AF765318

Plant Electrochemistry

The book comprehensively views the intersection between electrochemistry and botany. It attempts to inform people about the capacity of electrochemistry to provide information of interest to plant science. This includes electrophysiology, plant communication, and plant defense responses, as well as chemoecological and phylogenetic aspects. Conversely, the book also treats the capabilities of vegetal systems and vegetal-derived materials to be used in electrochemical applications for sensing, energy production and storage, and environmental issues.

Plant Electrochemistry
Physiology and Applications

Antonio Doménech-Carbó

CRC Press
Taylor & Francis Group
Boca Raton London New York

CRC Press is an imprint of the
Taylor & Francis Group, an **informa** business

Designed cover image: Photograph of leaves of Ginkgo biloba L. at the Botanic Garden of the University of Valencia. Phot. Josefa Prieto-Mossi.

First edition published 2025
by CRC Press
2385 NW Executive Center Drive, Suite 320, Boca Raton FL 33431

and by CRC Press
4 Park Square, Milton Park, Abingdon, Oxon, OX14 4RN

CRC Press is an imprint of Taylor & Francis Group, LLC

© 2025 Taylor & Francis Group, LLC

Reasonable efforts have been made to publish reliable data and information, but the author and publisher cannot assume responsibility for the validity of all materials or the consequences of their use. The authors and publishers have attempted to trace the copyright holders of all material reproduced in this publication and apologize to copyright holders if permission to publish in this form has not been obtained. If any copyright material has not been acknowledged please write and let us know so we may rectify in any future reprint.

Except as permitted under U.S. Copyright Law, no part of this book may be reprinted, reproduced, transmitted, or utilized in any form by any electronic, mechanical, or other means, now known or hereafter invented, including photocopying, microfilming, and recording, or in any information storage or retrieval system, without written permission from the publishers.

For permission to photocopy or use material electronically from this work, access www.copyright.com or contact the Copyright Clearance Center, Inc. (CCC), 222 Rosewood Drive, Danvers, MA 01923, 978-750-8400. For works that are not available on CCC please contact mpkbookspermissions@tandf.co.uk

Trademark notice: Product or corporate names may be trademarks or registered trademarks and are used only for identification and explanation without intent to infringe.

ISBN: 978-1-032-69729-1 (hbk)
ISBN: 978-1-032-69781-9 (pbk)
ISBN: 978-1-032-69782-6 (ebk)

DOI: 10.1201/9781032697826

Typeset in Times
by SPi Technologies India Pvt Ltd (Straive)

Dedication

To Leire, Gonzalo and Júlia, in the hope that they live in a green world

Contents

About the Author

Antonio Doménech-Carbó (Valencia, Spain, 1953) is Professor Emeritus at the Department of Analytical Chemistry, University of Valencia (PhD in Chemistry, 1989). His research is focused on solid-state electrochemistry with particular emphasis in the study of porous materials, vegetal matter and cultural heritage. He has supervised various PhD projects and directed several national research and development projects. For his research in cultural heritage, he has received the Demetrio Ribes Award from the Regional Government of Valencia in 2006 and 2019. He is the author of more than 300 articles in indexed journals, including one IUPAC (International Union of Pure and Applied Chemistry) technical report and several books; among them, *Electrochemical Methods in Archeometry, Conservation and Restoration* (Springer, 2009), *Electrochemistry of immobilized Particles and Droplets* (2nd ed. Springer, 2014), *Electrochemistry of Porous Materials* (Taylor & Francis, 2010; Chinese edition 2018; 2nd ed., 2021), and *Electrochemistry for Cultural Heritage* (Springer, 2023). Currently, he is a member of the editorial board of *ChemTexts* (Springer). He is the reviewer of the European Research Council and several national research agencies, and referee of more than 190 indexed journals.

Foreword

The electrochemical study of living organisms started just at the beginning of the birth of this scientific discipline in the frontier between the 18th and 19th centuries. Since then, electrochemistry and biology have been more or less directly coupled but always maintaining important connections. Over the last decades, the depth and variety of these connections have been enhanced and evolved from studies on plant electrophysiology to studies in energy production and storage. The purpose of the current monograph is to provide an approach to the interplay between electrochemistry and botany that may inform the botany community of the capabilities of electrochemistry and at the same time the electrochemistry community of the possibilities of using its discipline in the context of the vegetal world. Based on this general view, the contents of the book has been distributed among 15 chapters, each one focused on a concrete aspect where electrochemistry displays information of value for plant science.

For reasons of concision, the text has been written assuming that the essential biological concepts are known by the reader, i.e., the essentials of biochemistry and genetics, as well as plant anatomy and physiology. Eventually, the meaning of several "biological" terms will be briefly explained in the text. For the same reason, the discussion of the essentials topics on electrochemistry has been limited to the second half of the first chapter. The reader will find the basic cited literature used at the end of each chapter. This will be complemented by some additional literature. Given the enormous amount of publications in the field, this is necessarily short and, where available, concentrated in recent reviews.

The general scope adopted in the book results from a wide variety of influences, from scientific literature to informal talks with colleagues and students. Although the list of acknowledgments should be immense, the author wishes to express thanks to Fritz Scholz (Greifswald University) for fruitful discussions about the implementation of solid-state electrochemistry in botany and colleagues of the botanic garden of the University of Valencia, Elena Estrelles, Ana María Ibars, Josefa Prieto-Mossi, and Pilar Soriano, who have clarified concepts and introduced new problems of the vegetal world accessible to electrochemistry. The author wishes to thank the creativity and enthusiastic work of Adela Blasco, Gerardo Cebrián-Torrejón, Daiane Dias, Irene Domínguez, Géssica Domingos de Silveira, Yu Lee, Mariele Martini, Noemí Montoya, Anette Suleima Ortiz-Miranda, Sara Muñiz, Laura Osete-Cortina, Eva García-Martínez, and Francisco Manuel Valle-Algarra, as well as the continuous support of María Teresa Doménech-Carbó, Enrique García-España, José Vicente Gimeno-Adelantado, and Rufino Mateo-Castro. Naturally, errors and omissions are the exclusive responsibility of the author.

List of Acronyms

AFM	Atomic force microscopy
ADP	Adenosine diphosphate
AMP	Adenosine monophosphate
AP	Action potential
APB	Acid-producing bacteria
ATP	Adenosine triphosphate
BES	Bioelectrochemical systems
BSA	Bovine serum albumin
bRC	Bacterial photosynthetic reaction center
CE	Capillary electrophoresis
CPE	Constant phase element
CUPRAC	Cupric reducing antioxidant capacity
DMF	N,N'-Dimethylformamide
DMSO	Dimethylsulfoxide
DPhPC	1,2-diphytanoyl-*sn*–glycero-3-phosphocholine
DPPH	2,2-diphenyl-1-picrylhydrazyl radical
DPTL	2,3-di-O-phytanyl-*sn*-glycerol-1-tetraethylene glycol-DL-**α**-lipoic acid ester
dsDNA	Double-stranded deoxyribonucleic acid
ECL	Electrochemiluminiscence
EDL	Electric double-layer capacitor
EDTA	Ethylenediaminetetraacetic acid
EIS	Electrochemical impedance spectroscopy
ELISA	Enzyme-linked immunosorbent assay
EN	Electrochemical noise
EQCM	Electrochemical quartz crystal microbalance
FePD	Ethioporphyrin II
FLD	Fluorescence spectroscopy
FRAP	Ferric-reducing antioxidant power
FTO	Fluor-doped tin oxide electrode
gA	Gramicidin A
GCE	Glassy carbon electrode
GC-MS	Gas chromatography-mass spectrometry
HAT	Hydrogen atom transfer mechanism
HEPES	4-(2-hydroxyethyl)-1-piperazineethanesulfonic acid
HER	Hydrogen evolution reaction
HPLC	High-performance liquid chromatography
HPLC-MS	High-performance liquid chromatography-mass spectrometry
HTA	Hydroxyterephthalic acid
ISB	Iron-oxidizing bacteria
ISFET	Ion sensitive field effect transistor
ITO	Indium-doped tin oxide electrode
LCP	Monoolein liquid-crystalline cubic phase
LDL	Low-density lipoprotein
LEP	Local electrical potential
LIRB	Light-induced rhythmic bioelectrogenesis
MeCN	Acetonitrile
MFCs	Microbial fuel cells

MIC	Microbiologically influenced corrosion
NAD/NADH	Nicotinamide dinucleotide
NHE	Normal hydrogen electrode
NRB	Nitrate-reducing bacteria
OCP	Open-circuit potential
OER	Oxygen evolution reaction
ORAC	Oxygen radical absorbance capacity
ORR	Oxygen reduction reaction
PBS	Phosphate buffer saline
PSI	Photosystem I
PSII	Photosystem II
QRCE	Quasi-reference counter electrode
RAF	Radical adduct formation mechanism
SAMs	Self-assembled monolayers
SCE	Saturated calomel electrode
SECM	Scanning electrochemical microscopy
SEPT	Sequential electron proton transfer mechanism
SET-PT	Single electron transfer followed by proton transfer mechanism
SHE	Standard hydrogen electrode
SICM	Scanning ion conductivity microscopy
SOB	Sulfur-oxidizing bacteria
SOD	Superoxide dismutase
SP	System potential
SPLET	Sequential proton loss electron transfer mechanism
SRB	Sulfate-reducing bacteria
TAC	Total antioxidant capacity
TAP	Total antioxidant potential
TBA	Thiobarbituric acid
TBAP	n-Tetrabutylammonim perchlorate
TBARS	thiobarbituric acid-reactive substances
tBLM	Bilayer lipid membranes tethered to a metal electrode
TEAC	Trolox equivalence antioxidant capacity
TEMPO	2,2,6,6-tetramethylpiperidinyl-1-oxy radical
βTg	1-thio-β-D-glucose
THF	Tetrahydrofuran
TMPD	N,N,N',N'-tetramethyl-para-phenylene-diamine
TMQ	Tetramethylquinone
TOC	Total organic matter content
TPrA	Tri-n-propyl amine
TRAP	Total radical trapping antioxidant potential
VIMP	Voltammetry of immobilized particles
VOCs	Volatile organic compounds
VP	Variation potential

1 Overview

1.1 HISTORICAL PERSPECTIVE

The role of electricity in living organisms can be associated with the discovery of 'animal electricity' made by Luigi Galvani [1]. Several other authors, Bertholon in particular [2], thought previously about the possible biological effects of electrical phenomena in vegetables [3]. The subsequent development of studies around batteries, electrical currents, and electrochemistry (Volta, Ohm, Faraday, etc.) prompted the analysis of a variety of electrophysiological phenomena. In the middle of the 19th century, Du Bois-Reymond studied electrically hunting fishes and introduced the terms "action potential" and "action current" to describe the propagation of signals through nerves [4]. Several years later, Burdon-Sanderson recorded action potentials on leaves of the Venus flytrap (*Dionea muscipula* Ellis) in an event organized in 1873 by Charles Darwin [5, 6].

The recognition that living cells are surrounded by semipermeable membranes and the role of potassium ion concentration gradients in signal transmission were important achievements between the late 19th century and the 1920s [7]. In the 1930s, Umrath [8] and Osterhout [9] performed the first intracellular recordings of intracellular plant potentials. Research in this line led to the proposal of the ion-channel and ion-pump mechanisms [10]. In the 1960s, the discovery that "normal" plants had propagating action potentials such as the "sensitive" plants [11, 12] expanded notably the study of plant electrophysiology. Since then, the linking of electrical signals with respiration and photosynthesis [13, 14], pollination, and plant defense mechanisms have been widely studied [3].

In parallel, the connection between electrical potentials and photosynthetic activity was a subject of considerable interest. This matter was first addressed Haake in the late 19th century [15] when the sharp variations of the mesophyll voltage when passing from the dark to under illumination were discovered. The linkage between photoelectrical activity and photosynthesis was systematically developed in the 1970s [16, 17].

During the 20th century, electrochemistry experienced enormous growth concerning both its theoretical foundations and instrumental techniques and a range of practical applications. Potentiometric methods based on ion-selective electrodes (ISEs) and amperometric methods expanded their implementation in the vegetal world from the use of micropipette and microelectrode techniques [18, 19]. The development of the vibrating voltage probe for monitoring external voltage fields is related to the transmembrane passage of ions [20]. In turn, the creation of the patch-clamp technique [21] allowed for the *in situ* study of ion channels.

The expansion of voltammetric analysis from the seminal work on polarography due to Heyrovsky and his colleagues has been considerable since the 1950s [22]. This has exploited the development of pulsed techniques, new electrode materials, and an enormous variety of electrode modification strategies [23, 24] prompting *in vivo* analysis with sensors and biosensors [25]. Impedance techniques [26] and scanning electrochemistry techniques [27] have also increased the scope of electrochemistry. The analytical capabilities of electrochemical methods have been increased by the application of flow methods [28] and multivariate techniques of data analysis [29]. In the biological domain, these techniques have favored the metabolomic analysis [30].

Conjointly considered, all these matters depict a scenario in which electrochemistry and botany intersect deeply. Electrochemistry provides a theoretical background and a set of experimental techniques to interpret not only physiological processes in plants but also to provide taxonomic and phylogenetic information [31]. Additionally, electrochemistry yields relevant analytical information on plant composition, antioxidant capacities, nutritional properties, etc., of fruits and other plant parts. Conversely, the use of plants, and/or plant organelles, and/or plant components for

DOI: 10.1201/9781032697826-1

electroanalysis, electrosynthesis, and energy generation and storage is increasingly studied. These aspects will be treated in the next sections.

1.2 VEGETAL ELECTROCHEMISTRY

To some extent conventionally, this book can be divided into two great areas, "plant electrochemistry" (or "vegetal electrochemistry") and "plants in electrochemistry". The former corresponds to all aspects of vegetal physiology that can be described based on electrochemical concepts and the access of their understanding upon application of electrochemical methodologies. The second area of knowledge corresponds to the issues where vegetal matter in general enters into electrochemical methods of analysis, synthesis, and energy production and storage. In both cases, the research has been focused on green plants, but important groups such as algae and fungi can be similarly treated.

Biochemical processes in plants involve an enormous variety of reactions, one significant fraction of which corresponds to redox processes. Electron-transfer reactions in living organisms occur in aqueous media in conditions that differ from ordinary electrochemical experiments. The former involves outer-sphere and inner-sphere direct electron transfers between more or less complex molecular units; the second involves electron transfer between molecular units and an electronic conductor, the electrode. Despite this fundamental difference, most biochemical processes can be interpreted in electrochemical terms, and electrochemical techniques provide information on such processes. The determination of intracellular ionic concentrations [32] and the concomitant estimate of ion fluxes [33], as well as the analysis of membrane potentials and ion channels, focus on one of the essential issues in plant physiology [33]. Here, potentiometric techniques play an essential role. This has been enhanced by the development of patch-clamp devices, microelectrodes, and multi-electrode array technologies [35].

A second interesting aspect of the electrochemistry of plants is the study of molecular systems mimicking *in vitro* the processes occurring in biological media. The analysis of electrochemical pathways can provide, indirectly, mechanistic information on biochemical processes. On the other hand, most biochemical processes can be represented and modeled electrochemically. Figure 1.1 shows a paradigmatic example, corresponding to the equivalent circuits, based on the Hodgkin–Huxley proposal [36], used to describe an axon (a) and sieve tubes in plant phloem (b) [37]. In these schemes, the conductive branches represent ion channels and leak conductance with their different electromotive forces. See Table 1.1 to interpret the main components of equivalent circuits.

The aforementioned issues will be treated in Chapter 2, whereas another important aspect of plant physiology, the generation and propagation of electrical signals at different scales (ion

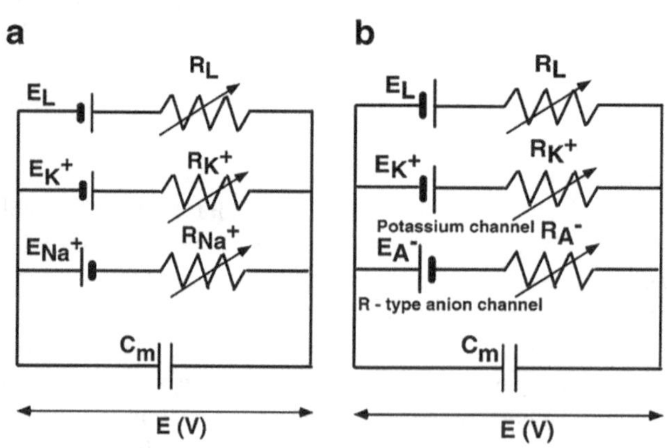

FIGURE 1.1 Equivalent circuits used to model (a) axon, and (b) sieve tubes in plant phloem. (Reproduced from ref. [37], with permission. Copyright Elsevier 2007.).

TABLE 1.1
Characteristics of the Three Basic Circuit Elements Used to Model Electrochemical Impedance Spectra

Circuit Element	Physical Quantity (Symbol, Units)	Impedance	Phase Angle/deg
Resistance	Ohmic resistance (R, Ω)	$Z_R = R$	0
Capacitor	Capacity or capacitance (C, F)	$Z_C = -j/C\omega$	–90
(Auto)induction	(Auto)induction (L, H)	$Z_L = jL\omega$	90

channels, cells, tissues, organs, and the whole-plant level) [38, 39], will be treated in Chapter 3. Electrochemical techniques are also used to extract information on plant composition, contaminants, and environmental relations. Important applications include food chemistry and agricultural chemistry. These matters will be treated in subsequent chapters.

1.3 PLANTS IN ELECTROCHEMISTRY

Research in applied electrochemistry has expanded the fields in which plants can be used for different purposes. These include studies concerning electroanalytical applications of plants (Chapter 13); studies on enzymatic, microbial, and plant fuel cells; and diverse arrangements for electrical energy storage and production that will be revised in Chapter 14.

Other electrochemical fields where plants or vegetal matter are directly involved are centered on corrosion and biological degradation processes, as well as topics such as the valorization of biochar. Several of these matters will be treated in Chapter 15.

1.4 ELECTROCHEMICAL METHODS

1.4.1 General Aspects

Generically, electrochemical methods involve charge transfer processes promoting chemical changes and/or informing on the chemical composition of the system under study. A significant part of these processes consists of the transfer of electrons through the interface between an electronic conductor, (electrode) and an ionic conductor (electrolyte). This is the case of, for instance, voltammetric and coulombimetric techniques. In other cases, the charge transfer through 'bulk' phases is studied. Conductimetry is a typical example. Here, we will focus attention on the techniques most frequently used in the study of plants.

In principle, all these methods involve the measurement of physical quantities – namely, difference of potential, current intensity, electrical charge, ohmic resistance, capacity, etc., in an experimental device, the electrochemical cell, under controlled conditions. Starting from a paradigmatic experiment, water electrolysis, we can consider the simplest electrochemical cell as constituted by two electrodes (Pt, Au, Hg, graphite, etc.) immersed into an aqueous solution having ionic conductivity. The electrical potential difference between the electrodes and the current passing through the cell can be measured with a voltmeter and an ammeter, respectively.

To acquire information on the chemical species existing in the electrolyte solution (this is the ordinary case for analytical purposes), we need to study the phenomena occurring at the surface of one electrode, the working electrode, whose (difference of) potential relative to a reference electrode is controlled. To minimize the current passing through this electrode (i.e., to disturb minimally the composition of the system under study), a third electrode (auxiliary electrode) of a large area is routinely used. The primary reference electrode is the standard hydrogen electrode (SHE). For practical purposes, secondary reference electrodes such as silver/silver chloride or calomel (Hg_2Cl_2/Hg) are used. These consist of the metal in contact with its insoluble (scarcely soluble) chloride salt in contact with a concentrated aqueous solution of NaCl or KCl. The potential of these reference electrodes relative to the SHE is well-known, being +0.2412 V for the (KCl) saturated calomel electrode (SCE) and +0.197 V for the (KCl) saturated Ag/AgCl electrode [40]. These reference electrodes are of application in aqueous solutions, but in nonaqueous solvents, a different set of electrodes has to be used. Here appears the problem of unknown junction potentials and, directly related to it, the absence of a universal scale of reference potentials. These matters will be treated in Chapter 4.

1.4.2 Electrodes and Electrochemical Processes

In principle, the working electrodes must be inert, i.e., they do not react with the components of the electrolyte solution and cannot be oxidized or reduced electrochemically under the operation conditions. Then, the ensemble electrode/electrolyte determines the interval of potentials accessible for analytical purposes. These are in general pH dependent. In the case of graphite electrodes in contact with 0.5 M H_2SO_4, the evolution of H_2 ca. −1.0 V vs. Ag/AgCl and the evolution of O_2 ca. 1.5 V vs. Ag/AgCl limit the accessible potential range. In the case of Pt and Au under equivalent conditions, the lower potential limit becomes ca. −0.2 V. The electrode oxidation also limits or seriously conditions the usable potential range. In neutral aqueous solutions, Pt and Au are oxidized at ca. 1.0 V forming a film of oxides onto the electrode surface while Hg is oxidized to Hg^{2+}/Hg_2^{2+} under equivalent conditions ca. 0.1 V.

A variety of electrode materials, shapes, and sizes can be used [26]. Microelectrodes (between 0.1 and 50 μm diameter) and ultramicroelectrodes (until 10 Å diameter) will be of particular interest in plant electrochemistry. Depending on the electrochemical technique used, the measurements are performed under quiescent or stirred solutions. Another possibility is the use of rotating electrodes. The electrolyte solution can be deaerated by bubbling N_2 or Ar. This procedure is aimed at removing the oxygen dissolved in the electrolyte. As seen in Chapter 5, O_2 is reduced electrochemically at moderately cathodic potentials and may react with the species involved in the processes of interest.

The electrochemical processes of interest will occur at the electrode/electrolyte solution interface. These consist of the transference of electrons through the interface accompanied by chemical changes in the same. Among other possibilities, these can consist of gas evolution (for example, $2H^+ + 2e^- \rightarrow H_2$), metal deposition (e.g., $Pb^{2+} + 2e^- \rightarrow Pb$), or simply the formation of new species in solution.

These processes are accompanied by the transport of charge through the electrolyte solution and the system of external conductors and can be coupled to chemical reactions with other species existing in the solution and/or the electrode. It is important to realize that the motion of ionic species through the electrolyte can be purely diffusive or convective. In electrophoretic techniques (usually ascribed to analytical separation techniques), ionic motion under large electric fields is used. However, in the majority of electrochemical techniques, this ionic migration is inconvenient, being avoided by the use of a supporting electrolyte (salts, acids, bases, buffers) in relatively large concentrations (typically, at least 0.10 M). This must be chemically and electrochemically inert under the conditions of operation.

1.4.3 Potentiometry

This technique is based on the measure of the difference of potential between the working electrode and the reference electrode under open-circuit conditions, i.e., when the current intensity passing through the system tends to zero [40]. The working electrode is responsive relative to the analyte under study. Under conditions of reversibility, the maximum difference of potential (or electromotive force) satisfies the Nernst equation. The typical application is the determination of ionic species in aqueous solution using ISEs. A schematic representation of the simplest potentiometric device is depicted in Figure 1.2. Here, the working electrode is connected through a voltmeter with high input (internal impedance) to the external reference electrode containing a standard internal solution, both being in contact with the problem solution. For a M^{n+} cation,

$$E = E_{pot} + \frac{RT}{nF} \ln a_{M^{n+}}. \tag{1.1}$$

Here, E represents the potential and a_{Mn+} the thermochemical activity of the metal ion. E_{pot} is a constant that includes the standard potential of the electrode, the reference electrode potential, and the junction potential between the different solutions. In diluted solutions of M^{n+}, the numerical values of the activities can be assumed to be equal to the numerical values of the concentrations, c_{Mn+}, and Eq. (1.1) can be approximated at 25°C by the expression (potentials in V),

$$E = E_{pot} + \frac{0.059}{n} \log c_{M^{n+}}. \tag{1.2}$$

This equation predicts a linear variation of the measured potential on the logarithm of concentration. As far as the divergence between the numerical values of thermochemical activities and concentrations increases with concentration, the E vs. $\log c_{Mn+}$ plots separate from linearity at high analyte concentrations. One possible alternative is to convert concentrations into activities using any of the suitable theoretical models. This practice increases the linearity interval but, of course, introduces some degree of arbitrary character [41].

Although not necessarily confined to ionic species, ISEs are of particular interest and extensive application. There are four basic types of ISEs of particular interest [42]: (i) glass electrodes (in particular, those sensitive to sensitive to H^+, Na^+, and NH_4^+), (ii) membrane electrodes in which an ion exchanger or a selective complexing agent is immobilized into a polymeric matrix, (iii) electrodes with sparingly soluble inorganic salts (LaF_3) or salt mixtures ($Ag_2S/AgCl$), and (iv) electrodes with enzymes embedded in a gel membrane. Different solid materials are used as potentiometric sensors for gases [43].

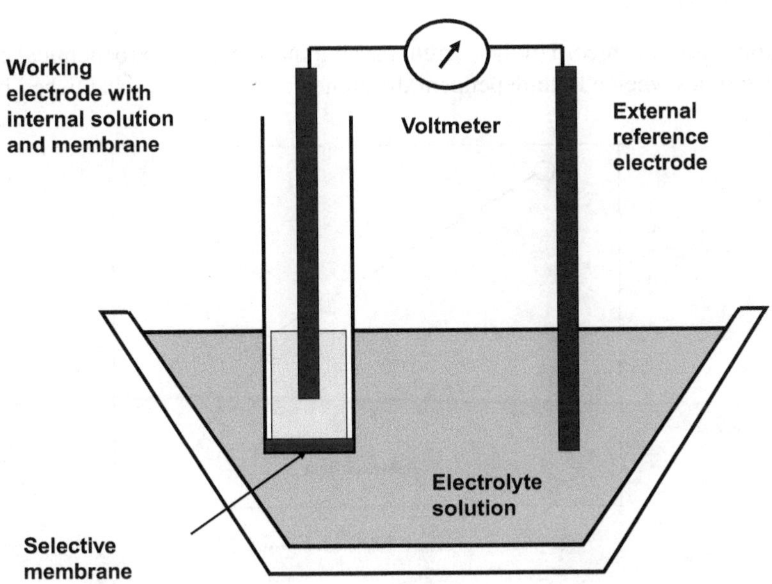

FIGURE 1.2 Schematics for a typical ISE.

In principle, the technique is also applicable when the electrochemical process is non-Nernstian, but there is a linear dependence between measured equilibrium potentials and logarithm of concentration. In this case, the slope of the E_{eq} vs. lonc_{Mn+} plots differs from 0.059/n V per decade, and the intercept has no thermochemical meaning.

The selectivity of ISEs is an essential property. Ideally, a selective potentiometric sensor should be responsive to only one ionic species. In practice, the ISE also responds to other interfering ions. The selectivity coefficient (k_{AB}) is a numerical measure of how well the ISE can discriminate the ion of interest relative to the interfering ion. When an electrode has equivalent responses to the two ions, then $k_{AB} = 1.0$. A log $K^{pot}{}_{A,B} < 1$ indicates a preference for the measuring ion relative to the interfering ion. This coefficient enters into the Nicolsky-Eisenman equation:

$$E = E_{pot} + \frac{RT}{z_A F} \ln\left[a_A + \sum_B K_{A,B}^{pot}\left(a_B\right)^{z_A/z_B}\right]. \tag{1.3}$$

In this equation, z_A and z_B are charge numbers, and a_A, a_B, are the activities of the primary ion, A, and of the interfering ion, B. This equation assumes a Nernstian response for all ions. The selectivity coefficients can be measured by separate solutions and mixed solutions methods, being the matched potential method, which in fact is independent on the Nicolsky-Eisenman equation, recommended by IUPAC [45].

ISEs possess a linearity (in logc_{Mn+} scale) interval at low concentrations. The detection limit is calculated from the intersection of the linear branch with the asymptote of the nonlinear region, as illustrated in Figure 1.3.

For potentiometric measurements in plant cells, it is necessary to measure the potential of the cellular membrane. For this purpose, double-barred microelectrode devices as schematized in Figure 1.4 are used. These contain the reference electrode that contacts the plant through a porous glass frit or an agar plug, an ion-selective barrel with a sensor plug in the tip, and a cell membrane potential barrel constituted by an Ag wire coated by AgCl in contact with 0.1 M KCl [19, 46].

1.4.4 Voltammetry

Voltammetry consists of the record of the current passing through an electrochemical cell containing two or three electrodes when a time-dependent difference of potential is established between them.

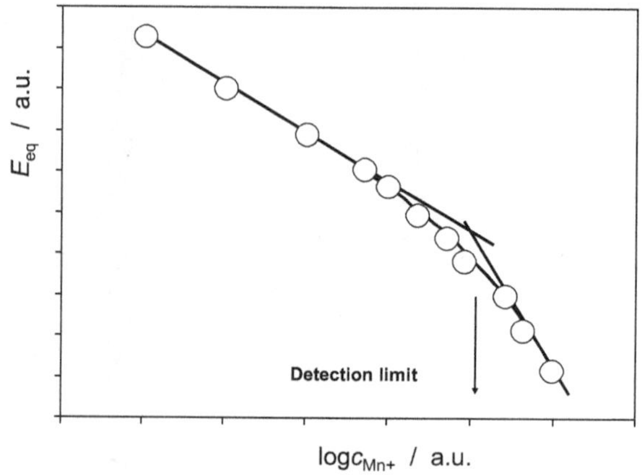

FIGURE 1.3 Scheme for the typical calibration graph of an ISE showing the procedure to calculate the detection limit.

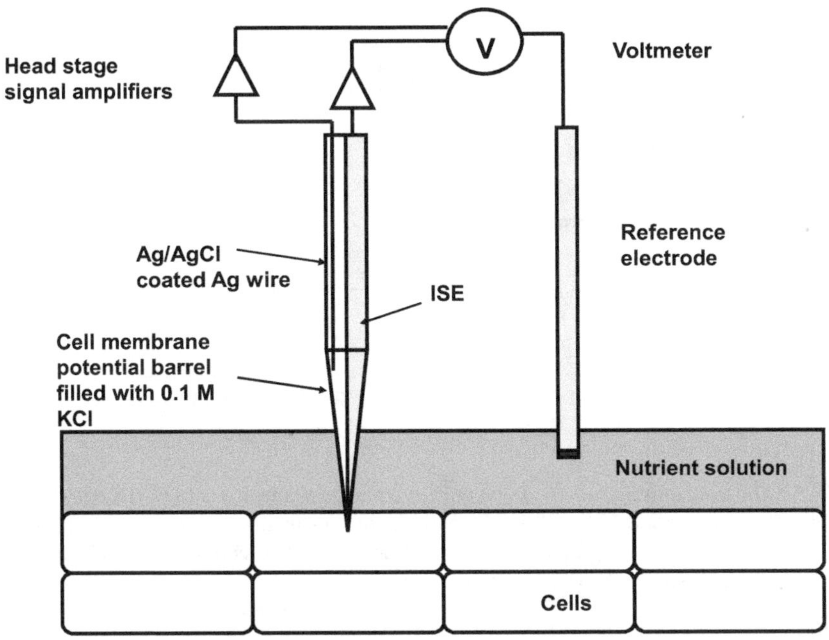

FIGURE 1.4 Scheme for a double-barreled, ion-selective microelectrode.

Usually, three-electrode arrangements are used. Here, the cell contains reference, working, and auxiliary electrodes connected to a potentiostatic device. The reference electrode provides a constant reference potential during the experiment, whereas the auxiliary or counterelectrode concentrates the gross of the current. The resulting current/potential representation is called the voltammogram. In most cases, the voltammograms are aimed at reflecting faradaic processes occurring at the working electrode/electrolyte solution interface. These are electron-transfer processes resulting in changes in the composition of the solution and/or the electrode, for instance, gas evolution or metal deposition. Depending on the type of excitation (time variation of the difference of potential), different voltammetric techniques are available. In linear potential scan techniques (linear and cyclic voltammetry), the potential varies linearly with time at a fixed potential scan rate v. In pulsed techniques (differential pulse voltammetry, square wave voltammetry, etc.), pulsed inputs of different amplitude and frequency are superimposed to a linear or staircase potential ramp. The same experimental arrangement is used in other techniques (chronocoulometry, chronoamperometry, etc.) and in conjunction with nonelectrochemical techniques, for instance, spectral measurements (spectroelectrochemistry).

In general, voltammetric experiments are carried out under quiescent (nonstirring) conditions in solutions containing a supporting electrolyte in a relatively large concentration (up to 0.05 M). In aqueous solution, NaCl, KCl, KNO_3, acetate, carbonate, and phosphate buffers, and HCl, H_2SO_4, NaOH, and KOH are used. Nonaqueous solvents such as dimethylsulfoxide (DMSO), N,N'-dimethylformamide (DMF), acetonitrile (MeCN), CH_2Cl_2, use other salts (Bu_4NPF_6, Et_4NClO_4, etc.) as supporting electrolytes. Ionic liquids [47] are increasingly used. The electrolyte solution is deaerated by bubbling nitrogen or argon to remove dissolved oxygen. The reason is that oxygen produces voltammetric signals and can react with other species existing in the solution and/or generated electrochemically. The reference electrodes must be adapted to the solvent; in aqueous electrolyte solutions, Ag/AgCl and Hg/Hg_2Cl_2 (calomel) are those typically used. The auxiliary electrode must have a large surface area; platinum wire and graphite rods are routinely used. As previously described, there is a variety of working electrodes, from metallic electrodes (Hg, Pt, Au, Ag, etc.) to carbon-based electrodes, mainly pyrolitic graphite and glassy carbon electrode (GCE).

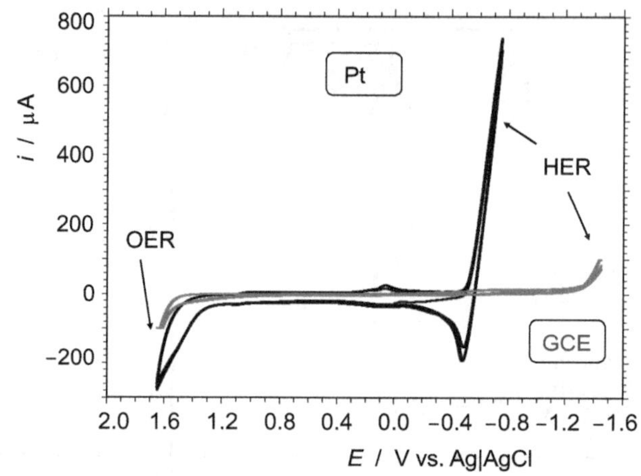

FIGURE 1.5 Cyclic voltammograms recorded at Pt (geometrical surface area 0.018 cm^2) and glassy carbon (GCE, geometrical surface area 0.071 cm^2) working electrodes on an air-saturated 0.25 M HAc/NaAc aqueous solution at pH 4.75. Potential scan initiated at 0.0 V in the negative direction. Potential scan rate 50 mV s^{-1}. Note that the axes are opposite to the IUPAC convention.

Indium-doped tin oxide (ITO) and fluor-doped tin oxide (FTO) are optically transparent electrodes used in photoelectrochemistry. Of particular interest are microelectrodes whose size ranges between a few micrometers to 100 nm [48].

It is important to realize that the voltammetric response is sensitive to the electrolyte solution composition and the nature of the electrode. This can be seen in Figure 1.5, where the cyclic voltammograms recorded at Pt and GCE on an aqueous acetate buffer solution at pH 4.75 are superimposed.

In both cases, the potential scan was initiated at 0.0 V vs. Ag/AgCl in the negative direction. At GCE, a rising current appears at ca. −1.4 V corresponding to the hydrogen evolution reaction (HER), formally represented as $2H^+$ (aq) + $2e^-$ → H_2 (g). When the applied potential reaches ca. 1.4 V, the oxygen evolution reaction (OER) takes place. This process, formally represented as $4OH^-$ (aq) → O_2 (g) + $2H_2O$ + $4e^-$, occurs at virtually the same potentials at Pt electrode. In contrast, the HER process starts at Pt electrode at much higher potentials than at GCE, ca. −0.5 V, thus revealing the importance of the electrode material. The voltammetric record is completed with other Pt-localized signals associated with metal surface and metal/electrolyte solution interface phenomena [49] and the reduction of dissolved oxygen (oxygen reduction reaction, ORR) at ca. −0.6 V.

The HER and OER processes mark the interval of potentials usable for studying the electrochemistry of other chemical species, but their voltammetric responses can be affected by electrode-associated signals. Figure 1.6 shows an example of the eventual complexity of the voltammograms when several electrochemical processes are superimposed. In this figure, the cyclic voltammograms recorded at Au electrode on an air-saturated $K_4Fe(CN)_6$ solution in aqueous acetate buffer at pH 4.75, are superimposed. The concentration of the hexacyanoferrate(II) salts is 1.0 mM, typical in conventional voltammetric experiments. In the potential range between 0.65 and −0.45 V vs. Ag/AgCl, the voltammograms show a well-defined couple of anodic (A_{Fe}) and cathodic (C_{Fe}) peaks at potentials of centered ca. 0.20 V. If the potential range is extended between 1.25 and −1.25 V, the voltammogram incorporates the large current for HER at ca. −0.8 V, preceded by the weak ORR signal at ca. −0.6 V. These are also accompanied by an anodic wave at 1.1 V (A_{Au}), which is followed, in the subsequent cathodic scan by a tall peak at 0.55 V (C_{Au}). The signal A_{Au} corresponds to the oxidation of Au to form a monolayer of gold oxides. The anodic process, after intermediate steps, is reversed in the process C_{Au} [50].

The oxidation of $Fe(CN)_6^{4-}$ can be represented as,

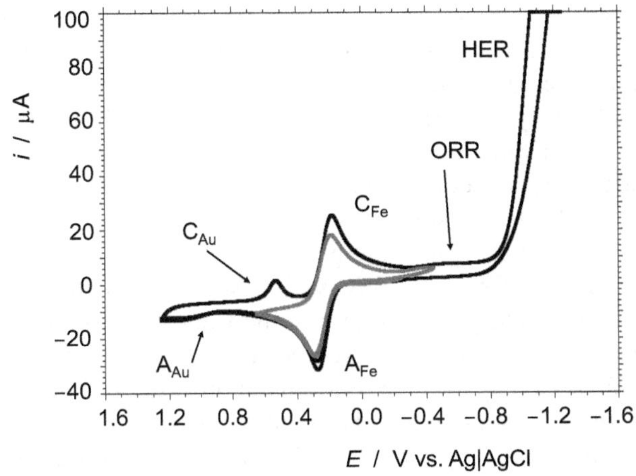

FIGURE 1.6 Cyclic voltammograms recorded at Au electrode (geometrical surface area 0.018 cm²) on an air-saturated 2.0 mM $K_4Fe(CN)_6$ plus 0.25 M HAc/NaAc aqueous solution at pH 4.75. Voltammograms recorded at two different potential ranges are superimposed. Potential scan initiated at −0.2 V in the positive direction. Potential scan rate 50 mV s⁻¹. Note that the axes are opposite to the IUPAC convention.

$$Fe(CN)_6^{4-}(aq) \rightarrow Fe(CN)_6^{3-}(aq) + e^-. \qquad (1.4)$$

This process takes place as an essentially reversible electron transfer through the electrode/electrolyte solution interface under conditions of diffusion control. Under these conditions, the voltammogram is characterized by the appearance of two peaks, cathodic and anodic, symmetrically placed relative to the midpeak potential, E_{mp}. This parameter, calculated as the half sum of the cathodic (E_{pc}) and anodic (E_{pa}) peak potentials, defines a formal potential of the couple, $E^{\circ\prime}$, which in principle can be taken as an approximate measure of the standard potential E°. Theoretically, E_{mp} is independent of the potential scan rate, and the theoretical peak-to-peak separation for an n-electron transfer equals $59/n$ mV at 25°C, also independent of the potential scan rate [26, 51].

The reversible, diffusion-controlled processes are theoretically characterized by the independence of the formal potential on the potential scan rate and the maintenance of a peak potential separation of $59/n$ mV at 25°C. The diffusion control implies the proportionality of the peak current on the square root of the potential scan rate, v. Peak currents are also proportional to the concentration of the electroactive species in the solution bulk, c, and to the square root of its diffusion coefficient, D, according to the Randles-Ševčík equation [26],

$$i_p = 0.4463nFSc\sqrt{nFDv/RT} \qquad (1.5)$$

valid for semi-infinite diffusion conditions at a planar electrode of surface area S.

Both anodic and cathodic peaks display an asymmetric profile. At the beginning of each wave, the current increases in an exponentially like manner by increasing (anodic peak) or decreasing (cathodic peak) the potential. After reaching the corresponding peak, the current decays in absolute value but more slowly than in the precedent region. In this second region, the electron-transfer rate is controlled by the diffusion of the electroactive ions toward the electrode surface, and the current is approximately proportional to the inverse of the square root of the potential.

The electrochemical cells in general display capacitive and resistive effects, which are superimposed on the faradaic ones. Among others, capacitive effects are associated with the double layer always produced in the region of the electrolyte solution vicine to the electrode surface [26]. As a result, the ideal voltammetric pattern previously described becomes distorted by uncompensated ohmic drops and capacitive effects. The most distinctive feature in the case of electrochemically reversible processes is the broadening of the voltammetric peaks and the increase of the separation between them, which increases on increasing the potential scan rate [26, 52]. Then, the reversibility is marked by the independence of the midpeak potential on the scan rate. An example of this situation is shown in Figure 1.7, where the cyclic voltammograms recorded at Pt electrode for an aqueous solution of $K_4Fe(CN)_6$ in 0.10 M KCl at four potential scan rates are superimposed. The separation between the cathodic and anodic peaks is enhanced by increasing potential scan rate v while the peak current increases. The essentially reversible character of the oxidation of $Fe(CN)_6^{4-}$ to $Fe(CN)_6^{3-}$ is suggested by the maintenance of the mid-peak potential at 0.20 V vs. Ag/AgCl, regardless of the value of the potential scan rate.

Most electrochemical processes are not reversible. The extreme case is the entirely irreversible electron transfer. Roughly speaking, the electron-transfer process is "slow" and the reverse reaction is blocked. As a result, the voltammogram consists of a unique cathodic (or anodic) peak without coupled anodic (or cathodic) peak. In principle, under conditions of diffusive control, the peak current also becomes proportional to the concentration of the electroactive species in the bulk solution and to the square root of the potential sweep rate and the diffusion coefficient. For an electrochemically irreversible n-electron transfer controlled by diffusion, the peak current recorded at a planar electrode will be,

$$i_p = 0.4958nFSc\sqrt{\alpha n_a FDv / RT}. \tag{1.6}$$

Classically, α is the electron-transfer coefficient, and n_a is the number of electrons transferred in the rate-determining step. It is recommended to replace the product αn_a by a unique coefficient β defined as the sum of the number of electrons n' transferred before the rate-determining step and the electron-transfer coefficient α of the latter [53].

In addition to "pure" electron-transfer processes with different degrees of reversibility, there are a variety of processes involving coupled chemical reactions. These are of particular importance in plant electrochemistry and will be treated in Chapter 4. The appearance of coupled chemical reactions can produce apparently irreversible voltammetric patterns such as that represented in Figure 1.8, corresponding to the oxidation of uric acid. The absence of coupled cathodic/anodic peaks does not necessarily indicate irreversible electron transfer. Accordingly, diagnostic criteria based on variation of electrochemical parameters with a potential scan rate [51] must be used.

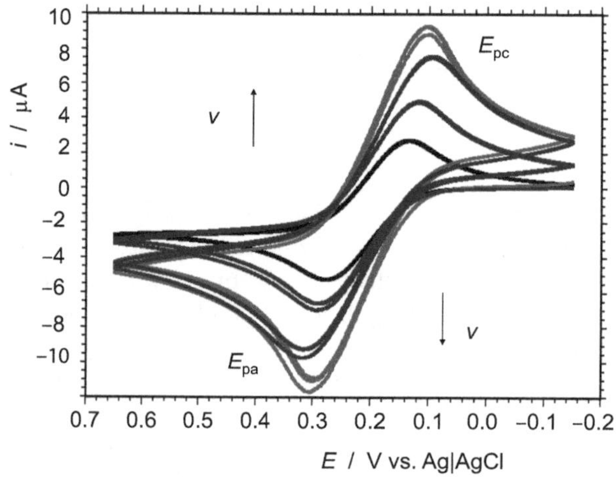

FIGURE 1.7 Cyclic voltammograms at Pt electrode of a 1.0 mM solution of $K_4Fe(CN)_6$ in 0.10 M KCl aqueous solution. Potential scan rates (v) of 10, 20, 50, and 100 mV s^{-1}.

Another interesting class of electrochemical processes is characterized by the persistent attachment of the electroactive species and its oxidation (or reduction) products to the electrode surface. The voltammetry of surface-confined species is characterized by symmetric peaks. In the reversible case, the cathodic and anodic peak potentials remain equal to the standard potential, and the peak current is proportional to the potential scan rate. Now, for a reversible surface-confined reaction, the peak current is [26]

$$i_p = n^2F^2sv\Gamma / 4RT, \quad (1.7)$$

where Γ is the initial surface concentration (mol cm^{-2}) of the electroactive species attached to the electrode.

1.4.5 Impedance Techniques

These are directed to determine the impedance parameters of an electrochemical cell when a sinusoidal potential input of low amplitude (5–50 mV) is applied. A three-electrode arrangement is in general use. Contrary to voltammetry, where in general electron-transfer processes involving changes in the chemical composition (faradaic processes) are studied, impedance techniques proceed, ideally, without chemical changes in the cell. This condition holds when the sinusoidal input is superimposed on the equilibrium potential of the system (frequently taken as the open-circuit potential). Of course, an arbitrary bias potential, E_{bias}, can be applied so that more or less intense faradaic processes can occur.

The most widely extended technique is electrochemical impedance spectroscopy (EIS). Here, the impedance parameters are determined for sinusoidal potential inputs of different frequencies (*f*), typically, between 0.001 and 10^5 Hz. The essential characteristic of all impedance measurements is that, in general, it appears a phase difference between the applied potential and the current [54]. This situation has to be described in terms of complex impedance:

$$Z = Z_{real} + jZ_{imag} \quad (j = \sqrt{-1}). \quad (1.8)$$

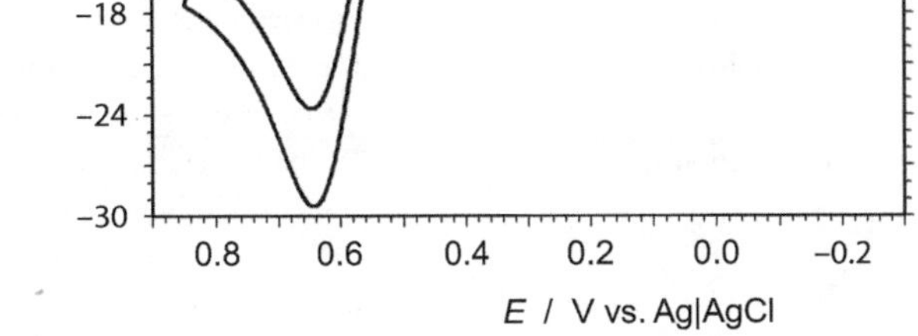

FIGURE 1.8 Cyclic voltammogram at GCE electrode of a 2.0 mM solution of urinc acid in 0.10 M potassium phosphate buffer at pH 7.0. Potential scan initiated at 0.0 V in the positive direction. Potential scan rate 50 mV s^{-1}. Note that the axes are opposite to the IUPAC convention.

Available instrumentation provides the absolute value of the total impedance, Z, of the circuit, and the phase angle (φ) between current and potential or, equivalently, the real and imaginary components (Z_{real}, Z_{imag}, respectively) of the total impedance.

These parameters depend on the frequency (or angular frequency, $\omega = 2\pi f$), and the representation of impedance parameters with frequency is called the impedance spectrum of the system. Strictly, impedance measurements constitute a "true" spectrum when the system satisfies the conditions imposed for a stationary system fulfilling the theory of linear systems constraints. In practice, this means that the system must be operated under stationary conditions, which can be derived from the application of Kramers-Kronig transforms [55].

The analysis of impedance spectra involves their modeling employing an equivalent circuit that reproduces experimental data in a statistically valid form. This equivalent circuit must be composed of elements (resistor, capacitors, autoinductions, etc.) that must have a definite physical meaning. The characteristics of the three basic circuit elements are summarized in Table 1.1. The (auto)induction element rarely appears in electrochemical studies.

The capacitance element can be considered a particular case of constant phase elements (CPEs) whose impedance can be expressed as

$$Z_{CPE} = Q_o\left(j\omega\right)^{-h}, \tag{1.9}$$

where Q_o is a constant, and $0 < h < 1$. The CPEs can be seen as nonideal capacitors. The ideal capacitor in Table 1.2 corresponds to $h = 1$ and $Q_o = 1/C$. The case $h = ½$ corresponds to another frequently used element, the Warburg element (W), whose impedance is generally expressed as $Z_W = W(j\omega)^{-1/2}$, W being the Warburg constant. This element is representative of diffusive effects in the cell.

The simplest equivalent circuit used to model conventional EIS is composed of a resistance in series with a parallel combination of a second resistance and a capacitor. This is the Randles circuit schematized as an inset in Figure 1.9. The constitutive elements are an ohmic resistance (R_s) representative of the resistance of the solution, connections, clamps, etc.; a charge transfer resistance (R_{ct})

TABLE 1.2
Summary of the Main Types of Operation of SECM

SECM Type	Operation	Description
Positive feedback	Tip reaction of the redox probe (R→O)	The probe tip approaches a conducting surface that favors the regeneration of R back to the tip
Negative feedback	Tip reaction of the redox probe R (R→O)	The probe tip approaches an inert surface that hinders diffusion of R to the tip by the substrate
Substrate generation/tip collection	Substrate reaction and tip reaction	Generate the redox species at the substrate and collect the species at the tip
Tip generation/substrate collection	Substrate reaction and tip reaction	Generate the redox species at the tip and collect the species at the substrate
Redox competition mode	Substrate and tip reaction	The substrate and the probe compete for the same electroactive species in solution
Direct mode	Tip reaction	The probe acts as a counter electrode. A substrate acts as a working electrode. There is a localized electric field between the substrate and the tip
Potentiometric	No faradaic reaction	The local potential is measured at the probe using an ion-selective electrode. No faradaic reaction occurs

Adapted from ref. [57], with permission. Copyright Elsevier 2020.

representative of the potential barrier associated with the transfer of electrons through the electrode/electrolyte interface; and the double-layer capacitance (C_{dl}). The latter element represents the capacitive effect due to the charge separation effect produced in the vicinity of the electrode surface. This results from the space distribution of ionic species, i.e., Helmholtz's double layer [40].

The real and imaginary components of the total impedance of the Randles circuit can be expressed as

$$Z_{real} = R_s + \frac{R_{ct}}{1 + R_{ct}^2 C_{dl}^2 \omega^2}, \quad (1.10)$$

$$Z_{imag} = -\frac{jR_{ct}^2 C_{dl} \omega}{1 + R_{ct}^2 C_{dl}^2 \omega^2}. \quad (1.11)$$

The resulting impedance spectrum can be represented in several ways. The Bode plots represents either the absolute value of the logarithm of total impedance, log|Z|, or the (minus) phase angle, -φ, vs. logarithm of frequency. The former graph consists of an inverted "s", as shown in Figure 1.9a. At the extreme of high frequencies, the total impedance tends to R_s, while at the extreme of low frequencies, it tends to $R_{ct} + R_s$. At intermediate frequencies, the Bode plot of log|Z| vs. log*f* approximates a straight line of slope −1. If the ideal capacitance C_{dl} is replaced by a CPE, the aforementioned intermediate region approximates a straight line of slope −*h*.

The −φ, vs. log*f* Bode plot presents a maximum at intermediate frequencies (Figure 1.9b). This marks the frequency at which the capacitive effects in the electrochemical cell are maximal. This frequency is related to the time constant of the RC unit. Finally, the Nyquist representation consists of the plot of $-Z_{imag}$ vs. Z_{real}. In the case of the Randles circuit (Figure 1.10), it consists of a depressed semicircle whose diameter in the real impedance axis can be taken as the charge transfer resistance.

Most impedance spectra cannot be modeled using the Randles circuit so that a variety of models have been proposed to describe experimental impedance spectra. Examples will be seen in the next chapters. An interesting operational view is that, given the large range of frequencies used in EIS

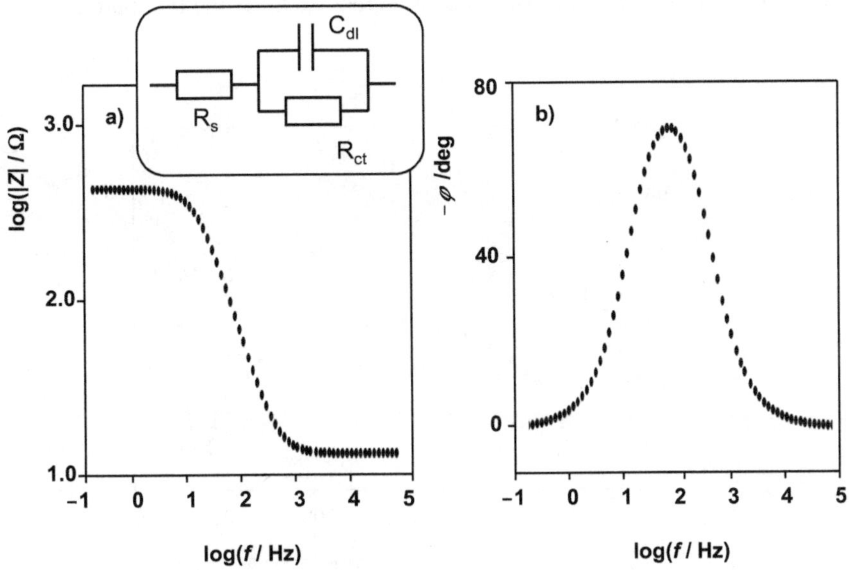

FIGURE 1.9 Bode plots of impedance spectra modeled by the Randles circuit (inset) taking R_s = 15 Ω, R_{ct} = 400 kΩ, and C_{dl} = 30 μF. (a) log|Z| vs. logf; (b) −φ vs. logf representations.

measurements, impedance data offer the possibility to record different relaxation phenomena, in turn resulting in different time constants. These are reflected in successive loops in Nyquist plots and successive (minus)phase angle maxima or inflections in Bode plots.

The EIS technique has been extensively used in the analysis of surfaces where impedance parameters are usually normalized to surface area units. Then, resistances are expressed as Ω cm^2 and capacitances as F cm^{-2}.

1.4.6 Scanning Electrochemical Microscopy

Scanning electrochemical microscopy (SECM) can be considered a chemical microscopy technique. It originates as a technique of surface analysis inspired by scanning electron microscopy, scanning tunneling microscopy, and atomic force microscopy. SECM is based on the record of the current flowing through a microelectrode or ultramicroelectrode tip (usually 1–25 µm diameter), which moves relative to a surface, termed the substrate, both in contact with a suitable electrolyte [56]. The tip can be horizontally and vertically displaced at a close distance from the substrate using stepper and piezoelectric motors. There is the possibility of applying adjustable excitation potentials to the substrate (substrate potential, E_S) and the tip (tip potential, E_T) employing a low-current bipotentiostat. The electrode arrangement is completed with reference and auxiliary electrodes.

Although it can be used to simply determine the topography of the problem surface, the SECM technique enables us to determine the "electrochemical topography" of the surface, i.e., the local concentration of electroactive species. The simplest operation mode is schematized in Figure 1.11. Here, the tip is immersed into a solution containing a redox probe such as $[Fe(CN)_6]^{3-}/[Fe(CN)_6]^{4-}$, ferrocene, or dissolved oxygen. The tip potential is maintained at a fixed value where the redox probe is oxidized or reduced under diffusive control and the current flowing through the tip is measured. Under these conditions, the tip current depends on the distance to the surface and the conducting nature of the substrate.

It is relevant to note that the diffusion of the redox probe on the tip microelectrode is hemispherical or cylindrical. When the distance between the tip and the surface is around a few tip diameters, the diffusion of the redox probe from the bulk solution to the tip is blocked by the surface of the substrate. If the tip approximates to an insulating surface, the tip current decreases monotonically because of the inhibition of radial diffusion (negative feedback). If the tip approaches an electrical

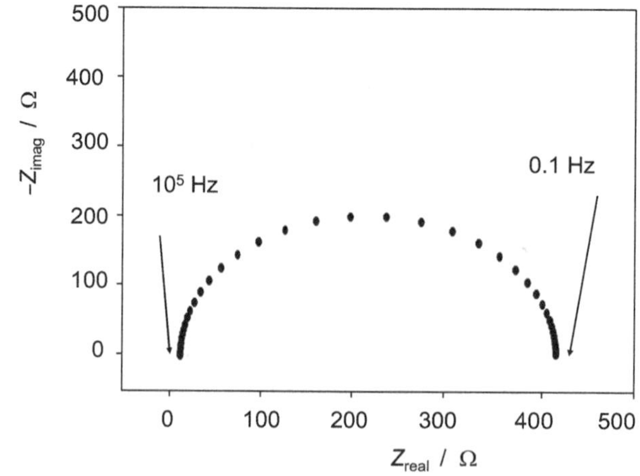

FIGURE 1.10 Nyquist plot ($-Z_{imag}$ vs. Z_{real}) of impedance spectra modeled by the Randles circuit in Figure 1.9 taking R_s = 15 Ω, R_{ct} = 400 kΩ, and C_{dl} = 30 µF.

conductor, the tip current increases (positive feedback). The corresponding tip current vs. tip-substrate distance graphs are the so-called approach curves [56].

Under the aforementioned conditions, if the tip is scanned horizontally at a fixed distance over a rough surface of homogeneous composition, the variations of the tip current provide a topographic map of the surface. If the surface is heterogeneous in composition, the tip current reflects the local variations in topography and conductivity. There are other possible operation modes – namely, tip generation/substrate collection, substrate generation/tip collection, redox competition, amperometric, potentiometric, etc. The main types of SECM operation are summarized in Table 1.2 and schematized in Figure 1.11 [57].

As a result, a variety of applications in biological systems have been reported. These include imaging of living cells, concentration profiles of oxygen evolved in photosynthetic electron transport in plant leaves [58], molecular absorption, and metabolite distribution in biofilms [59], among others.

Figure 1.12 illustrates one example of monitoring complex biochemical reaction pathways using SECM. This figure shows the color maps of double-stranded deoxyribonucleic acid fibers (dsDNA) deposited onto Pt electrode in contact with air-saturated 0.10 M Bu_4NPF_6/DMSO [60]. The images correspond to a study aimed to assess the reactivity of chartreusin, an aromatic polyketide glycoside isolated from *Streptomyces chartreusis* Leach et al., which exhibits antibacterial and antitumor activity. Here, the dissolved oxygen acts as a redox probe exploiting its one-electron essentially reversible reduction (at ca. −0.85 V vs. Ag/AgCl) in aprotic media [61]. In the experiments in Figure 1.12, the tip potential was fixed at a value (E_T = −0.70 V vs. Ag/AgCl), close to the midpeak potential of the $O_2/O_2^{\cdot-}$ couple. When no potential input was applied to the substrate (Figure 1.12b), the dsDNA fibbers appeared as negative feedback features, whereas the Pt surface directly exposed to the electrolyte appeared as a positive feedback background.

Applying a potential of −1.0 V to the substrate, dissolved oxygen is increasingly reduced at the Pt surface and the local concentration of superoxide radical anion increases. As a result, the positive feedback region becomes sharply contrasted (Figure 1.12b). The addition of chartreusin to the solution determines significant changes in the SECM pattern. When no potential input was applied to the substrate (Figure 1.12c), the negative feedback features were prolonged, suggesting that chartreusin

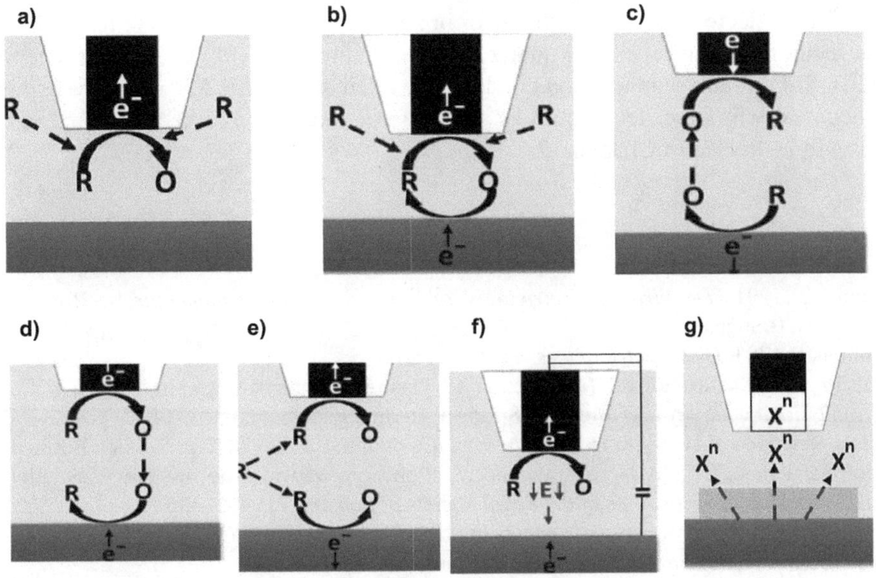

FIGURE 1.11 Main modes of SECM operation. The oxidation process Ox → Rd + e⁻ acts as a redox probe. (Adapted from ref. [57], with permission. Copyright Elsevier 2020.).

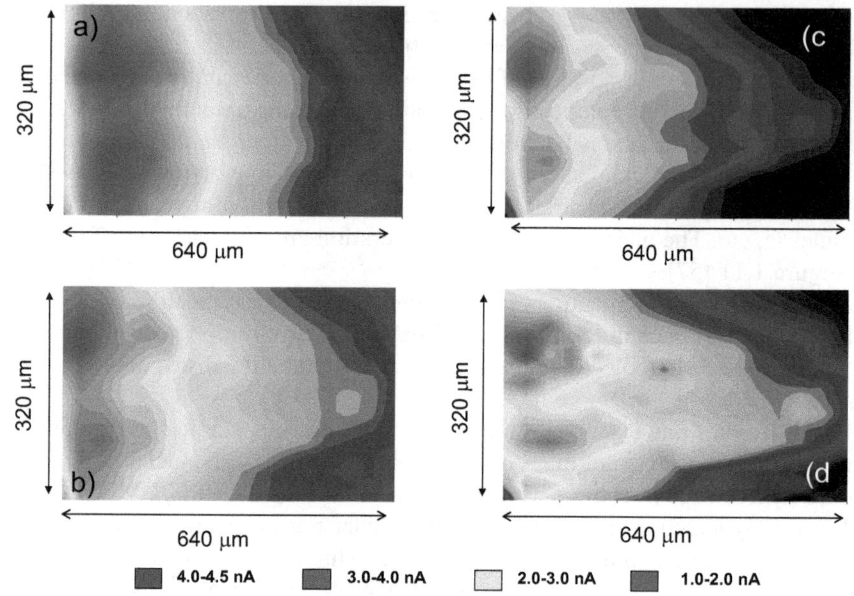

FIGURE 1.12 SECM color maps (here presented as different intensities of gray) for a deposit of dsDNA fibers attached to a Pt electrode in contact with air-saturated: (a, b) 0.10 M Bu_4NPF_6/DMSO solution, and (c, d) 0.10 M Bu_4NPF_6/DMSO plus 0.03 mM chartreusin solutions. (a, c) $E_S = 0.00$ V; $E_T = -0.70$ V; (b, d) $E_S = -1.00$ V; $E_T = -0.70$ V. (Adapted from ref. [60], with permission. Copyright the Royal Society of Chemistry 2017.).

interacts with the dsDNA fibers partially blocking the exposed Pt surface. When a potential of −1.0 V is applied to the substrate (Figure 1.12d), not only dissolved oxygen but also chartreusin is reduced. Under these conditions, the profile of the negative feedback features is clearly modified. This suggests that there is an interaction between the reduced form of chartreusin and the superoxide radical anion. These results were interpreted in terms of significant differences in the radical scavenging activity of chartreusin in the presence and absence of dsDNA [60].

There are other electrochemical methods for imaging and monitoring redox reactivity at the cellular scale, including electrochemical quartz crystal microbalance (EQCM), electrochemiluminescence (ECL), SECM, and scanning ion conductivity microscopy (SICM), as well as combinations between them and with other techniques as atomic force microscopy (AFM) [57, 62, 63]. Several approaches will be treated in Chapter 12.

REFERENCES

[1] Galvani, L. 1791. *De viribus electricitatis in motu musculari commentaries*. Bononiae Instituti Scientiarum, Bologna.
[2] Bertholon, P. 1783. *De l'electricite des vegetaux*. Paris.
[3] Stahlberg, R. 2006. Historical Introduction to Plant Electrophysiology, in Volkov, A.G. Ed. *Plant Electrophysiology Theory and Methods*. Springer, Berlin-Heidelberg, pp. 3–14.
[4] Du Bois-Reymond, E.H. 1848. *Untersuchungen uber thierische Elektrizitat*. Reimer, Berlin.
[5] Burdon-Sanderson, J. 1873. Note on the electrical phenomena which accompany stimulation of leaf of *Dionea muscipula*. *Proceedings of the Royal Society of London*. 21: 495–496.
[6] Darwin, C. 1896. The power of movements in plants. Appleton, New York.
[7] Bernstein, J. 1912. *Elektrobiologie*. Thieme, Braunschweig.
[8] Umrath, K. 1930. Untersuchungen uber Plasma und Plasamstromung a Characeen. IV. Potentialmessungen an *Nitella mucronata* mit besonderer Berucksichtigung der Erregungserscheinungen. *Protoplasma*. 9: 576–597.

[9] Osterhout, W.J.V. 1936. Electrical phenomena in large plant cells. *Physiological Reviews*. 16: 216–237.
[10] Higinbotham, N. 1973. Electropotentials of plant cells. *Annual Review of Plant Physiology*. 24: 25–46.
[11] Gunar I.I.; Sinykhin, A.M. 1962. A spreading wave of excitation in higher plants. *Proceedings of the Academy of Sciences USSR (Botany)*. 142: 214–215.
[12] Karmanov, V.G.; Lyalin, O.O.; Mamulashvili, G.G. 1972. The form of action potentials and cooperativeness of the excited elements in stems of winter squash. *Soviet Plant Physiology*. 19: 354–420.
[13] Gunar, I.I.; Sinykhin, A.M. 1963. Functional significance of action currents affecting the gas exchange of higher plants. *Soviet Plant Physiology*. 10: 219–226.
[14] Koziolek, C.; Grams, T.E.; Schreiber, U.; Matyssek, R.; Fromm, J. 2003. Transient knockout of photosynthesis mediated by electrical signals. *New Phytologist*. 161: 715–722.
[15] Haake, O. 1892. Uber die Ursachen elektrischer Strome in Pflanzen. *Flora*. 75: 455–487.
[16] Higinbotham, N. 1973. Electropotentials of plant cells. *Annual Review of Plant Physiology*. 24: 25–46.
[17] Spanswick, R.M. 1974. Evidence for an electrogenic pump in *Nitella translucens*. I. Control of the light-stimulated component of the membrane potential. *Biochimica and Biophysica Acta*. 332: 387–398.
[18] Messerli, M.A.; Robinson, K.R.; Smith, P.J.S. 2006. Electrochemical Sensor Applications to the Study of Molecular Physiology and Analyte Flux in Plants, in Volkov, A.G. Ed. *Plant Electrophysiology Theory and Methods*. Springer, Berlin-Heidelberg, pp. 72–107.
[19] Lew, R.R. 2006. Use of Double Barrel Micropipettes to Voltage-Clamp Plant and Fungal Cells, in Volkov, A.G. Ed. *Plant Electrophysiology Theory and Methods*. Springer, Berlin-Heidelberg, pp. 138–154.
[20] Jaffe, L.F.; Nuccitelli, R. 1974. An ultrasensitive vibrating probe for measuring steady extracellular currents. The *Journal of Cell Biology*. 63: 614–628.
[21] Hamill OP, Marty A, Neher E, Sakmann B, Sigworth FJ (1981) Improved patch-clamp techniques for high-resolution current recording from cells and cell-free membrane patches. *Pflugers Archiv*. 391: 85–100.
[22] Scholz, F. 2015. Voltammetric techniques of analysis: the essentials. *ChemTexts*. 1: 17.
[23] Díaz-Cruz, J.M.; Serrano, N.; Pérez-Ràfols, C.; Ariño, C.; Esteban, M. 2020. Electroanalysis from the past to the twenty-first century: challenges and perspectives. *Journal of Solid State Electrochemistry*. 24: 2653–2661.
[24] Smutok, O.; Katz, E. 2024. Electroanalytical instrumentation—how it all started: history of electrochemical instrumentation. *Journal of Solid State Electrochemistry*. 28: 683–710.
[25] Xiao, T.; Wu, F.; Hao, J.; Zhang, M.; Yu, P.; Mao, L. 2017. In Vivo Analysis with Electrochemical Sensors and Biosensors. *Analytical Chemistry*. 89: 300–313.
[26] Bard, A.J.; Faulkner, L.R. 2001. *Electrochemical Methods Fundametals and Applications*, 2nd ed. Wiley, New York.
[27] Lasia, A. 2014. *Electrochemical Impedance Spectroscopy and its Applications*. Springer, Berlin-Heidelberg.
[28] Timofeeva, I.; Nugbienyo, L.; Pochivalov, A.; Vakh, C.; Shishov, A.; Bulatov, A. 2021. Flow-based methods and their applications in chemical analysis. *ChemTexts*. 7: artic. 24.
[29] Díaz-Cruz, J.M.; Esteban, M.; Ariño, C. 2020. *Chemometrics in Electroanalysis*. Monographs in Electrochemistry series, Scholz, F. Edit. Springer, Berlin-Heidelberg.
[30] Shockcor, J.P.; Holmes, E. 2002. Metabonomic applications in toxicity screening and disease diagnosis. *Current Topics in Medicinal Chemistry*. 2: 35–51.
[31] Doménech-Carbó, A.; Ibars, A.M.; Prieto-Mossi, J.; Estrelles, E.; Scholz, F.; Cebrián-Torrejón, G.; Martini, M. 2015. Electrochemistry-based chemotaxonomy in plants using the voltammetry of microparticles methodology. *New Journal of Chemistry*. 39: 7421–7428.
[32] Miller, A.J.; Zhen, R.-G. 1991. Measurement of intracellular nitrate concentration in Chara using nitrate-selective microelectrodes. *Planta*. 184: 47–52.
[33] Henriksen, G.H.; Bloom, A.J.; Spanswick, R.M. 1990. Measurement of net fluxes of ammonium and nitrate at the surface of barely roots using ion-selective microelectrodes. *Plant Physioogy*. 93: 271–280.
[34] Sumikama, T.; Saito, S.; Ohmine, I. 2006. Mechanism of Ion Permeation in a Model Channel: Free Energy Surface and Dynamics of K^+ Ion Transport in an Anion-Doped Carbon Nanotube. *Journal of Physical Chemistry B* 110: 20671–20677.
[35] Masi, E.; Ciszak, M.; Stefano, G.; Renna, L.; Azzarello, E.; Pandolfi, C.; Mugnai, S.; Balŭska, F.; Arecchi, F.; Mancuso, S. 2009. Spatiotemporal dynamics of the electrical network activity in the root apex. *Proceedings of the National Academy of Sciences of USA*. 106: 4048–4053.

[36] Hodgkin, A.L.; Huxley, A.F. 1952. A quantitative description of membrane current and its application to conduction and excitation in nerve. *Journal of Physiology*. 117: 500–544.

[37] Volkov, A.G.; Lang, R.D.; Volkova-Gugeshashvili, M.I. 2007. Electrical signaling in *Aloe vera* induced by localized thermal stress. *Bioelectrochemistry*. 71: 192–197.

[38] Fromm, G. 2006. Long-Distance Electrical Signaling and Physiological Functions in Higher Plants, in Volkov, A.G. Ed. *Plant Electrophysiology Theory and Methods*. Springer, Berlin-Heidelberg.

[39] Li, J.-H.; Fan, L.-F.; Zhao, D.-J.; Zhou, Q.; Yao, J.-P.; Wang, Z.-Y.; Huang, L. 2021. Plant electrical signals: A multidisciplinary challenge. *Journal of Plant Physiology*. 261: artic. 153418.

[40] Bard, A.J.; Inzelt, G.; Scholz, F. Eds. 2008. *Electrochemical Dictionary*. Springer, Berlin-Heidelberg.

[41] Doménech-Carbó, A.; Navarro, P.; Arán, V.J.; Muro, B.; Montoya, N.; García-España, E. 2010. Selective electrochemical discrimination between dopamine and phenethylamine-derived psychotropic drugs using electrodes modified with an acyclic receptor containing two terminal 3-alkoxy-5-nitroindazole rings. *Analyst*. 135: 1449–1455.

[42] Kahlert, H. 2022. Potentiometry, in Scholz, F. Ed. *Electroanalytical Methods*. Springer, Berlin.

[43] Doménech-Carbó, A. 2021. *Electrochemistry of Porous Materials*. Taylor & Francis, Boca Raton.

[44] Umezawa, Y.; Umezawa, K.; Sato, H. 1995. Selectivity coefficients for ion-selective electrodes: recommended methods for reporting $K^{pot}{}_{A,B}$ values (IUPAC Technical Report). *Pure and Applied Chemistry*. 67: 507–518.

[45] Umezawa, Y.; Bühlmann, P.; Umezawa, K.; Tohda, K.; Amemiya, S. 2000. Potentiometric selectivity coefficients of ion-selective electrodes. Part I. Inorganic cations (IUPAC Technical Report). *Pure and Applied Chemistry*. 72: 1851–1856

[46] Miller, A.J.; Wells, D.N. 2006. Electrochemical methods and Measuring Transmembrane Ion Gradients, in Volkov, A.G. Ed. *Plant Electrophysiology Theory and Methods*. Springer, Berlin-Heidelberg, pp. 138–154.

[47] Ohno, H. 2005. *Electrochemical aspects of Ionic Liquids*. Wiley-Interscience, New York.

[48] Štulík, K.; Amatore, C.; Holub, K.; Mereček, V.; Kutner, W. 2000. Microelectrodes. Definitions, characterization, and applications (IUPAC Technical Report). *Pure and Applied Chemistry*. 72: 1483–1492.

[49] Dubois, N.; Grimaud, A. 2019. The hydrogen evolution reaction: from material to interfacial descriptors. *Chemical Science*. 10: 9165–9181.

[50] Diaz-Morales, O.; Calle-Vallejo, F.; de Munck, C.; Koper, M.T. 2013. Electrochemical water splitting by gold: evidence for an oxide decomposition mechanism. *Chemical Science*. 4: 2334–2343.

[51] Nicholson, R.S.; Shain, I. 1964. Theory of Stationary Electrode Polarography Single Scan and Cyclic Methods Applied to Reversible, Irreversible, and Kinetic Systems. *Analytical Chemistry*. 36: 706–723.

[52] Nicholson, R.S. 1965. Some Examples of the Numerical Solution of Nonlinear Integral Equations. Analytical Chemistry. 37: 667–671.

[53] Guidelli, R.; Compton, R.G.; Feliu, J.M.; Gileadi, E.; Lipkowski, J.; Schmickler, W.; Trasatti, S. 2014. Defining the transfer coefficient in electrochemistry: An assessment (IUPAC Technical Report). *Pure and Applied Chemistry*. 86: 245–258.

[54] Retter, H.; Lohse, H. 2005. Electrochemical Impedance Spectroscopy, in Scholz, F. Ed. *Electroanalytical Methods*, Springer, Berlin.

[55] Macdonald, D.D.; Sikora, A.; Engelhardt, G. 1998. Characterizing electrochemical systems in the frequency domain. *Electrochimica Acta*. 43: 87–107.

[56] Bard, A.J.; Mirkin, M.V. Eds. 2001. *Electrochemical Microscopy*, Taylor & Francis, Boca Raton.

[57] Shi, X.; Qing, W.; Marhaba, T.; Zhang, W. 2020. Atomic force microscopy - Scanning electrochemical microscopy (AFM-SECM) for nanoscale topographical and electrochemical characterization: Principles, applications and perspectives. *Electrochimica Acta*. 332: artic. 135472.

[58] Tsionky, M.; Cardon, Z.G.; Bard, A.J.; Jackson, R.B. 1997. *Plant Physiology*. 113: 895–901.

[59] Caniglia, G.; Kranz, C. 2020. Scanning electrochemical microscopy and its potential for studying biofilms and antimicrobial coatings. *Analytical and Bioanalytical Chemistry*. 412: 6133–6148.

[60] Doménech-Carbó, A.; Cebrián-Torrejón, G.; Montoya, N.; Ueberschaar, N.; Scotti, M.T.; Benfodda, Z.; Hertweck, C. 2017. Electrochemical monitoring of ROS generation by anticancer agents: the case of chartreusin. *RSC Advances*. 7: 45200–45210.

[61] Sawyer, D.T.; Chiericato, G.; Angelis, C.T.; Nanni, E.J.; Tsuchiya, T. 1982. Effects of Media and Electrode Materials on the Electrochemical Reduction of Dioxygen. *Analytical Chemistry*. 54: 1720–1724.

[62] Huang, K.; Wang, Y.H.; Zhang, H.; Wang, T.Y.; Liu, X.H.; Liu, L.; Jiang, H. 2023. Application and outlook of electrochemical technology in single-cell analysis. *Biosensors & Bioelectronics*. 242: artic. 115741.

[63] Bedendi, G.; De Moura Torquato, L.D.; Webb, S.; Cadoux, C.; Kulkarni, A.; Sahin, S.; Maroni, P.; Milton, R.D.; Grattieri, M. 2022. Enzymatic and Microbial Electrochemistry: Approaches and Methods. *ACS Measurement Science*. 2: 517–541.

ADDITIONAL LITERATURE

HISTORICAL LITERATURE

Biedermann, W. 1895. *Elektrophysiologie*. Fischer, Jena.

Blinks, L.R. 1949. The source of the bioelectric potentials in large plant cells. *Proceedings of the National Academy of Sciences of USA*. 35: 566–575.

Bose, J.C. 1906. *Plant response as a means of physiological investigation*. Longman Green, London.

Heyrovsky, J. 1922. Elektrolysa se rtuťovou kapkovou kathodou. *Chemické Listy*. 16: 256–264.

Lemstrom, S. 1902. *Elektrokultur*. Springer, Berlin Heidelberg New York.

Nernst, W. 1889. Die elektromotorische Wirksamkeit der Ionen. *Zeitschrift für Physikalische Chemie*. 4: 129–181.

Randles, J.E.B. 1948. A cathode ray polarograph. Part II.—The current-voltage curves. *Transactions of the Faraday Society*. 44: 327–338.

Ševčík, A. 1948. Oscillographic polarography with periodical triangular voltage. *Collection of Czechoslovak Chemical Communications*. 13: 349–377.

Stern, K. 1924. *Elektrophysiologie der Pflanzen*. Springer, Berlin Heidelberg New York.

Umrath, K. 1932. Der Erregungsvorgang bei *Nitella mucronata*. *Protoplasma*. 17: 258–300.

GENERAL DISCUSSIONS ON PLANT ELECTROPHYSIOLOGY

Kitasato, H. 2003. Membrane potential genesis in *Nitella* cells, mitochondria and thylakoids. *Journal of Plant Research*. 116: 401–418.

Pickard, B.G. 1973. Action potentials in higher plants. *The Botanical Review*. 39: 172–201.

Trewawas, A. 2003. Aspects of plant intelligence. *Annals of Botany*. 92: 1–20.

ELECTROCHEMICAL TECHNIQUES

Batchelor-McAuley, C.; Kätelhön, E.; Barnes, E.O.; Compton, R.G.; Laborda, E.; Molina, A. 2015. Recent Advances in Voltammetry. *ChemistryOpen*. 4: 224–260.

Compton, R.G.; Sokolov, S.V. 2024. Electrochemistry needs electrochemists: "goodbye to rotating discs". *Journal of Solid State Electrochemistry*. 28: 1041–1047.

Doménech-Carbó, A.; Koshevoy, I.O.; Montoya, N.; Pakkanen, T.A. 2010. Electrochemical anion sensing using electrodes chemically modified with Au(I)-Cu(I) heterotrimetallic alkynyl cluster complexes containing ferrocenyl groups. *Analytical and Bioanalytical Chemistry*. 397: 2013–2022.

Doménech-Carbó, A.; Doménech-Carbó, M.T. 2023. *Electrochemistry for Cultural Heritage*, Monographs in Electrochemistry series, F. Scholz, L. Péter Eds. Springer, Berlin-Heidelberg.

Fabry, P.; Seibert, E. 1997. Electrochemical Sensors, in Gellings, P.J.; Bouwmeester, H.J.M. Eds. *CRC handbook of solid state electrochemistry*. CRC Press, Boca Raton.

Gomes Cavalheiro, É.T.; Brett, C.M.A.; Oliveira-Brett, A.M. Fatibello-Filho, O. 2012. Bioelectroanalysis of pharmaceutical compounds. *Bioanalytical Reviews*. 4: 31–53.

Huang, L.; Li, Z.; Lou, Y.; Cao, F.; Zhang, D.; Li, X. 2018. Recent Advances in Scanning Electrochemical Microscopy for Biological Applications. *Materials*. 11: artic. 1389.

Inzelt, G. 2014. Crossing the bridge between thermodynamics and electrochemistry. From the potential of the cell reaction to the electrode potential. *ChemTexts*. 1: 1.

Lovrić, M. 2002. Square-wave voltammetry, in Scholz, F., Ed. *Electroanalytical Methods*. Springer, Berlin.

Marken, F.; Neudeck, A.; Bond, A.M. 2000. Cyclic Voltammetry, in Scholz, F. Ed. *Electroanalytical methods*. Springer, Berlin.

Mirceski, V.; Skrzypek, S.; Stojanov, L. 2018. Square-wave voltammetry. *ChemTexts*. 4: 17.

Polcari, D.; Dauphin-Ducharme, P.; Mauzeroll, J. 2016. *Chemical Reviews*. 116: 13234–13278.
Pyun, S.-I. 2021. Thermodynamic aspects of energy potential systems with focus on osmotic potential and selectively permeable membrane (Donnan) systems including two applications of the Donnan potential. *ChemTexts*.7: 20.
Scholz, F.; Schröder, U.; Gulabowski, R.; Doménech-Carbó, A. 2014. *Electrochemistry of Immobilized Particles and Droplets*, 2nd ed. Springer, Berlin-Heidelberg.
Seeber, R.; Zanardi, C.; Inzelt, G. 2015. Links between electrochemical thermodynamics and kinetics. *ChemTexts*. 1: 18.
Squissato, A.L.; Munoz, R.A.A.; Banks, C.E.; Richter, E.M. 2020. An Overview of Recent Electroanalytical Applications Utilizing Screen-Printed Electrodes Within Flow Systems. *ChemElectroChem*. 7, 2211–2221.
Sin, P.; Laforge, F.O.; Mirkin, M.V. 2007. Scanning electrochemical microscopy in the 21st century. *PhysChemChemPhys*. 9: 802–823.
Wang, J. 2023. *Analytical Electrochemistry*, 4th ed. VCH Publ., New York.

Electrochemical Processes

Brown, A.P.; Anson, F.C. 1977. Cyclic and Differential Pulse Voltammetric Behavior of Reactants Confined to the Electrode Surface. *Analytical Chemistry*. 49: 1589–1595.
Meng, Y.; Aldous, L.; Belding, S.R.; Compton, R.G. 2012. The Hydrogen Evolution Reaction in a Room Temperature Ionic Liquid: Mechanism and Electrocatalyst Trends. *PhysChemChemPhys*. 14: 5222–5228.
Mom, R.; Frevel, L.; Velasco-Vélez, J.-J.; Plodinec, M.; Knop-Gericke, A.; Schlögl, R. 2019. The Oxidation of Platinum under Wet Conditions Observed by Electrochemical X-Ray Photoelectron Spectroscopy. *Journal of the American Chemical Society*. 141: 6537–6544.
Sarabia, F.J.; Sebastián-Pascual, P.; Koper, M.T.M.; Climent, V.; Feliu, J.M. 2019. Effect of the Interfacial Water Structure on the Hydrogen Evolution Reaction on Pt(111) Modified with Different Nickel Hydroxide Coverages in Alkaline Media. *ACS Applied Materials and Interfaces*. 11: 613–623.

2 Electrophysiology I

2.1 SEVERAL GENERAL CONSIDERATIONS ON PLANT PHYSIOLOGY

Plants are autotrophic organisms requiring the supply of light, carbon dioxide, water, and minerals from the environment to survive. Their physiology is based on the occurrence of a series of complex oxidation-reduction processes in cells. These processes occur in aqueous solution and are coupled to different transport phenomena. The differences in pressure and solute concentration gradients determine water transport in the vascular strands. The transport of water and nutrients from the soil to the xylem and from xylem to the substomatal cavity is termed short-distance transport. This is complemented by the long-distance transport through the xylem to the foliar parts of the plant [1].

Given the essential role of water in plant physiology, the so-called water potential was introduced to express the free energy of water per volume unit (J m^{-3}). Equivalently, water potential, ψ_w, can be conceived as a pressure and expressed in megapascal (MPa). Roughly speaking, this quantity reflects the ability of a plant to absorb water, the higher the water potential of the plant, the lower its ability to absorb water. In a living cell, ψ_w results from the sum of four components, pressure potential (ψ_p), solute/osmotic potential (ψ_s), gravitational potential (ψ_g), and matric potential (ψ_m):

$$\psi_w = \psi_p + \psi_s + \psi_g + \psi_m \tag{2.1}$$

The pressure potential represents the pressure needed to stop water entrance into a cell. Also known as the turgor pressure, this quantity represents the difference between inside and outside hydrostatic pressures across the plasma membrane and the cell wall. The solute (or osmotic) potential represents the reduction of water potential due to the presence of solutes corresponding to the well-known osmotic pressure in physical chemistry, proportional to the effective (accounting for ionization) concentration of solutes.

When a plant cell is in contact with an isotonic (i.e., having the same osmotic pressure) solution, the cell remains flaccid (see Scheme in Figure 2.1a). If the cell is in contact with a hypotonic solution (Figure 2.1b), water ingress from the solution into the cell, which becomes turgid, and the plasma membrane is connected to the cell wall. Now, the cell experiences equal and opposite pressures through the wall (wall pressure). If the cell is in contact with a hypertonic solution, there is a net water loss from the vacuoles. Now, the plasma membrane is connected to the cell wall through a series of thin (Hechtian) strands (Figure 2.1c) [1].

Gravitational potential is associated with the gravitation effect on water columns (i.e., hydrostatic pressure) and mainly affects the motion in tall trees. Matric potential is associated with water adsorption by hydrophilic surfaces of plants, giving rise to the phenomenon of imbibition. This potential is only significant in dehydrated cells and tissues such as seeds and desert plants.

Water transport between cells (short-distance transport) can be made via diffusion, osmosis, or mass (or bulk) flow. The latter determines the water transport in the xylem and phloem elements. The long-distance transport of xylary elements distributes water and nutrients through the plant. The ending phase of water transport is transpiration, consisting of water evaporation in the aerial parts of the plant. This process occurs through the stomata that are open during the day for gaseous exchange. Ultimately, the loss of water via transpiration acts as a driving force for water absorption in the roots and transport through the plant [1].

DOI: 10.1201/9781032697826-2

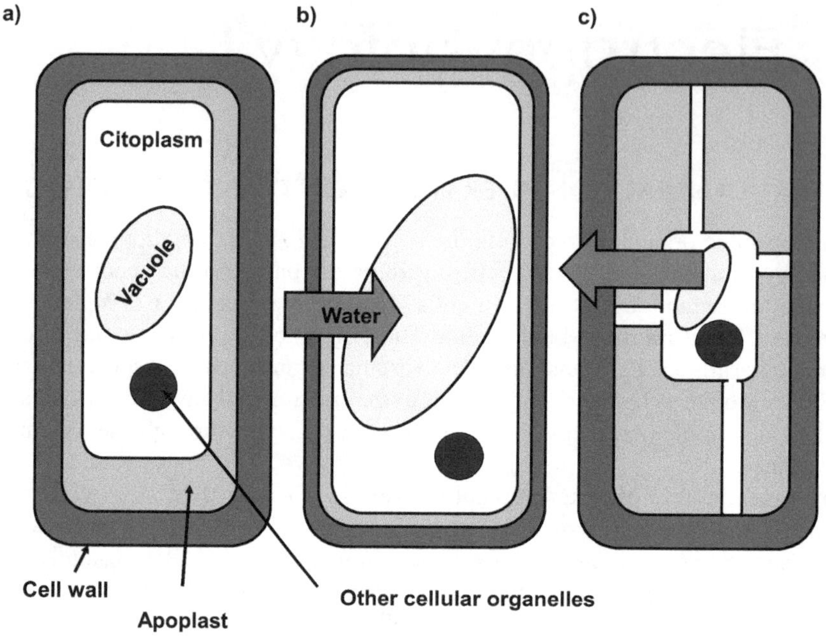

FIGURE 2.1 Schematic representation of a plant cell in contact with an (a) isotonic, (b) hypotonic, and (c) hypertonic solution.

2.2 TRANSPORT THROUGH CELL MEMBRANES

Cell membranes are selectively permeable barriers that separate the cell from the immediate environment. These are constituted as bilayer lipid barriers of a hydrophobic nature so that the ability of a given chemical species to permeate membranes is related to their hydrophobicity. The permeability of plasma membranes is high for molecules such as CO_2, N_2, and O_2, and for small nonpolar and some polar molecules (for example, ethanol). However, the permeability is low for charged species, small polar molecules (water, in particular), and large molecules such as sugars and nucleotides [2, 3].

The bilayer is constituted by phospholipids that have hydrophobic tails facing between them and polar heads facing the intra- and extracellular fluids. The cell membrane also contains glycolipids, and cholesterol-bound integral (strongly bound) and peripheral (weakly bound) proteins. Several integral proteins can cross the membrane (transmembrane proteins) [2, 3].

The transport of ions and other nutrients through the membrane is made by specific transport proteins embedded in the membrane. The corresponding transport according to the concentration gradient of a given species is referred to as facilitated diffusion. This transport is performed via carriers (or uniporters) and ion channels. This is often called passive transport. When proteins move the species against its electrochemical gradient, active transport occurs. This type of transport is divided into primary and secondary. Primary active transport is made by "pumps" that extract the necessary energy from the hydrolysis of adenosine triphosphat (ATP) or pyrophosphate. Secondary active transporters (or cotransporters) utilize ion gradients generated by pumps associated with the transport of other species against their electrochemical gradients [3, 4].

Cellular transporters can be divided into four groups – namely, pores, channels, carriers, and pumps. Their function is the maintenance of cellular composition (homeostasis) or the control of specific cell functions. The expression of several of these proteins is specific to tissues or membrane domains, whereas others are expressed in most cell membranes.

Pores are open communications between both sides of the membrane. Channels are communications that can be closed stochastically by conformational changes (gatting) and then become impermeable. Carriers are communications that are not simultaneously accessible to both sides. Pumps are equivalent to carriers coupled to a metabolic energy source [3]. It is convenient to underline that the major pathway for water transport across membranes is through water channel proteins, and aquaporins, and only a minor fraction of water is transported by simple diffusion.

Passive transport is driven by chemical or electrochemical gradients encompassing differences in concentration, electrical potential, and/or pressure between the two sides of the membrane. In the case of the transport of an ion (i) of charge z_i, the difference in the electrochemical potential, $\Delta\mu_i$, can be expressed under isothermal conditions as [3],

$$\Delta\mu_i = z_i E_m F + RT \ln \frac{c_i^{inn}}{c_i^{out}} + V_i \Delta p \tag{2.2}$$

where E_m is the membrane potential, c_i denotes the concentrations of the species i outside and inside the cell membrane, V_i is the molar volume of i, Δp is the transmembrane hydrostatic pressure difference, and the other symbols have their customary meaning. In the case of noncharged species, $z = 0$, the first term of the aforementioned equation vanishes. If the pressure difference is negligible, Eq. (2.2) can be rewritten to express the difference of electrical potential, ΔE, as

$$\Delta E = \frac{RT}{F} \ln \frac{c_i^{inn}}{c_i^{out}}. \tag{2.3}$$

Figure 2.2a depicts a simplified diagram representative of the transport of species i across a membrane that separates two phases. In the absence of pores or channels, the transport process involves

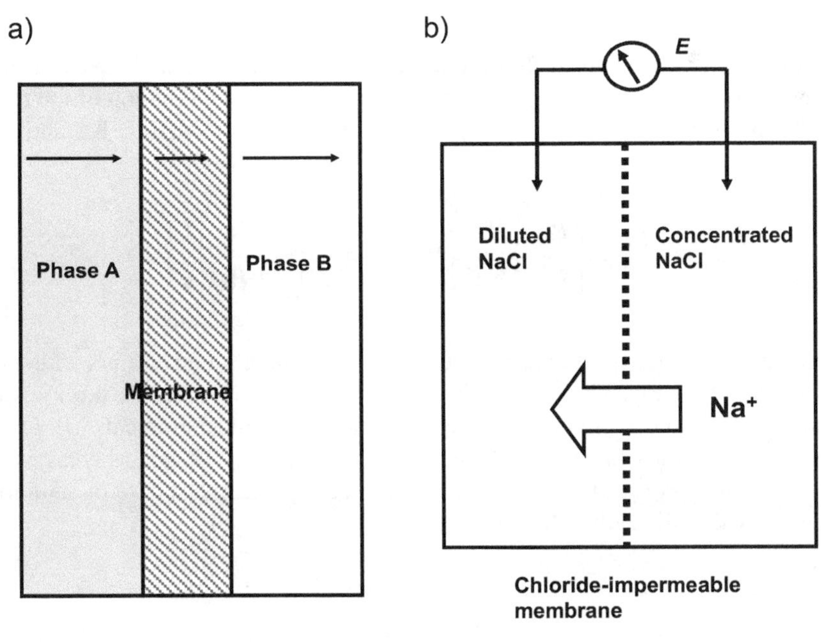

FIGURE 2.2 Simplified representations of (a) transport of a species i across a lipophilic membrane that separates two aqueous phases and (b) electrochemical equilibrium established between both sides of a semipermeable membrane that separates NaCl solutions of different concentrations.

the diffusion of i through phases A and B, the diffusion through the membrane, and the establishment of partition equilibrium between the hydrophilic and the lipophilic (membrane) phases [3]. In an experiment in which the solutions A and B are stirred so that the respective concentrations of i-species remain constant, and under steady-state conditions, the net flux of i-species, J_i (mol cm^{-2} s^{-1}), across the membrane can be estimated as [3],

$$J_i = -\frac{D_i K_i}{\delta} \Delta c_i = -P_i \Delta c_i, \tag{2.4}$$

where δ (cm) is the membrane thickness, D_i (cm^2 s^{-1}) is the diffusion coefficient of the i-species, K_i the partition coefficient (representative of the two-phase water-lipophilic membrane equilibrium), and Δc_i the concentration difference between the phases A and B. The D_i, K_i, and δ parameters can be condensed in a unique phenomenological parameter, P_i, the permeability of i-species (cm s^{-1}).

Figure 2.2b shows a scheme to illustrate the establishment of electrochemical equilibrium through a semipermeable membrane [3]. Here, two compartments containing NaCl solutions with different concentrations are separated by a chloride-impermeable membrane. Then, Na^+ ions move through the membrane because of the existence of a concentration gradient between the compartments. As a result, the membrane acquires opposite charges on both sides because of the presence of a tiny excess of Na^+ ions on one compartment and a tiny excess of Cl^- ions on the other. This causes a transmembrane (difference of) potential that acts as a driving force compensating the chemical (due to the difference of concentrations) driving force.

Electrodiffusive ion transport is a mediated process that, in the case of large pores filled with an aqueous electrolyte, can be assimilated to diffusive transport. Under several simplifying assumptions, the Goldman-Hodgkin-Katz equation applies [4, 5]:

$$J_i = -\frac{P_i z_i E_m F}{RT} \left[\frac{c_i^{out} - c_i^{inn} \exp(E_m F / RT)}{1 - \exp(E_m F / RT)} \right], \tag{2.5}$$

where, as before, E_m represents the difference between the intracellular and the extracellular potential. Under zero-current conditions, Eq. (2.5) gives this quantity as a function of the permeabilities and concentrations of all ions. As an example, for the permeation of Na^+, K^+, and Cl^- ions, we obtain [3],

$$E_m = -\frac{RT}{F} \ln \left[\frac{P_{Na+}[Na^+]_{inn} + P_{K+}[K^+]_{inn} + P_{Cl-}[Cl^-]_{inn}}{P_{Na+}[Na^+]_{out} + P_{K+}[K^+]_{out} + P_{Cl-}[Cl^-]_{out}} \right]. \tag{2.6}$$

Protein-mediated passive transport is carried out by pores, channels, and carriers without employing any metabolic energy supply. This means that the total energy involved in the transport process equals (or is lower than) the energy disposable in the electrochemical gradients [3]. In the diffusive transport, the flux increases linearly with the concentration (Eq. (2.4)), whereas in the carrier-mediated passive transport saturation kinetics is involved. The simplest case of saturation kinetics can be described by the Michaelis-Menten kinetics:

$$J_i = -\frac{c_i J_i(\max)}{c_i + K_m}, \tag{2.7}$$

where J_i(max) represents the maximum (saturation) flux and K_m is a constant whose meaning is that it represents the concentration at which the flux equals J_i(max)/2.

2.3 TRANSMEMBRANE POTENTIALS AND ION FLUXES

As briefly described in Chapter 1, the determination of transmembrane potentials plays a prominent role in the study of plant physiology [1]. The main electrochemical approach consists of the determination of ionic concentrations in cells. For this purpose, ion-selective microelectrodes (ISEs) measure the difference of electrical potential between the inside and outside of the cell.

Measurements require a solid anchoring of the plant material, ideally preserving its normal state. Avoiding the dissection of the plant material is desired as far as this operation leads to local wounding which alters, via gene expression [2], the composition of the plant material. During the measurements, the tissue can be perfused with nutrient solution to prevent large local concentration gradients of ions in the vicinity of the cells. The position of the microelectrodes is fixed with the help of micromanipulators for cellular impalement. The operations are usually performed under a microscope with the objective of long working distance.

Potentiometric measurements provide the sum of the potential difference due to the intracellular ion activity and that due to the voltage across the cell membrane. This requires the use of two single electrodes or a double-barreled electrode where the ion-sensing electrode is combined with a cell voltage-measuring electrode, as was previously schematized in Figure 1.4.

The identification of the internal cell compartment (cytoplasm or vacuole) in which the microelectrode is located may offer difficulties. One alternative is the use of triple-barreled microelectrodes yielding pH and Ca^{2+} determination. The cytoplasm maintains relatively constant concentrations of hydrogenions and Ca^{2+} (pH 7.2, Ca^{2+} 100 nM) in contrast with large variations in the tonoplast [6].

Noninvasive determination of ionic fluxes can also be achieved using slowly vibrating ISEs located in different positions in the solution outside the plant tissues. The concentration gradients determined in this way permit the calculation of the net fluxes of ions throughout the membrane [7, 8].

As an alternative to potentiometric measurements using ISEs, amperometric sensing can also be used. This offers high sensitivities but usually low selectivity. In both techniques, the effect of temperature, ionic strength, and the presence of interferent species, in particular the confounding effect exerted by metabolic inhibitors, have to be taken into account [9, 10]. As in all electrochemical techniques involving solid electrodes, electrode conditioning (cleaning and activation) is crucial to obtain repeatable and reproducible results.

The scope of potentiometric sensors has been extended to the study of ion fluxes in vivo using self-referencing ion vibrating electrode techniques. Figure 2.3 shows an image of the experimental

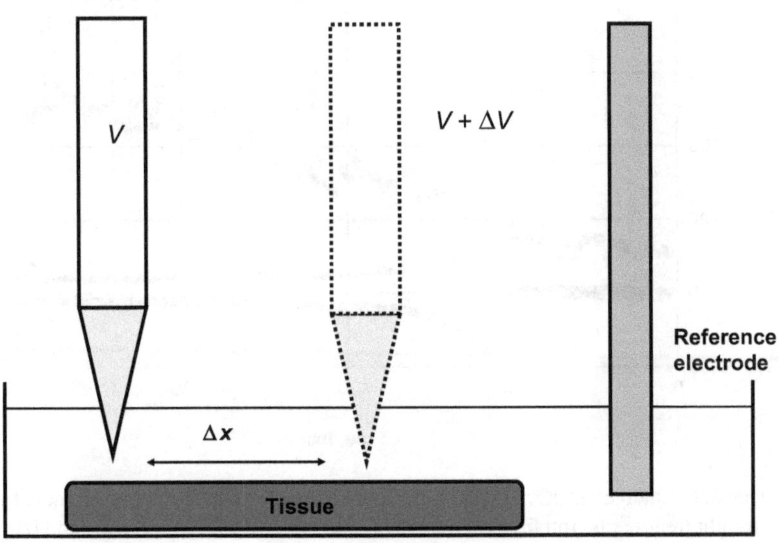

FIGURE 2.3 Scheme of the electrode arrangement for the potentiometric measurement of ion fluxes on plant tissues.

arrangement used for *in vivo in situ* determination of ion fluxes on plant tissues using ISEs with an ion-selective barrier. The position of the ISE is controlled with the aid of a microscope and an electric micro-manipulator, all the arrangements being placed within a Faraday cage on an anti-vibration platform. The electrode is moved through a known distance, and the potential change is measured by an electrometer.

Figure 2.4 shows the transient changes in the fluxes of K^+, H^+, and Ca^{2+} ions of soybean plants subjected to polyethylene glycol-induced drought treatment [11]. The ion fluxes were measured from leaf mesophyll cells (A) and root epidermal cells (B). Root and leaf segments were preincubated in a KCl plus $CaCl_2$ control solution, and potentiometric measurements were carried out for 10 min in the control solution to ensure steady initial values before adding 50 g L^{-1} polyethylene glycol for another 30 min.

The first step is the calibration of the ISE upon immersion into KCl solutions of known concentration. As a result, a linear E vs. $\log[K^+]$ plot is obtained. Ideally, the slope of this representation should be Nernstian (59 mV $decade^{-1}$ at 298 K). After this step, the calibration of ISE response to K^+ ion fluxes is performed by placing the electrode in a KCl solution of concentration c_B and

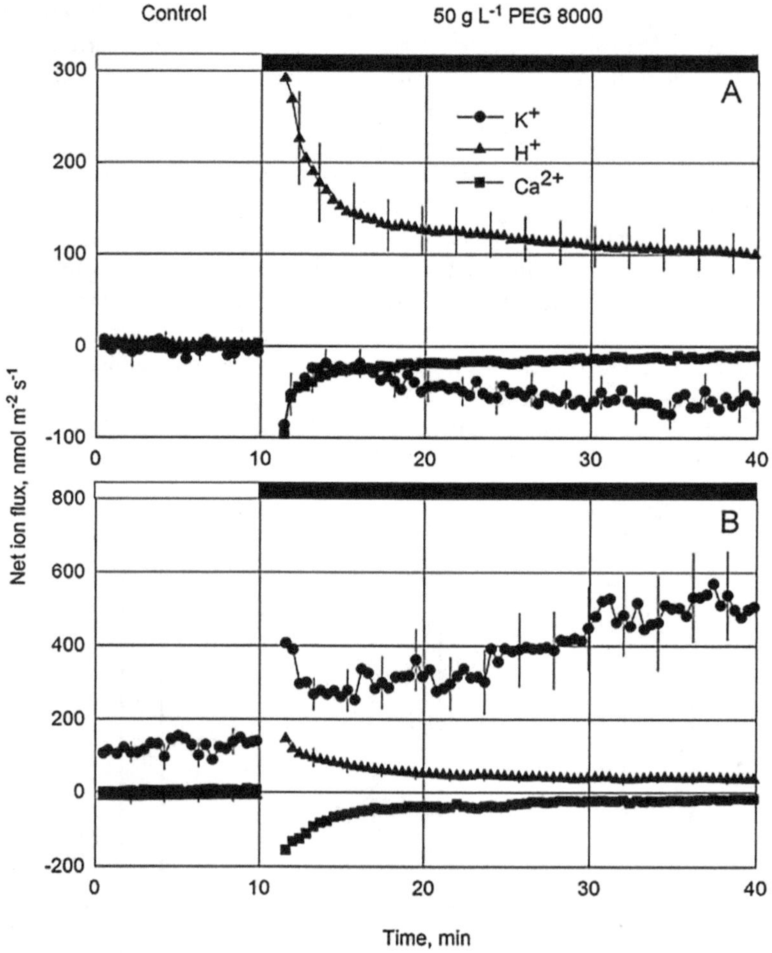

FIGURE 2.4 Transient changes in K^+, H^+ and Ca^{2+} ion fluxes of soybean plants subjected to polyethylene glycol-induced drought treatments. Ion fluxes measured from (A): leaf mesophyll cells and (B): root epidermal cells. Roots were collected from 10-day-old hydroponically grown seedlings, and leaves were sampled from well-water glasshouse-grown plants. (Reproduced from ref. [11], with permission. Copyright Elsevier 2014.).

incorporating a K^+ source. The measured steady concentration c_i can be related to the distance x between the ISE and the K^+ source as [9]

$$c_i \approx c_B + \frac{K}{x}. \tag{2.8}$$

where K is a proportionality constant. The measured potential difference, ΔE, can be approxiated by the relationship [12]

$$\Delta E = -\frac{K(SL)a}{2.302\left(c_B x^2 + Kx\right)}, \tag{2.9}$$

which applies to planar diffusion. Depending on the microelectrode geometry, cylindrical or hemispherical diffusion must be considered [13]. In this equation, (SL) is the Nernstian slope of the conventional ISE calibration and a is the vibration amplitude. The ion flux in the x-direction can be evaluated as

$$J_i = D\frac{\Delta c_i}{\Delta x} \approx D\frac{2.302\Delta V c_B}{\Delta x(SL)}. \tag{2.10}$$

The potential difference measured between positions (or "poles") A, B, can be related to ion concentrations similar to Eq. (2.3),

$$\Delta E = \frac{RT}{F}\ln\frac{c_i^A}{c_i^B}, \tag{2.11}$$

assuming that the slope of the calibration graph is the same. To compensate for the effect of the background solution of concentration c_B, one can take [14]

$$\Delta c_i = c_i^A - c_i^B = c_B\left[10^{\frac{\Delta E}{(SL)}} - 1\right]. \tag{2.12}$$

In the case of amperometric sensing, it is assumed that the current measured under fixed electrochemical conditions (I) is proportional to the concentration of the i-ion; i.e.,

$$I = g c_i. \tag{2.13}$$

After a classical calibration experiment, the proportionality constant, g, can be obtained as the slope of the linear I vs. c_i representation. Then, the difference in concentration between the two poles will be

$$\Delta c_i = \frac{\Delta I}{g} = \frac{I_A - I_B}{g}. \tag{2.14}$$

I_A, I_B, being the amperometric currents measured when the electrode was placed in the respective A, B positions [14]. As in potentiometric sensing, interference and matrix effects, including analyte buffering must be accounted for.

Figure 2.5 shows the time variation of drought-induced fluxes of K^+ (A), H^+ (B), and Ca^{2+} (C) ions in leaf mesophyll cells of soybean plants measured over 20 days of drought treatments with polyethylene glycol [11]. There are significant differences in the influx/efflux of the different ionic species depending on the severity of the drought conditions and time.

Figure 2.6 illustrates an application of ion flux measurement for monitoring salt shock effects. The time variation of the net flux of K^+, Ca^{2+}, and H^+ at apical regions of non-inoculated (NI) and *Paxillus involutus* Batchs ex Fr.-inoculated roots of *Populus* × *canescens* Aiton is represented in this figure [15]. The roots of *P.* × *canescens* were cultured in the absence and in the presence of *P. involutus* strains MAJ and NAU for ten days. A NaCl shock was performed by adding NaCl stock solution into the measuring chamber until a final concentration of 100 mM. Ion flux measurements were carried out at the apex (ca. 300 μm of the root tip). Steady fluxes for each ion before the salt shock

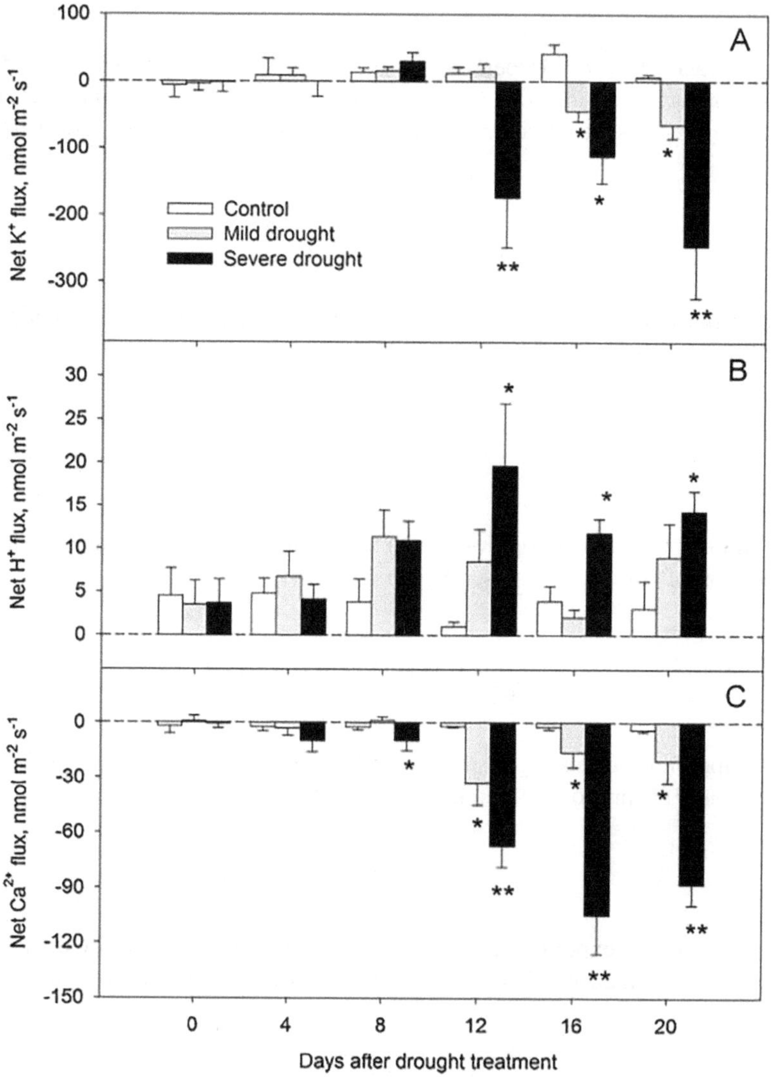

FIGURE 2.5 Drought-induced fluxes of K^+ (A), H^+ (B) and Ca^{2+} (C) ions in leaf mesophyll cells of soybean plants. The ionic fluxes were measured over 20 days of drought treatments with polyethylene glycol. Leaves were collected 0, 4, 8, 12, 16, and 20 days after drought. (Reproduced from ref. [11], with permission. Copyright Elsevier 2014.).

differ significantly from fluxes after the saline shock, but there are significant differences between the response of K^+, Ca^{2+}, and H^+. Such data are relevant in the context of studies concerning the enhancement of tolerance to salinity of wood species in the presence of highly salt-tolerant ectomycorrhizal (EM) fungi [16].

2.4 ION CHANNELS

Ion transport through plant membranes is involved in many cellular biological processes – namely, nutrient and metabolite acquisition and distribution, pH and ionic homeostasis, turgor generation, energy transduction and signaling, and waste excretion [9]. Ionic equilibrium in plant cells is achieved via ATPases on the plasma membrane, which maintains an electrochemical gradient by proton, and via the flux of other ions – namely potassium, calcium, and chloride through ion channels and transporters. Vacuolar pyrophosphatases (PPases) operate as a third "pump". As a result, the cell cytosol is maintained at a practically constant and neutral pH, whereas vacuoles and external apoplast become acidic, ca. pH 5.5 [3, 10]. It is pertinent to remark that plant cells differ from animal cells in handling ionic fluxes. In plant cells, gradients in H^+ concentration across the plasma membrane generate the so-called proton motive force that drives ionic transport. Animal cells maintain a gradient of Na^+ and K^+ using the sodium motive force to drive ionic transport. Major cation channels in plants include K^+- and Ca^{2+}-selective channels, whereas anion channels are (except in a few cases such as malate ion) much less selective, allowing the flux of Cl^-, NO_3^-, and carboxylates [2].

Cation channels consist of several transmembrane alpha helices (also termed transmembrane spans or domains) that form pores. Channel activation (opening) or deactivation (closing) is caused by the rearrangement of transmembrane spans driven by different activators or inhibitors. These include membrane voltage, hydrogenions and cations, G-proteins, ATP, cyclic nucleotides, hormones, ROS, and amino acids but also other factors such as stretching or gravity [17].

Numerous studies have been performed on model systems, typically, bilayer lipid membranes tethered to a metal electrode (tBLMs). These systems are prepared from thiolipids with a terminal sulfur group, accompanied by a spacer (or tether) oligopeptide or polyethylene glycol molecules. The sulfur group ensures a strong attachment to gold and mercury electrodes, while the spacers accommodate an aqueous subphase that mimics the properties of cell cytosol [18].

Figure 2.7 illustrates the structures of one such system constituted by a tBLM where the inner leaflet consisted of a 2,3-di-O-phytanyl-*sn*-glycerol-1-tetraethylene glycol-DL-α-lipoic acid ester (DPTL) thiolipid and the outer leaflet 1,2-diphytanoyl-*sn*–glycero-3-phosphocholine (DPhPC) [19]. Electrochemical studies were conducted on β-Tg monolayer self-assembled on the gold (111) monocrystals previously covered by a 1-thio-β-D-glucose (β-Tg) monolayer. This was self-assembled simply by immersion of the monocrystal gold electrode in a methanolic solution of β-Tg.

Gramicidin A (gA) is a linear peptide containing 15 alternating L- and D-amino acids [20]. gA forms ion channels in the bilayer lipid membrane prompting transport of monovalent cations [21]. The primary structure of gA can be described as a right-handed β-helix with a β-sheet hydrogen bond pattern. Depending on the environment, monomeric, dimeric, and double-stranded conformations can be adopted. These are labeled β^m, with m being the total number of residues per turn. In the lipid bilayer, the hydrogen bonds between the tryptophan side chains and the lipid polar head groups orient the gA helix forming dimeric units. The exterior region of such units is hydrophobic, whereas the inner hydrophilic region allows the formation of a column of water and ion channel properties [19].

Figure 2.7 also depicts the equivalent circuit used to model EIS measurements. Here, R_{sol} represents the solution resistance and R_m the resistance of the bilayer lipid membrane. The capacitive effect of this bilayer is represented by a constant-phase element (CPE_m), whereas the impedance effect of the β-Tg spacer is modeled by a second constant-phase element (CPE_{sp}) [19, 22].

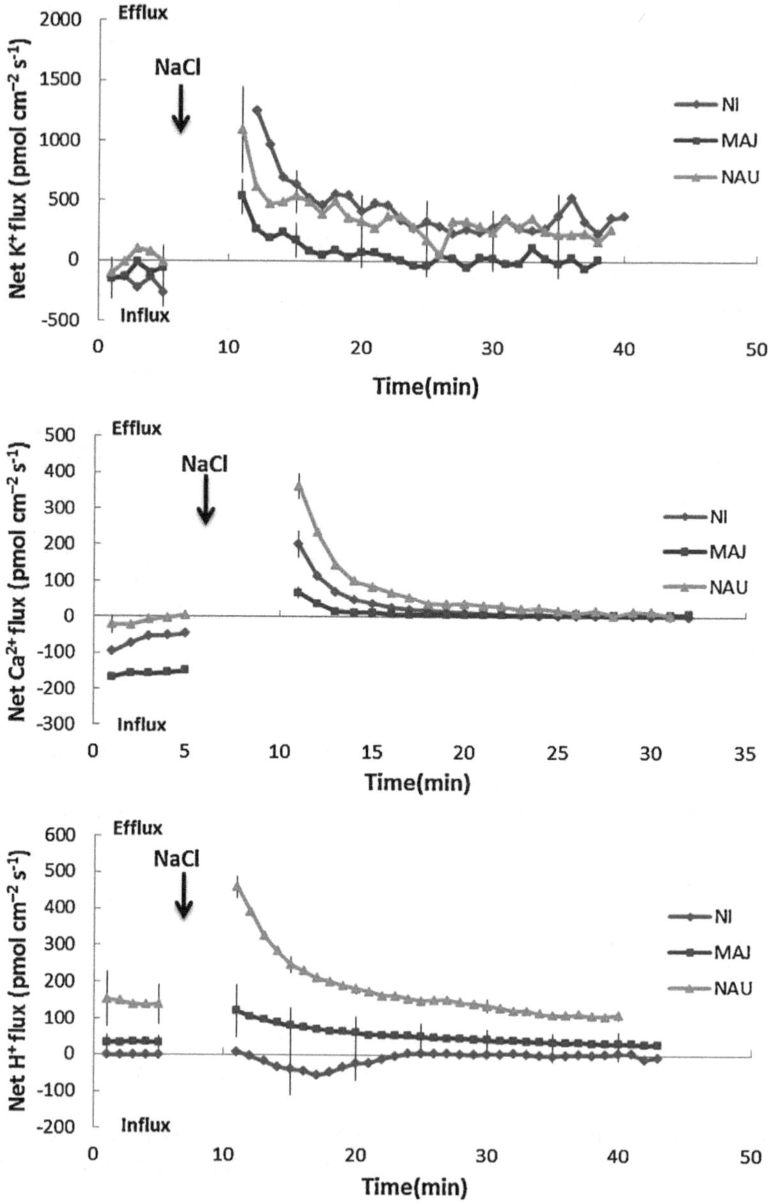

FIGURE 2.6 Monitoring of salt shock effects on transient K^+, Ca^{2+}, and H^+ ion fluxes at apical regions of NI (rhombs) and ***Paxillus involutus***-inoculated roots of ***Populus*** × ***canescens***. The roots of ***P.*** × ***canescens*** were cultured without or with the ***P. involutus*** strains MAJ (squares) and NAU (triangles) for ten days. Ion fluxes were measured potentiometrically at the apex (ca. 300 μm from the root tip) before and after the addition of NaCl into the measuring chamber reaching a final concentration of 100 mM. Each point is the mean of four individual plants, and bars represent the standard error. (Adapted from ref. [15], with permission. Copyright Elsevier 2014.).

Figure 2.8 depicts the Bode plots obtained for DPhPC/gA bilayer membrane on a gold electrode in contact with 0.1 M NaF [19]. The logarithm of the absolute value of total impedance (log|Z|) vs. logarithm of frequency (log*f*) representation (A) shows an almost linear inversed s-shaped profile with light differences between data recorded at bias potentials between 0.0 and −0.5 V vs. Ag/AgCl.

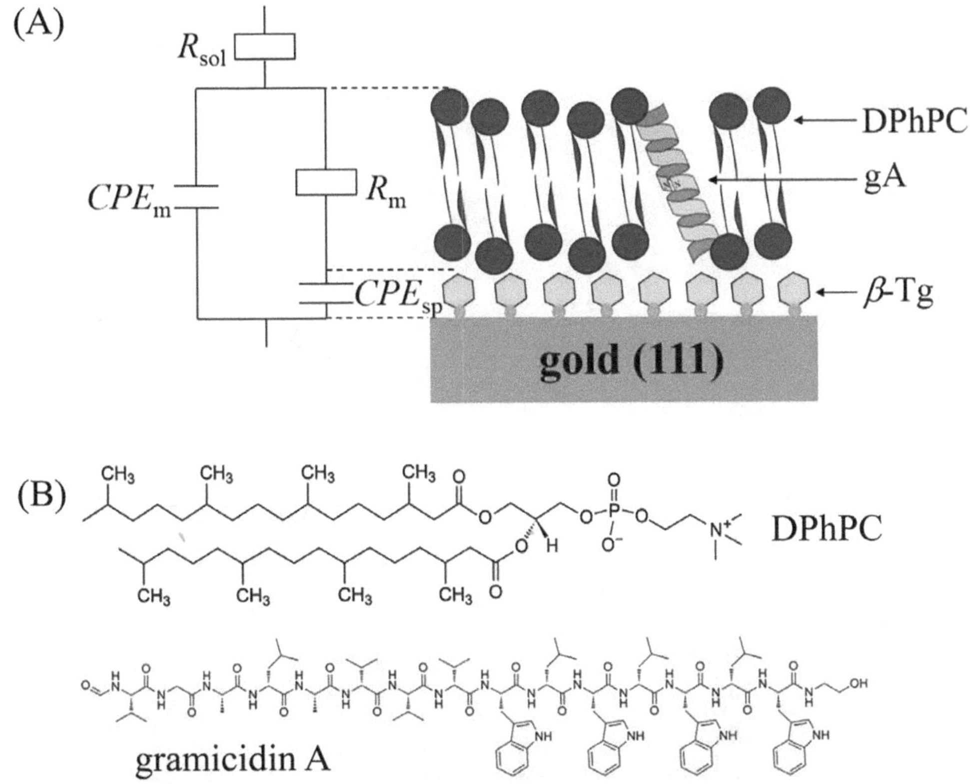

FIGURE 2.7 Scheme of the DPhPC/gA bilayer membrane on β-Tg-modified gold (111) monocrystal electrode accompanied by the equivalent circuit used to model EIS measurements (A) and structural formulas of DPhPC (B) and gA (C). (Adapted from ref. [19], with permission. Copyright Elsevier 2021.).

The (minus) phase angle vs. log*f* plots (B) shows a plateau at ca. −80 deg at frequencies between 0.1 and 100 Hz, thus denoting the predominance of capacitive effects.

As can be seen in the figure, the impedance spectra are satisfactorily reproduced by the equivalent circuit in Figure 2.7A. Using this equivalent circuit, the values of the different impedance elements were obtained in the absence and the presence of gA. Pertinent data are shown in Figure 2.9, where the membrane capacitance (Q_m) and membrane resistance (R_m) are plotted as a function of the electrode potential [19, 23]. The top horizontal axis plots the difference between the bias potential and the potential of zero free charge on the gold electrode ($E - E_{pzfc}$). This difference approximates the transmembrane potential across the DPhPC/gA bilayer. The values of Q_m are similar with and without gA, although slightly larger in the presence of the peptide. This can be interpreted by assuming that water is transported through gA channels [24]. As a result, water concentration in the bilayer increases, thus enhancing the interface capacity.

In contrast, the bilayer resistance without gA is significantly larger than the resistance with gA at all potentials. This feature can be interpreted by assuming that gA forms ion channels in the bilayer transporting Na^+ ions. The result is the decrease of R_m.

Figure 2.10 shows a schematic representation of the classical model of protein-mediated membrane transport for a monovalent cation M^+ [18]. The transporter protein in a conformational state 1 (PRt_1) incorporates the cation M^+ through the first aqueous/membrane interface forming the corresponding adduct (PRt_1-M^+). It suffers a new conformational change (Prt_2-M^+) that permits the release of the metal cation in the second aqueous/membrane interface.

Under steady-state conditions, mediated transport involves the establishment of a heterogeneous equilibrium between the ion (for simplicity, M^+) and the carrier (S) through the membrane (mem)/water (aq) interface [25]:

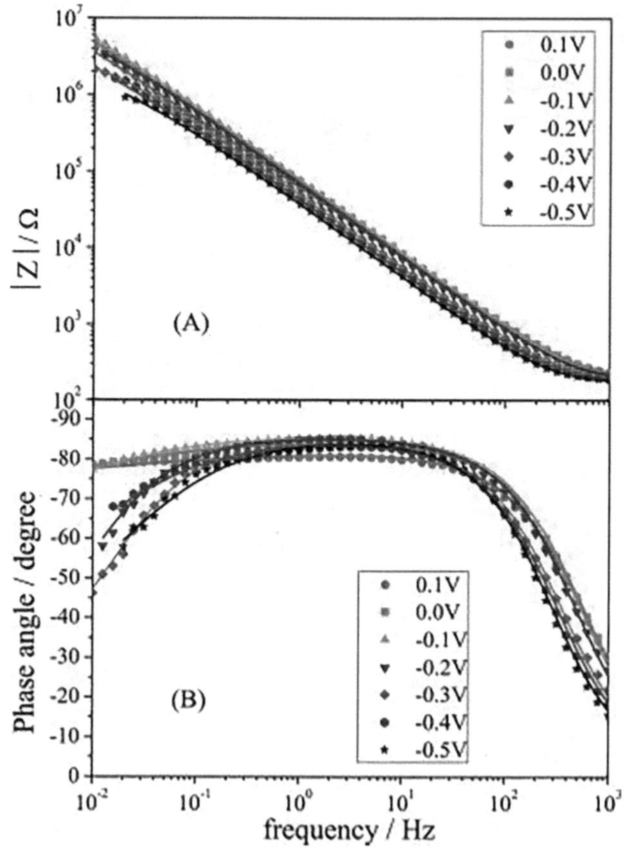

FIGURE 2.8 Bode plots of the logarithm of total impedance (A) and the (minus)phase angle (B) vs. logarithm of frequency for DPhPC/gA bilayer membrane on a gold electrode in contact with 0.1 M NaF. Data at different bias potentials are superimposed. (Adapted from ref. [19], with permission. Copyright Elsevier 2021.).

$$M^+(aq) + S(mem) \leftrightarrows M^+ - S(mem). \tag{2.15}$$

Let k_1, k_{-1}, be the rate constants for the direct and reverse processes, respectively. Under conditions of zero current, the equilibrium represented by [25]

$$K = \frac{k_1}{k_{-1}} = \frac{\Gamma_{MS}}{\Gamma_S c_M} \tag{2.16}$$

is attained. In this equation, c_M represents the concentration of M^+ in the aqueous phase and Γ_S, Γ_{MS}, the interfacial concentrations (mol cm^{-2}) of the carrier and the carrier-metal ion adduct, respectively. Additionally, a second adduct formation equilibrium must take place in the aqueous phase. The corresponding equilibrium constant will be

$$K_W = \frac{c_{MS}}{c_S c_M}, \tag{2.17}$$

where c_S, c_{MS}, are the concentrations of the free carrier and the M^+-carrier adduct in the aqueous solution phase. Introducing the membrane thickness, δ, the carrier interfacial concentration will be [25]

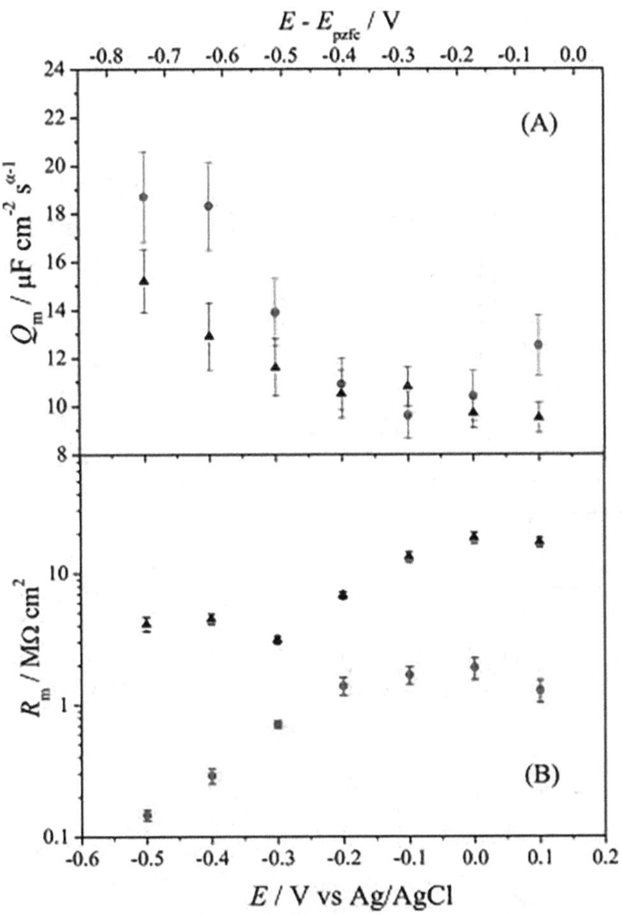

FIGURE 2.9 Plots of membrane capacitance (Q_m) and membrane resistance (R_m) vs. the bias potential obtained by fitting EIS data of DPhPC bilayer membrane on gold electrode in contact with 0.1 M NaF in the absence (black triangles) and presence (red squares) of gA to the equivalent circuit in Figure 2.7. The top horizontal axis plots the difference between the potential and the potential of zero-charge ($E - E_{pzfc}$). (Adapted from ref. [19], with permission. Copyright Elsevier 2021.).

$$\Gamma_S = \frac{\delta}{2}\frac{c_S + c_{MS}}{1 + c_M K_W}. \tag{2.18}$$

2.5 ELECTROGENIC TRANSPORT

Most biochemical processes in plants can be described in terms of multielectron reactions between reactants placed in an aqueous phase and a hydrophobic phase. These systems can be modeled based on the redox chemistry of reactants between two immiscible solvents and involve redox mediation and the concomitant cascade electron transfer pathways. Figure 2.11 illustrates the example of the oxidation of nicotinamide adenine dinucleotide reduced form (NADH) by oxygen at the n-octane-water interface in the presence of ethioporphyrin II (FeEP) and tetramethylquinone (TMQ) [26]. The overall process determines the oxidation of NADH by oxygen catalyzed by FeEP and TMQ, which cycle between the oxidized (quinone), and its reduced $TMQH_2$ forms. This view is consistent with data for cytochrome c oxidase that adheres to the octane/water interface when an enzyme

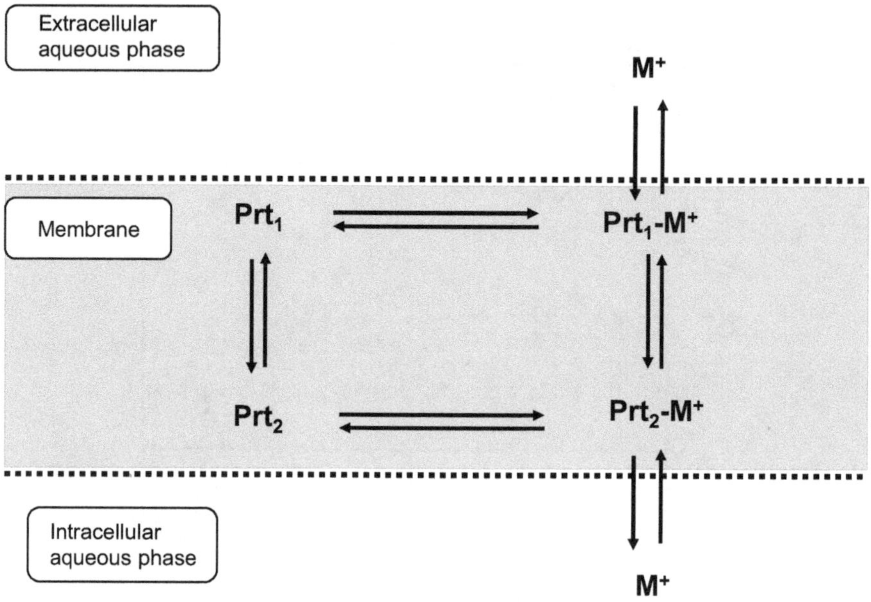

FIGURE 2.10 Scheme for the classical model of protein-mediated membrane transport for a monovalent cation M^+. (Adapted from ref. [18], with permission. Copyright Elsevier 2003.).

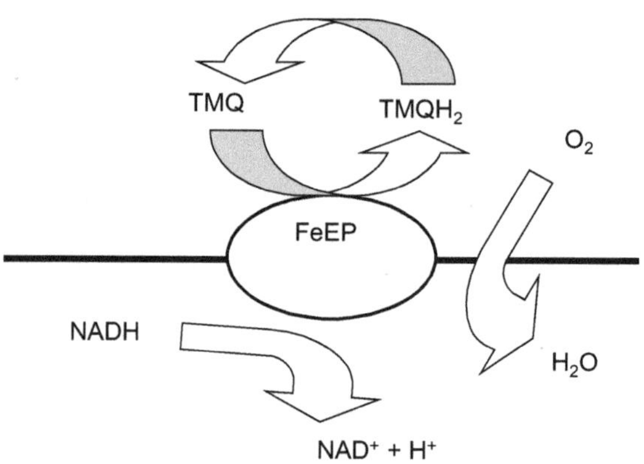

FIGURE 2.11 Schematic representation of redox processes involved in NADH oxidation at the ***n***-octanol/water interface mediated by ethioporphyrin II (FeEP) and tetramethylquinone (TMQ). (Adapted from ref. [26], with permission. Copyright Elsevier 1998.).

suspension is added to the system. This type of process involves coupled transport of protons and electrons. For example, cytochrome oxidases in the mitochondrial membrane can transport up to eight protons across the membrane per four electrons when catalyzing the water-to-oxygen reduction. Four protons cross the mitochondrial membrane and the other four remain bound to the reaction complex [26].

The first approach to the thermochemistry of these electron transfer processes departs from the Born equation of the free Gibbs energy of solvation of an ion, assumed to be a nonpolarizable sphere uniformly charged:

$$\Delta G_{\text{solv}} = \frac{z^2 e^2}{8\pi r \varepsilon_0}\left(1 - \frac{1}{\varepsilon}\right), \tag{2.19}$$

where z is the formal charge of the ion, r is its radius, ε_o is the vacuum dielectric permittivity, and ε is the relative dielectric permittivity (the "dielectric constant") of the solvent. Accordingly, the free energy of ion transfer from solvent A to solvent B, $\Delta G_{A \to B}$, will be

$$\Delta G_{A \to B} = \frac{z^2 e^2}{8\pi r \varepsilon_o}\left(\frac{1}{\varepsilon_A} - \frac{1}{\varepsilon_B}\right), \tag{2.20}$$

where ε_A, ε_B, are the respective relative dielectric permittivities of the solvents. The more detailed analysis involves the introduction of energetic terms associated with solvent reorganization and other terms. From the thermochemical point of view, the condition of electroneutrality must hold in both liquid phases so that electron transfer must be balanced by ion transfer [27]. Theoretical approaches include kinetic studies based on Marcus theory [26], theories of electron transfer [28], and quantum chemical calculations [29]. Such systems can be treated as highly selective interfacial catalytic systems [26].

An interesting aspect is that the estimate of solvation Gibbs free energies implies, in principle, extra-thermochemical considerations [30]. However, individual free energies of ion transfer between two solvents can be determined in three-phase electrochemical systems [31]. These include the solid-state electrochemistry of ion-insertion solids and immobilized droplets [32]. From the electrochemical point of view, a general problem of electron transfer processes between two immiscible fluid phases is the absence of a solvent-independent electrode potential scale [30], eventually approximated from ion-permeable solids [33]. Recent developments on the separation of ionic and electronic contributions to the free energy change of the overall electrochemical process [30, 34, 35] could also be applied.

Figure 2.12 [36] shows an example of an electrochemical study on the functional activity of membrane proteins and electrogenic pumps. Here, the transmembrane pump Na^+/K^+-ATPase was mimicked using a monoolein liquid-crystalline cubic phase (LCP). LCP forms the walls of a lipid bilayer encompassing a network of aqueous channels. The channels favor protein contact with the aqueous environment, whereas the lipid fraction stabilizes the hydrophobic part of the enzyme. The voltammograms in Figure 2.12 were recorded at a glassy carbon electrode (GCE) modified with Na^+/K^+-ATPase incorporated into LCPs film using different buffered electrolyte solutions containing ammonium phosphomolybdate as a redox probe. In the absence of Na^+ or K^+ ions in the electrolyte, no voltammetric signals were recorded. However, in the presence of such ions, the voltammogram shows oxidation peaks for molybdenum-centered oxidation processes, thus signaling the Na^+/K^+-ATPase activity [36].

2.6 APPROACH TO ENZYME ELECTROCHEMISTRY

Enzymes represent the majority of natural catalysts controlling essential biological functions in living organisms. Among them, oxidoreductases catalyze the transfer of electrons between two or more substrates. Their function can be correlated to electrochemical systems. Enzymatic electrochemistry can be seen from two perspectives: via electrochemical modeling and monitoring of enzymatic processes and using enzymes as electrode modifiers for electrocatalytic purposes extended to analytical and energy production applications [37].

Although other kinetic equations apply to describe enzymatic reactions (in particular, in macromolecular environments) [38], enzymatic processes are generally described in terms of the Michaelis-Menten (or Henry-Michaelis-Menten) kinetics. This consists of two steps. In the first step, the substrate (S) reacts with the enzyme (E), forming an enzyme substrate (ES) complex. In the second step, the complex forms a product P regenerating the enzyme

$$E + S \to ES \to E + P. \tag{2.21}$$

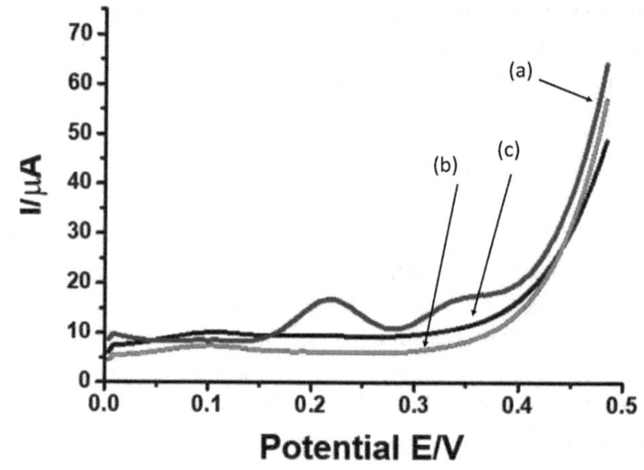

FIGURE 2.12 ATP-hydrolyzing activity of Na^+/K^+-ATPase incorporated into LCP film on a GCE. Square wave voltammograms of the buffer solution containing ammonium phosphomolybdate complex (a), buffer without NaCl (b), and KCl (c). Potential scan initiated at 0.0 V vs. Ag/AgCl in the positive direction, square wave amplitude 20 mV, frequency 20 Hz. (Adapted from ref. [36], with permission. Copyright Elsevier 2019.).

Let k_1, k_{-1}, be the rate constants of the direct and inverse first-step reaction and k_2 the rate constant for the second step. It is assumed that the enzyme concentration is much less than the substrate concentration, that the second process follows first-order kinetics, and that the rate of breakdown of ES equals the rate of formation of ES (steady-state assumption). Then, the initial reaction rate in terms of substrate concentration, c_S, satisfies the relationship:

$$-\frac{dc_S}{dt} = \frac{k_2 c_E c_S}{K_M + c_S}, \tag{2.22}$$

where c_S and c_E are the concentrations of substrate and enzyme, and K_M, the Michaelis-Menten constant, equals $(k_2 + k_{-1})/k_1$. The product k_2c_E represents the maximum reaction rate. When $K_M >> c_S$, i.e., at low substrate concentrations, the overall reaction approaches a first-order kinetics. At high substrate concentrations ($K_M << c_S$), the overall reaction approaches zero-order kinetics.

Chronoamperometry can be used as a sensitive technique for studying enzymatic catalysis. Figure 2.13a shows the current/time curves recorded to monitor the ATP-hydrolyzing activity of Na^+/K^+-ATPase incorporated into LCP film [36] whose voltammetric response was illustrated in Figure 2.12. In these experiments, a constant potential input was applied to promote the oxidation of the phosphomolybdate complex, which acts as a redox probe. In the absence of enzyme in the LCP but in the presence of Na^+ and K^+ ions, a typical current decay was obtained. The current becomes enhanced when Na^+/K^+-ATPase is incorporated into the LCP film. The current/time curve resulting after subtraction (Figure 2.13b) displays a well-defined maximum at short times. This chronoamperometric feature, also appearing in nucleation-driven electrodeposition processes, can be associated with stepped conformational changes in channel-forming peptides [39].

A simple case can be modeled assuming that there occurs a first-order chemical reaction between the mediator and a given reactant. It will be assumed that the substrate reduces the redox mediator (whose total concentration c is assumed to be constant) with a homogeneous rate constant k_c and that the mediator is reoxidized at the electrode surface with an electrochemical rate constant k_e. This constant equals $DA/\delta V$, where D is the diffusion coefficient of the mediator, A is the electrode area, δ the steady thickness of the diffusion layer, and V is the sample volume [40]. Let γ be the fraction

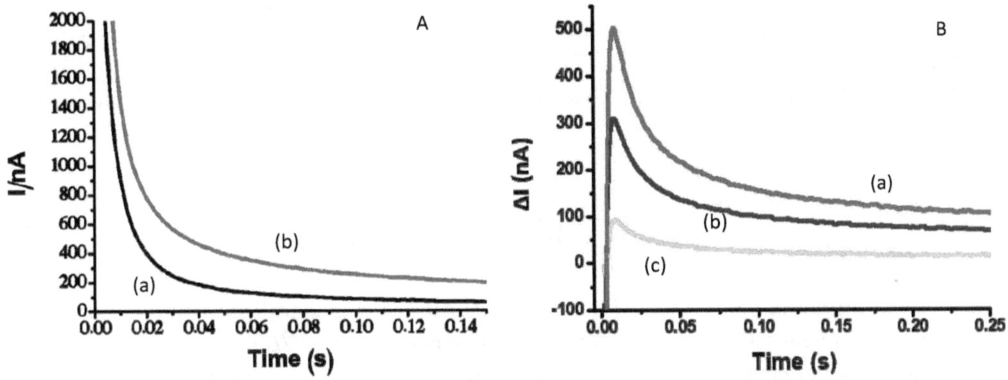

FIGURE 2.13 Chronoamperometric monitoring of ATP-hydrolyzing activity of Na^+/K^+-ATPase incorporated into LCP film. A: Current/time curves at LCP (a) and Na^+/K^+-ATPase incorporated into LCP (b) in contact with a buffer solution containing ammonium phosphomolybdate complex with Na^+ and K^+ ions; potential step 20 mV. B: Difference curves applying potential steps of 10 mV (a), 20 mV (b), and 50 mV (c). (Adapted from ref. [36], with permission. Copyright Elsevier 2019.).

of redox mediator in its oxidized form present under steady-state conditions. This quantity verifies the relationship

$$k_e(1-\gamma)c = k_c\gamma c. \tag{2.23}$$

Then,

$$\gamma = \frac{k_e}{k_e + k_c}. \tag{2.24}$$

Under non-steady-state conditions, and assuming first-order kinetics, the fraction of oxidized form of the redox probe, $\beta(t)$, satisfies

$$\frac{d\beta(t)}{dt} = k_e(1-\beta(t)) - k_c\beta(t). \tag{2.25}$$

The corresponding electrochemical current i at a time t will be

$$i = \frac{nFADc}{\delta}(1-\beta(t)). \tag{2.26}$$

Finally, introducing a dimensionless time constant, τ, the chronoamperometric current will be

$$i = \frac{nFADc}{\delta}(1-\gamma)\left[1-e^{-\tau/(1-\gamma)}\right]. \tag{2.27}$$

This general equation permits the distinction of four different limiting cases. For example, if the chemical step is rate-determining ($k_c << k_e$), the current becomes proportional to $1 - e^{-\tau(1-\gamma)}$. In the inverse case ($k_c >> k_e$), $\gamma \rightarrow 1$ and the current tends to vanish. An analog treatment has been developed for the case of Michaelis-Menten kinetics, now leading to more complex analytical expressions [40].

Another modeling considers the situation where the enzyme catalyst is immobilized within a support matrix introducing the consideration of diffusion in thin bounded film [41]. In the case of planar diffusion of a substrate S within a thin layer of thickness L, immobilized on a support electrode, containing the catalyst in concentration c_{cat}. The diffusion equation can be expressed as

$$D\frac{d^2c_S}{dx^2} = \frac{k_c c_{cat} c_S}{K_M + c_S}, \tag{2.28}$$

where D is the diffusion coefficient of the mobile substrate. A generalized solution has been provided in terms of a generalized reaction/diffusion parameter, γ (= kL^2/D); a saturation parameter, α, equal to the ratio between the reactant concentration at the layer/solution interface, b; and the Michaelis-Menten constant, K_M. The normalized current response, y, is [41]

$$y = \alpha\sqrt{\frac{\gamma}{1+\alpha}}\tanh\left[\sqrt{\frac{\gamma}{1+\alpha}}\right]. \tag{2.29}$$

Four limiting cases can be distinguished depending on the values of the aforementioned generalized parameters. When the catalytic reaction is slower than substrate diffusion (case I) and $\gamma << 1$, the reaction occurs uniformly through the film, and the current is first order relative to substrate concentration, catalyst concentration, and layer thickness.

$$i = nFA\left(k_c / K_M\right)bLc_{cat}. \tag{2.30}$$

When $\gamma >> 1$, the reaction between substrate and catalytic sites is fast, and the layer thickness is large compared to the kinetic length. Then, the reaction occurs concentrated in a thin layer of thickness X outside the film. Here (case II), the chronoamperometric current becomes

$$i = nFA\left(k_c / K_M\right)bXc_{cat}. \tag{2.31}$$

When $\alpha >> 1$ and $\gamma << 1$, the catalytic reaction can be considered as saturated. The substrate concentration in the film is larger than the Michaelis-Menten constant for the catalytic reaction, and the chronoamperometric current (case III) can be expressed as

$$i = nFAk_cLc_{cat}. \tag{2.32}$$

Under these circumstances, the current is first order with respect to catalyst and layer thickness but of zero order with respect to substrate concentration. The rate-determining step of the overall reaction process involves the break of the catalyst/substrate adduct throughout the film.

The case IV corresponds to $\gamma >> 1$ and $\alpha >> 1$. In this case, the catalytic kinetics is rapid, and the substrate concentration in the layer is larger than K_M. The current is now given by the relationship

$$i = nFA\left(k_c / \sqrt{K_M}\right)Xc_{cat}\sqrt{b}; \tag{2.33}$$

i.e, the current is half order with respect to substrate concentration and first order to the catalyst. The reaction is concentrated in a thin reaction layer adjacent to the film/solution interface. Under these

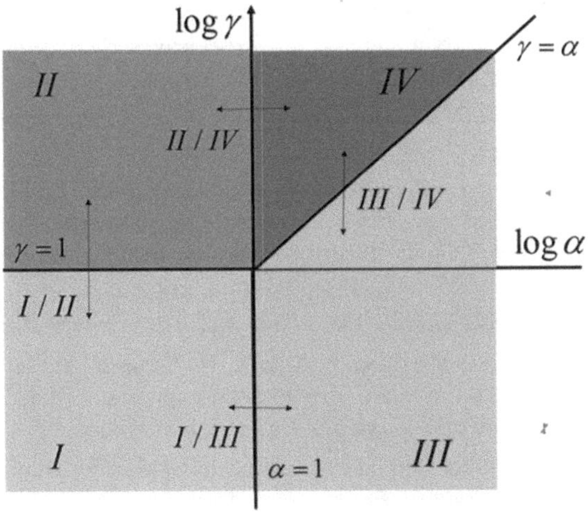

FIGURE 2.14 Kinetic case diagram illustrating four kinetically limiting situations corresponding to the theoretical description of electrocatalytic processes in thin film electrodes. (Adapted from ref. [41], with permission. Copyright Elsevier 2020.).

conditions, the external regions of the film become saturated, and the inner regions of the film remain unsaturated. Figure 2.14 depicts a scheme summarizing the aforementioned limiting cases. Here, cases II and IV are separated by the line corresponding to $\gamma = \alpha$.

2.7 CONCLUDING REMARKS

The matters previously treated do not exhaust the variety of physiological processes in plants whose study can be performed using electrochemical techniques. Several other biochemical processes, including essential ones such as photosynthetic function and specific enzymatic activities, will be treated in subsequent chapters. The relevant point to underline here is that electrochemical (mainly potentiometric, voltammetric, and amperometric) techniques possess high capabilities for monitoring essential physiological processes in living organisms. Upon appropriate theoretical modeling, electrochemical techniques can significantly contribute to improving our knowledge of important aspects of plant physiology.

REFERENCES

[1] Kathpalia, R.; Bhatla, S.C. 2018. Plant Water relations, in Bhatla, S.C.; Lal, M.A. *Plant Physiology, Development and Metabolism*. Springer, Singapore, pp. 3–36.

[2] Bhatla, S.C. 2018. Water and Solute Transport, in Bhatla, S.C.; Lal, M.A. *Plant Physiology, Development and Metabolism*. Springer, Singapore, pp. 83–115.

[3] Reuss, L. 2008. Mechanisms of Ion Transport Across Cell Membranes and Epithelia, in *Seldin and Giebisch's The Kidney*. Elsevier, Amsterdam, pp. 35–56.

[4] Goldman, D.E. 1943. Potential, impedance, and rectification in membranes. *The Journal of General Physiology*. 27: 37–60.

[5] Hodgkin, A.L.; Katz, B. 1949. The effect of sodium ions on the electrical activity of the giant axon of the squid. *Journal of Physiology (London)*. 108: 37–77.

[6] Miller, A.J.; Wells, D.M. 2006. Electrochemical Methods and Measuring Transmembrane Ion Gradients, in Volkov, A.G. Ed. *Plant Electrophysiology Theory and Methods*. Springer, Berlin-Heidelberg, pp. 15–34.

[7] Delessert, C.; Wilson, I.W.; van der Straeten, D.; Dennis, E.S.; Dolferus, R. 2004. Spatial and temporal analysis of the local response to wounding in *Arabidopsis* leaves. *Plant Molecular Biology*. 55:165–181.

[8] Lucas, W.J.; Kochian, L.V. 1986. Ion transport processes in corn roots: An approach utilizing microelectrode techniques, in Gensler, W.G. Ed. *Advanced agricultural instrumentation: design and use*. Nijhoff, Dordrecht, pp 402–425.

[9] Kunkel, J.G.; Cordeiro, S.; Xu, Y..; Shipley, A.M.; Feijó, J.A. 2006. Use of Non-Invasive Ion-Selective Microelectrode Techniques for the Study of Plant Development, in Volkov, A.G. Ed. *Plant Electrophysiology Theory and Methods*. Springer, Berlin-Heidelberg, pp. 109–137.

[10] Shabala, S. 2006. Non-Invasive Microelectrode Ion Flux Measurements In Plant Stress Physiology, in Volkov, A.G. Ed. Plant *Electrophysiology Theory and Methods*. Springer, Berlin-Heidelberg, pp. 35–71.

[11] Mak, M.; Babla, M.; Xu, S.-C.; O'Carrigan, A.; Liu, X.-H.; Gong, Y.-M.; Holford, P.; Chen, Z.-H. 2014. Leaf mesophyll K^+, H^+ and Ca^{2+} fluxes are involved in drought-induced decrease in photosynthesis and stomatal closure in soybean. *Environmental and Experimental Botany*. 98: 1–12.

[12] Kühtreiber, W.M.; Jaffe, L.F. 1990. Detection of extracellular calcium gradients with a calcium-specific vibrating electrode. *The Journal of Cell Biology*. 110: 1565–1573.

[13] Henriksen, G.H.; Raman, D.R.; Walker, L.P.; Spanswick, R.M. 1992. Measurement of net fluxes of ammonium and nitrate at the surface of barley roots using ion-selective microelectrodes. *Plant Physiology*. 99: 734–747.

[14] Messerli, M.A.; Robinson, K.R.; Smith, P.J.S. 2006. Electrochemical Sensor Applications to the Study of Molecular Physiology and Analyte Flux in Plants, in Volkov, A.G. Ed. *Plant Electrophysiology Theory and Methods*. Springer, Berlin-Heidelberg, pp. 73–107.

[15] Ma, X.; Sun, M.; Sa, G.; Zhang, Y.; Li, J.; Sun, J.; Shen, X.; Polle, A.; Chen, S. 2014. Ion fluxes in *Paxillus involutus*-inoculated roots of *Populus × canescens* under saline stress. *Environmental and Experimental Botany*. 108: 99–108.

[16] Luo, Z.; Li, K.; Gai, Y.; Göbel, C.; Wildhagen, H.; Jiang, X.; Feußner, I.; Rennenberg, H.; Polle, A. 2011. The ectomycorrhizal fungus (*Paxillus involutus*) modulates leaf physiology of poplar towards improved salt tolerance. *Environmental and Experimental Botany*. 72: 304–311.

[17] Demidchik, V.; Sokolik, A.; Yurin, V. 2006. Electrophysiological Characterization of Plant Cation Channels, in Volkov, A.G. Ed. *Plant Electrophysiology Theory and Methods*. Springer, Berlin-Heidelberg, pp. 173–185.

[18] Naumann, R.; Walz, D.; Schiller, S.M.; Knoll, W. 2003. Kinetics of valinomycin-mediated K^+ ion transport through tethered bilayer lipid membranes. *Journal of Electroanalytical. Chemistry*. 550: 241–252.

[19] Su, Z.; Goodall, B.; Leitch, J.J.; Lipkowski, J. 2021. Ion transport mechanism in gramicidin A channels formed in floating bilayer lipid membranes supported on gold electrodes. *Electrochimica Acta*. 375: artic. 137892.

[20] Reiter, R.; Zaitseva, E.; Baaken, G.; Halimeh, I.; Behrends, J.C.; Zumbuehl, A. 2019. Activity of the Gramicidin A ion channel in a lipid membrane with switchable physical properties. *Langmuir*. 35: 14959–14966.

[21] Yamaguchi, T.; Kitazumi, Y.; Kano, K.; Shirai, O. 2020. Permselectivity of Gramicidin A channels based on single-channel recordings. *Electroanalysis*. 32: 1093–1099.

[22] Valincius, G.; Meškauskas, T.; Ivanauskas, F. 2012. Electrochemical impedance spec- troscopy of tethered bilayer membranes. *Langmuir*. 28: 977–990.

[23] Su, Z.; Shodiev, M.; Leitch, J.J.; Abbasi, F.; Lipkowski, J. 2018. Role of transmembrane potential and defects on the permeabilization of lipid bilayers by alamethicin, an ion-channel-forming peptide. *Langmuir*. 34: 6249–6260.

[24] Portella, G.; Polupanow, T.; Zocher, F.; Boytsov, D.A.; Pohl, P.; Diederichsen, U.; de Groot, B.L. 2012. Design of peptide-membrane interactions to modulate single-file water transport through modified gramicidin channels. *Biophysical Journal*. 103: 1698–1705.

[25] Stark, G.; Ketterer, B.; Benz, R.; Läuger, P. 1971. The Rate Constants of Valinomycin-Mediated Ion Transport through Thin Lipid Membranes. *Biophysical Journal*. 11: 981–994.

[26] Volkov, A.G. 1998. Redox reactions at liquid hydrocarbon/water interfaces: biophysical aspects. *Electrochimica Acta*. 44: 139–153.

[27] Kakiuchi, T. 1995. Free energy coupling of electron transfer and ion transfer in two immiscible fluid systems. *Electrochimica Acta*. 40: 2999–3003.

[28] Schmickler, W. 1997. Electron-transfer reactions across liquid|liquid interfaces. *Journal of Electroanalytical Chemistry*. 428: 123–127.

[29] Osakai, T.; Ebina, K. 1996. Quantum chemical approach to the Gibbs energy of ion transfer between two immiscible liquids. *Journal of Electroanalytical Chemistry*. 412: 1–9.

[30] Trasatti, S. 1986. The absolute electrode potential: an explanatory note. *Pure and Applied Chemistry*. 58: 955–966.

[31] Scholz, F.; Doménech-Carbó, A. 2019. The thermodynamics of insertion electrochemical electrodes – a team play of electrons and ions across two separate interfaces. *Angewandte Chemie International Edition*. 58: 3279–3284.

[32] Scholz, F.; Schröder, U.; Gulaboski, R.; Doménech-Carbó, A. 2014. *Electrochemistry of Immobilized Particles and Droplets*, 2nd edn, Springer, Berlin-Heidelberg.

[33] Doménech-Carbó, A. 2012. Solvent-Independent Electrode Potentials of Solids Undergoing Insertion Electrochemical Reactions: Part I. Theory. *Journal of Physical Chemistry C*. 116: 25977–25983.

[34] Cisternas, R.; Kahlert, H.; Wulff, H.; Scholz, F. 2015. The electrode responses of a tungsten bronze electrode differ in potentiometry and voltammetry and give access to the individual contributions of electron and proton transfer. *Electrochemistry Communications*. 56: 34–37.

[35] Doménech-Carbó, A.; Koshevoy, I.O.; Montoya, N. 2016. Separation of the Ionic and Electronic Contributions to the Overall Thermodynamics of the Insertion Electrochemistry of Some Solid Au(I) Complexes. *Journal of Solid State Electrochemistry*. 20: 673–681.

[36] Zatloukalova, M.; Nazaruk, E.; Bilewicz, R. 2019. Electrogenic transport of Na^+/K^+-ATPase incorporated in lipidic cubic phases as a model biomimetic membrane. *Electrochimica Acta*. 310: 113–212.

[37] Bedendi, G.; De Moura Torquato, L.D.; Webb, S.; Cadoux, C.; Kulkarni, A.; Sahin, S.; Maroni, P.; Milton, R.D.; Grattieri, M. 2022. *ACS Measurement Science*. 2: 517–541.

[38] Schnell, S.; Turner, T.E. 2004. Reaction kinetics in intracellular environments with macromolecular crowding: simulations and rate laws. *Progress in Biophysics & Molecular Biology*. 85: 235–260.

[39] Guidelli, R.; Becucci, L. 2016. Mechanism of voltage-gated channel formation in lipid Membranes. *Biochimica and Biophysica Acta*. 1858: 748–755.

[40] Longatte, G.; Guille-Collignon, M.; Lemaître, F. 2017. Electrocatalytic Mechanism Involving Michaelis–Menten Kinetics at the Preparative Scale: Theory and Applicability to Photocurrents from a Photosynthetic Algae Suspension With Quinones. *ChemPhysChem*. 18: 2643–2650.

[41] Lyons, M.E.G. 2020. Understanding the kinetics of catalysed reactions in microheterogeneous thin film electrodes. *Journal of Electroanalytical Chemistry*. 872: artic. 114278.

ADDITIONAL LITERATURE

Transmembrane Potentials

Hamill, O.P.; Marty, A.; Neher, E.; Sakmann, B.; Sigworth, F.J. 1981 Improved patch-clamp techniques for high-resolution current recording from cells and cell-free membrane patches. *Pflugers Archiv*. 391: 85–100.

Henriksen, G.H.; Bloom, A.J.; Spanswick, R.M. 1990. Measurement of net fluxes of ammonium and nitrate at the surface of barely roots using ion-selective microelectrodes. *Plant Physiology*. 93: 271–280.

Kochian, L.V.; Shaff, J.E.; Kuhtrieber, W.M.; Jaffe, L.; Lucas, W.J. 1992. Use of extracellular, ionselective, vibrating microelectrode system for the quantification of K^+, H^+ and Ca^{2+} fluxes in maize roots and maize suspension cells. *Planta*. 188: 601–610.

Newman, I.A. 2001. Ion transport in roots: measurement of fluxes using ion-selective microelectrodes to characterize transporter function. *Plant Cell & Environment*. 24: 1–14.

Ion Fluxes

Ahmadi, F.; Samadi, A.; Sepehr, E.; Rahimi, A.; Shabala, S. 2023. Potassium homeostasis and signaling as a determinant of *Echinacea* species tolerance to salinity stress. *Environmental and Experimental Botany*. 206: artic. 105148.

Kohler, B.; Hills, A.; Blatt, M.R. 2003. Control of guard cell ion channels by hydrogen peroxide and abscisic acid indicates their action through alternate signaling pathways. *Plant Physiology*. 131: 385–388.

Li, X.; Ye, G.; Shen, Z.; Li, J.; Hao, D.; Kong, W.; Wang, H.; Zhang, L.; Chen, J.; Guo, H. 2023. Na^+ and K^+ homeostasis in different organs of contrasting *Zoysia japonica* accessions under salt stress. *Environmental and Experimental Botany*. 214: artic. 105455.

Miller, G.A.D.; Suzuki, N.; Ciftci-Yilmaz, S.; Mittler, R.O.N. 2010. Reactive oxygen species homeostasis and signaling during drought and salinity stresses. *Plant, Cell and Environment*. 33: 453–467.

Xue, L.; Zhao, D.-J.; Wang, Z.-Y.; Wang, X.-D.; Wang, C.; Huang, L.; Wang, Z.-Y. 2016. The calibration model in potassium ion flux non-invasive measurement of plants *in vivo in situ*. *Information Processing in Agriculture*. 3: 76–82.

Ion Channels

Ehrenstein, G.; Lecar, H. 1977. Electrically gated ion channels in lipid bilayers. *Quarterly Reviews of Biophysics*. 10: 1–34.

Menestrina, G.; Ropele, M. 1989. Voltage-dependent gating properties of the channel formed by *E. Coli* hemolysin in planar lipid membranes. *Bioscience Reports*. 9: 465–473.

Schwarz, G. 1978. On the physico-chemical basis of voltage-dependent molecular gating mechanisms in biological membranes. *The Journal of Membrane Biology*. 43: 127–148.

Electrogenic Transport

Born, M. 1920. Volumen und Hydratationswärme der Ionen. *Zeitschrift für Physik* 1: 45–48.

Doménech-Carbó, A.; Koshevoy, I.O.; Montoya, N.; Pakkanen, T.A. 2011a. Estimation of free energies of anion transfer from solid state electrochemistry of alkynyl-based Au(I) dinuclear and Au(I)-Cu(I) cluster complexes containing ferrocenyl groups. *Electrochemistry Communications*. 13: 96–98.

Doménech-Carbó, A.; Montoya, N.; Scholz, F. 2011b. Estimation of individual Gibbs energies of cation transfer employing the insertion electrochemistry of solid Prussian blue. *Journal of Electroanalytical Chemistry*. *657*: 117–122.

Doménech-Carbó, A.; Koshevoy, I.O.; Montoya, N.; Karttunen, A.J.; Pakkanen; T.A. 2011c. Determination of individual Gibbs energies of anion transfer and excess Gibbs energies using an electrochemical method based on insertion electrochemistry of solid compounds. *Journal of Chemical & Engineering Data*. 56: 4577–4586.

Doménech-Carbó, A.; Koshevoy, I.O.; Montoya, N.; Pakkanen, T.A.; Doménech-Carbó, M.T. 2012a. Solvent-Independent Electrode Potentials of Solids Undergoing Insertion Electrochemical Reactions: Part II. Experimental Data for Alkynyl-diphosphine Dinuclear Au(I) Complexes Undergoing Electron Exchange Coupled to Anion Exchange. *Journal of Physical Chemistry C*. 116: 25984–25992.

Doménech-Carbó, A. Scholz, F.; Montoya, N. 2012b. Solvent-Independent Electrode Potentials of Solids Undergoing Insertion Electrochemical Reactions: Part III. Experimental Data for Prussian Blue Undergoing Electron Exchange Coupled to Cation Exchange. *Journal of Physical Chemistry C*. 116: 25993–25999.

Tissandier, M.D.; Cowen, K.A.; Feng, W.Y.; Gundlach, E.; Cohen, M.H.; Earhart, A. D.; Coe, J.V. Free Energy of Transfer of Hydrated Ion Clusters from Water to an Immiscible Organic Solvent. *Journal of Physical Chemistry A*. 102: 7787–7794.

Volkov, A.G.; Gugeshashvili, M.I.; Deamer, D.W. 1995. Energy conversion at liquid/liquid interfaces: artificial photosynthetic systems. *Electrochimica Acta*. 40: 2849–2868.

Kinetics

Albery, W.J.; Cass, A.E.G.; Shu, Z.X. 1990. Inhibited enzyme electrodes. Part 1: Theoretical model. *Biosensors & Bioelectronics*. 5: 367–378.

Dharmalingam, K.M.; Veeramuni, M. 2019. Akbari-Ganji's Method (AGM) for solving non-linear reaction - Diffusion equation in the electroactive polymer film. *Journal of Electroanalytical Chemistry*. 844: 1–5.
Marcus, R.A. 1956. On the Theory of Oxidation-Reduction Reactions Involving Electron Transfer. I. *Journal of Chemical Physics*. 24: 966–978.
Marcus, R.A. 1991. Theory of electron-transfer rates across liquid-liquid interfaces. 2. Relationships and application. *Journal of Physical Chemistry*. 95: 2010–2013.
Menten, L.; Michaelis, M.I. 1913. Die kinetik der invertinwirkung. *Biochemistry*. 49: 333–369.
Rasmussen, M.; Minteer, S.D. 2014. Thylakoid direct photobioelectrocatalysis: utilizing stroma thylakoids to improve bio-solar cell performance. *PhysChemChemPhys*.16: 17327–17331.

3 Electrophysiology II

3.1 INTRODUCTION

Although there was a long tradition concerning the study of electrical signaling in plants [1], the theoretical and experimental background increased significantly since the 1980s [2]. These signals are generated in plants in response to stimuli (either internal or external) and have in common the involvement of changes in membrane potential. These include different electrical signals differing by shape, velocity, and constancy [1]. From a general perspective, electrical signals in plants can be divided into four groups, local electrical potentials (LEPs), action potentials (APs), variation potentials (VPs), and system potentials (SPs) [3].

The generation, propagation, and induction of plant electrical signals involve differential gene expression [4]. The primary cause of such signals is the change in transmembrane potential originated by changes in the electrochemical gradient inside and outside the plasma membrane. These are mainly associated to changes the concentration of K^+, Ca^{2+}, Cl^- , and H^+, [5]. The former are interchanged through the cell cytoplasm mainly through genetically encoded ion channels, whereas H+ is mainly transported through proton pumps and transporters [3].

Such signals permit a fast systemic information transmission so that the entire plant is informed when any perturbation has been produced in any region of the same. Thus, the wounded cells send a signal to both adjacent and distant cells, displaying a rapid response in terms of the inhibition of translation elongation/termination, the prevention of ribosome movement along mRNA and translation of viral mRNA, and the activation of genes involved in the defense response [6].

Plants possess most of the chemistry associated with neuromotoric system in animals. This includes ion channels, acetylcholine (neurotransmitter), cellular messengers (e.g., calmodulin), cellular motors (actin, myosin), and sensors for external mechanical, luminous, thermal, etc., stimuli [7].

From an evolutionary point of view, although electrical signaling is highly developed in pluricellular organisms, both plants and animals inherited their basic neuronal capabilities from their bacterial ancestors, further evolving in different directions [8]. For instance, the pressure-sensitive ion channels were characterized in *Escherichia coli* [9]. In the course of evolution, plants colonized dry land and their excitability and signal transmission capabilities were used to develop increasingly sophisticated survival tactics [3].

3.2 LOCAL ELECTRICAL POTENTIALS

LEP consists of a subthreshold response induced by the change in environmental factors. This potential lacks the refractory period and is strictly "'local", i.e., cannot propagate from the original location of the stimulus to other parts of the plant. It depends on the type of cells at the stimulus. In the case of moderate light stimulation, LEP of mesophyll cells is mainly associated with Ca^{2+} influx, while LEP of guard cells is driven by H^+, Cl^- efflux, and K^+ influx [3, 10].

3.3 ACTION POTENTIALS

AP is a short-term electrical pulse induced by non-damaging stimuli that propagates over a long distance, maintaining constant amplitude and constant velocity. APs can be considered self-propagating signals whose magnitude is independent of the intensity of the stimulus. APs have absolute and relative refractory periods [3].

Plant APs although, in principle, are similar to signaling in animal cells of several tissues, such as epithelium, differ from APs in neurons. Here, these are generated via Na^+/K^+ exchange, while

DOI: 10.1201/9781032697826-3

plant APs appear to be supported by the relatively large difference between the rest membrane potential and the cytoplasm and rely on $Ca^{2+}/K^{+}/Cl^{-}$ fluxes [2].

These consist of potential inputs characterized by rapid information transmission through plasmodesmata, traveling with constant velocity and amplitude [4]. In vascular plants, long-distance AP transmission occurs through the phloem, mainly via the sieve tubes, although lateral transmission may also occur. The transmission rate usually ranges between 0.01 and 0.2 m s^{-1} but often reaches values similar to those of APs in nerves, up to 30 m s^{-1} [11]. As a first step, the plasma membrane becomes depolarized, forming a receptor potential. The change in transmembrane potential produces a wave of depolarization that propagates over the cell membrane and along the conductive bundles of tissue. When the membrane is depolarized beyond a certain threshold potential, the AP is generated.

Figure 3.1 shows the typical experimental arrangement used for AP measurement [12]. The measurements are performed inside a Faraday cage by inserting two reversible nonpolarizable electrodes (usually Ag/AgCl) in the stem, root, or any plant leaf. In the figure, light is the external stimulus but can be replaced by thermal, mechanical, chemical, electrical, etc., external inputs. Since APs appear as potential pulses of short duration, it is required that the signal must be sampled at a rate greater than the highest frequency component in the signal, f_s. The threshold frequency, given by the Shannon sampling theorem, is the so-called Nyquist rate, corresponding to $f_s/2$.

In the absence of external stimuli, the potential difference between the electrodes remains constant as can be seen in Figure 3.2, corresponding to Ag/AgCl electrodes inserted in different positions on leaves of *Aloe vera* L. [13]. Figure 3.3 shows the time variation of the potential difference measured in the same plant after the application of a cold shock produced by a small ice block placed on a leaf [13]. APs look like sharp, asymmetrical potential pulses of duration ca. 0.004 s and amplitudes between 0.01 and 0.04 V. The propagation velocity of APs can be estimated by dividing the distance between the Ag/AgCl electrodes by the time between maximum AP values. In the experiment in Figure 3.3, an average (seven plants) propagation velocity of 132 ± 12 m s^{-1} was determined [13].

Remarkably, the electrical impulses propagate along all the leaves of the plant with constant amplitude and velocity, but the amplitude of the potential pulse on a treated leaf decreases with

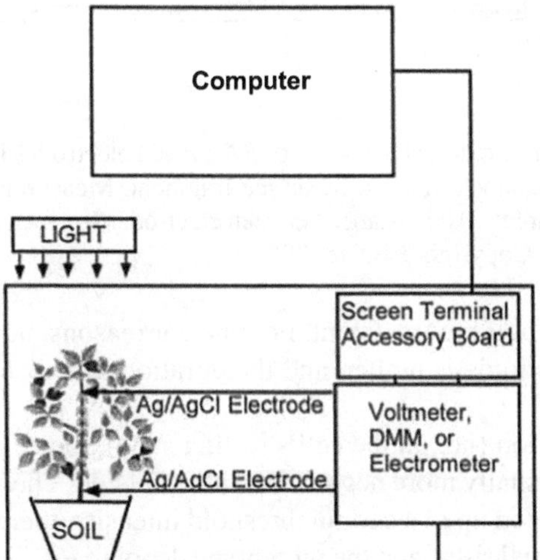

FIGURE 3.1 Experimental arrangement used to measure APs under the application of light stimuli. These can be replaced by mechanical, thermal, etc., stimuli. (Adapted from ref. [12], with permission. Copyright Elsevier 2000.).

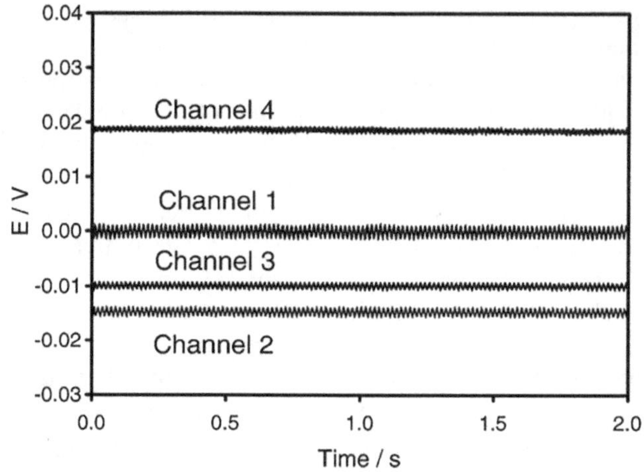

FIGURE 3.2 Potential difference recorded between two Ag/AgCl electrodes inserted in different positions on leaves of ***Aloe vera*** in the absence of external stimuli. Measurements were recorded at 500,000 scans s^{-1}. The distance between electrodes for each channel was 1 cm. (Reproduced from ref. [13], with permission. Copyright Elsevier 2007.).

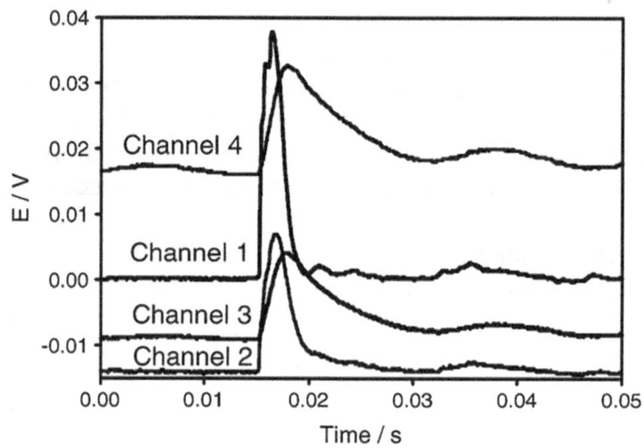

FIGURE 3.3 Potential difference recorded between two Ag/AgCl electrodes inserted in different positions on leaves of ***Aloe vera*** after a cold shock induced by an ice fragment. Measurements were recorded at 100,000 scans s^{-1} and 400,000 scans $sample^{-1}$. The distance between electrodes for each channel was 1 cm. (Reproduced from ref. [13], with permission. Copyright Elsevier 2007.).

increasing distance from the point of treatment. For unclear reasons, near the point where the stimulus was applied, the AP amplitude is higher, and the duration time is lower than these same values on other leaves [13].

In the absence of stimulation (i.e., in the cell's resting state), the electrical potential of the inside of the cellular membrane is usually more negative than the outside. Then, the cell membrane is polarized. When the cell is stimulated up to a certain threshold intensity, there is a cascade of ion channels to open: the membrane potential rises, and the membrane depolarizes. Then, there is an efflux of positive ions, and the membrane potential returns to its original value (i.e., the cell returns to its resting state). Before returning to this state, there is a brief period of hyperpolarization during which the membrane potential drops below its resting potential [14]. Figure 3.4 shows a schematic representation of AP spikes, including the depolarization, repolarization, and hyperpolarization periods. Absolute and relative refractory periods where the membrane remains "insensitive" are also marked.

The current view [14] of this sequence of processes departs from the resting cell with Ca^{2+} and Cl^- ions kept apart from the electrochemical equilibrium. As a result of non-damaging stimulation, membrane depolarization is initiated by the activation of Ca^{2+}-dependent permeable anion channels. This activation produces the influx of calcium into the cytosol, the concomitant activation of Cl^- channels and the efflux of Cl^-. As a result, the concentration of calcium ions in the cytoplasm increases, activating the voltage-dependent K^+ channels and anion channels yielding, indirectly, a K^+ efflux. When the plasma membrane returns progressively to its resting potential, repolarization occurs. The reduction of membrane depolarization involves the suppression of Ca^{2+} influx and cancels stimulation for Cl^- flux. This is accompanied by the triggering of K^+ efflux via activation of voltage-gated K^+ channels.

3.4 VARIATION POTENTIALS

Also termed slow-wave potentials, they consist of a long-time potential wave, induced by damaging stimuli, whose amplitude and duration are related to the intensity of the stimulus. In contrast with APs that follow an "all-or-none" law, VPs depend on stimulus type, distance from zone of damage, and plant species [15]. VPs lack a refractory period and can be propagated over a long distance. In contrast with APs, the amplitude and velocity of VPs decrease gradually [3]. VP is a long-term electrical signal, with durations of minutes or longer. Another distinctive characteristic of VPs is their ability to pass through dead or poisoned tissues.

VP pulses have irregular shapes and can include two components, a long-term depolarization over the course of a minute to several minutes and often a fast impulse depolarization, similar to AP signals [16]. The long-term component of VPs has amplitudes ranging in a wide interval (ca. tens mV) and low-velocity propagation (ca. mm s^{-1}).

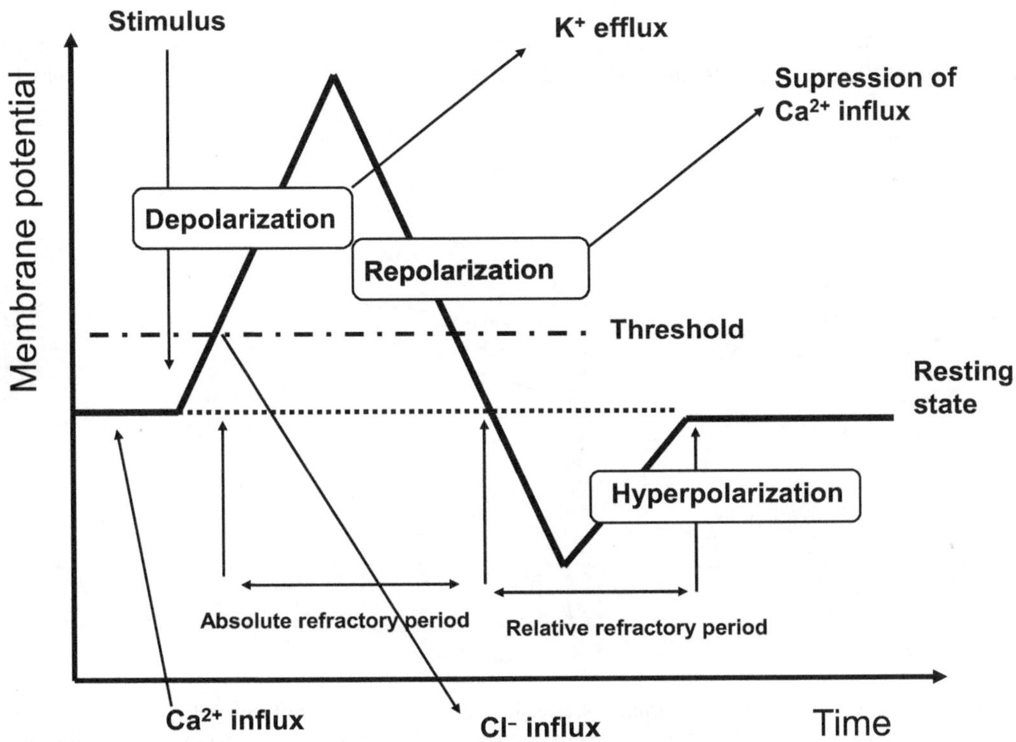

FIGURE 3.4 Scheme for the changes in membrane potential accompanying APs.

Figure 3.5 depicts the VPs induced by burning on a leaf tip of wheat measured by three electrodes positioned at different distances from the burning apical zone (2–3 mm) submitted to open-fire for 2 s [15]. The propagation velocity of the signal decreased from 0.40 cm s^{-1}, as measured by an electrode placed at a distance of 3 cm from the burn site, to 0.07 cm s^{-1}, as measured by another electrode situated at 15 cm from this site. As can be seen in the figure, the amplitude of the VP signal also decreases with the distance from the burning point.

Different basic hypotheses have been proposed to describe the mechanism of VP propagation in plants. The most widely accepted view attributes the long-term depolarization to the local electrical response induced by the propagation of a specific factor. This factor may be a hydraulic wave [17, 18] or a chemical agent (a "wound substance"), a revisited [16, 19] "old" idea [15].

This second approach can be modeled considering the one-dimensional diffusion of a substance based on Fick's law:

$$\frac{\partial c(x,t)}{\partial t} = D\frac{\partial^2 (c(x,t)}{\partial x^2}. \quad (3.1)$$

In this equation, $c(x,t)$ represents the concentration of the substance at a time t at a distance x from the origin of coordinates, and D is the diffusion coefficient characteristic of the substance. This equation can be applied to the propagation of a wounding substance along the x-axis. Assuming that the concentration in the damage point is c_o and that the substance diffuses from this point located at $x = 0$, Eq. (3.1) leads to [20]

$$c(x,t) = \frac{c_0}{2\sqrt{\pi Dt}}\exp\left[-\frac{x^2}{4Dt}\right]. \quad (3.2)$$

Assuming that the signaling evolves at a given point when the concentration of the wound substance in this point reaches a critical threshold value c_{crit}, the position of that point (x_{crit}) will be given by [15]

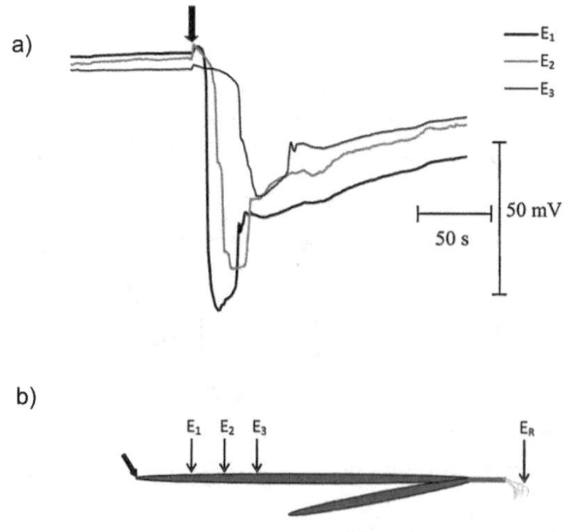

FIGURE 3.5 (a) VP induced by the burning of a leaf tip of wheat (*Triticum aestivum* L.) and (b) schematics for electrode positioning. E1, E2, and E3 are electrodes separated 3, 5 and 7 cm, respectively from the zone of burning (arrow); ER is the reference electrode. (Adapted from ref. [15], with permission. Copyright Elsevier 2012.).

$$x_{\text{crit}} = \sqrt{-4Dt \ln\left(\frac{2c_{\text{crit}}\sqrt{\pi Dt}}{c_o}\right)}. \tag{3.3}$$

This equation has meaning only for $2(c_{\text{crit}}/c_o)(\pi Dt)^{1/2} < 1$. An effective diffusion coefficient of the wounding substance, D_{eff}, can be calculated from experiments with radioactive tracers, as [15]

$$D_{\text{eff}} = \frac{x_2^2 - x_1^2}{4t}, \tag{3.4}$$

where t is the time after wounding, x_1 is the distance between the bounding region and the first point at which radioactivity was measured. The second distance, x_2, is that corresponding to a second point at which the activity was reduced by e times relative to the first measurement point. In the case of wounding, the estimated values of D_{eff} are around 10^{-2} cm^2 s^{-1} [15], clearly larger than the typical values for ions in aqueous solution (around 10^{-6} cm^2 s^{-1}). This large apparent diffusion coefficient could result from the nonlaminar water flow in the xylem and the induction of turbulent diffusion [15]. This is an important factor for considering VP response to damage could also be connected with the spread of hydraulic waves (see the following).

The membrane potential can be treated as stationary, being expressed as (see also Section 2.2) [15]

$$E_m = \frac{g_{K+}E_{K+} + g_{Cl-}E_{Cl-} + g_{Ca2+}E_{Ca2+} + g_{PH}E_{PH} + g_{PCa}E_{PCa} + g_{Sy}E_{Sy}}{g_{K+} + g_{Cl-} + g_{Ca2+} + g_{PH} + g_{PCa} + g_{Sy}}. \tag{3.5}$$

In this equation, g_{K+}, g_{Cl-}, g_{Ca2+}, are electrical conductivities of the K^+, Cl^-, and Ca^{2+} channels, whereas g_{PH}, g_{PCa}, g_{Sy} are the conductivities of the proton, Ca^{2+}-ATPases, and $2H^+/Cl^-$ symporter, respectively. The corresponding E-symbols denote the respective equilibrium potentials. The conductivity coefficients can be calculated from the flux of the respective species, J_J, and its corresponding equilibrium potential, E_J, as [15]

$$g_J = F\frac{J_J}{E_m - E_J}. \tag{3.6}$$

A simplified diagram showing an equivalent electric circuit used to schematize the ion mechanism for membrane potential is presented in Figure 3.6 [10]. The membrane potential includes the contribution of ionic reversal potentials (here limited to K^+, Cl^-, and Ca^{2+} channels) each one in series with a resistive element representative of the respective ionic maximum conductance. Capacitive effects are represented by membrane capacitance.

The hydraulic model of VP propagation considers that there is a direct pressure increase in xylem vessels and apoplast under damage [21], and an indirect pressure increase caused by osmotically active chemicals released from damaged cells [22, 23]. A simplified model assumes that the xylem vessel can be taken as a tube where the Poiseuille equation applies [23]:

$$p = p_o + (p_o - p_L)\frac{x}{L}, \tag{3.7}$$

where p is the stationary pressure at a distance x from the extreme of the tube, p_o is the pressure at $x = 0$, and p_L the pressure at the extreme of the tube. We can apply this equation to the xylem tubes in the absence of damage. When a damaging stimulus is applied in $x = 0$, the pressure increases. Assuming that there is no pressure increase at the end of the tube relative to the undamaged system, the equation for the pressure increase p' will be

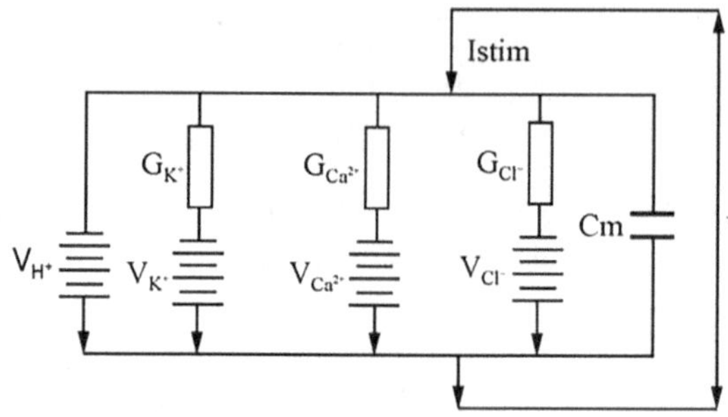

FIGURE 3.6 Simplified equivalent circuit schematizing the ion mechanism for membrane potential. V is the membrane potential and C_m is its capacitance. I_{stim} is the stimulus current; V_{H+}, V_{K+}, V_{Ca2+}, V_{Cl-} represent the ionic reversal potentials, and G_{K+}, G_{Ca2+}, G_{Cl-} the respective ionic maximum conductance. (Reproduced from ref. [10], with permission. Copyright Elsevier 2000.).

$$p' = p_o'\left(1 - \frac{x}{L}\right), \tag{3.8}$$

where p_o' is the pressure increment at $x = 0$. The simplest proposed model assumes a linear variation of the pressure with the time after application of the stimulus. If the total duration of the response is t_r, Eq. (3.8) can be converted into pressure/time equations; three variants have been proposed, assuming that the pressure increment at the origin increases linearly with time, remains constant, or decreases linearly with time [23]:

$$p_o' = p_m'\left(1 - \frac{t}{t_r}\right), \tag{3.9}$$

$$p_o' = \frac{1}{2} p_m', \tag{3.10}$$

$$p_o' = p_m' \frac{t}{t_r}, \tag{3.11}$$

valid for $t \le t_r$. In these equations, p_m' is the maximum pressure increment at $x = 0$. Accordingly, the total pressure change can be determined as the area under the pressure increment/time functions, Q, that is the same in all three variants, i.e.,

$$Q = \frac{1}{2} p_m' t_r. \tag{3.12}$$

The model leads to predicting the dependence of the propagation velocity, v_p, on the distance from the damaged region as a function of the ratio between Q and the threshold pressure change, Q_{crit}, necessary for the initiation of the VP. The respective expressions for the aforementioned variants are [23]

$$v_p = \frac{2L}{t_r G}\left(1 - \frac{G}{1 - x/L}\right)^{1/2} (1 - x/L)^2, \tag{3.13}$$

$$v_p = \frac{L}{t_r G}(1 - x / L)^2, \quad (3.14)$$

$$v_p = \frac{2L}{t_r G^{1/2}}(1 - x / L)^{3/2}, \quad (3.15)$$

being $G = Q/Q_{crit}$.

The sequence of processes determining the generation of VPs starts with the activation of mechanically sensitive or ligand-dependent cation channels by the effect of the damaging stimulus. There is a depolarization phase with an increase in the concentration of Ca^{2+} in the cytosol accompanied by activation of Ca^{2+}-dependent anion channels. Chloride depletion in the cytosol leads to depolarization phase formation and activation of voltage-dependent K^+ channels. Now, long-time repolarization is initiated through the combined effect of the aforementioned ions and the protons. This results in long-time irregular repolarization [3].

3.5 SYSTEM POTENTIALS

SPs are induced by stimulation of the injured leaves with specific species (Na^+, Mg^{2+}, K^+, Ca^{2+}, glutamic acid). SPs appear as potential pulses whose amplitude and duration are related to the intensity of the stimulus. SPs can be propagated over long distances, but the amplitude and propagation velocity gradually decrease [3, 24, 25]. SPs are related to the activation/deactivation of the proton pump caused by the change in hydraulic pressure. The mass efflux of H^+ produces the hyperpolarization of the plasma membrane, whose inactivation results in de-hyperpolarization.

3.6 SIGNALING

In discussing the hypothesis of the wounding substance for VP generation, we bypassed an important problem: The wound substance has not been unambiguously identified [23]. Several signaling compounds such as H_2O_2, systemin, jasmonic acid, abscisic acid, and glutamate, among others, have been discussed as possible wound substances [15, 26, 27].

These substances act, in principle, as "second messengers", small intracellular molecules that relay information within the cell. A change in their concentration induced by a given stimulus results in a specific physiological response [28]. Recent studies suggest that these substances act not only for intracellular but also for intercellular signaling [29].

Signaling chemical species such as Ca^{2+}, reactive oxygen species (ROS), and reactive nitrogen species (RNS) form signatures in the plant's immune and wound responses. It is known that following pathogen infections or physical damage, there is a rapid production of ROS. This process, the so-called oxidative burst, is an important signature promoting local immune response and cell-to-cell communication [30].

There is a complex signaling regulation systems. For example, both Ca^{2+} and ROS signaling regulate each other and are balanced to regulate immunity responses in plants [31]. Figure 3.7 illustrates the complex series of relationships proposed to describe the Ca^{2+} dependence of ROS signaling and the ROS dependence of Ca^{2+} signaling [31]. Relevant aspects to mention are that the Ca^{2+} signal possibly enhances ROS influx by activating aquaporins, while the Ca^{2+} signal possibly activates catalase enzymes, which act as ROS scavengers. At your time, hydrogen peroxide and hydroxyl radicals activate Ca^{2+} permeable channels [31].

In this scenario, the classical electrochemical methods centered on the measurement of ionic fluxes and membrane potentials provide valuable tools for better understanding biochemical processes, as summarized in Chapter 2. Complementary electrochemical methods aimed at studying signaling processes and ROS chemistry will be treated in Chapters 9 and 10.

3.7 PLANT COMMUNICATION

Plants can communicate between them using different pathways. These include volatile organic compounds (VOCs), root grafting, the rhizosphere (root ball), and mycorrhizal networks [32]. In addition, there is communication via electrostatic and electromagnetic [33] and "mechanic" (acoustic or sound) vibrations [34].

Nonhydraulic signaling between roots and shoots of plants provides a sensitive link between soil changes and physiological responses via electrical signaling [35]. This is favored by the relatively high conductivity of water-impregnated soils. In trees, communication over long distances may be achieved through phloem-transmitted electrical signals [7].

The synchronous electrostimulation of plants from two different points is being used as an electrochemical technique to study plant communication [32]. The electrostimulation of plants can induce plant movement, activation of ion channels and ion transport, activation of enzymatic systems, and gene expression. These are accompanied by the generation of action and electrotonic potentials [33]; the synchronous electrostimulation technique is based on the "collision" of electrotonic potentials that propagate with either a positive or negative pole in front of them. In extreme cases, this results in amplification or annihilation of the signals.

Figures 3.8 and 3.9 show the experimental results on electrotonic signal conduction in tomato (*Lycopersicon esculentum* Mill. cv Cosmonaut Volkov) plants [33]. To avoid root-to-root connections or mycorrhizal networks between plants, the soil between plants was separated by a nonpenetrable barrier. The experimental arrangement was close to that schematized in Figure 3.1 but here incorporating two Pt-electrodes connected to a function generator for electrostimulation. These electrodes and the usual Ag/AgCl electrodes were inserted into the plant vascular boundless.

After the insertion of the electrodes, the plants were allowed to rest until a stable potential difference was obtained between the Ag/AgCl electrodes. During the day, and in the absence of any external stress, the potential difference between these electrodes remained stable. This situation changed significantly when a periodic potential input of ±1V between Pt-electrodes was applied. Now, the response varies with the type of stimulus involving a phase difference between the stimulus and the response, and decreases exponentially with the distance.

This situation can be modeled using an equivalent circuit consisting of a resistor (with resistance R) and a capacitor (with capacity C) in series (RC series circuit). Here the output voltage is directly proportional to the rate of variation of the difference between input (V_{out}) and output (V_{in}) voltages [33]:

$$V_{\text{out}}(t) = RC\frac{\mathrm{d}}{\mathrm{d}t}\left(V_{\text{in}}(t) - V_{\text{out}}(t)\right). \tag{3.16}$$

When the rate of variation of the output potential is much lower than the rate of variation of the input potential, it is possible to approximate:

$$V_{\text{out}}(t) = RC\frac{\mathrm{d}V_{\text{in}}(t)}{\mathrm{d}t}. \tag{3.17}$$

This corresponds to the so-called electrical differentiator [33]. Thus if the input is sinusoidal of amplitude A and angular frequency ω ($V_{in}(\mathrm{t}) = A\sin(\omega t)$), then

$$V_{\text{out}}(t) = RC\frac{\mathrm{d}A\sin(\omega t)}{\mathrm{d}t} = RCA\omega\cos(\omega t). \tag{3.18}$$

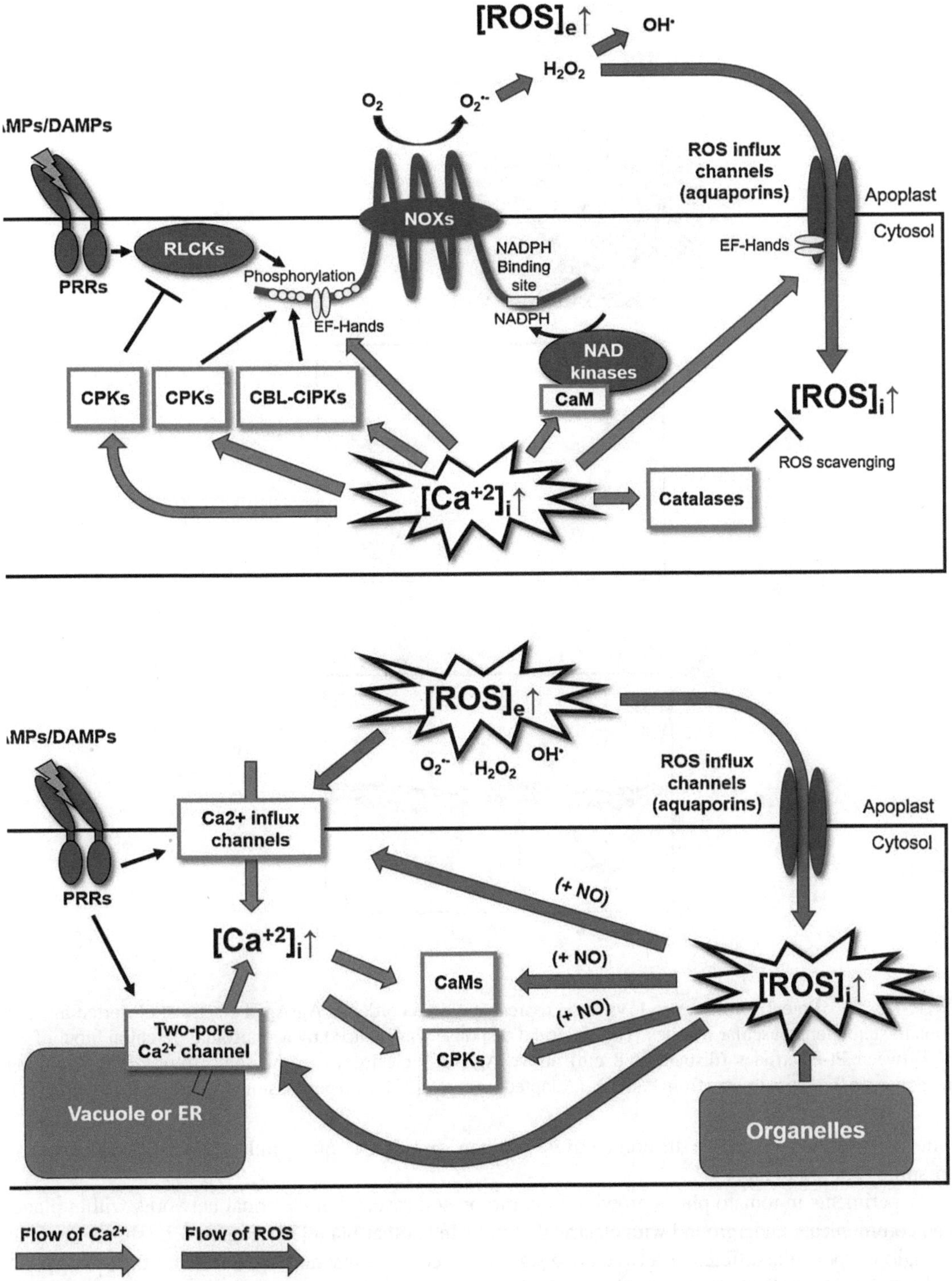

FIGURE 3.7 Schemes for processes proposed to describe (a) Ca^{2+}-dependent ROS signaling and (b) ROS-dependent Ca^{2+} signaling. For a detailed interpretation of such diagrams, see ref. [31]. (Adapted from this ref., with permission. Copyright Elsevier 2019.).

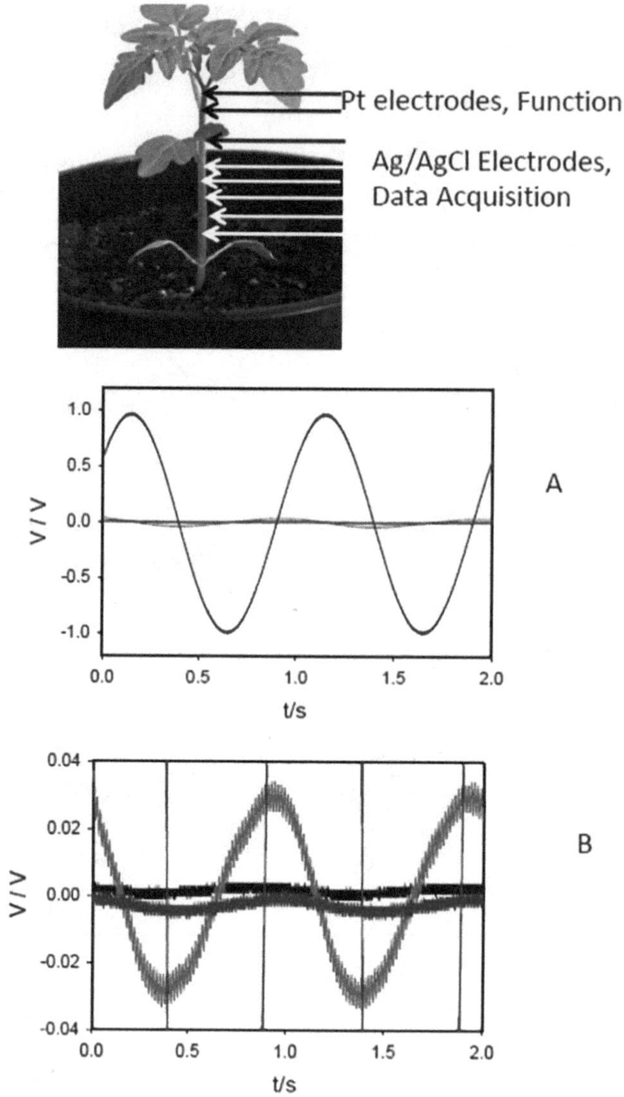

FIGURE 3.8 Potential difference V vs. time representation recorded at Ag/AgCl electrodes inserted into the tomato stem along vascular bundles. The electrical response was induced by a sinusoidal potential input of ±1 V between Pt-electrodes (distance 0.8 cm) above Ag/AgCl electrodes in the same stem. High-resolution responses in B were extracted from Panel A. (Adapted from ref. [33], with permission. Copyright Elsevier 2018.).

This corresponds to a phase difference of 90°. When stimuli are square pulses, the response consists of spikes analog to APs.

Experiments in tomato plants growing in common soil reveal that electrical networks within plants can communicate underground with electrical circuits in another plant [33]. Figure 3.10 shows the time variation of potential difference between two Ag/AgCl electrodes inserted into the stem of a tomato plant induced by the application of sinusoidal and triangular potential inputs to a second tomato plant in the same pot [33]. At the frequency used in this experiment, 500 Hz, no phase differences were detected, but such differences appear at lower frequencies. This system can be modeled using an electrical circuit constituted by RC units corresponding to cell-to-cell communication within the same plant stem and the propagation of the signal through the plant stem [33]. In the model, root-to-root signal propagation is represented by two resistive elements, but more complicated modeling can be proposed [36, 37].

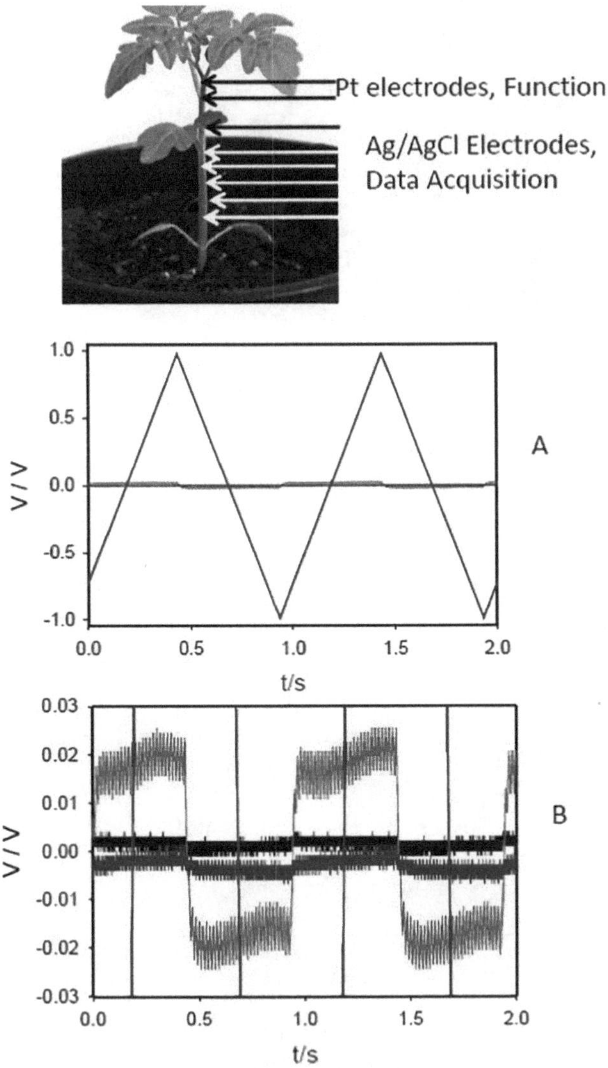

FIGURE 3.9 Time variation of the potential difference *V* recorded in a tomato plant using the same arrangement as that in Figure 3.7 using a ±1V triangular wave from a function generator. High-resolution responses in B were extracted from Panel A. (Adapted from ref. [33], with permission. Copyright Elsevier 2018.).

Through-soil communication of plants may involve, as previously noted, root grafting, the rhizosphere (root ball), and mycorrhizal networks [32]. The rhizosphere can be seen as a medium that hosts a diverse array of microorganisms interacting with plants through chemical signals. These signals are produced in response to specific stimuli and activate regulatory mechanisms [38]. The entire ecological niche can be referred to as a microbiome and includes bacteria, fungi, archaea, oomycetes, viruses, and lower and higher eukaryotes [38, 39]. From the electrochemical point of view, analysis of rhizosphere processes also deals with soil and biochar analysis [40]. This last can be defined as a carbon-rich material produced when crop and wood residues break down thermally in the presence of little oxygen. Here, zeta and streaming potentials [41] and cation exchange capacity [42] are electrochemical parameters to be evaluated [38]. The application of electrochemical techniques to study communication in this context is a promising field of research. Here, solid-state techniques, which have been applied to soil analysis (see Chapter 13), are also of potential interest in this regard.

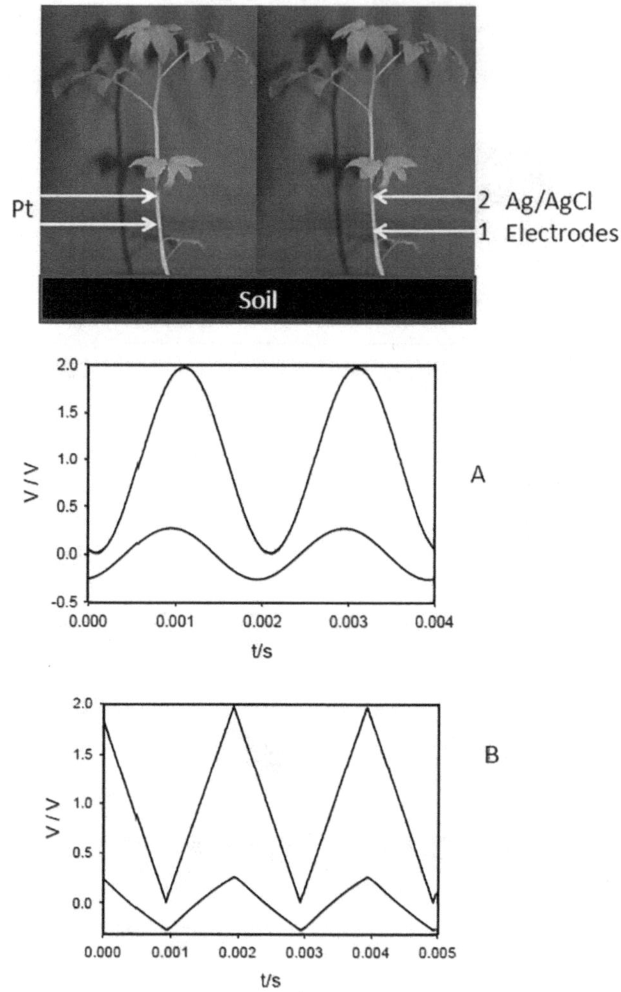

FIGURE 3.10 Potential difference between two Ag/AgCl electrodes inserted into the stem of a tomato plant induced by the application of 2 V sinusoidal (a) or triangular (b) potential input of frequency 500 Hz to a second tomato plant in the same pot. Distance between Pt-electrodes 0.2 cm; distance between lower Pt and Ag/AgCl electrodes and soil was 4 cm. Measurements were carried out at 500,000 scans/s with a low pass filter at 250,000 scans/s. (Adapted from ref. [33], with permission. Copyright Elsevier 2018.).

3.8 ELECTROMAGNETIC FIELDS

The study of electrophysiology of plants was aided by techniques involving the measurement of magnetic effects associated with electric currents. The electrosuperconducting quantum interference devices (SQUID), available since the 1970s [43], have been increasingly used in the study of electrophysiological signals in living organisms [44]. Although experimental arrangements are conditioned by the cryogenic arrangement needed for achieving the semiconducting behavior in sensing devices, a notable amount of research has been carried out. Most have been conducted on "giant" internodal cells of *Chara corallina* Willdenow and other green algae [44].

The application of external static electric and magnetic fields, as well as time-dependent electromagnetic fields, has also been widely studied. Experiments with seedlings of different plant species have detected effects on the growth of their primary roots proliferative activity and cell reproduction in the meristem, changes in the distribution of condensed chromatin and nucleolus compactization in nuclei, accumulation of lipid bodies, among others, in the presence of weak magnetic fields [45]. Magnetic fields influence the regulation of plant functions and growth, and appear to increase tolerance against different types of environmental stress [46]. The effects of electromagnetic fields appear to be more intense at low frequencies [47], whereas the effect of high-frequency fields on living organisms remains controversial [48].

A variety of physicochemical phenomena have been invoked to interpret this phenomenology. Among them, oscillator and cyclotron resonance models, quantum effects (interference, coherent quantum excitation), free radical reactions, phase transitions, and modifications of chemical kinetics have been proposed [45]. Possibly, a direction to be exploited in future research could deal with magnetoelectrochemistry. Classically, the application of magnetic fields to electrochemical cells yields a magnetoconvective effect influencing the transport properties of paramagnetic ions [49]. Other effects (energy level splitting in radicals and diamagnetic orientation of aromatic molecules) have been described [50]. The paramagnetic and magnetoconvective effects may influence intra- and intercellular ionic fluxes but also condition molecular preorganization processes associated with chemical or electrochemical reactions [51].

3.9 ENVIRONMENTAL APPLICATIONS

Conjointly considered, plant electrical potentials display valuable information about physiological processes in plants occurring at different levels and on plant response to external stimuli. In this context, periodic potentials can be used for monitoring the response of plants to different types of environmental stress. Figure 3.11 illustrates a case in which light-induced rhythmic bioelectrogenesis (LIRB) was used to characterize the salt sensitivity of wheat cultivars [52]. In this experiment, seedlings were placed in a chamber with a layer of a solid agarose cell using two Ag/AgCl electrodes. The leaf tip surface of a wheat seedling contacts the solid agarose gel on the undersurface of the lid and the root is in contact with the solid agarose gel in the bottom of the chamber. Figure 3.11 depicts the variation of the electrical potential difference between the leaf tip surface coupling region

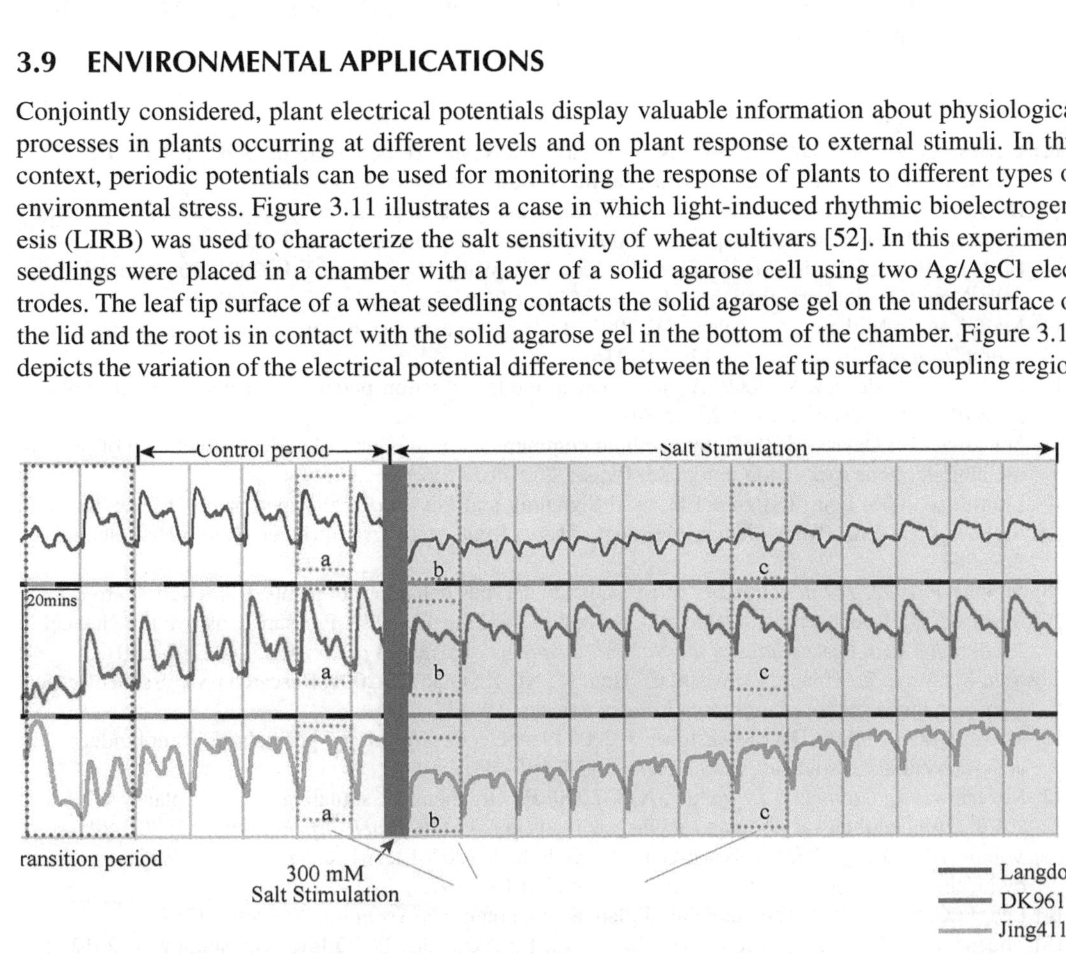

FIGURE 3.11 LIRB waveform for durum wheat "Langdon" (upper area) and wheat "DeKang 961" (middle area), wheat "Jing 411" (lower area). (a) Last complete light-dark cycle wave before continuous salt stimulation. (b) First complete light-dark cycle wave during the salt stress period. (c) Seventh light-dark cycle wave during the salt stress period. The brown vertical bar and the associated arrow indicate the salt stimulation; the interference caused by the lid change was removed. (Adapted from ref. [52], with permission. Copyright Elsevier 2019.).

and the root coupling region for wheat "DeKang 961," wheat "'Jing 411," and durum wheat "Langdon." The seedlings were submitted to light/dark stimulation cycles of 10 min under nonsaline, and saline (300 mM NaCl) conditions. The graph shows different periodic waveforms for the three different specimens with significant differences between control and saline-stressed probes. Such differences were satisfactorily correlated to chlorophyll fluorescence measurements [52].

3.10 CONCLUDING REMARKS

The electrical activity of plants is of crucial importance for the regulation of a great variety of physiological processes in plants. Plant electrical signals propagate over long distances within plants to trigger different physiological responses. The cell-to-cell transference of information is complemented by the so-called second messengers who convert extracellular signals into intracellular information to be decoded into the appropriate intracellular biochemical processes. All this signaling machinery involves ionic fluxes and variations in membrane potentials, thus underlying the importance of an electrochemical view of plant physiology. Electrochemistry offers an increasingly growing set of techniques capable of yielding information on plant physiology and the relations between plants and the environment.

REFERENCES

[1] Davies, E. 2006. Electrical Signals in Plants: Facts and Hypotheses, in Volkov, A.G. Ed. *Plant Electrophysiology Theory and Methods*. Springer, Berlin-Heidelberg, pp. 407–422.

[2] Davies, E. 1987. Action potentials as multifunctional signals in plants: a hypothesis attempting to unify apparently disparate wound responses. *Plant, Cell and Environment*. 10: 623–631.

[3] Li, J.-H.; Fan, L.-F.; Zhao, D.-J.; Zhou, Q.; Yao, J.-P.; Wang, Z.-Y.; Huang, L. 2021. Plant electrical signals: A multidisciplinary challenge. *Journal of Plant Physiology*. 261: artic. 153418.

[4] Maffei, M.E.; Mithofer, A.; Boland, W. 2007. Before gene expression: early events in plant-insect interaction. *Trends in Plant Science*. 12: 310–316.

[5] Sukhov, V.; Vodeneev, V. 2009. A mathematical model of action potential in cells of vascular plants. *Journal of Membrane Biology*. 232: 59–67.

[6] Stankovic, B.; Davies, E. 1997. Intercellular communication in plants: electrical stimulation of proteinase inhibitor gene expression in tomato. *Planta*. 202: 402–406.

[7] Fromm, J. 2006. Long-Distance Electrical Signaling and Physiological Functions in Higher Plants, in Volkov, A.G. Ed. *Plant Electrophysiology Theory and Methods*. Springer, Berlin-Heidelberg, pp. 269–286.

[8] Simons, P. 1992. *The action plant*. Movement and nervous behaviour in plants. Blackwell, Oxford.

[9] Martinac, B.; Buechner, M.; Delcour, A.H.; Adler, J.; Kung, C. 1987. Pressure-sensitive ion channel in *Escherichia coli*. *Proceedings of the National Academy of Sciences of the USA*. 84: 2297–2301.

[10] Yan, X.; Wang, Z.; Huang, L.; Wang, C.; Hou, R.; Xu, Z.; Qiao, X. 2009. Research progress on electrical signals in higher plants. *Progress in Natural Science*. 19: 531–541.

[11] Volkov, A.G.; Collins, D.J.; Mwesigwa, J. 2000 Plant electrophysiology: pentachlorophenol induces fast action potentials in soybean. *Plant Science*. 153: 185–190.

[12] Mwesigwa, J.; Collins, D.J.; Volkov, A.G. 2000. Electrochemical signaling in green plants: effects of 2,4-dinitrophenol on variation and action potentials in soybean. *Bioelectrochemistry*. 51: 201–205.

[13] Volkov, A.G.; Lang, R.D.; Volkova-Gugeshashvili, M.I. 2007. Electrical signaling in *Aloe vera* induced by localized thermal stress. *Bioelectrochemistry*. 71: 192–197.

[14] Lee, J.; Calvo, P. 2023. The potential of plant action potentials. *Synthese*. 202: artic. 176.

[15] Vodeneev, V.; Orlova, A.; Morozova, E.; Orlova, L.; Akinchits, E.; Orlova, O.; Sukhov, V. 2012. The mechanism of propagation of variation potentials in wheat leaves. *Journal of Plant Physiology*. 169: 949–954.

[16] Vodeneev, V.; Akinchits, E.; Sukhov, V. 2015. Variation potential in higher plants: Mechanisms of generation and propagation. *Plant Signaling & Behavior*. 10: artic. e1057365.

[17] Malone, M.; Stankovic, B. 1991. Surface potentials and hydraulic signals in wheat leaves following localized wounding by heat. *Plant, Cell and Environment*.14: 431–436.

[18] Mancuso, S. 1999. Hydraulic and electrical transmission of wound-induced signals in *Vitis vinifera*. *Australian Journal of Plant Physiology*. 26: 55–61.

[19] Evans, M.J.; Morris, R.J. 2017. Chemical agents transported by xylem mass flow propagate variation potentials. *The Plant Journal*. 91: 1029–1037.

[20] Codling, E.A.; Plank, M.J.; Benhamou, S. 2008. Random walk models in biology. *Journal of Royal Society Interface*. 5: 813–834.

[21] Stahlberg, R.; Cleland, R.E.; van Volkenburgh, E. 2006. Slow wave potentials—A propagating electrical signal unique to higher plants, in *Communication in Plants. Neuronal Aspects of Plant Life*. Baluška, F.; Mancuso, S.; Volkmann, D. Eds. Springer: Berlin/Heidelberg, New York, pp. 291–308.

[22] Malone, M. 1994. Wound-induced hydraulic signals and stimulus transmission in *Mimosa pudica* L. *New Phytologist*. 128: 49–56.

[23] Sukhova, E.; Akinchits, E.; Gudkov, S.V.; Pishchalnikov, R.Y.; Vodeneev, V.; Sukhov, V. 2021. A Theoretical Analysis of Relations between Pressure Changes along Xylem Vessels and Propagation of Variation Potential in Higher Plants. *Plants*. 10: artic. 372.

[24] Vodeneev, V.; Mudrilov, M.; Akinchits, E.; Balalaeva, I.; Sukhov, V. 2018. Parameters of electrical signals and photosynthetic responses induced by them in pea seedlings depend on the nature of stimulus. *Functional Plant Biology*. 45: 160–170.

[25] Felle, H.H.; Zimmermann, M.R. 2007. Systemic signalling in barley through action potentials. *Planta*. 226: 203–214.

[26] Choi, W.G.; Miller, G.; Wallace, I.; Harper, J.; Mittler, R.; Gilroy, S. 2017. Orchestrating rapid long-distance signaling in plants with Ca^{2+}, ROS and electrical signals. *The Plant Journal*. 90: 698–707.

[27] Toyota, M.; Spencer, D.; Sawai-Toyota, S.; Jiaqi, W.; Zhang, T.; Koo, A.J.; Howe, G.A.; Gilroy, S. 2018. Glutamate triggers long-distance, calcium-based plant defense signaling. *Science*. 361: 1112–1115.

[28] Plieth, C. 2016. Calcium, metaphors, and zeitgeist in plant sciences, *Plant Physiology*. 171: 1790–1793.

[29] Gilroy, S.; Białasek, M.; Suzuki, N.; Górecka, M.; Devireddy, A.R.; Karpiński, S.; Mittler, R. 2016. ROS, calcium, and electric signals: key mediators of rapid systemic signaling in plants. *Plant Physiology*. 171: 1606–1615.

[30] Miller, G.; Schlauch, K.; Tam, R.; Cortes, D.; Torres, M.A.; Shulaev, V.; Dangl, J.L.; Mittler, R. 2009. The plant NADPH oxidase RBOHD mediates rapid systemic signaling in response to diverse stimuli. *Science Signaling*. 2: ra45–ra45.

[31] Marcec, M.J.; Gilroy, S.; Poovaiah, B.W.; Tanaka, K. 2019. Mutual interplay of Ca^{2+} and ROS signaling in plant immune response. *Plant Science*. 283: 343–354.

[32] Volkov, A.G.; Toole, S.; WaMaina, M. 2019. Electrical signal transmission in the plant-wide web. *Bioelectrochemistry*. 129: 70–78.

[33] Volkov, A.G.; Shtessel, Yu.B. 2018. Electrical signal propagation within and between tomato plants. *Bioelectrochemistry*. 124: 195–205.

[34] Mishra, R.C.; Ghosh, R.; Bae, H. 2016. Plant acoustics: in the search of a sound mechanism for sound signaling in plants. *Journal of Experimental Botany*. 67: 4483–4494.

[35] Davies, W.J.; Zhang, J. 1991. Root signals and the regulation of growth and development of plants in drying soil. *Annual Review of Plant Physiology and Plant Molecular Biology*. 42: 55–76.

[36] Johnson, D.; Gilbert, L. 2015. Interplant signaling through hyphal networks. *New Phytologist*. 205: 1448–1453.

[37] Volkov, A.G.; Shtessel, Yu.B. 2016. Propagation of electrotonic potentials in plants: experimental study and mathematical modeling. *AIMS Biophysics*. 3: 358–378.

[38] Solomon, W.; Janda, T.; Molnár, Z. 2024. Unveiling the significance of rhizosphere: Implications for plant growth, stress response, and sustainable agriculture. *Plant Physiology and Biochemistry*. 206: artic. 108290.

[39] Berg, G.; Smalla, K. 2009. Plant species and soil type cooperatively shape the structure and function of microbial communities in the rhizosphere. *FEMS Microbiology Ecology*. 68: 1–13.

[40] Farhangi-Abriz, S.; Ghassemi-Golezami, K. 2023. Improving electrochemical characteristics of plant roots by biochar is an efficient mechanism in increasing cations uptake by plants. *Chemosphere*. 313. artic. 137365.

[41] Li, Z.Y.; Liu, Y.; Zheng, Y.Y.; Xu, R.K. 2015. Zeta potential at the root surfaces of rice characterized by streaming potential measurements. *Plant Soil*. 386: 237–250.

[42] Rengel, Z.; Robinson, D.L. 1989. Determination of cation exchange capacity of ryegrass roots by summing exchangeable cations. *Plant Soil*. 116: 217–222.

[43] Zimmerman, J.E.; Thiene, P.; Harding, J.T. 1970. Design and operation of stable RF-biased superconducting point-contact quantum devices, and a note on properties of perfectly clean metal contacts. *Journal of Applied Physics*. 41: 1572–1580.

[44] Tronteli, Z.; Thiel, G.; Jazbinsek, V. 2006. Magnetic Measurements in Plant Electrophysiology, in Volkov, A.G. Ed. *Plant Electrophysiology Theory and Methods*. Springer, Berlin-Heidelberg, pp. 187–218.

[45] Belyavskaya, N.A. 2004. Biological effects due to weak magnetic fields on plants. *Advances in Space Research*. 34: 1566–1574.

[46] Radhakhrisnan, R. 2019. Magnetic field regulates plant functions, growth and enhances tolerance against environmental stresses. Physiology *Molecular Biology Plants*. 25: 1107–1119.

[47] Goldsworthy, A. 2006. *Effects of Electrical and Electromagnetic Fields on Plants and Related Topics*, in Volkov, A.G. Ed. *Plant Electrophysiology Theory and Methods*. Springer, Berlin-Heidelberg, pp. 247–422.

[48] Vian, A.; Davies, E.; Gendraud, M.; Bonnet, P. 2016. Plant Responses to High Frequency Electromagnetic Fields. *BioMed Research International*. 2016: artic. 1830262.

[49] Fahidy, T.Z. 1990. The statistical indeterminacy of the magnetic field effect on electrolytic mass transport. *Electrochimica Acta*. 35: 929–932.

[50] Waskaas, M.; Kharkats, Y.I. 1999. Magnetoconvection Phenomena: A Mechanism for Influence of Magnetic Fields on Electrochemical Processes. *Journal of Physical Chemistry B*. 104: 4876–4883.

[51] Doménech-Carbó, A.; Costero, A.M.; Bañuls, M.J.; Aurell, M.J. 2004. Magnetoelectrochemical modulation of pre-organization processes in a 4,40-dinitrobiphenyl azacrown macrocyclic lactam. *Electrochemistry Communications*. 6: 908–912.

[52] Wang, Z.-Y.; Qin, X.-H.; Li, J.-H.; Fan, L.-F.; Zhou, Q.; Wang, Y.-Q.; Zhao, X.; Xie, C.-J.; Wang, Z.-Y.; Huang, L. 2019. Highly reproducible periodic electrical potential changes associated with salt tolerance in wheat plants. *Environmental and Experimental Botany*. 160: 120–130.

ADDITIONAL LITERATURE

Historical Literature

Adrian, E.D.; Bronk, D.W. 1928. The discharge of impulses in motor nerve fibres. I. Impulses in single fibres of the phrenic nerve. *Journal of Physiology*. 66: 81–101.

Bose, J.C. 1925. Transmission of stimuli in plants. *Nature*. 115: 457–457.

Burdon Sanderson, J. 1873. Note on the electrical phenomena which accompany irritation of the leaf of *Dionaea muscipula*. *Proceedings of the Royal Society of London A Mathematical and Physical Sciences*. 21: 495–496.

Gaffey, C.T.; Mullins, L.J. 1958. Ion fluxes during the action potential in *Chara*. *Journal of Physiology*. 144: 505–524.

Hope, A.B. 1961. Ionic relations of cells of Chara corallina. V. The action potential. *Australian Journal of Biological Sciences*. 14: 312–322.

Hormann, G. 1898. *Studien uber die Protoplasmastromung bei den Characacean*. Fischer, Jena.

Ricca, U. 1916. Soluzione d'un problema di fisiologia: la propagazione di stimulo nella Mimosa. *Nuovo Giornale di Botanico Italiano*. 23: 51–170.

Sinukhin, A.M.; Britikov, E.A. 1967. Action potentials in the reproductive system of plant. *Nature*. 215: 1278–1280.

Umrath, K. 1929. Uber Erregungsleitung bei hoheren Pflanzen. *Planta*. 7: 174–207.

Local Electrical Potentials

Marten, I.; Deeken, R.; Hedrich, R.; Roelfsema, M.R.G. 2010. Light-induced modification of plant plasma membrane ion transport. *Plant Biology*. 12: 64–79.

Wang, C.; Huang, L.; Wang, Z.Y.; Qiao, X.J. 2010. Monitoring and analysis of electrical signals in water-stressed plants. *New Zealand Journal of Agricultural Research*. 50: 823–829.

Action Potentials

Choi, W.-G.; Hilleary, R.; Swanson, S.J.; Kim, S.-H.; Gilroy, S. 2016. Rapid, long-distance electrical and calcium signaling in plants. *Annual Review of Plant Biology*. 67: 287–307.

Eschrich, W.; Fromm, J.; Evert, R.F. 1988. Transmission of electric signals in sieve tubes of zucchini plants. *Botanica Acta*. 101:327–331.

Huber, A.E.; Bauerle, T.L. 2016. Long-distance plant signaling pathways in response to multiple stressors: The gap in knowledge. *Journal of Experimental Botany*. 67: 2063–2079.
Pawlikowska-Pawlega, B.; Król, E.; Trębacz, K.; Gawron, A. 2009. Genistein and changes of resting and action potentials in *Conocephalum conicum*. *Journal of Plant Physiology*. 166: 712–719.

Variation Potentials

Evans, M.J.; Morris, R.J. 2017. Chemical agents transported by xylem mass flow propagate variation potentials. *The Plant Journal*. 91: 1029–1037.
Stahlberg, R.; Cosgrove, D.J. 1997. The propagation of slow wave potentials in pea epicotyls. *Plant Physiology*. 113: 209–217.

System Potentials

Sukhov, V.; Akinchits, E.; Katicheva, L.; Vodeneev, V. 2013. Simulation of variation potential in higher plant cells. *Journal of Membrane Biology*. 246: 287–296.
Zimmermann, M.R.; Maischak, H.; Mithofer, A.; Boland, W.; Felle, H.H. 2009. System potentials, a novel electrical long-distance apoplastic signal in plants, induced by wounding. *Plant Physiology*. 149: 1593–1600.

Siignaling

Hlavinka, J.; Nožková-Hlaváčková, V.; Floková, K.; Novák, O.; Nauš, J. 2012. Jasmonic acid accumulation and systemic photosynthetic and electrical changes in locally burned wild type tomato, ABA-deficient sitiens mutants and sitiens pre-treated by ABA. *Plant Physiology and Biochemistry*. 54: 89–99.
Herde, O.; Peña-Cortés, H.; Fisahn, J. 1995. Proteinase inhibitor II gene expression induced by electrical stimulation and control of photosynthetic activity in tomato plants. *Plant and Cell Physiology*. 36: 737–742.
Pelagio-Flores, R.; Ruiz-Herrera, L.F.; López-Bucio, J. 2016. Serotonin modulates Arabidopsis root growth via changes in reactive oxygen species and jasmonic acid-ethylene signaling. *Physiologia Plantarum*. 158: 92–105.
Song, Y.Y.; Zeng, R.S.; Xu, J.F.; Li, J.; Shen, X.; Yihdego, W.G. 2010. Interplant communication of tomato plants through underground common mycorrhizal networks. *PLoS One*. 5: artic. e13324.
Wan, J.; Zhang, P.; Sun, L.; Li, S.; Wang, R.; Zhou, H.; Wang, W.; Xu, J. 2018. Involvement of reactive oxygen species and auxin in serotonin-induced inhibition of primary root elongation. *Journal of Plant Physiology*. 229: 89–99.

Plant Communication

Faure, D.; Vereecke, D.; Leveau, J.H.J. 2009. Molecular communication in the rhizosphere. *Plant Soil*. 321: 279–303.

Electromagnetic Fields

Binhi, V.N.; Rubin, A.B. 2022. Theoretical concepts in magnetobiology after 40 years of research. *Cells*. 11: artic. 274.
Nyakane, N.E. Markus, E.D.; Sedibe, M.M. 2019. The Effects of Magnetic Fields on Plants Growth: A Comprehensive Review. *International Journal of Food Engineering*. 5: 79–87.
Selaga, T.; Selaga, M. 1996. Response of Pinus sylvestris L needles to electromagnetic fields. Cytological and ultrastructural aspects. *Science of the Total Environment*. 180: 65–73.
Smith, S.D.; McLeod, B.R.; Liboff, A.R. 1993. Effects of SR tuning 60 Hz magnetic fields on sprouting and early growth of *Raphanus sativus*. *Bioelectrochemistry and Bioenergetics*. 32: 67–76.

Environmental Applications

Chatterjee, S.K.; Ghosh, S.; Das, S.; Manzella, V.; Vitaletti, A.; Masi, E.; Santopolo, L.; Mancuso, S.; Maharatna, K. 2014. Forward and inverse modelling approaches for prediction of light stimulus from electrophysiological response in plants. *Measurement*. 53: 101–116.
Xie, J.; Wu, Y.; Xing, D.; Li, Z.; Chen, T.; Duan, R.; Zhu, X. 2021. A comparative study on the circadian rhythm of the electrical signals of *Broussonetia papyrifera* and *Morus alba*. *Plant Signaling & Behavior*. 16: artic. e1950899.

4 Electrochemistry of Plants

4.1 INTRODUCTION

Vegetal matter contains an enormous variety of chemical components. Apart from structural components lignins and celluloses, and those constituting the genoma, a variety of very-long-chain aliphatics fatty acids, aldehydes, primary and secondary alcohols, ketones, and alkanes of chain lengths C20–C36, as well as C38–C70 alkyl esters, as well as triterpenoids, tocopherols, or aromatic compounds can be identified [1]. One important fraction is formed by primary metabolites, compounds that are involved in the fundamental metabolic pathways for growth and development. These are accompanied by secondary metabolites. These generally perform nonfundamental but important functions in the plants, for instance, defense against predation.

Given the nutritional and biomedical utilities of many secondary metabolites, much research has been devoted to isolate, identify, and characterize their chemical and biochemical properties. Not all metabolites are electroactive, as well as they do not provide voltammetric responses under the conditions in which they are isolated. A significant fraction, however, is electrochemically active, so electrochemical methods constitute an important source of information about their properties.

In this chapter, we will examine the electrochemistry of the most widely studied families of secondary metabolites, most of polyphenolic nature [2]: quinones, flavonoids and other compounds based on benzopyran unit, phenolic acids and ester derivatives, lignins, tannins, lignans, stilbenes, curcuminoids, and chalcones. For this purpose, it is convenient to clarify several basic concepts around the characterization of the electrochemical reactivity of these compounds.

4.2 THERMOCHEMICAL ASPECTS, REVERSIBILITY

We are interested here in the processes where a given chemical species in the solution phase interchanges electrons with an electrode. This is an electronic conductor whose chemical composition is not altered in the experiment. The electrolytic solution, usually but not necessarily aqueous, contains the electroactive species of interest accompanied by an electrochemically silent salt in large concentration. This is the supporting electrolyte ensuring the high ionic conductivity of the solution and making negligible ionic migration effects [3]. The electrolyte solution frequently buffered at a given pH, is in general deoxygenated. As discussed in Chapter 1, the electrochemical process involves, in the simplest case, the one-electron reduction or oxidation of the problem species yielding a second species in the solution phase. The reduction of an oxidized form, A^{ox}, to a reduced form, A^{rd}, can be represented as

$$A^{ox}{}_{solv} + ne^- \rightarrow A^{rd}{}_{solv}, \tag{4.1}$$

where the subscript solv denotes solvated species in the electrolyte solution, and the charges of A^{ox} and A^{rd} are omitted for simplicity. This is a faradaic process because there is a change in the composition of the system. This kind of process may involve the evolution of gases (for instance, $2H^+ + 2e^- \rightarrow H_2$) or the deposition (or dissolution) of a solid (for instance, $Cu^{2+} + 2e^- \rightarrow Cu$). For our purposes, the relevant point to emphasize is that the electrochemical process can be formally treated as a heterogeneous chemical reaction in which one of the reagents, the electron, must pass through the electrode/electrolyte interface [4]. This process will occur at a finite rate depending (at least) on the energetic barrier established between the aforementioned two phases. This can be expressed using an electrochemical rate constant whose value will depend on the difference of potential established between the working electrode and a reference electrode.

 DOI: 10.1201/9781032697826-4

The process will be considered electrochemically reversible when the electrochemical rate constant is high enough to assume that the process is driven by thermochemical considerations [5, 6]. In these circumstances, the Nernst equations apply and the equilibrium potential of the system in arbitrary conditions, E_{eq}, can be expressed as

$$E_{eq} = E^{\circ}\left(A^{ox} / A^{rd}\right) + \frac{RT}{nF}\ln\frac{a\left(A^{ox}\right)}{a\left(A^{rd}\right)}. \tag{4.2}$$

Here, $E^{\circ}(A^{ox}/A^{rd})$ represents the standard potential of the couple, $a(A^{ox})$, $a(A^{rd})$, represent the thermochemical activities of the oxidized and reduced forms in the electrolyte solution; all other symbols have their customary meaning. In general, this Nernst equation can be rewritten by replacing activities with concentrations and passing the Neperian logarithm to the decimal logarithm. At 25 °C, the resulting equation is:

$$E_{eq} = E^{\circ}\left(A^{ox} / A^{rd}\right) + \frac{0.059}{n}\log\frac{\left[A^{ox}\right]}{\left[A^{rd}\right]}. \tag{4.3}$$

Ideally, the standard potential of the A^{ox}/A^{rd} couple in aqueous solution corresponds to the equilibrium potential of the electrochemical cell constituted by the standard (or normal) hydrogen electrode (SHE) and the problem couple (which includes an inert electrode), with all species being in the activity unit at the standard pressure. Notice that currently, this is 1 bar = 10^5 Pa. Earlier, the standard pressure was taken as 1 atm = 1.01325 bar. This causes a negligible difference in the potential of SHE of 0.169 mV [7]. Assuming, by simplicity, that $n = 1$, the cell reaction is

$$A^{ox}{}_{aq} + (1/2)H_{2\,gas} \rightarrow H^{+}{}_{aq} + A^{rd}{}_{aq}, \tag{4.4}$$

whose standard Gibbs energy change is

$$\Delta G^{o}{}_{cell} = \Delta G^{o}{}_{H^{+}aq} + \Delta G^{o}{}_{Ardaq} - \Delta G^{o}{}_{Aoxaq} - (1/2)\Delta G^{o}{}_{H_2gas}. \tag{4.5}$$

Here, $\Delta G^{\circ}{}_{i}$ represents the standard Gibbs energy of formation of the *i*-species. By convention, $\Delta G^{\circ}{}_{H+aq} = 0$ and $\Delta G^{\circ}{}_{H2gas} = 0$ so that, introducing the Nernst equation relating electrode potentials and Gibbs energy ($\Delta G^{\circ} = -nFE^{\circ}$), one can write [8],

$$E^{o}{}_{cell} = E^{\circ}\left(A^{ox} / A^{rd}\right) = (1/F)\left(\Delta G^{o}{}_{Ardaq} - \Delta G^{o}{}_{Aoxaq}\right). \tag{4.6}$$

Accordingly, the standard potential of the A^{ox}/A^{rd} couple in an aqueous solution relative to the SHE is directly related to the standard Gibbs energy of formation of the oxidized and reduced species.

In the case of nonaqueous solvents, the situation is complicated by the existence of a liquid junction potential whose value is in principle unknown. To correlate electrochemical measurements in different solvents, a solvent-independent scale of standard redox is required, but, conversely, the determination of junction potentials requires just the disposal of a solvent-independent redox potential scale. This problem of circularity requires the introduction of extra-thermodynamic assumptions to be faced [9].

One way to access this problem is to consider the standard Gibbs energy of electron transfer in the gas phase ($\Delta G^{\circ}{}_{g}$) and the energies of solvation of the oxidized and reduced forms, $\Delta G^{\circ}{}_{solv}(A^{ox})$, $\Delta G^{\circ}{}_{solv}(A^{rd})$. These quantities are summarized in the thermochemical cycle in Figure 4.1.

Then, the standard Gibbs energy of electron transfer in the solvent, ΔG°_{e}, will satisfy the relationship [10]

$$\Delta G^{o}{}_{e} = \Delta G^{o}{}_{g} + \Delta G^{o}{}_{solv}\left(A^{rd}\right) - \Delta G^{o}{}_{solv}\left(A^{ox}\right) - \Sigma_{e}, \tag{4.7}$$

where Σ_e represents the Gibbs energy of electron transfer from the vacuum to the metal constituting the electrode. The electron is at zero energy in the vacuum scale (assumed to be at rest at an infinite distance), and ΔG°_{g} corresponds to the difference between the Gibbs energies of the formation of A^{ox} and A^{rd} in the gas phase [8]:

$$\Delta G^{o}{}_{g} = \Delta G^{o}{}_{Ardg} - \Delta G^{o}{}_{Aoxg}. \tag{4.8}$$

Ultimately, this problem deals with the definition of solvent-independent electrode potentials [10] and the determination of individual solvation energies of ions [11]. As shown in Table 4.1 [12–15], these can show significant differences as is also reflected in the considerable differences in the autoprotolysis constants (K_{auto}) of amphiprotic solvents, those predominantly used in electrochemistry. A representative list is summarized in Table 4.2 [16].

For many practical purposes, the problem is focused on converting electrode potentials determined relative to an arbitrary reference electrode in a nonaqueous solvent into potential relative to the aqueous SHE. One possible approach is to refer the potentials to the previously mentioned vacuum scale where the most adopted value for the standard potential of the SHE is $E^{\circ}_{NHE} = -4.44$ eV. Then, the standard potential of the A^{ox}/A^{rd} one-electron couple in the nonaqueous solvent S relative to the aqueous SHE, $E^{\circ}(A^{ox}/A^{rd})_{NHE}$, can be calculated as [8]

$$E^{\circ}\left(A^{ox}/A^{rd}\right)_{NHE} = (1/F)\left[\Delta G^{o}{}_{solv}\left(A^{ox}\right) - \Delta G^{o}{}_{solv}\left(A^{rd}\right) - \Delta G^{o}{}_{g} - 4.44\right]. \tag{4.9}$$

4.3 MECHANISTIC CONSIDERATIONS

The description of the electrochemical response in general requires the consideration of not only thermochemical but also kinetic aspects. As described in Chapter 1, the kinetics of electron transfer determines the occurrence of electrochemical processes with different degrees of reversibility. A second crucial aspect is the possibility of multiple electron transfers and coupled chemical reactions. The electron transfer (E) can be preceded, accompanied, and/or followed by chemical reactions (C)

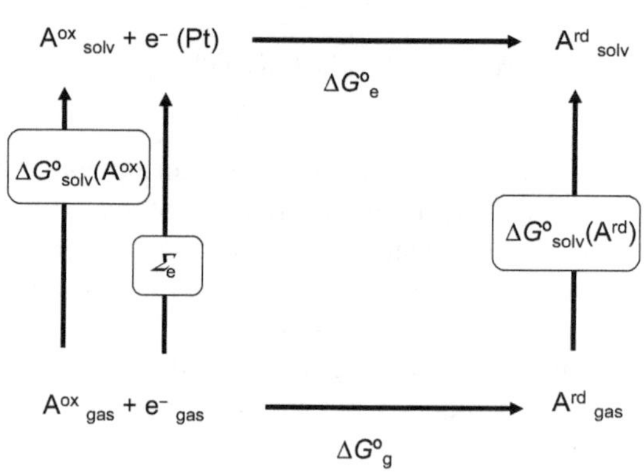

FIGURE 4.1 Thermochemical cycle for the reduction of the A^{ox}/A^{rd} couple in a given solvent using a Pt working electrode.

TABLE 4.1
Gibbs Free Energy of Solvation of Several Cations and Anions in Water, Methanol and Acetonitrile [12–15]

Ion	ΔG°_{aq} / kJ mol–1	ΔG°_{MeOH} / kJ mol–1	ΔG°_{MeCN} / kJ mol–1
Li^+	–529	–524	–520
Na^+	–424	–415	–416
K^+	–352	–340	–345
F^-	–465	–439	–400
Cl^-	–340	–327	–295
Br^-	–315	–303	–279

TABLE 4.2
Autoprotolysis Constants (K_{auto}) of Some Amphiprotic Solvents at 25°C [7, 16]

Solvent	Conjugate cation	Conjugate anion	log(K_{auto}/ mol² L–²)
Sulfuric acid	$H_3SO_4^+$	HSO_4^-	3.33
Water	H_3O^+	OH^-	14.00
Ethanol	$H_5C_2\text{-}OH_2^+$	$H_5C_2\text{-}O^-$	18.88
N,N-Dimethylacetamide	$(H_3C)COH\text{-}N(CH_3)_2^+$	$(H_2C)CON(CH_3)_2^-$	23.95
Acetone	$(H_3C)_2C{=}OH^+$	$(H_3C)(H_2C)C\text{-}O^-$	32.5
Dimethyl sulfoxide	$(H_3C)_2SOH^+$	$(H_3C)(H_2C)CO^-$	33.3
Acetonitrile	$(H_3C)CNH^+$	$(H_2C)CN^-$	≥ 33.3

between the parent electroactive species and/or its products of oxidation/reduction with other species existing in the solution or adhering to the electrode surface. In electrochemical terminology, EC, CE, ECE, ECEC, CECE, etc., processes may occur. This complex scenario is completed by the possibility of catalysis, adsorption/desorption, gas evolution, solid deposition/dissolution, electropolymerization, disproportionation/ comproportionation, etc., processes [3, 17].

In a variety of cases of interest in organic electrochemistry, protonation and deprotonation processes are coupled to the electron-transfer steps, and disproportionation/comproportionation processes may be involved [18]. A widely studied case is the reduction of *o*-and *p*-quinones (Q). In aprotic media, their reduction occurs in two successive one-electron steps at separate potentials (see Figure 4.2a).

These processes correspond to the successive formation of the semiquinone radical anion and the dianion, schematized in Figure 4.3.

Under conditions of electrochemical reversibility, the equilibrium constant for the disproportionation process, K_{dis},

$$2Q^{\cdot-} \leftrightarrows Q + Q^{2-} \tag{4.10}$$

can be calculated from the equilibrium potentials of the $Q/Q^{\cdot-}$ and $Q^{\cdot-}/Q^{2-}$ couples as

$$\ln K_{dis} = F\frac{E^{\circ}\left(Q^{\cdot-}/Q^{2-}\right) - E^{\circ}\left(Q/Q^{\cdot-}\right)}{RT}. \tag{4.11}$$

When $E^{\circ}(Q/Q^{\cdot-}) > E^{\circ}(Q^{\cdot-}/Q^{2-})$, the disproportionation constant is low. Under conditions of reversibility, the voltammogram will show two one-electron couples. When $E^{\circ}(Q^{\cdot-}/Q^{2-})$ approaches

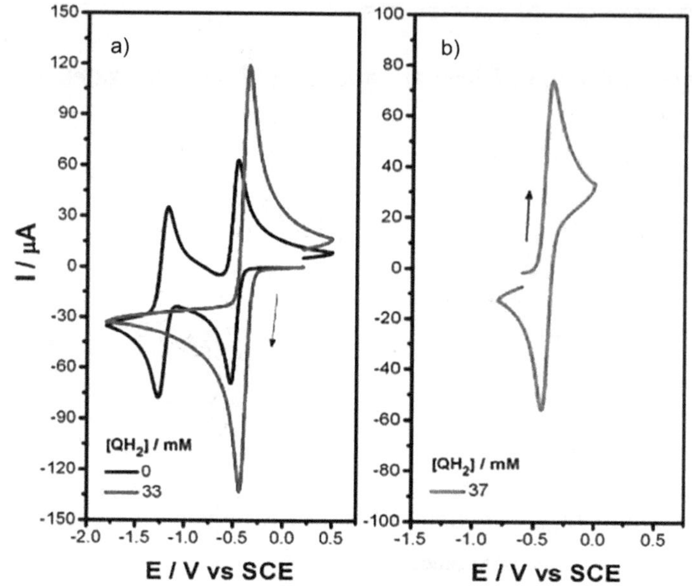

FIGURE 4.2 Cyclic voltammetry on glassy carbon electrode of 0.2 M ***n***-Bu_4NPF_6/MeCN of (a) 2 mM ***p***-benzoquinone in the presence of 0 and 33 mM of hydroquinone and (b) 37 mM of hydroquinone + 4 mM of ***n***-Bu_4NOH. Potential scan rate 100 mV s^{-1}. (Adapted from ref. [19], with permission. Copyright Elsevier 2016.).

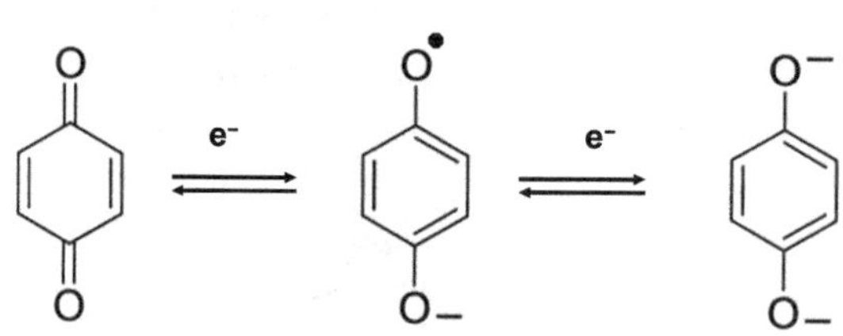

FIGURE 4.3 Scheme for the electrochemical reduction of p-benzoquinone in aprotic media.

$E°(Q/Q^{-})$ or even when $E°(Q^{-}/Q^{2-}) > E°(Q/Q^{-})$, the voltammetric signals merge to give a unique two-electron wave. In principle, the mechanism involves the sequential transfer of the electrons, but a concerted, two-electron transfer is conceived as possible [8].

In the presence of proton-donor species, as well as in protic solvents, the situation becomes more complex because different protonated species can coexist. Frequently, a unique voltammetric couple is recorded (see Figure 4.2a, b). This corresponds to the overall two-electron reduction:

$$Q + 2H^{+} + 2e^{-} \rightarrow H_2Q. \tag{4.12}$$

This voltammetric response has to be interpreted in terms of the successive transfer of two protons and two electrons in either ECEC or CECE sequences [20, 21]. Figure 4.4 shows the general scheme for the electrochemical reduction of *p*-benzoquinone.

Figure 4.5 superimposes the cyclic voltammogrmas at GCE of *p*-hydroquinone (H_2Q) in unbuffered water solution before and after the addition of 1 and 2 equivalents of OH^{-}. The initial voltammogram shows an oxidation peak at 0.5 V and a separated reduction peak at ca. 0.02 V. The addition of an equivalent of OH^{-}, formally leading to a solution of HQ^{-}, results in a significant modification of the current/potential profile, now consisting of the initial anodic peak and an apparently reversible

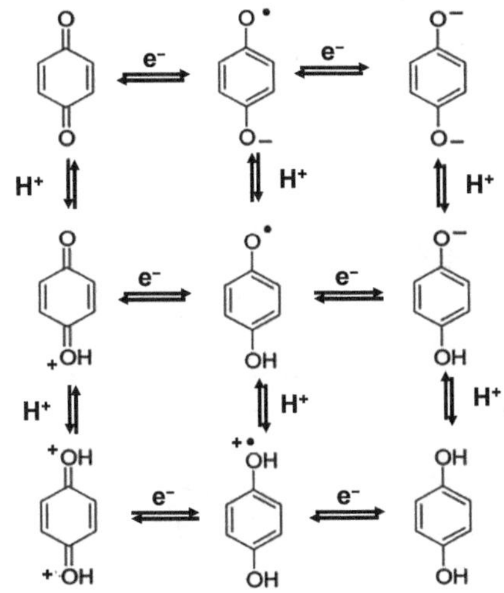

FIGURE 4.4 Scheme for the electrochemical reduction of *p*-benzoquinone in protic media.

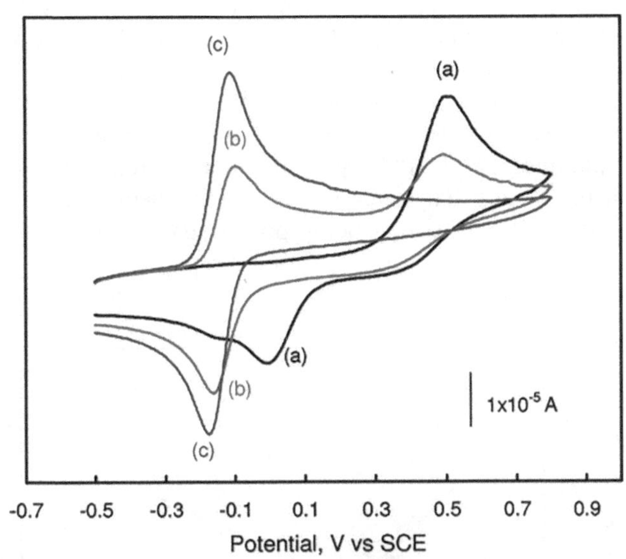

FIGURE 4.5 Cyclic voltammograms at GCE of ***p***-hydroquinone in unbuffered water solution before (a) and after the addition of 1 (b) and 2 (c) equivalents of OH^- from $Ca(OH)_2$. Potential scan rate 100 mV s^{-1}. (Adapted from ref. [21], with permission. Copyright the American Chemical Society 2007.).

couple at ca. 0.15 V. The addition of a second equivalent of OH^-, formally leading to a Q^{2-} solution, determines a new change in the voltammogram. Here, a unique, apparently reversible two-electron couple appears at ca. 0.15 V.

Attributing this couple to the Q to Q^{2-} oxidation, the equilibrium potential, in principle represented by the midpeak potential recorded in the voltammograms will be, under the usual assumptions (298 K, replacing activities by concentrations) [20]:

$$E_{eq} = E^{\circ}\left(Q/Q^{2-}\right) + \frac{0.059}{2}\log\frac{[Q]}{\left[Q^{2-}\right]}. \tag{4.13}$$

$$E_{eq} = E^{\circ}\left(Q/Q^{2-}\right) + \frac{0.059}{2}\log\left(1 + \frac{\left[H^{+}\right]}{K_{a2}} + \frac{\left[H^{+}\right]^{2}}{K_{a1}K_{a2}}\right). \tag{4.14}$$

Here, K_{a1} and K_{a2} are the first and second acidity constants of *p*-hydroquinone, corresponding to the equilibria $H_2Q \leftrightarrows HQ^- + H^+$ and $HQ^- \leftrightarrows Q^{2-} + H^+$, respectively. In buffered solutions, the voltammetric response can be fitted to Eq. (4.14), as shown in Figure 4.6 for solutions of the sodium salt of 2-anthraquinonesulfonic acid (AQS).

In this figure, the pH variation of midpeak potentials measured in conditions such as in Figure 4.5 in AQS solutions in buffered and unbuffered media at fixed ionic strength is depicted. Data in buffered solutions vs. saturated calomel electrode (SCE) can be fitted reasonably to Eq. (4.14) taking $E°(AQS/AQS^{2-})$ −0.600 V vs. SCE, $pK_{a1} = 7.68$ and $pK_{a2} = 10.92$, but data in unbuffered solutions diverge from this equation. This feature can be attributed, generically, to the variation in the effective concentration of the species in the vicinity of the electrode surface.

Following the square scheme in Figure 4.4, assuming that all protonation equilibria are fast, each overall electron transfer can be treated as a simple electron transfer. The apparent formal potential depends on the corresponding quantity of each electron-transfer step, the pKa's, the pH, and an apparent standard heterogeneous electron-transfer rate constant, $k°$. This last quantity in turn depends on the pKa's and the pH [20]. This means that the kinetics of the electron-transfer process influence the voltammetric response. In addition, this response can also be influenced by other kinetic factors (for instance, disproportionation/comproportionation reactions). The sum of these factors determines that the voltammetric record is dependent on the concentration of the quinone, the pH, the type and concentration of the supporting electrolyte and its buffering capacity. Notice that the presence of other species in the solution can promote parasitic reactions, with all the processes being potentially conditioned by the adsorption of species onto the electrode surface [19].

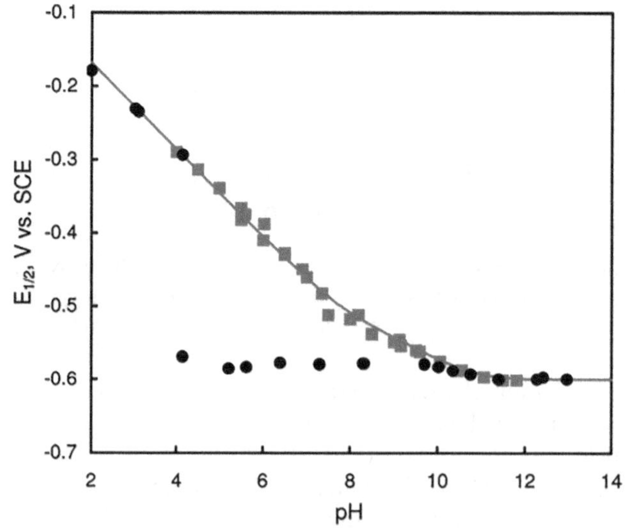

FIGURE 4.6 Variation of cyclic voltammetric midpeak potentials on the pH for 0.5 mM AQS solutions in buffered (squares) and unbuffered (circles) aqueous media of fixed ionic strength (0.2 M KCl + NaOH). The continuous line corresponds to the fit of experimental data to Eq. (4.14) taking $E°(AQS/AQS^{2-})$ −0.600 V vs. SCE, $pK_{a1} = 7.68$, and $pK_{a2} = 10.92$. (Adapted from ref. [21], with permission. Copyright the American Chemical Society 2007.).

Under favorable conditions, voltammetric midpeak potentials possess thermochemical significance and can be correlated with electron and proton affinities (see Figure 4.1) estimated from quantum mechanical calculations (Figure 4.7). Here, gas phase electron affinities from quantum mechanics calculations are correlated to midpeak potentials for the $Q/Q^{\cdot -}$ couple for a series of quinones [22].

Another frequent type of electrochemical reaction involves the participation of water as a reagent. A well-known case is the oxidation of phenols to *o*- and *p*-quinones, illustrated in Figure 4.8 [2, 23]. Here, the cyclic voltammogram (first and second potential scans are superimposed) at GCE of a phenol solution at pH 7.0 is accompanied by the corresponding scheme. The main oxidation wave at ca. 0.7 V vs. SCE corresponds to the one-electron, one-proton oxidation of phenol to phenoxide radical. This process is followed by the reaction with water giving rise to a mixture of *o*- and *p*-benzoquinone. The concentration of such species increases in successive potential scans as evidenced by the appearance of two couples at equilibrium potentials of ca. 0.32 and 0.42 V corresponding to the quinone to hydroquinone transformations previously described in this section.

4.4 SOLID-STATE ELECTROCHEMISTRY

Until now, the described electrochemistry involved species in the solution phase in both their oxidized and reduced forms. A second type of electrochemistry can be applied to the vegetal world: the solid-state electrochemistry. This is based on the record of the electrochemical response of nonconducting solids in contact with electrolyte solutions where they are sparingly soluble. The most direct antecedent was the voltammetric study of organic compounds embedded into carbon paste electrodes [26]. The carbon paste was formed by binding graphite powder with a nonconducting (nujol oil, paraffin oil) or conducting (aqueous solutions, ionic liquids) liquid. This analytical approach was significantly expanded from the development of the voltammetry of immobilized particles (VIMP) [27]. Here, an aliquot, if necessary, at the nanogram level, of the solid material is abrasively transferred to the surface of an inert electrode, typically paraffin-impregnate graphite.

The VIMP methodology has been applied to a variety of materials [28] and can be seen as a technique for solid-state analysis at the sub-micro and nano levels [29]. This technique is not limited to

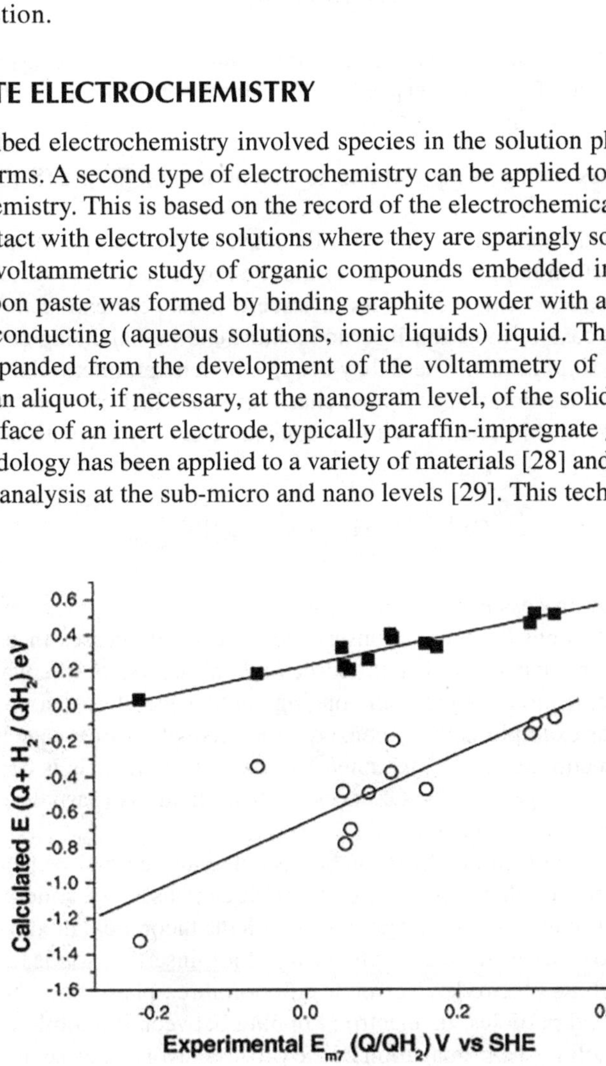

FIGURE 4.7 Correlation between gas phase electron affinities from quantum mechanics calculations and midpeak potentials for the $Q/Q^{\bullet -}$ couple determined for a series of quinones. Calculations were performed at the B3LYP/6-31G(d,p) (squares) and HF/6-31G(d,p) (circles) levels of theory. (Reproduced from ref. [22], with permission. Copyright Elsevier 2006.).

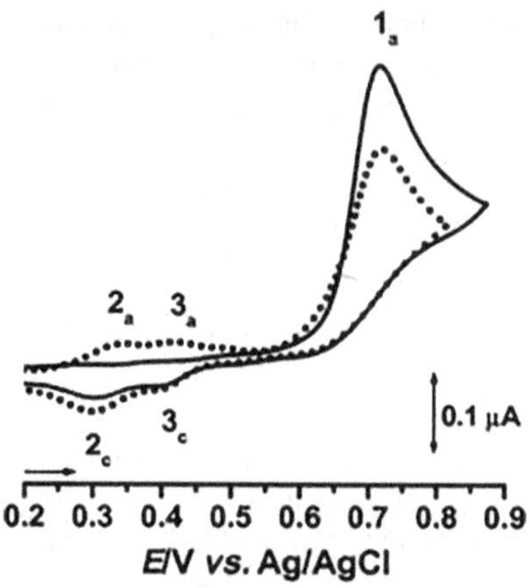

FIGURE 4.8 Cyclic voltammograms at a GCE, in 30 μM phenol in pH 7.0 showing the first (continuous line) and second (points line) scans. Potential scan rate 50 mV s^{-1}. Inset: phenol oxidation processes. (Adapted from ref. [23], with permission. Copyright Elsevier 2011.).

voltammetric techniques, being of application to potentiometry [30], EIS [31], and SECM [32], among other techniques.

The simplest case is the topotactic solid-to-solid conversion of ion-permeable solids. In this type of process, there is electron hopping between immobile redox centers in the solid coupled, by reasons of charge conservation, with the ingress or issue of ions permeating the solid from or to the electrolyte [27]. In the case of most ion-insertion solids, charge-balancing cations/anions enter (or egress) into (from) the solid lattice, where they occupy, in general, interstitial positions.

In the organic chemistry domain, the most characteristic VIMP pattern consists of the proton-assisted reduction of a solid in contact with an aqueous electrolyte solution. This process can be represented as

$$\{\text{Ox}\}_{\text{solid}} + n\text{H}^{+}{}_{\text{aq}} + ne^{-} \rightarrow \{\text{Rd}\}_{\text{solid}}. \tag{4.15}$$

The electron-transfer process is necessarily coupled to the ingress (or issue) of hydrogen ions in (from) the parent solid to ensure charge conservation. Redox processes in organic solids such as described in Eq. (4.15) involve a significant modification of the system of covalent bonds. This may involve a drastic reorganization of hydrogen bonding, and, hence, they can imply serious structural reorganizations. A clear example is the *p*-benzoquinone reduction to *p*-catechol. If occurring as a solid-to-solid transformation, a solid constituted by nonpolar molecules is converted into one containing intermolecular hydrogen bonds. Of course, other situations (partially soluble solids, redox processes with phase changes) can occur [27].

The electrochemistry involved in VIMP can be treated as a three-phase electrochemistry because it can be modeled in terms of charge transfer between three phases, the solid particle, the base inert electrode, and the electrolyte solution. In agreement with the theoretical model developed by Lovrić, Scholz, Oldham et al. to describe the electrochemistry of ion-insertion solids [27], the redox reaction initiates at the particle/base electrode/electrolyte solution three-phase boundary. Then, the process expands through the solid particles via electron hopping between immobile redox centers coupled to ion diffusion in the solid. The propagation of the process involves, as schematized in Figure 4.9, the transfer of electrons through the particle/base electrode interface and the transfer of protons (ions in general) through the particle/electrolyte solution interface.

When a topotactic proton-assisted, solid-to-solid reduction occurs (Figure 4.9a), the reaction front advances through the solid particle progressively converting the oxidized solid into its

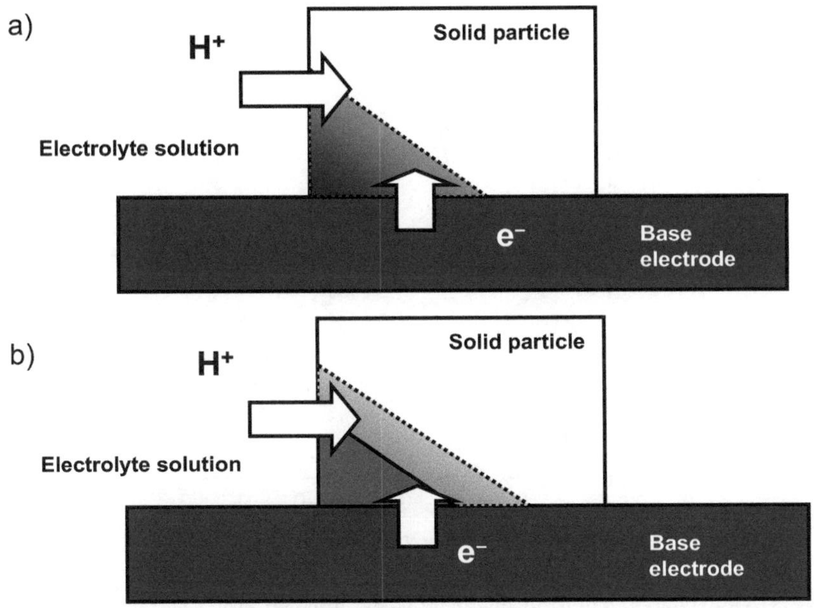

FIGURE 4.9 Scheme for solid-state electrochemistry processes involving the proton-assisted reduction of a cuboid solid particle. (a) Topotactic solid-to-solid transformation, (b) solid-to-solid transformation with phase segregation assuming that an intermediate layer (see text) is formed. The reaction fronts are denoted by dotted lines.

(protonated) reduced form. When the process involves phase segregation (Figure 4.9b), it can be assumed the reaction front advances through the solid forming an intermediate layer between the original, oxidized phase and the new reduced phase.

The VIMP technique can be directly applied to study plant composition from extracts of leaves, roots, etc., in organic solvents. After evaporation of the solvent in air, microparticulate deposits of organic compounds are formed on solid electrodes (typically, glassy carbon). These are put into contact with aqueous electrolyte solutions, and the voltammetric response is recorded. Figure 4.10 shows an example of this procedure applied to the deposits from ethanolic extracts of "Pholia magra" (*Cordia ecalyculata* Vell., order Labiales, fam. Boraginaceae) and *Centella asiatica* L. Urb. (order Apiales, fam. Apiaceae) immersed into air-saturated potassium phosphate buffer at neutral pH [32]. When the potential was scanned from 0.0 V vs. Ag/AgCl in the positive direction, a series of apparently irreversible anodic peaks (A_{pph}) was recorded in the potential range 0.2–1.2 V. Since the electrolyte solution was not deaerated, the cathodic signals are dominated by a gross wave at ca. −0.40 V (C_{ox}), which corresponds to the reduction of dissolved oxygen. The anodic signals A_{pph} can be attributed to the oxidation of polyphenolic plant components (lignins, flavonoids, etc.) in the solid state.

The solid-state nature of these processes was assessed by SECM using a $K_4Fe(CN)_6$ solution in aqueous acetate buffer as a redox probe [32, 33]. Figure 4.11 shows the color maps recorded when the potential applied to the tip (E_T) produces the oxidation of $Fe(CN)_6^{4-}$ ions under diffusion-controlled conditions. If no potential input was applied to the microparticulate substrate (E_S = 0.00 V) the insulating particles of polyphenolic compounds appear as (dark) irregularly rounded negative feedback features (Figure 4.11a). When a potential of 0.80 V was applied to the substrate, the oxidative processes took place starting from the particle/base electrode/electrolyte solution three-phase junction. The redox process results in the alteration of the profile of the negative feedback features that can be observed in Figure 4.11b.

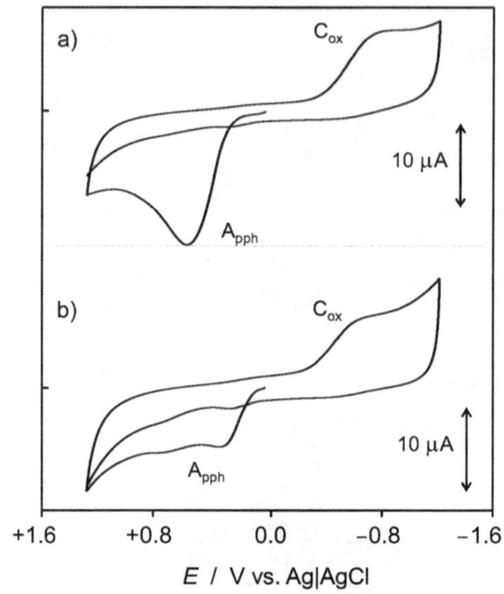

FIGURE 4.10 Cyclic voltammograms of microparticulate films on glassy carbon electrode prepared from ethanolic extracts of (a) *Cordia ecalyculata Vell.* and (b) *Centella asiatica* L. Urb. immersed into air-saturated 0.10 M aqueous potassium phosphate buffer at pH 7.00. Potential scan rate 50 mV s^{-1}. Semi-derivative data convolution was applied to increase peak resolution. (Adapted from ref. [32], with permission. Note that the axes are opposite to the International Union of Pure and Applied Chemistry (IUPAC) convention. Copyright the Royal Society of Chemistry 2015.).

Solid-sate electrochemical processes can proceed also as reductive or oxidative dissolutions and eventually precipitations when the reduced or oxidized product is sparingly soluble. Figure 4.12 shows the atomic force microscopy (AFM) imaging of an aggregate of crystals of morin, a flavonoid, attached to a graphite plate in contact with acetic/acetate buffer at pH 4.75. Under these conditions, morin is oxidized at potentials ca. 0.4 V vs. Ag/AgCl (see Chapter 5). The initial image shows a series of fibrous, elongated crystals. After the application of an oxidative potential input of 0.45 V for 5 min, the AFM images reveal the progressive corrugation of the crystals accompanied by their apparent dissolution and the appearance of minor irregular crystals. These features can be interpreted in terms of the oxidative dissolution of morin yielding a soluble intermediate subsequently deposited as a crystalline compound.

The voltammetric analysis of microparticulate deposits on graphite electrodes obtained upon evaporation of plant extracts (from leaves, stems, roots, seeds, etc.) permits the record of well-defined electrochemical responses [33–35]. Applications of this methodology for analytical, chemo-ecological and phylogenetic studies will be treated in subsequent chapters.

4.5 IN SITU ELECTROCHEMISTRY

Another extended approach to the study of plants is the performance of analysis in situ inserting the electrodes in leaves, stems, fruits, etc. Two-electrode and three-electrode arrangements can be used for electrochemical measurements exploiting microelectrode fabrication techniques. These have dimensions of tens of micrometers or less, down to submicrometer range, and a variety of geometries, including microrings, microdisks, microbands, microspheres, microhemispheres, and different electrode arrays [36, 37].

The first aspect to consider is the mass transport in solution near the electrode surface. In an electrochemical process uncomplicated by parallel reactions, adsorption effects, etc., the

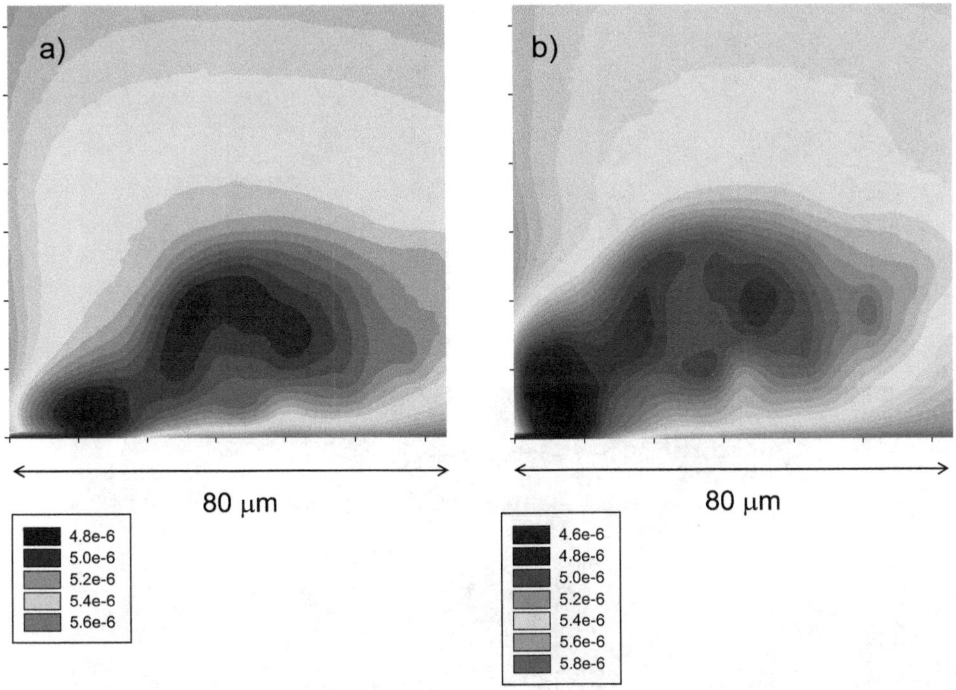

FIGURE 4.11 SECM map colors of a microparticulate deposit on Pt substrate of an ethanolic extract of an herbal sample. Electrolyte: 2.0 mM $K_4Fe(CN)_6$ plus 0.25 M HAc/NaAc aqueous buffer at pH 4.75. (a) E_T = 0.30 V; E_S = 0.00 V; (b) E_T = 0.30 V; E_S = 0.80 V. (Reproduced from ref. [32], with permission. Copyright the Royal Society of Chemistry 2015.).

voltammetric current signal is proportional to the flux of the electroactive species toward the solution-electrode interface. In an ordinary experiment, the potential scan is initiated at a value at which no electrode reaction occurs. When sufficient cathodic or anodic potential is reached, the concentration of the electroactive species diminishes in the vicinity of the electrode and a concentration gradient develops over a certain distance from the electrode surface into the bulk of the solution. As previously noted (see Chapter 1), the current is proportional to the concentration gradient at the electrode surface.

The solution volume within which the diffusional flux of the electroactive species takes place is termed the "diffusion layer". Then, the flux Φ of the electroactive species toward the electrode is described by the product

$$\Phi = \frac{Dc}{\delta}, \tag{4.16}$$

where D is the diffusion coefficient of the substance, c bulk concentration of the same, and δ the diffusion layer thickness. For planar semi-infinite diffusion, the diffusion layer thickness is a time-dependent quantity given by the relationship

$$\delta = \sqrt{\pi D t}. \tag{4.17}$$

For nonplanar electrodes, the diffusion layer thickness depends on the shape and size of the electrode. For electrodes with radial symmetry, characterized by a radius r, the diffusive behavior (hemispherical, cylindrical, etc.) will depend on the δ/r ratio. Interestingly, at a sufficiently short time, the electrode behaves as an infinitely large planar electrode.

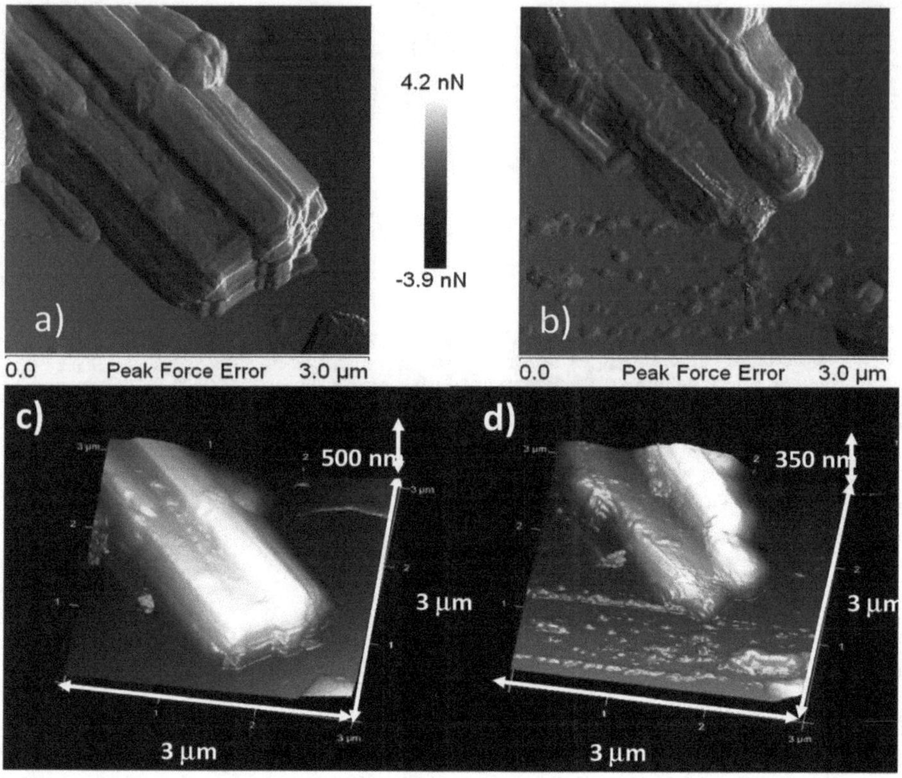

FIGURE 4.12 AFM amplitude error channel graphs (a, b) and topographic images (c, d) of a crystal of morin attached to a graphite plate in contact with 0.25 M sodium acetate buffer, pH 4.75 (a, c) before, and (b, d) after application of an oxidative potential step at 0.45 V during 5 min.

Apart from their inherent capability for measuring small solution volumes, microelectrodes offer several interesting characteristics [37, 38]:

- A steady state for a faradaic process is attained rapidly.
- The faradaic-to-charging current ratio increases by decreasing the electrode area. This facilitates the use of fast potential scanning.
- The ohmic drops are low as the measured currents are low.
- When microelectrode arrays are used, the gap between the microelectrodes has to be considered. Here, the signal-to-noise ratio is enhanced when the diffusion layers overlap totally. The interactions between the individual electrodes permit the regeneration of the electroactive species through redox cycling.

Of course, the working electrodes can be modified to acquire specific electroanalytical capabilities. Figure 4.13 shows a simple scheme for in situ measurement in plants [39]. Here, the electrode arrangement includes a ZnO nanodisk microelectrode electrodeposited on a spiny-shaped indium-doped tin oxide (ITO) glass plate modified with Cu, Zn-SOD for the detection of superoxide anion radical [39].

Figure 4.14 depicts a schematic representation of a three-electrode arrangement applied to fruits [39]. This type of voltammetric analysis can be applied to detect specific plant components [40] and pesticides [41], as well as to humidity, nutrients, and volatile organic components, among others

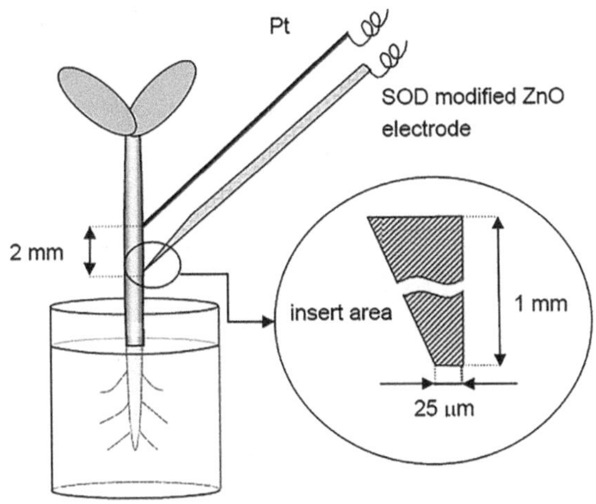

FIGURE 4.13 Scheme for in situ voltammetric measurements in plants. (Adapted from ref. [38], with permission. Copyright Elsevier 2015.).

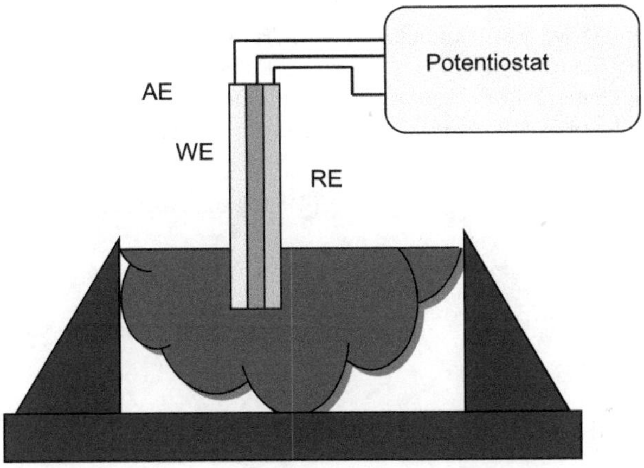

FIGURE 4.14 Schematic representation of the three-electrode arrangement used to measure the voltammetric response in tomato fruits. (Adapted from ref. [39], with permission. Copyright Elsevier 2015.).

[42]. Electrochemical measurements can also be used to monitor physiological processes in plants and plant organelles, including response to oxidative burst [43], signaling transduction [44] to gene expression [45].

Figure 4.15 exemplifies the application of in situ electrochemistry for the discrimination of tomato varieties [39]. Here, the voltammograms reflect the oxidation of ascorbic acid and polyphenolic components of the fruit producing variety-sensitive current/potential profiles.

Figure 4.16 shows another type of in situ voltammetric sensing applied to different parts of a germinated pea seedling [47]. The sampling process was made with a punch extracting several milligrams of vegetal matter from the root tip and the embryo. The sample was transferred onto the surface of a multiwalled carbon nanotubes–modified carbon tape deposited onto the ITO electrode. The device incorporates auxiliary and reference electrodes and was covered by a disk of filter paper. The seed samples were adhered here, and the voltammetric response was recorded as shown in Figure 4.16. This response was dominated by the anodic peaks assigned to the oxidation of indole-3-acetic acid and salicylic acid [46].

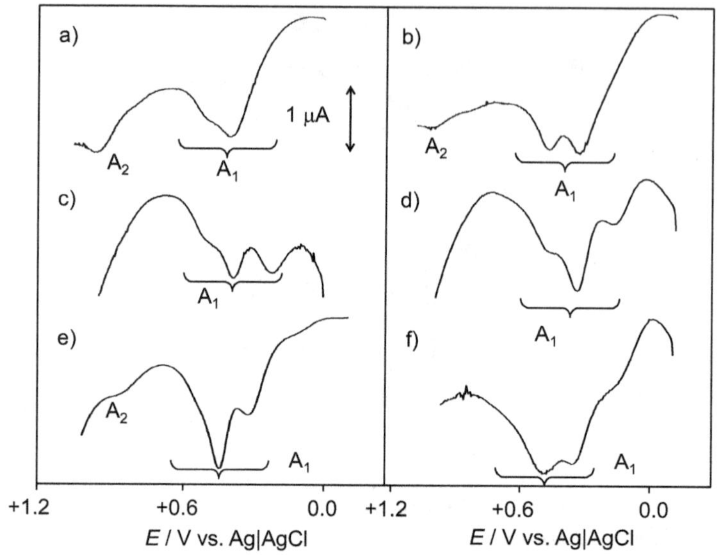

FIGURE 4.15 In situ square wave voltammograms at GCE of tomato specimens of the varieties (a) Pitenza, (b) Ventero, (c) Vernal, (d) Tigre, (e) Delizia, (f) Shirem. Potential scan initiated at –0.05 V in the positive direction; potential step increment 4 mV; square wave amplitude 25 mV; frequency 5 Hz. (Reproduced from ref. [39], with permission. Note that the axes are opposite to the IUPAC convention. Copyright Elsevier 2015.).

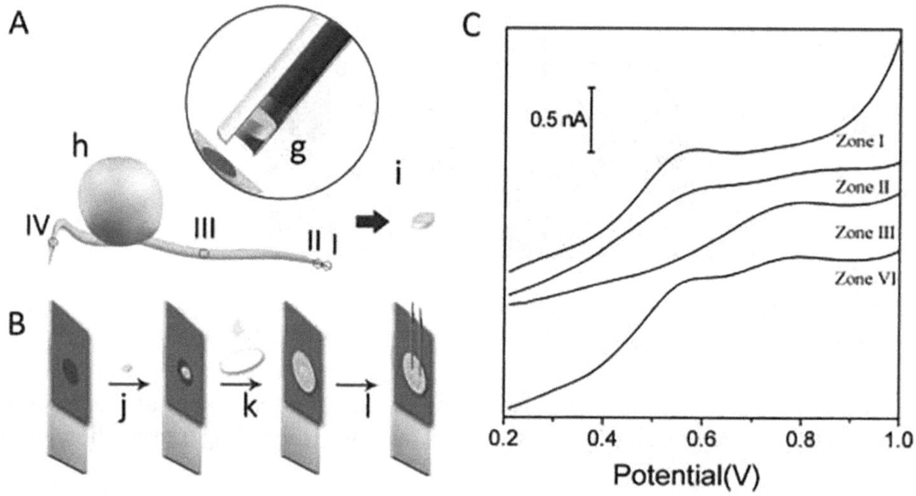

FIGURE 4.16 Scheme for the process of sampling with a paper-based electrochemical sensor (a) and its application on the surface of working electrodes (b). Differential pulse voltammetric curves (c) of zones I, II, III, IV of the pea seedling (h). The tiny parts were sampled with a sample punch (g). The obtained part was then put on the surface of a modified electrode (j). The buffer solution with a volume of 10 μL was then dropped and covered by a piece of filter paper (k). A clasp with a platinum wire and an Ag/AgCl wire was used to provide the counter electrode and the reference electrode (l). (Adapted from ref. [46], with permission. Copyright Elsevier 2017.).

During the last decade, a variety of sensing devices have been developed for in situ analysis based on potentiometric, voltammetric, amperometric, and impedimetric techniques. In particular, wearable sensors capable of being adhered to the surface of plant organs are under intensive research [47]. These approaches will be treated in the chapter specifically devoted to analytical issues of plant electrochemistry.

4.6 BIOFILMS AND ORGANELLES

The philosophy of in situ electrochemistry can be extended to measurements at biofilms and cellular suspensions. Figure 4.17 depicts the differential pulse voltammetry curves recorded at a gold-based microsensor in biofilms of *Pseudomonas aeruginosa* Migula aimed to detect pyocyanin [48]. *P. aeruginosa* is an opportunistic bacterial pathogen, responsible for various acute and chronic infections, and pyocyanin is a molecular marker of specific virulence. In this study [48], the effect of curcumin, an antibacterial agent, on the formation of *P. aeruginosa* biofilms was monitored electrochemically by introducing biofilm suspensions into the microchannels of the gold-based chip. The voltammograms show an oxidation signal at −0.35 V vs. Ag/AgCl, whose intensity is representative of the pyocyanin concentration. Data in Figure 4.17 superimpose the voltammograms obtained in the absence and in the presence of curcumin at different times and the change of pyocyanin concentration and biofilms biomass over time in both cases with and without the presence of curcumin.

This philosophy can be extended to organelles separated from cells. As an example, protoplasts isolated from oilseed rape leaf [45]. The protoplasts were immobilized with the aid of a gelatin-based adherent on the stage of an inverted fluorescent microscope placed in a Faraday cage. Under the microscope, the microelectrode tip, consisting of a carbon fiber microdisk electrode modified with Pt nanoparticles, was positioned close to the isolated protoplast. This was submitted to oxidative burst with the aid of a second micromanipulator, and the amperometric

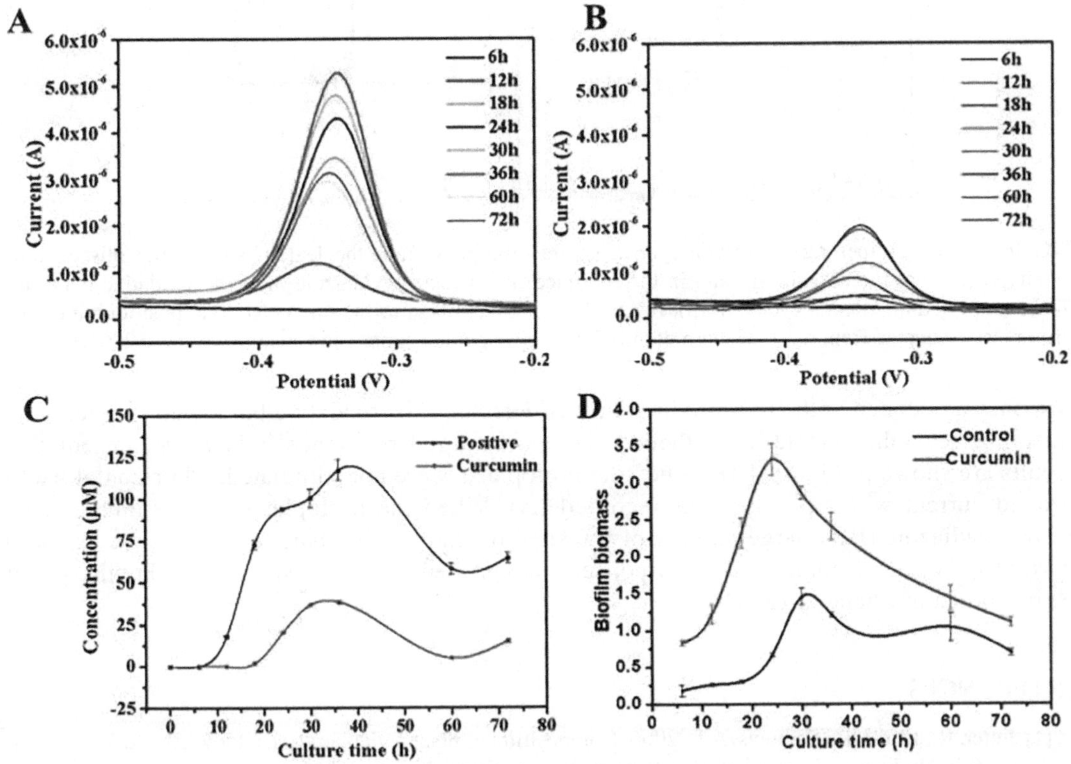

FIGURE 4.17 Differential pulse voltammetry curves recorded by in situ electrochemical monitoring of pyocyanin during the formation process of *Pseudomonas aeruginosa* biofilms. Voltammograms were obtained (A) without curcumin at different times and (B) in the presence of curcumin at different times. Curves (C) correspond to the change of pyocyanin concentration over time with and without the presence of curcumin. Curves (D) represent the change of biofilm biomass over time with and without the presence of curcumin. (Adapted from ref. [48], with permission. Copyright Elsevier 2021.).

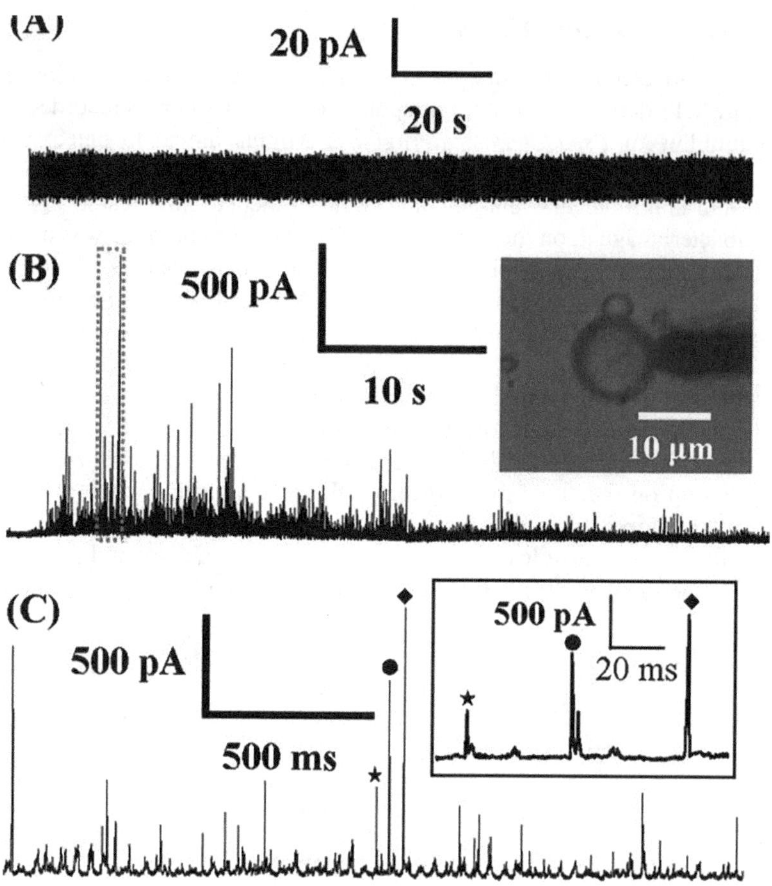

FIGURE 4.18 Amperometric response on a single protoplast from the leaf of susceptible oilseed rape genotype (215) in the absence (a) and in the presence (b) of benzothiadiazole inducing oxidative burst. (c) Three amplified microburst events in (b). Inset: photographic image of the microelectrode positioning on the protoplast. (Adapted from ref. [43], with permission. Copyright the American Chemical Society 2009.).

response associated with H_2O_2 production was detected. The oxidative burst was simulated by injecting benzothiadiazole, a synthetic systemic acquired resistance inducer. Representative results are shown in Figure 4.18. When the protoplasts were not stimulated, a horizontal background current without spikes was recorded (A). When the protoplasts were stimulated by benzothiadiazole (B), a large number of transient oxidative microburst events appear as sharp current spikes in the amperometric response. An expanded view of in situ electrochemistry will be presented in Chapters 12 and 13.

REFERENCES

[1] Jetter, R.; Kunst, L.; Samuels, A.L. 2006. Composition of plant cuticular waxes. In: Riederer, M.; Müller, C., Eds. *Biology of the plant cuticle*. Blackwell Publ, Oxford, pp 182–215.

[2] Chiorcea-Paquim, A.-M.; Enache, T.A.; De Souza Gil, E.; Oliveira-Brett, A.M. 2020. Natural phenolic antioxidants electrochemistry: Towards a new food science methodology. Comprehensive *Reviews in Food Science and Food Safety*. 19: 1680–1726.

[3] Bard, A.J.; Faulkner, L.R. 2001. *Electrochemical Methods Fundametals and Applications*, 2nd ed. Wiley, New York.

[4] Scholz, F. 2015. Voltammetric techniques of analysis: the essentials. *ChemTexts*. 1: artic. 17.
[5] Inzelt, G. 2014. Crossing the bridge between thermodynamics and electrochemistry. From the potential of the cell reaction to the electrode potential. *Chem Texts*. 1: artic.11
[6] Inzelt, G. 2015. Links between electrochemical thermodynamics and kinetics. *Chem Texts*. 1: artic.18.
[7] Bard, A.J.; Inzelt, G.; Scholz, F. Eds. 2008. *Electrochemical Dictionary*. Springer, Berlin-Heidelberg.
[8] Evans, D.H. 2008. One-Electron and Two-Electron Transfers in Electrochemistry and Homogeneous Solution Reactions. *Chemical Reviews*. 108: 2113–2144.
[9] Trasatti, S. 1986. The absolute electrode potential: An explanatory note. *Pure and Applied Chemistry*. 58: 966–966.
[10] Doménech-Carbó, A. 2012. Solvent-Independent Electrode Potentials of Solids Undergoing Insertion Electrochemical Reactions: Part I. Theory. *Journal of Physical Chemistry C*. 116: 25977–25983.
[11] Scholz, F.; Doménech-Carbó, A. 2019. The Thermodynamics of Insertion Electrochemical Electrodes—A Team Play of Electrons and Ions across Two Separate Interfaces. *Angewandte Chemie Internacional Edition*. 58: 3279–3284.
[12] Tissandier, M.D.; Cowen, K.A.; Feng, W.Y.; Gundlach, E.; Cohen, M.H.; Earhart, A.D.; Coe, J.V. 1998. *Journal of Physical Chemistry A*. 102: 7787.
[13] Moyer, S. A.; Bonnesen, P. V. 1997, in *Supramolecular Chemistry of Anions*; Bianchi, A., García-España, E., Bowman-James, K., Eds. Wiley, New York; pp 1–44.
[14] Doménech-Carbó, A.; Koshevoy, I.O.; Montoya, N.; Karttunen, A.J.; Pakkanen, T.A. 2011. Determination of individual Gibbs energies of anion transfer and excess Gibbs energies using an electrochemical method based on insertion electrochemistry of solid compounds. *Journal of Chemical & Engineering Data*. 56: 4577–4586.
[15] Doménech-Carbó, A.; Montoya, N.; Scholz, F. 2011. Estimation of individual Gibbs energies of cation transfer employing the insertion electrochemistry of solid Prussian blue. *Journal of Electroanalytical Chemistry*. 657, 117–122.
[16] Reinhart, C. 2004. *Solvents and solvent effects in organic chemistry*. Wiley-VCH, Weinheim.
[17] Savéant, J.-M. 2006. *Elements of Molecular and Biomolecular Electrochemistry*. Wiley-Interscience, New York.
[18] Costentin, C. 2008. Electrochemical Approach to the Mechanistic Study of Proton-Coupled Electron Transfer. *Chemical Reviews*. 108: 2145–2179.
[19] Gamboa-Valero, N.; Astudillo, P.D.; González-Fuentes, M.A.; Leyva, M.A.; Rosales-Hoz, M.D.J.; González, F.J. 2016. Hydrogen bonding complexes in the quinone-hydroquinone system and the transition to a reversible two-electron transfer mechanism. *Electrochimica Acta*. 188: 602–610.
[20] Laviron, E. (1984) Electrochemical reactions with protonations at equilibrium: Part X. The kinetics of the *p*-benzoquinone/hydroquinone couple on a platinum electrode. *Journal of Electroanalytical Chemistry*. 164: 213–227.
[21] Quan, M.; Sanchez, D.; Wasylkiw, M.F.; Smith, D.K. 2007. Voltammetry of Quinones in Unbuffered Aqueous Solution: Reassessing the Roles of Proton Transfer and Hydrogen Bonding in the Aqueous Electrochemistry of Quinones. *Journal of the American Chemical Society*. 129: 12847–12856.
[22] Cape, J.L.; Bowman, M.K.; Kramer, D.M. 2006. Computation of the redox and protonation properties of quinones: Towards the prediction of redox cycling natural products. *Phytochemistry*. 67: 1781–1788.
[23] Enache, T.A.; Oliveira-Brett, A.M.. 2011. Phenol and parasubstituted phenols electrochemical oxidation pathways. *Journal of Electroanalytical Chemistry*. 655: 9–16.
[24] Jetter, R.; Kunst, L.; Samuels, A.L. 2006. Composition of plant cuticular waxes. In: Riederer, M.; Müller, C., Eds. *Biology of the plant cuticle*. Blackwell Publ, Oxford, pp 182–215.
[25] Kuwana, T.; French, W.G. 1964. Electrooxidation or Reduction of Organic Compounds into Aqueous Solutions Using Carbon Paste Electrode. *Analytical Chemistry*. 36: 241–242.
[26] Scholz, F.; Meyer, B. 1998. Voltammetry of solid microparticles immobilized on electrode surfaces, in *Electroanalytical Chemistry, A Series of Advances*. Bard, A.J.; Rubinstein, I., Eds, Marcel Dekker, New York, vol. 20, pp 1–86.
[27] Scholz, F.; Schröder, U.; Gulabowski, R.; Doménech-Carbó, A. 2014. *Electrochemistry of Immobilized Particles and Droplets*, 2nd ed. Springer, Berlin-Heidelberg.
[28] Doménech-Carbó, A.; Labuda, J.; Scholz, F. 2013. Electroanalytical chemistry for the analysis of solids: characterization and classification (IUPAC Technical Report). *Pure and Applied Chemistry*. 85: 609–631.

[29] Doménech-Carbó, A.; Navarro, P.; Arán, V.J.; Muro, B.; Montoya, N.; García-España, E. 2010. Selective electrochemical discrimination between dopamine and phenethylamine-derived psychotropic drugs using electrodes modified with an acyclic receptor containing two terminal 3-alkoxy-5-nitroindazole rings. *Analyst*. 135: 1449–1455.
[30] Redondo-Marugán, J.; Piquero-Cilla, J.; Doménech-Carbó, M.T.; Ramírez-Barat, B.; Al Sekhaneh, W.; Capelo, S.; Doménech-Carbó, A. 2017. Characterizing archaeological bronze corrosion products intersecting electrochemical impedance measurements with voltammetry of immobilized particles. *Electrochimica Acta*. 246: 269–279.
[31] Doménech-Carbó, A.; Doménech-Carbó, M.T.; Silva, M.; Valle-Algarra, F.M.; Gimeno-Adelantado, J.V.; Bosch-Reig, F.; Mateo-Castro, R. 2015. Screening and mapping pigments in paintings using scanning electrochemical microscopy (SECM). *Analyst*. 140: 1065–1075.
[32] Martini, N.; de Carvalho, L.M.; Blasco-Blasco, A.; Doménech-Carbó, A. 2015. Screening and authentication of herbal formulations based on microextraction-assisted voltammetry of microparticles. *Analytical Methods*. 7: 5740–5747.
[33] Doménech-Carbó, A.; Ibars, A.M.; Prieto-Mossi, J.; Estrelles, E.; Scholz, F.; Cebrián-Torrejón, G.; Martini, M. 2015. Electrochemistry-based chemotaxonomy in plants using the voltammetry of microparticles methodology. *New Journal of Chemistry*. 39: 7421–7428.
[34] Doménech-Carbó A.; Ibars, A.M.; Prieto-Mossi, J.; Estrelles, E.; Doménech-Carbó, M.T.; Ortiz-Miranda, A.S.; Lee, Y. 2017. Access to phylogeny from voltammetric fingerprints of seeds: the *Asparagus* case. *Electroanalysis*. 29: 643–650.
[35] Domínguez, I.; Doménech-Carbó, A. 2015. Screening and authentication of tea varieties based on microextraction-assisted voltammetry of microparticles. *Sensors and Actuators B*. 210: 491–499.
[36] Štulik, K.; Amatore, C.; Holub, K.; Mareček, V.; Kutner, W. 2000. Microelectrodes, definitions, characterization, and applications (IUPAC technical Report). *Pure and Applied Chemistry*. 72: 1483–1492.
[37] Wightman, R.M.; Wipf, D.O. 1989. Voltammetry at Ultramicroelectrodes, in *Electroanalytical Chemistry*, Bard, A.J., Ed., Vol. 15, Marcel Dekker, New York and Basel, pp. 267–353.
[38] Deng, Z.; Rui, Q.; Yin, X.; Liu, H.; Tian, Y. 2008. In Vivo Detection of Superoxide Anion in Bean Sprout Based on ZnO Nanodisks with Facilitated Activity for Direct Electron Transfer of Superoxide Dismutase. *Analytical Chemistry*. 80: 5839–5846.
[39] Doménech-Carbó, A.; Domínguez, I.; Hernández, P.; Gavara, R. 2015. Electrochemical tomato (*Solanum lycopersicum* L.) characterization using contact probe *in situ* voltammetry. *Food Chemistry*. 127: 318–325.
[40] Liu, Y.; Li, M.; Li, H.; Wang, G.; Long, Y.; Li, A.; Yang, B. 2019. In Situ Detection of Melatonin and Pyridoxine in Plants Using a CuO–Poly(L-lysine)/Graphene-Based Electrochemical Sensor. *ACS Sustainable Chemistry & Engineering*. 7: 19537–19545.
[41] Umapathi, R.; Ghoreshian, S.M.; Sonwal, S.; Rani, G.M.; Huh, Y.S. 2022. Portable electrochemical sensing methodologies for on-site detection of pesticide residues in fruits and vegetables. *Coordination Chemistry Reviews*. 453: artic. 214305.
[42] Kim, M.-Y.; Lee, K.H. 2022. Electrochemical Sensors for Sustainable Precision Agriculture—A Review. *Frontiers in Chemistry*. 10: artic. 848320.
[43] Ai, F.; Chen, H.; Zhang, S.-H.; Liu, S.-Y.; Wei, F.; Dong, X.-Y.; Cheng, J.-K.; Huang, W.-H. 2009. Real-Time Monitoring of Oxidative Burst from Single Plant Protoplasts Using Microelectrochemical Sensors Modified by Platinum Nanoparticles. *Analytical Chemistry*. 81: 8453–8458.
[44] Doménech-Carbó, A.; Dias, D.; Donnici, M. 2021. In vivo electrochemical monitoring of signaling transduction of plant defense against stress in leaves of Aloe vera L. *Electroanalysis*. 33: 1024–1032.
[45] Pandey, R.; Teig-Sussholz, O.; Schuster, S.; Avni, A.; Sacham-Diamand, Y. 2018. Integrated electrochemical Chip-on-Plant functional sensor for monitoring gene expression under stress. *Biosensors & Bioelectronics*. 117: 493–500.
[46] Sun, L.-J.; Xie, Y.; Yan, Y.-F.; Yang, H.; Gu, H.-Y.; Bao, N. 2017. Paper-based analytical devices for direct electrochemical detection of free IAA and SA in plant samples with the weight of several milligrams. *Sensors and Actuators B*. 247: 336–342.
[47] Lee, G.; Wei, Q.; and Zhu, Y. 2021. Emerging Wearable Sensors for Plant Health Monitoring. *Advanced Functional Materials*. 31: artic. 2106475.
[48] Liu, L.; Cao, X.; Ma, W.; Chen, L.; Li, S.; Hu, B.; Xu, Y. 2021. In-situ and continuous monitoring of pyocyanin in the formation process of *Pseudomonas aeruginosa* biofilms by an electrochemical biosensor chip. *Sensors and Actuators B*. 327: artic. 128945.

ADDITIONAL LITERATURE

Thermochemical Considerations

Bryantsev, V.S.; Diallo, M.S.; Goddard III, W.A. 2008. Calculation of Solvation Free Energies of Charged Solutes Using Mixed Cluster/Continuum Models. *Journal of Physical Chemistry B*. 112: 9709–9719.

Cisternas, R.; Kahlert, H.; Wulff, H.; Scholz, F. 2015. The electrode response of a tungsten bronze electrode differ in potentiometry and voltammetry and give access to the individual contributions of electron and proton transfer. *Electrochemistry Communications*. 56: 34–37.

Doménech-Carbó, A.; Koshevoy, I.O.; Montoya, N.; Pakkanen, T.A.; Doménech-Carbó, M.T. 2012. Solvent-Independent Electrode Potentials of Solids Undergoing Insertion Electrochemical Reactions: Part II. Experimental Data for Alkynyl-diphosphine Dinuclear Au(I) Complexes Undergoing Electron Exchange Coupled to Anion Exchange. *Journal of Physical Chemistry C*. 116: 25984–25992.

Doménech-Carbó, A.; Scholz, F.; Montoya, N. 2012. Solvent-Independent Electrode Potentials of Solids Undergoing Insertion Electrochemical Reactions: Part III. Experimental Data for Prussian Blue Undergoing Electron Exchange Coupled to Cation Exchange. *Journal of Physical Chemistry C*. 116: 25993–25999.

Doménech-Carbó, A.; Koshevoy, I.O.; Montoya, N. 2016. Separation of the ionic and electronic contributions to the overall thermodynamics of the insertion electrochemistry of some solid Au(I) complexes. *Journal of Solid State Electrochemistry*. 20: 673–681.

Gritzner, G. 1990. Polarographic half-wave potentials of cations in nonaqueous solvents. *Pure and Applied Chemistry*. 62: 1839–1858.

Izutsu, K. 2011. Liquid Junction Potentials between Electrolyte Solutions in Different Solvents. *Analytical Sciences*. 27: 658–694.

Pandey, M.D.; Martí-Centelles, V.; Burguete, M.I.; Montoya, N.; Luis, S.V.; García-España, E.; Doménech-Carbó, A. 2016. Bisferrocenyl-functionalized pseudopeptides: access to separated ionic and electronic contributions for electrochemical anion sensing. *RSC Advances*. 6: 35257–35266.

Trasatti, S. 1990. The "absolute" electrode potential- the end of the history. *Electrochimica Acta*. 35: 269–271.

Tsao, R. 2010. Chemistry and biochemistry of dietary polyphenols. *Nutrients*. 2: 1231–1246.

Theoretical Aspects of Voltammetry

Compton, R.G.; Banks, C.E. 2011. *Understanding Voltammetry*, 2nd ed., Imperial College Press, London.

Laviron, E. 1979. General expression of the linear potential sweep voltammogram in the case of diffusionless electrochemical systems. *Journal of Electroanalytical Chemistry*. 101: 19–28.

Lopez-Tenes, M.; Gonzalez, J.; Molina, A. 2014. Two-Electron Transfer Reactions in Electrochemistry for Solution-Soluble and Surface-Confined Molecules: A Common Approach. *Journal of Physical Chemistry C*. 118: 12312–12324.

Marcus, R.A. 1965. On the theory of electron-transfer reactions. VI. Unified treatment for homogeneous and electrode reactions. *Journal of Chemical Physics*. 43: 679–701.

Mirčesky, V.; Lovirć, M. 2001. Ohmic drop effects in square-wave voltammetry. *Journal of Electroanalytical Chemistry*. 497: 114–124.

Mirceski, V.; Gulaboski, R.; Lovric, M.; Bogeski, I.; Kappl, R.; Hoth, M. 2013. Square-Wave Voltammetry: A Review on the Recent Progress. *Electroanalysis*. 25: 2411–2422.

Myland, J.C.; Oldham, K.B. 2005. Quasireversible cyclic voltammetry of a surface confined redox system: a mathematical treatment. *Electrochemistry Communications*. 7: 282–287.

Polcyn, D.S.; Shain, I. 1966. Multistep Charge Transfers in Stationary Electrode Polarography. *Analytical Chemistry*. 38: 370–375.

Nicholson, R.S. 1965. Theory and Application of Cyclic Voltammetry for Measurement of Electrode Reaction Kinetics. *Analytical Chemistry*. 37: 1351–1355.

Nicholson, R.S.; Shain, I. 1965. Theory of Stationary Electrode Polarography for a Chemical Reaction Coupled between Two Charge Transfers. *Analytical Chemistry*. 37: 178–190.

Seeber, R.; Zanardi, C.; Inzelt, G. 2015. Links between electrochemical thermodynamics and kinetics. *ChemTexts*. 1: artic. 18.

Solid State Electrochemistry

Bauer, D.; Gaillochet, M.P. 1974. Etude du comportement de la pate de carbone a compose electroactif incorporé. *Electrochimica Acta*. 19: 597–606.

Brainina, Kh.Z.; Vidrevich, M.B. 1981. Stripping analysis of solids. *Journal of Electroanalytical Chemistry*. 121: 1–28.
Chen, A.; Rogers, E.I.; Compton, R.G. 2010 Abrasive stripping voltammetric studies of lignin and lignin model compounds. *Electroanalysis*. 22: 1037–1044.
Doménech-Carbó, A. 2004. A model for solid-state voltammetry of zeolite-associated species. *Journal of Physical Chemistry B*. 108: 20471–20478 .
Doménech-Carbó, A. 2015. Theoretical scenarios for the electrochemistry of porous silicate-based materials: an overview. *Journal of Solid State Electrochemistry*.19: 1887–1903.
Doménech-Carbó, A. 2020. *Electrochemistry of Porous Materials*, 2nd edit. Taylor & Francis, Boca Raton.
Doménech-Carbó, A.; Doménech-Carbó, M.T. 2023. *Electrochemistry for Cultural Heritage*. Monographs in Electrochemistry series, Scholz, F.; Peter, L., Eds. Springer, Berlin-Heidelberg.
He, J.-B.; Zhou, Y.; Meng, F.-S. 2009 Time-derivative cyclic voltabsorptometry for voltammetric characterization of catechin film on a carbon-paste electrode: one voltammogram becomes four. *Journal of Solid State Electrochemistry*. 13: 679–685.
Janeiro, P.; Oliveira-Brett, A.M. 2005 Solid state electrochemical oxidation mechanisms of morin in aqueous media. *Electroanalysis*. 17: 733–738.
Laviron, E. 1979. General expression of the linear potential sweep voltammogram in the case of diffusionless electrochemical systems. *Journal of Electroanalytical Chemistry*. 101: 19–28.
Lovrić, M.; Hermes, M.; Scholz, F. 1998. The effect of the electrolyte concentration in the solution on the voltammetric response of insertion electrodes. *Journal of Solid State Electrochemistry*. 2: 401–404.
Lovrić, M.; Scholz, F. 1997. A model for the propagation of a redox reaction through microcrystals. *Journal of Solid State Electrochemistry*. 1: 108–113.
Lovrić M, Scholz F 1999 A model for the coupled transport of ions and electrons in redox conductive microcrystals. *Journal of Solid State Electrochemistry*. 3: 172–175.
Nitschke, L.; Henrion, G.; Damaschun, F.; Scholz, F. 1989. A New technique to Study the Electrochemistry of Minerals. *Naturwissenschaften*. 76: 167–168.
Oldham, K.B. 1998. Voltammetry at a three-phase junction. *Journal of Solid State Electrochemistry*. 2: 367–377.
Scholz, F.; Nitschke, L.; Henrion, G. 1989. A New procedure for Fast Electrochemical Analysis of Solid Materials. *Naturwissenschaften*. 76: 71–72.
Scholz, F.; Nitschke, L.; Henrion, G. 1989. Identification of Solid Materials with a New Electrochemical technique - The Abrasive Stripping Voltammetry. *Fresenius' Zeitschrift für Analytische Chemie*. 334: 56–58.
Schröder, U.; Oldham, K.B.; Myland, J.C.; Mahon, P.J.; Scholz, F. 2000. Modelling of solid state voltammetry of immobilized microcrystals assuming an initiation of the electrochemical reaction at a three-phase junction. *Journal of Solid State Electrochemistry*. 4: 314–324.
Schultz. F.A.; Kuwana, T. 1965. Electrochemical studies of organic compounds disolved in carbon-paste electrodes. *Journal of Electroanalytical Chemistry*. 10: 95–103.

In Situ Electrochemistry of Plants

Gao, J.; Li, H.; Li, M.; Wang, G.; Long, Y.; Li, P.; Li, C.; Yang, B. 2021. Polydopamine/graphene/MnO_2 composite-based electrochemical sensor for in situ determination of free tryptophan in plants. *Analytica Chimica Acta*. 1145: 103–113.
Gieling, T.; Vanstraten, G.; Janssen, H.; Wouters, H. 2005. ISE and Chemfet Sensors in Greenhouse Cultivation. *Sensors and Actuators B*. 105: 74–80.
Ren, Q.-Q.; Huang, X.-R.; Liu, G.-C.; Jun, O.-Y.; Li, M.-T.; Chen, H.; Zhao, Y.-D.; Chen, W. 2015. A field-compatible technique using an electrochemical sensing microbundle for real-time and simultaneous in vivo measurement of hydrogen peroxide, nitric oxide, and pH under drought stress. *Sensors and Actuators B*. 220: 743–748.
Sun, L.-J.; Xie, Y.; Yan, Y.-F.; Yang, H.; Gu, H.-Y.; Bao, N. 2017. Paper-based Analytical Devices for Direct Electrochemical Detection of Free IAA and SA in Plant Samples with the Weight of Several Milligrams. *Sensors and Actuators B*. 247: 336–342.
Wilson, G.S.; Johnson, M.A. 2008. In-vivo electrochemistry: What can we learn about living systems? *Chemical Reviews*. 108: 2462–2481.
Zhao, F.; He, J.; Li, X.; Bai, Y.; Ying, Y.; Ping, J. 2020. Smart Plant-Wearable Biosensor for In-Situ Pesticide Analysis. *Biosensors & Bioelectronics*. 170, artic. 112636.

5 Electrochemistry of Vegetal Components I

5.1 INTRODUCTION

The vegetal matter is composed of a variety of chemical compounds. These include polymeric species with structural functions such as cellulose and lignins, compounds responsible for essential biological (metabolic, genetic) functions, primary and secondary metabolites, compounds responsible for signal transduction, defense, and plant communication, etc. [1]. A significant fraction of these compounds is electroactive under "ordinary" electrochemical conditions in aqueous solution or solution in organic solvents. Accordingly, electrochemical methods can be used to obtain information on the composition and functionality of such compounds. The *in vitro* reactivity can be directly accessed under these ordinary conditions, but *in vivo* reactivity can also be studied using electrochemical methods. The scope of electrochemistry of vegetal components is being expanded by the development of the solid state and in situ techniques presented in the precedent chapter. In the current one, the electrochemical behavior of the most extensively studied families of vegetal components will be described.

5.2 PHENOLIC COMPOUNDS

Higher plants synthesize and accumulate a variety of phenolic compounds. These include simple phenolics (phenolic acids and ester derivatives and coumarins) and polyphenols. This second group includes many components, including flavonoids, proanthocyanidins, stilbenes, lignans, and lignins. These families of compounds have a common carbon C6-C3 phenylpropanoid unit. The biosynthesis of most polyphenols is associated with the deamination of phenylalanine and tyrosine amino to cinnamic acids, within the phenylpropanoid pathway. The attachment of one or more –OH groups to the phenyl ring originates an enormous variety of compounds [2, 3].

5.2.1 Flavonoids and Related Compounds

The flavonoids are a wide group of plant components (more than 4,000 natural flavonoids have been identified) that can be defined as benzopyran derivatives constituted by a phenyl ring (ring A) fused with a pyran ring (ring C), in addition to another phenyl ring (ring B) substituted at position 2 of ring C. This structural unit (the flavan unit) incorporates indoxyl, metoxyl, and/or glucosides as substituents.

Depending on their structure, flavonoids can be divided into seven major types illustrated in Figure 5.1: flavones (flavonones), flavonols (flavon-3-ols), flavanones, flavanonols (3-hydroxyflavanones), flavanols, anthocyanidins and anthocyanins, and isoflavonoids. Their electrochemistry is mainly determined by the number and position on the flavan backbone of the –OH groups. The pairs of –OH groups in *o*- and *p*- positions (*o*- and *p*-catechol units) oxidize to the corresponding *o*- and *p*-quinones vie two-proton, two-electron transfer processes. These processes, particularly favored for catechol groups in the B ring, occur as ECEC and/or CECE pathways, which frequently look like apparently reversible voltammetric couples.

Figure 5.2 shows a series of cyclic voltammograms recorded at different pHs on a solution of quercetin in aqueous phosphate buffer [4]. The voltammetric response approximates that expected for a diffusion-controlled reversible oxidation process, as suggested by the linear variation of the

DOI: 10.1201/9781032697826-5

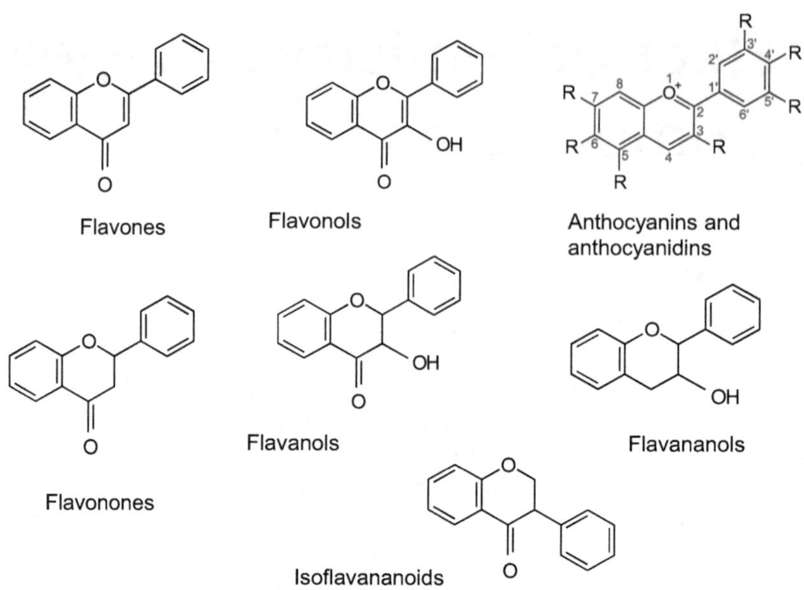

FIGURE 5.1 Structural backbone of the principal families of flavonoids.

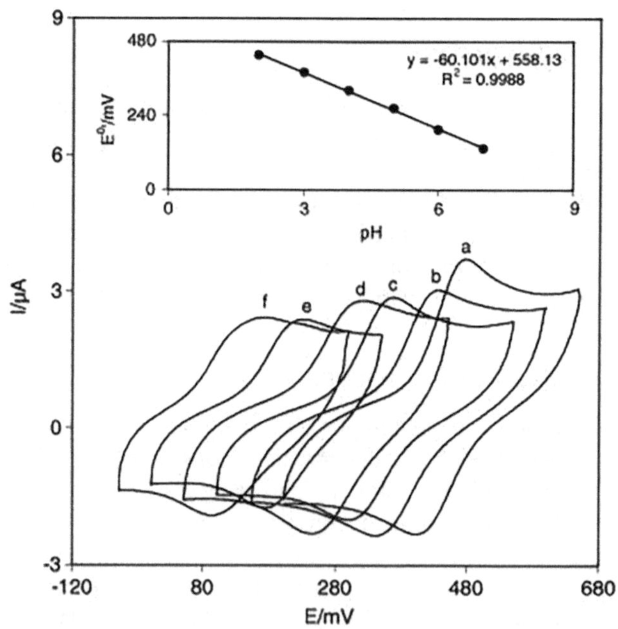

FIGURE 5.2 Cyclic voltammograms at glassy carbon electrode of a 0.3 mM solution of quercetin in 0.10 M phosphate buffer at different pH 2 (a), 3 (b), 4 (c), 5 (d) 6 (e), and 7 (f). Potential scan rate 100 mV s^{-1}. Inset: Variation of the mid-peak potential with the pH. (Adapted from ref. [4], with permission. Copyright Elsevier 2005.).

peak current of the square root of the potential scan rate [4]. The variation of the mid-peak potential with the pH (inset in Figure 5.2) is reasonably linear with a slope close to the theoretical value (59 mV $decade^{-1}$) expected for a reversible process involving the transfer of an equal number of protons and electrons.

Eventually, the ECEC or CECE pathways representing the "true" electrochemistry associated with the oxidation of catechol units may be interrupted by the occurrence of dimerization or polycondensation reactions of the semiquinone radical intermediate. The electrochemical oxidation of flavonoids, however, in general, implies more complex pathways. Figure 5.3 compares the cyclic voltammetric response of quercetin (3,3',4',5,7-pentahydroxyflavone), luteolin (3',4',5,7-tetrahydroxyflavone), and apigenin (3',5,7-trihydroxyflavone) in MeCN solution [5]. These flavonoids differ in the number and distribution of hydroxyl groups and exhibit different voltammetric profiles.

For our purposes, the relevant point to emphasize is that depending on the solvent and other experimental conditions, several oxidation products can be formed not only involving the *o*-catechol unit in ring B. In the case of quercetin, the first oxidation peak corresponds to the *o*-catechol to *o*-quinone oxidation in ring B, but a variety of compounds can be obtained after exhaustive electrolysis as schematized in Figure 5.4 [5].

Figure 5.5 shows the cyclic voltammetry of morin (2',3,4',5,7-pentahydroxyflavone) in MeCN solution at various scan rates [6]. Here, the main oxidation process involves the hydroxyl to ketone conversion of one –OH group of the ring B and the –OH unit of the ring C (see Figure 5.6a). In the case of catechin, the semiquinone form in ring B resulting from the one-electron, one-proton initial oxidation is in equilibrium with a tautomeric, which could result in different final products (see Figure 5.6b) [7].

The oxidation of *m*-diphenol (resorcinol) groups is similar to that of monophenols and usually involves an EC mechanism, which involves an initial one-electron, one-proton process yielding a semiquinone radical. This is rapidly followed by dimerization reactions or reactions with water forming different products [2, 3, 8]. These processes are schematized in Figure 5.7. These processes can be multiplied to give rise to polymeric species. Electropolymerization processes can eventually produce undesired electrode poisoning effects or, on the contrary, be exploited for analytical purposes [9].

Solid-state electrochemistry follows equivalent patterns [10, 11]. Interestingly, reduction processes, poorly studied in the solution phase, can be recorded. This is illustrated in Figure 5.8 for microparticulate deposits of luteolin attached to paraffin-impregnated graphite electrodes in contact with an aqueous acetic/acetate buffer. When the potential scan is initiated in the positive direction

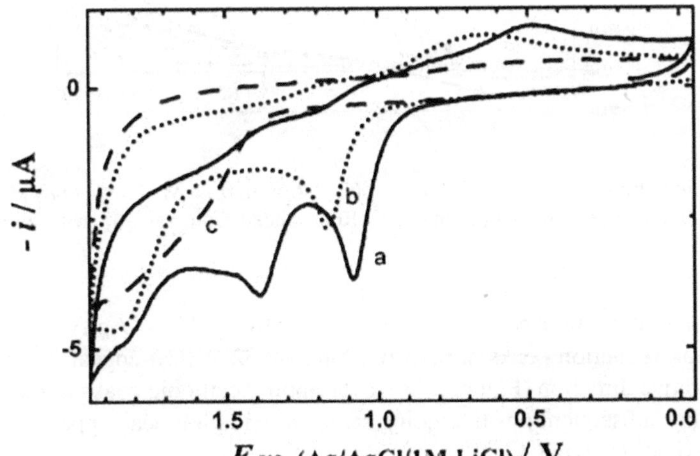

FIGURE 5.3 Cyclic voltammograms at glassy carbon microelectrode of 0.41 mM quercetin (a), 0.27 mM luteolin (b), and 0.27 mM apigenin solutions in 0.10 M Bu_4NPF_6/MeCN solution. Potential scan rate 0.25 mV s^{-1}. (Reproduced from ref. [5], with permission. Copyright Elsevier 2011.).

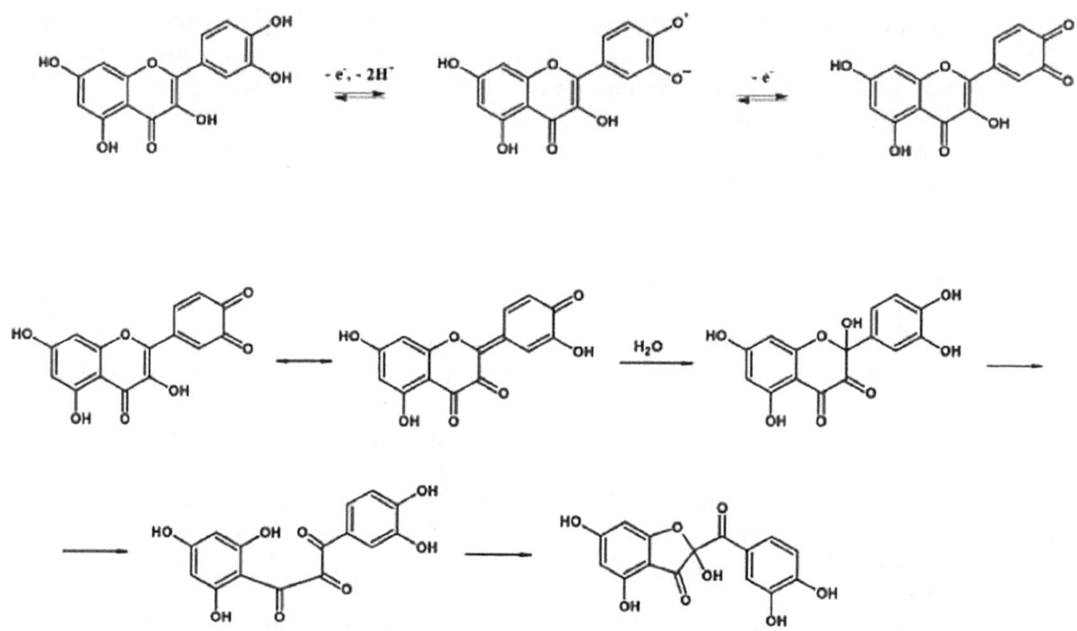

FIGURE 5.4 Species involved in the oxidation of quercetin in MeCN. (Adapted from ref. [5], with permission. Copyright Elsevier 2011.).

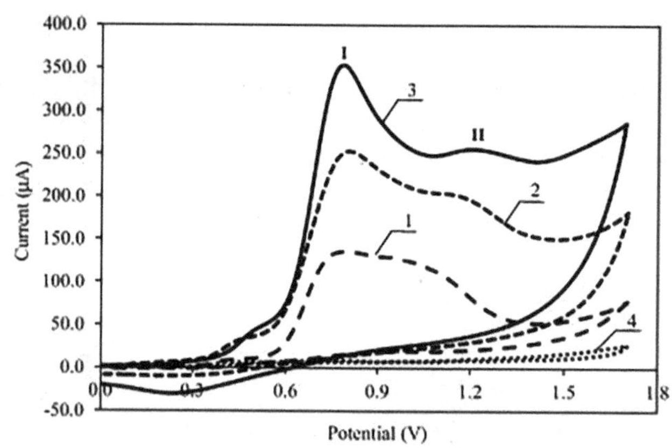

FIGURE 5.5 Cyclic voltmmograms of 2.0 mM morin solution in 0.10 M Bu_4NClO_4/MeCN at potential scan rates of 10 (1), 50 (2), 100 (3), and 10 (4) mV s^{-1}. (Reproduced from ref. [6], with permission. Copyright Elsevier 2014.).

(Figure 5.8a), a main oxidation peak (A_1) is recorded at 0.30 V vs. Ag/AgCl. In the subsequent cathodic scan, weak reduction peaks at 0.15 (C_1) and −0.32 V (C_2) appear. When the potential is scanned in the negative direction (Figure 5.8b), overlapping cathodic peaks at −0.50 (C_3) and −0.75 V (C_4) appear. In the subsequent positive-going scan, an oxidation peak appears at 0.10 V (A_2) preceding the main oxidation signal A_1.

This voltammetry can be interpreted based on literature dealing with flavonoid electrochemistry in the solution phase [4–7, 12–15], attributing the process A_1 to the two-proton, two-electron oxidation of the *o*-catechol group. This process is probably followed by the reaction with water and ring opening as schematized in Figure 5.9. Process C_1 can be assigned to the reversal of the

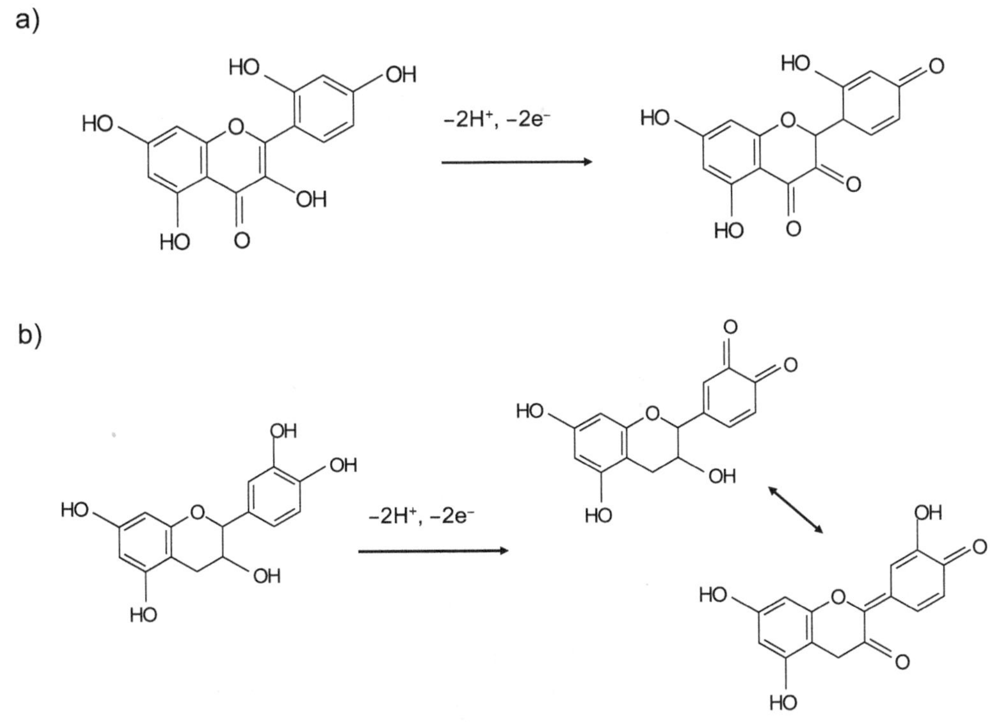

FIGURE 5.6 Scheme representative of the main oxidation processes of (a) morin and (b) catechin.

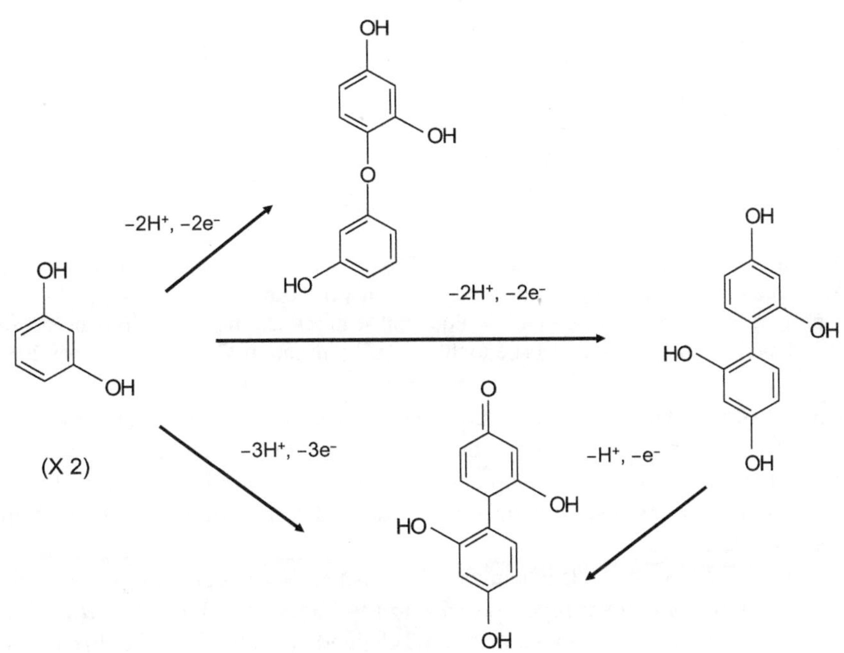

FIGURE 5.7 Electrochemical pathways for the oxidation of resorcinol.

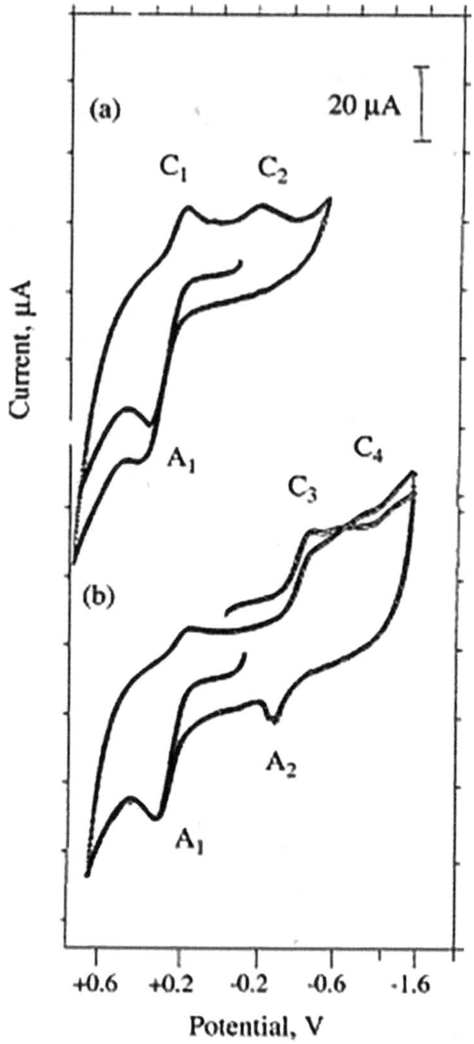

FIGURE 5.8 Cyclic voltammograms of luteolin-modified graphite electrodes in contact with 0.25 M HAc + 0.25 M NaAc aqueous buffer. Potential initiated at 0.0V in (a) the positive direction and (b) the negative direction and prolonged to potentials ca. −0.85 V. Potential scan rate 20 mV s^{-1}. (Reproduced from ref. [10], with permission. Note that the axes are opposite to the IUPAC convention. Copyright Elsevier 2005.).

aforementioned process, whereas process C_2 would correspond to the *o*-quinone to *o*-catechol reduction of the "open" derivative. The cathodic processes C_3 and C_4 can be represented, tentatively, as the successive two-proton, two-electron processes schematized in Figure 5.9 [10]. Notice that this complex electrochemistry involves not only solid-to-solid transformations but also dissolution-mediated processes.

The coexistence of different electrochemical pathways enriches – but complicates – the voltammetric behavior of polyphenolic compounds. Steric hindrance and diastereomization effects can also influence the redox behavior, as well as the pH-dependent stability of the different flavonoids. In addition, it has to be taken into account that most polyphenolic compounds have significant chelating capacity for metal ions [16].

The electrochemistry of other benzopyran derivatives, chromones, coumarins (and their derivatives: styrylchromones and neoflavonoids), and xanthones (see Figure 5.10) offers characteristics similar to that of flavonoids [17].

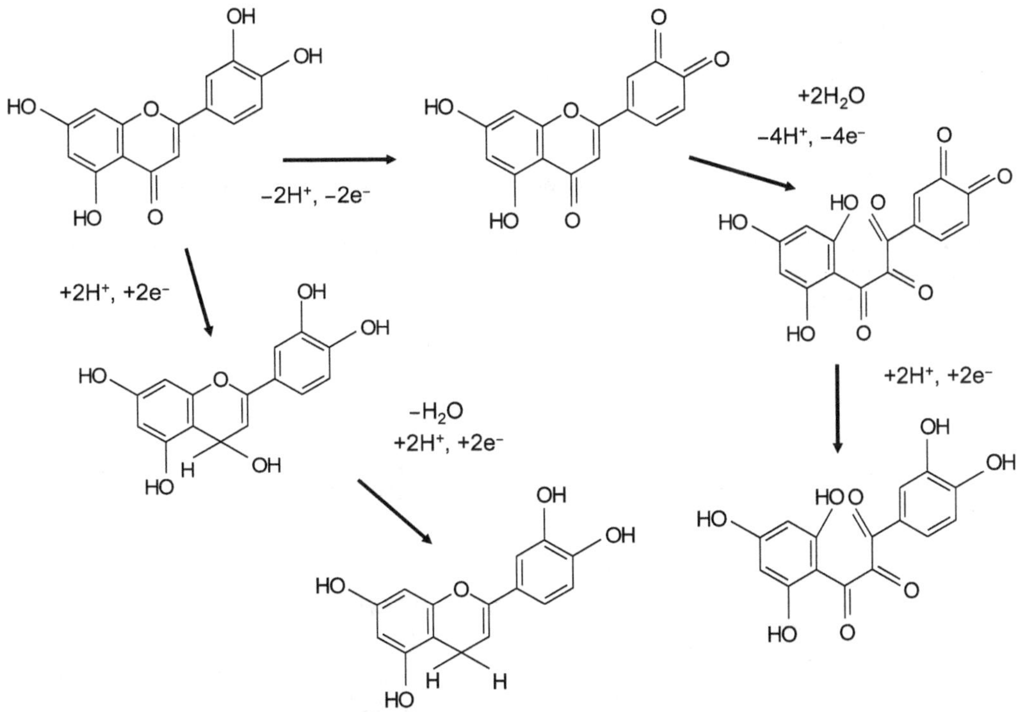

FIGURE 5.9 Tentative scheme representing the electrochemical processes associated with the solid-state electrochemistry of luteolin. (Adapted from ref. [10], with permission. Copyright Elsevier 2005.).

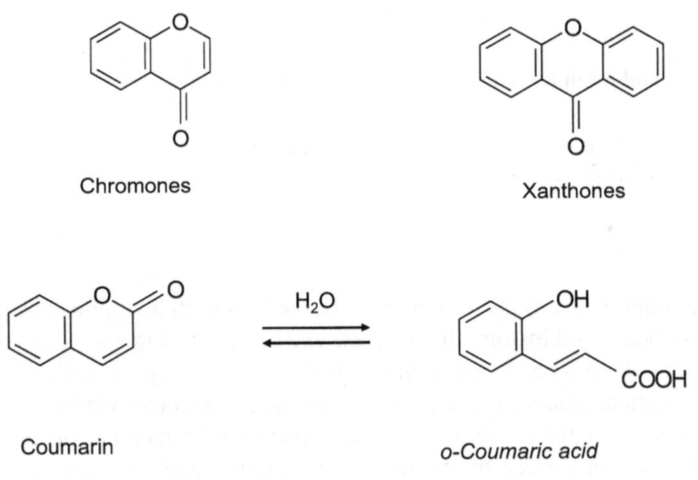

FIGURE 5.10 Chemical structures of chromones, xanthones, coumarin/coumaric acid.

5.2.2 Phenolic Acids and Ester Derivatives

Phenolic acids represent another abundant class of natural compounds with antioxidant properties. This group includes the derivatives of hydroxybenzoic and hydroxycinnamic (coumaric) acids. Figure 5.11 depicts the chemical structure of salicylic acid and caffeic acid, examples, respectively of hydroxybenzoic- and hydroxycinnamic-type acids. The biosynthesis of hydroxycinnamic acids gives rise to hydroxycinnamoyl ester derivatives. The electrochemistry of these compounds follows patterns directly related to those previously described. The electrochemistry of compounds such as

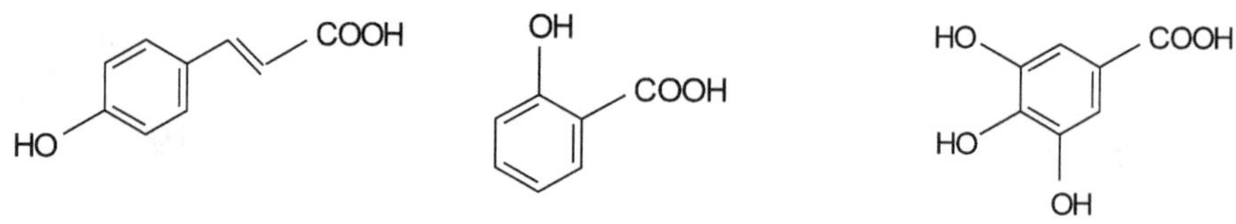

Caffeic acid Salicylic acid Gallic acid

FIGURE 5.11 Chemical structures of caffeic acid (hydroxycinnamic acid-type), salicylic acid, and gallic acid (hydroxybenzoic acid-type).

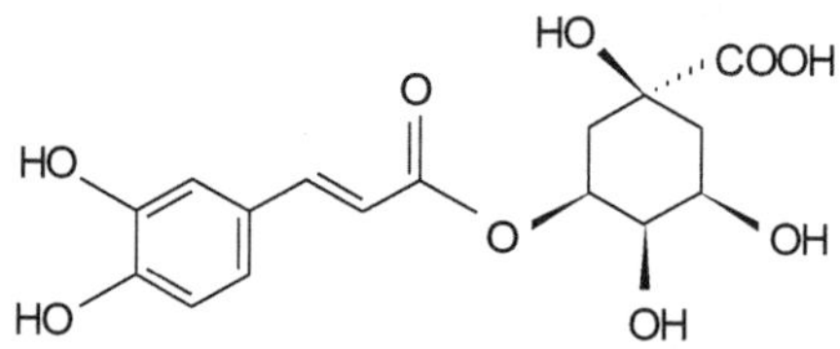

Chlorogenic acid

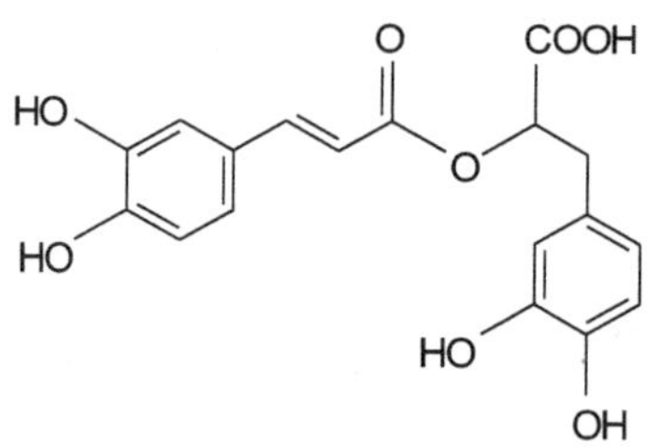

Rosmarinic acid

FIGURE 5.12 Structural formulas of chlorogenic acid and rosmarinic acid.

chlorogenic acid and rosmarinic acid (Figure 5.12) which contain *o*-catechol groups is dominated by the oxidation to the corresponding *o*-quinones [18].

5.3 QUINONES

These form several important groups of plant metabolites. As with flavonoids, apart from biological functions, they have been used historically as dyes. The basic structures are depicted in Figure 5.13. Prenyl-substituted *p*-quinones are found in mitochondria, many bacteria, and/or chloroplasts. These include ubiquinones, menaquinones, and plastoquinones. The electrochemistry of these families of compounds is dominated by the reduction of the quinone motif, which has been previously studied. This group is usually accompanied by methyl, hydroxyl, and other substituents so that the corresponding oxidative patterns accompany the reduction of the *o*-quinone unit, as has been studied in solid state in contact with aqueous electrolytes [19, 20].

The electrochemistry can be complicated by the occurrence of tautomerization, cyclicization, and cycle opening reactions. The cyclic voltammetry of lapachol (2-hydroxy-3-(3-methyl-2-butenyl)-1,4-naphtho-quinone) in protic media is shown in Figure 5.14. On initiating the potential scan at 0.0 V vs. Ag/AgCl in the negative direction (Figure 5.14a), a well-defined reduction peak C_1 is recorded at ca. −0.50 V. This is followed, in the subsequent anodic scan, by a shoulder near −0.25 V (A_1) and a well-defined oxidation peak at 0.0 V. This peak precedes a second intense anodic peak at 0.90 V (A_2). In a second cathodic scan, the peak C_1 is preceded by a

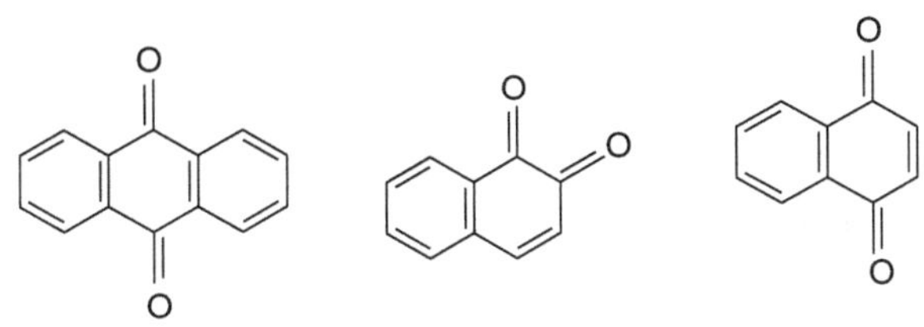

FIGURE 5.13 Chemical structure of (a) anthraquinones, (b) 1,2-naphtoquinones, and (c) 1,4-naphtoquinones.

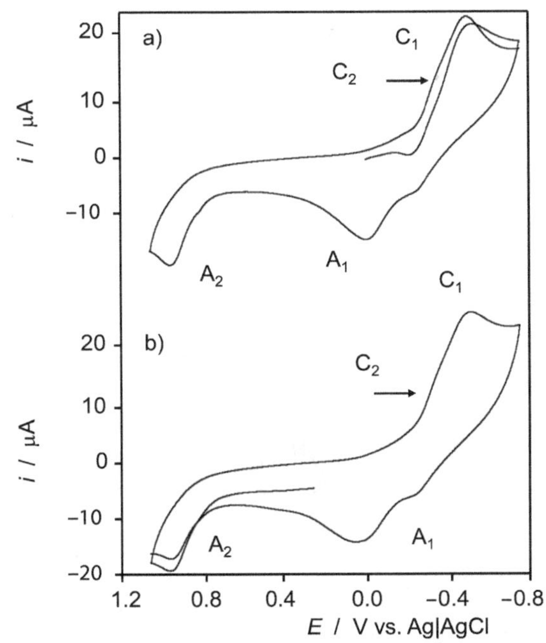

FIGURE 5.14 Cyclic voltmmograms at GCE of a 0.5 mM solution of lapachol in 50:50 (v/v) MeOH/water mixture containing 0.10 M HAc/NaAc. Potential scan initiated at (a) 0.0 V in the negative direction, (b) 0.20 V in the positive direction. Potential scan rate 50 mV s^{-1}.

shoulder ca. −0.40 V (C_2). This shoulder also appears if the potential scan is initiated at 0.20 V in the positive direction (Figure 5.14b). Here, the peak A_2 appears in the initial positive-going potential scan. These features denote that the parent naphthoquinone can be either reduced (process C_1) or oxidized (process A_2) and that process C_2 can be assigned to a species generated during oxidation step A_2. The interpretation of this voltammetric pattern is illustrated in Figure 5.15. The couple C_1/A_1 can be attributed to the two-proton, two-electron *o*-quinone to *o*-catechol conversion, in principle, via ECEC or CECE mechanisms. The process A_2 corresponds to the (apparently) irreversible oxidation of lapachol to dehydrolapachol, which in turn experiences a two-proton, two-electron reduction in the process C_2 [21].

The electrochemistry of compounds forming the vitamin K complex, phylloquinone, menaquinone, and menadione, has been repeatedly studied. Due to their lipophilic character, electrochemical studies have been focused on aprotic solvents. Here, the usual two one-electron reduction processes yielding the anion radical and the dianion are obtained [22]. The electrochemical behavior in aqueous media has been studied in 1,2-dimyristoyl-sn-glycero-3-phosphocholine monolayers on the surface of mercury electrodes [23] and stabilized in micellar media [24].

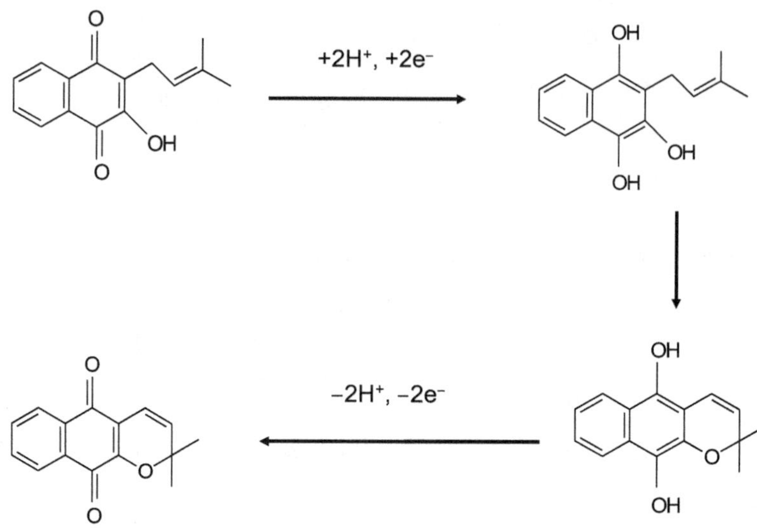

FIGURE 5.15 Scheme for the cyclic voltammetry of lapachol in Figure 5.14 according to [21].

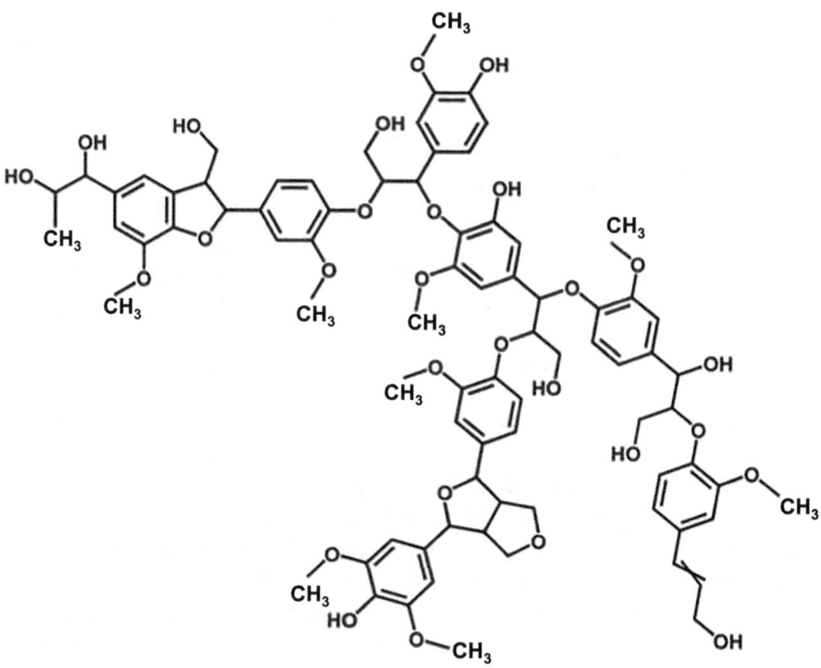

FIGURE 5.16 Chemical structure of lignin organosolv.

5.4 LIGNINS, LIGNANS, AND TANNINS

Lignins are complex polymeric components located in the secondary cell wall of higher plants that, via interactions with cellulose and hemicelluloses, provide mechanical support and defense against pathogenic attacks. Lignins can be considered as composed of three basic phenyl- propanoid units – namely, *p*-hydroxyphenyl, guaiacyl, and syringyl, derived from the corresponding monolignols (*p*-coumaryl, coniferyl, and sinapyl alcohols [25]. Figure 5.16 depicts the chemical structure of lignin organosolv.

The abundant presence of hydroxy and methoxy units bound to aromatic centers ensures the occurrence of electrochemical oxidation processes in line with previously studied families.

The polyphenolic structure of lignins is responsible for their antioxidant properties, favored by the π-polyconjugation. Their antioxidant capacity, however, is diminished by the heterogeneity of polymeric structure. The electrochemistry of lignins is focused on the oxidation of the free phenolic groups and the *o*-methoxy groups. These are eventually accompanied by the oxidation of the double bonds between the outermost carbon atoms in the side chain. Figure 5.17 depicts the molecular schemes for the electrochemical oxidation processes of (a) syringic acid [26] and (b) β-O-4 coniferyl alcohol dimer [27] in aqueous media which are representative of lignin structural units. These oxidation processes in general look like irreversible voltammetric waves in the same region of potentials as other polyphenolic compounds. The voltammetry of microparticulate deposits from ethanolic extracts of lignins has been applied for cultural heritage studies [27, 28].

Lignans are produced by oxidative coupling of two phenylpropanoid units, but the molecular structure presents considerable variability. Tannins are also polymeric polyphenolic components of plants also characterized by antioxidant properties. The electrochemistry of these families of compounds has been, however, the subject of relatively few studies [2].

5.5 STILBENES, CURCUMINOIDS, CHALCONES, ESSENTIAL OILS, AND OTHERS

Stilbenes, curcuminoids, and chalcones are structurally characterized by two phenolic rings linked by an alkyl chain. Resveratrol is the most representative stilbene whose electrochemistry (in both the *cis* and *trans* forms) has been widely studied [29]. Chalcones can be seen as open-chain flavonoids where the two aromatic rings are joined by an aliphatic three-carbon α,β-unsaturated carbonyl

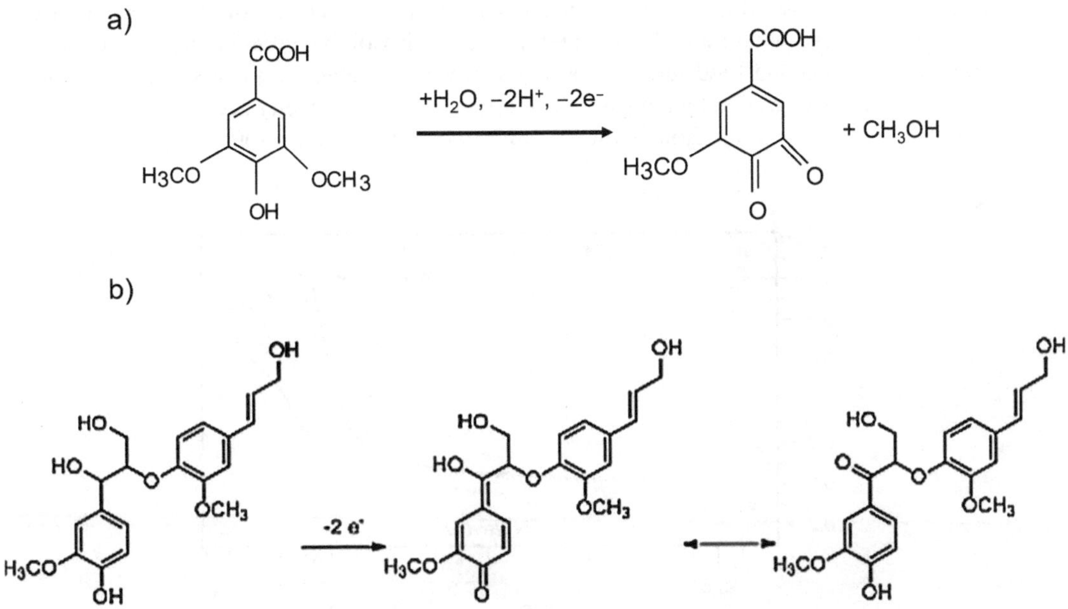

FIGURE 5.17 Scheme for the electrochemical oxidation processes of (a) syringic acid and (b) β-O-4 coniferyl alcohol dimmer. (Adapted from refs. [26–28], with permission. Copyright the Royal Society of Chemistry, 2017, 2019 and Elsevier 2015.).

system. The electrochemistry of curcumin and several analogs has been studied in solution phase [30] and solid state [31]. In all cases, the oxidation of phenolic units centers the observed electrochemistry.

Essential oils consist of mixtures of compounds, mainly terpenes with repeated isoprene (C_5H_8) units mostly incorporating phenolic functions. Typical examples are eugenol, guaiacol, thymol, and vanillin. Figure 5.18 shows the cyclic voltammetric response of vanillin in MeCN solution [31]. In the initial positive-going potential scan, a prominent anodic peak ca. 1.1 V vs. ferrocenium/ferrocene couple was recorded. This signal (wave I) was followed by a weak cathodic peak (wave II) in the reverse scan. This voltammetric response has been interpreted in terms of the sequence of processes represented schematically in Figure 5.19 [32]. The electrochemistry in aqueous solution follows a similar pattern [33], but the overall electrochemical pathway in nonaqueous media can be considerably more complex [32]. The solid-state electrochemistry of vanillin in contact with aqueous electrolytes has been also studied [34].

Vanillin, as a known product of the degradation of lignins, has been proposed as a marker compound for dating purposes in the archaeological domain. This is based on the determination of the rate of loss of vanillin from lignin, which follows a first-order rate law [35]. Given the electroactive character of vanillin and lignins, voltammetric data could be potentially usable for dating purposes [36].

Brazilein and brazilin are tetracyclic components of a natural pigment, Brazilwood, constituted by two aromatic rings, one pyrone, and one a five-membered unsaturated ring whose backbone can be derived from an isoflavonoid structure. These compounds can be seen as components of a redox couple, like that formed by hematoxylin and haematein (see Figure 5.20) which are the main coloring agents of logwood (or litmus Campeche). The solid-state electrochemistry of these compounds reflects the proton-assisted interconversions centered on the phenolic units [10].

5.6 ASCORBIC ACID

The electrochemistry of ascorbic acid has been extensively studied because of its biological importance: the L-enantiomer is vitamin C. This is a ubiquitous hydrosoluble vitamin that can be found in many biological systems and foodstuffs, in particular fresh vegetables and fruits. The solutions of ascorbic acid are easily degraded by atmospheric oxygen, a process accelerated by light, and heavy metal cations. Ascorbic has a γ-lactone structure with two dissociable protons (pK_a 44.0 and 11.3)

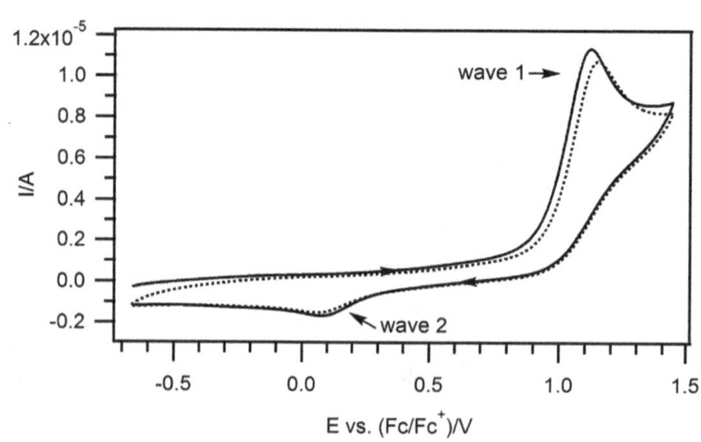

FIGURE 5.18 Cyclic voltammogram at Pt electrode (2 successive scans) of 2 mM vanillin solution in 0.2 M Bu_4NPF_6/MeCN. Potential scan rate 100 mV s^{-1}. Continuous line: first scan; points line: second scan. (Reproduced from ref. [32], with permission. Copyright Elsevier 2016.).

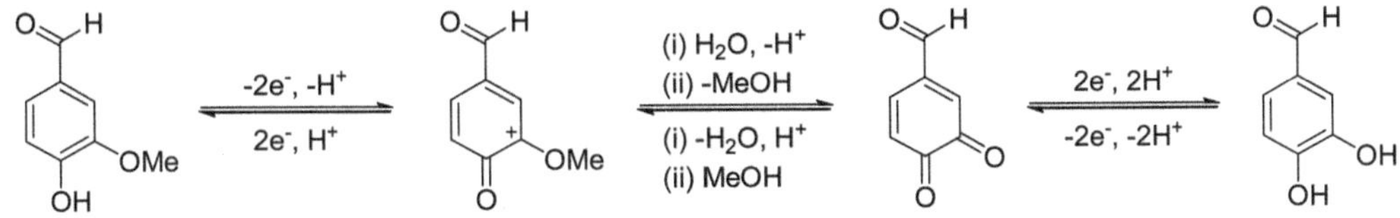

FIGURE 5.19 The scheme proposed for the description of the voltammetry of vanillin in MeCN solution. Experimental data (continuous lines) and simulated voltammogram (see text, dotted lines). (Reproduced from ref. [32], with permission. Copyright Elsevier 2016.).

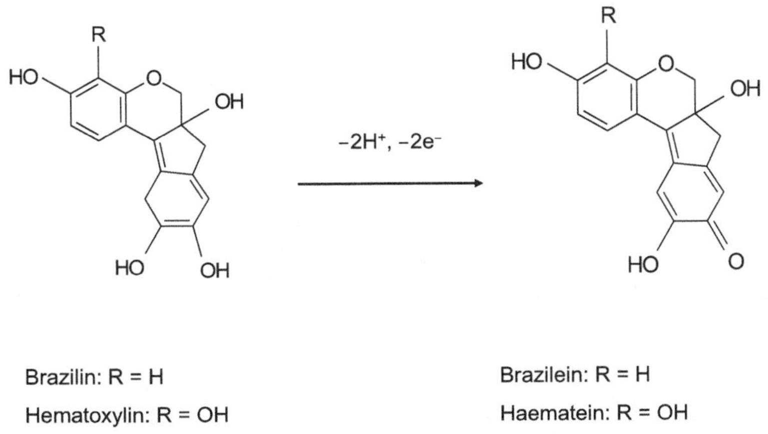

FIGURE 5.20 Structure and electrochemical interconversion pathway of the brazilin/brazilein and hematoxylin/haematein couples.

[37]. Then, in an aqueous solution at biological pHs, the ascorbate ion becomes the predominating form.

In an aqueous solution at carbon electrodes, ascorbic acid shows an (apparently) irreversible oxidation wave. This process has been widely used for analytical purposes using a plethora of modified electrodes. Figure 5.21 shows a series of cyclic voltammograms recorded in a solution of ascorbic acid in phosphate buffer saline (PBS) at unmodified GCE (a), GCE decorated with gold nanoparticles (b), and GCE modified with polyindole film decorated with gold nanoparticles (c). The voltammograms recorded for the aforementioned electrode modifier deposited on GCE and anodized aluminum oxide are compared in (d) [38]. One can see that the voltammetric profile is sensitive to the concentration of ascorbic acid and the electrode modification.

The electrochemical process corresponds to the two-proton, two-electron oxidation to dehydroascorbic acid, followed by a fast irreversible reaction opening the lactone ring, resulting in the formation of 2,3-diketogluconic acid. This is adsorbed on the electrode, thus resulting in electrode fouling [39, 40]. Figure 5.22 depicts the corresponding scheme.

At gold electrodes, the ascorbate ion is electrochemically oxidized to a diketolactone, which is rapidly dehydrated to dehydroascorbic acid. This rearranges to another ene-diol, which is further oxidized at higher potentials [39].

As will be treated in the chapter devoted to electroanalytical aspects, there are many studies dealing with the electrochemical determination of ascorbic acid in fruits, beverages, etc. In most cases, the analytical method involves chemically modified electrodes searching for high sensitivity and selectivity. A critical view on analytical strategies is, however, convenient because the oxidation of ascorbic acid is frequently treated as a single diffusion-controlled process. This is an oversimplification of the "true" multistep mechanism.

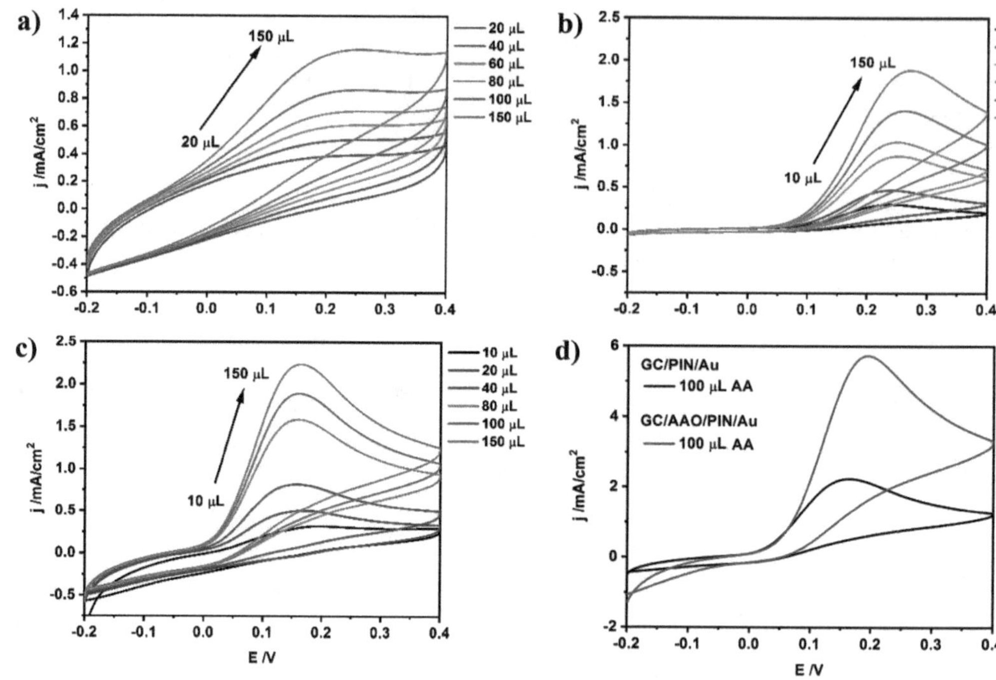

FIGURE 5.21 Cyclic voltammograms of 10 mM ascorbic acid solution in PBS at scan rate 20 mV s^{-1} on the (a) GCE, (b) GCE decorated with gold nanoparticles, (c) GCE modified with as-deposited polyindole decorated with gold nanoparticles, and (d) the above (GC/PIN/Au) compared with the same nanocomposite deposited on anodized aluminum oxide (GC/AAO/PIN/Au). Concentration of ascorbic acid in the solution of, 0.5 mM (10 μL), 1 mM (20 μL), 2 mM (40 μL), 6 mM (60 μL), 8 mM (80 μL), 10 mM (100 μL), 15 mM (150 μL). (Adapted from ref. [38], with permission. Copyright Elsevier 2020.).

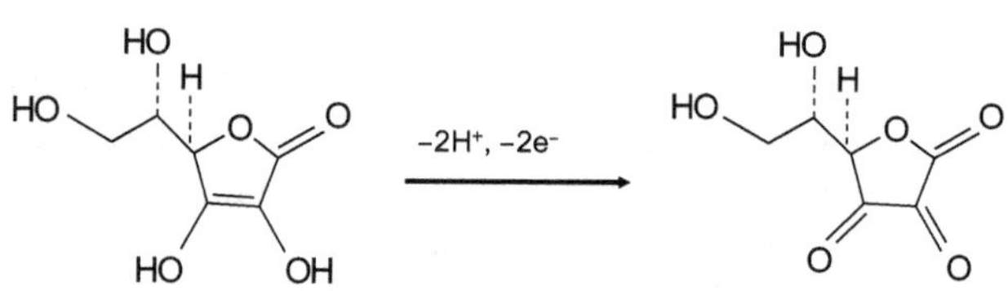

FIGURE 5.22 Scheme for the electrochemical oxidation of ascorbic acid to dehydroascorbbic acid.

5.7 CAROTENOIDS

Carotenoids are widely distributed among plants, serving as light harvesting and photoprotection by preventing the formation of damaging singlet oxygen. In addition, carotenoids act as energy transfer to chlorophylls for the performance of photosynthesis due to their efficient absorption of light energy in spectral regions weakly absorbed by chlorophyll [41]. Carotenoids are characterized by a long backbone chain of conjugated carbon-carbon double bonds. More than 600 compounds of this family have been isolated from plants. These compounds, however, are quite sensitive to light, oxygen, and acid attack. Lycopene represents the basic, acyclic skeleton of carotenoids, consisting of eight connected isoprene units forming a carbon chain with 11 conjugated double bonds.

Because of solubility, the electrochemistry of carotenoids has been studied in organic solvents. Figure 5.23 shows the cyclic voltammograms recorded at Pt electrode for a β-carotene solution in CH_2Cl_2 [42]. This response, representative of most carotenoids, corresponds to a two-electron

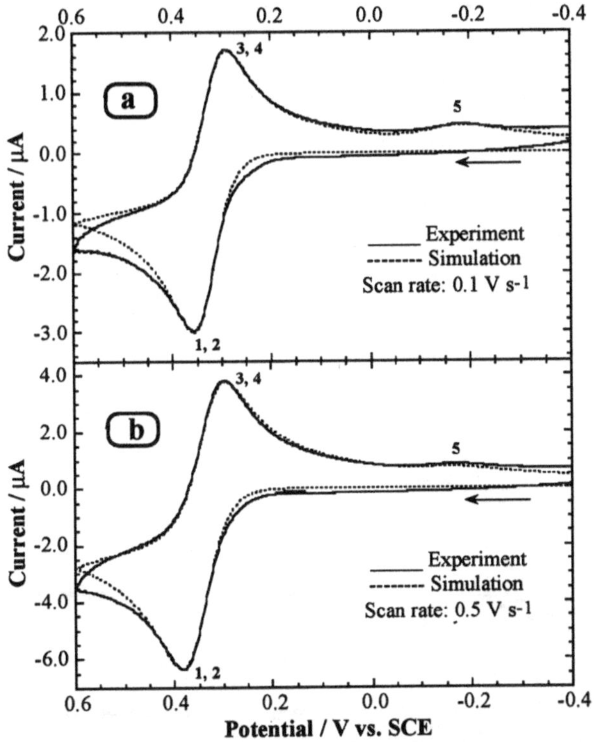

FIGURE 5.23 Cyclic voltammograms at Pt electrode of 1.0 mM *β*-carotene solution in 0.1 M $(Bu_4N)PF_6$/CH_2Cl_2. Potential scan rate: (a) 100 and (b) 500 mV s^{-1}. (Adapted from ref. [42], with permission. Copyright Elsevier 2000.).

oxidation process that results from the merging of two successive one-electron waves (EE pathway) [43, 44]. These waves correspond to the successive formation of a cation radical and a dication, and are eventually mediated by comproportionation/disproportionation and deprotonation processes [45]. The voltammetric record and the electrochemical pathway are sensitive to solvent properties and can become more complicated. In aprotic solvents a cation radical is produced in the first oxidation step but in solvents such as tetrahydrofuran (THF) nonradical cations are formed [42]. Astaxanthin, having a high antioxidant capacity, is characterized by two conjugated carbonyl groups. The voltammetry of a microparticulate deposit resulting from the evaporation of an acetone solution of this compound in contact with aqueous electrolytes reveals the appearance of the two-proton, two-electron reduction of carbonyl groups to hydroxyl units [46]. There is also a reductive electrochemistry of carotenoids in aprotic solvents, leading to radical anions and dianions (see next section) [47].

The extensive electron delocalization in carotenoids facilitates their active role in electron transfer processes in which the cation radical is an integral part of charge-separated states involving porphyrins and quinones [42]. Carotenoids can also act as cross-linking units on electrode surfaces under the effect of transient oxygen radicals [43].

5.8 TOCOPHEROLS, CALCIFEROLS, AND RETINOL

These groups of substances are constituents of the vitamins A, D, and E. These are natural compounds synthesized by plants that have different structural backbones. Figure 5.24 illustrates the cyclic voltammetric response of retinol and several related compounds (including *β*-carotene) in

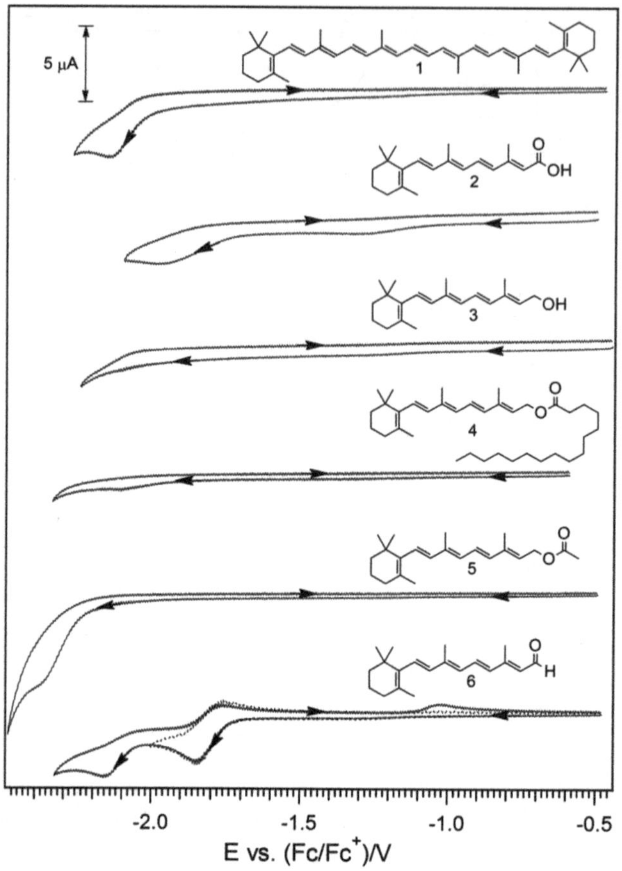

FIGURE 5.24 Initial cathodic scan cyclic voltammograms at GCE of 1 mM solutions of β-carotene (1), retinoic acid (2), retinol (3), retinyl palmitate (4), retinyl acetate (5), and retinal (6) 0.2 M Bu_4NPF_6/ CH_2Cl_2. Potential scan rate 100 mV s^{-1}. (Reproduced from ref. [47], with permission. Copyright the American Chemical Society 2014.).

CH_2Cl_2 corresponding to their reductive electrochemistry. This consists of two successive one-electron transfers yielding successively an anion radical and a dianion, but the peak separation varies significantly between different compounds (eventually merging on a unique wave), and there is a possibility of coupled chemical reactions [47].

The oxidative electrochemistry of retinol and related compounds can be compared in Figure 5.25. Except for β-carotene, which exhibits a reversible two-electron oxidation, these compounds display (apparently) irreversible oxidation processes coupled with chemical reactions resulting in a complex behavior [47]. A similar oxidative electrochemistry has been reported for ergocalciferol [48].

α-Tocopherol (αTOH) acts as an antioxidant inside the lipid phase of biomembranes reacting with an oxidized radical site (LOO•) as [49]

$$\alpha \text{TOH} + \text{LOO}\bullet \rightarrow \alpha \text{TO}\bullet + \text{LOOH}, \tag{5.1}$$

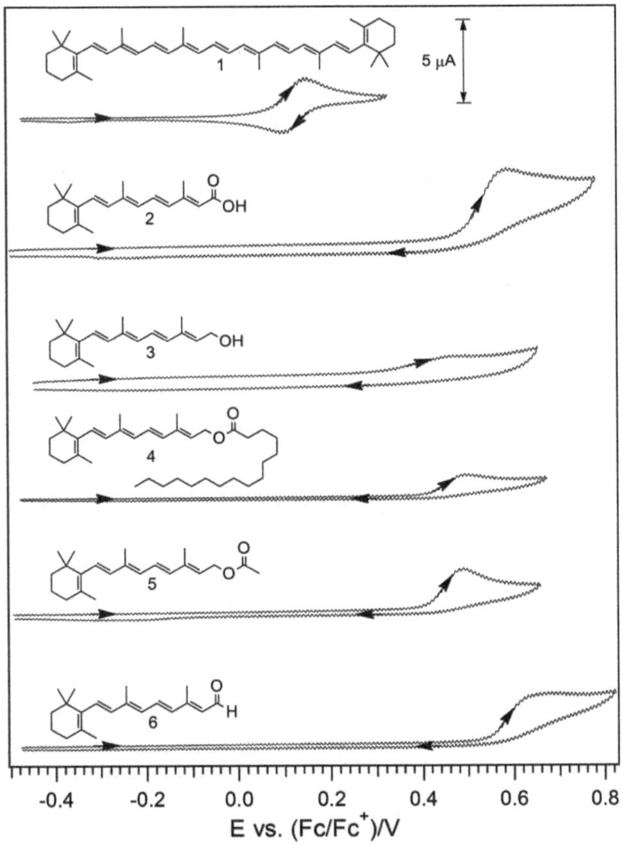

FIGURE 5.25 Initial anodic scan cyclic voltammograms at GCE of 1 mM solutions of β-carotene (1), retinoic acid (2), retinol (3), retinyl palmitate (4), retinyl acetate (5), and retinal (6) 0.2 M Bu_4NPF_6/ CH_2Cl_2. Potential scan rate 100 mV s^{-1}. (Reproduced from ref. [47], with permission. Copyright the American Chemical Society 2014.).

$$\alpha TO\bullet + LOO\bullet \rightarrow (TO)OOL. \tag{5.2}$$

This antioxidant behavior agrees with rich oxidative electrochemistry that has been studied in aprotic media, lecithin microdroplets deposited onto GCE in contact with aqueous media [49], and hydro-alcoholic solutions [50]. Figure 5.26 shows voltammograms representative of the electrochemistry of α-tocopherol microdroplets/films and α-tocopherol films coated with a multilayer of lecithin in contact with aqueous solutions at different pHs [49]. This is an example of how the solid-state electrochemistry "philosophy" [51] can be applied to obtain information on biochemical processes.

The electrochemical processes in Figure 5.26 can be interpreted as schematized in Figure 5.27 [49]. The initial oxidation process (A_1) consists of a two-electron, one-proton oxidation forming the phenoxonium cation, which is subsequently hydroxylated and deprotonated to yield the corresponding quinone via ring opening. This process involves the breaking of the chromanol ring structure, a

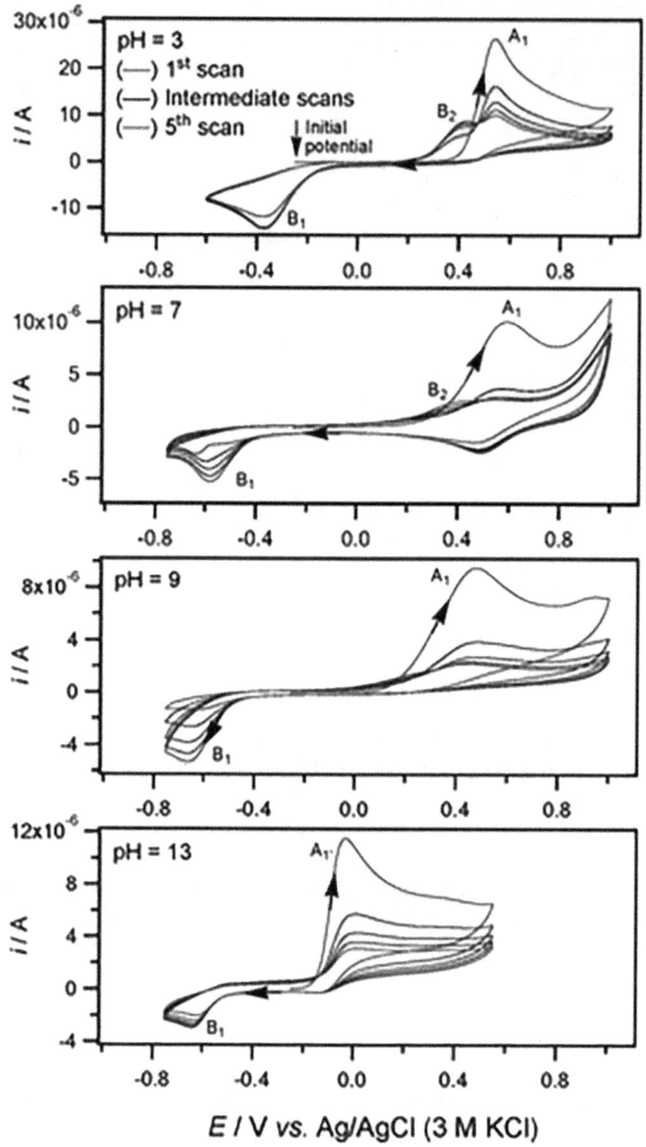

FIGURE 5.26 Cyclic voltammograms obtained at microdroplets/films of *α*TOH/lecithin and αTOH films coated with a multilayer of leci deposited onto a GCE in contact, citric acid-phosphate buffer solutions (pH 3 and 7), and Britton-Robinson buffer solutions (pH 9 and 13). Potential scan rate 100 mV s^{-1}. (Adapted from ref. [49], with permission. Copyright the American Chemical Society 2009.).

chemically irreversible process under mild electrochemical conditions. In sufficiently alkaline media, αTOH forms the phenolate anion, which is easily oxidized, and the A_1 signal shifts to less positive potentials. In turn, the B_1/B_2 processes correspond to the *p*-quinone/*p*-catechol interconversion. The electrochemistry in aprotic solvents offers the possibility of stabilizing intermediate radical species and the occurrence of dimerization processes [49].

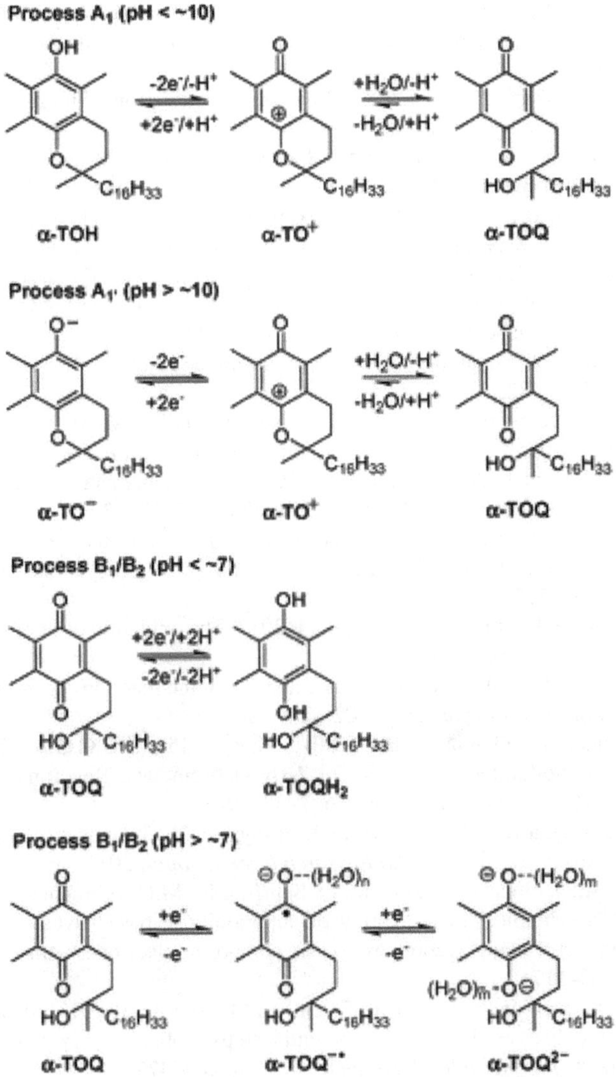

FIGURE 5.27 Processes involved in the electrochemical oxidation/reduction of αTOH films attached to electrode surfaces in contact with aqueous solutions. (Reproduced from ref. [49], with permission. Copyright the American Chemical Society 2009.).

REFERENCES

[1] Jetter, R.; Kunst, L.; Samuels, A.L. 2006. Composition of plant cuticular waxes. In: Riederer, M.; Müller, C., Eds. *Biology of the plant cuticle*. Blackwell Publ, Oxford, pp 182–215.

[2] Chiorcea-Paquim, A.-M.; Enache, T.A.; De Souza Gil, E.; Oliveira-Brett, A.M. 2020. Natural phenolic antioxidants electrochemistry: Towards a new food science methodology. Comprehensive *Reviews in Food Science and Food Safety*. 19: 1680–1726.

[3] Enache, T.A.; Oliveira-Brett, A.M. .2011. Phenol and parasubstituted phenols electrochemical oxidation pathways. *Journal of Electroanalytical Chemistry*. 655: 9–16.

[4] Zare, H.R.; Namazian, M.; Nasirizadeh, N. 2005. Electrochemical behavior of quercetin: Experimental and theoretical studies. *Journal of Electroanalytical Chemistry*. 584: 77–83.

[5] Sokolová, R.; Degano, I.; Ramešová, Š.; Bulíčková, J.; Hromadová, M.; Gál, M.; Fiedler, J.; Valášek, M. 2011. The oxidation mechanism of the antioxidant quercetin in nonaqueous media. *Electrochimica Acta*. 56: 7421–7427.
[6] Masek, A.; Chrzescijanska, E.; Zaborski, M. 2014. Electrooxidation of morin hydrate at a Pt electrode studied by cyclic voltammetry. *Food Chemistry*. 148: 18–23.
[7] Timbola, A.K.; de Souza, C.D.; Giacomelli, C.; Spinelli, A. 2006. Electrochemical oxidationof quercetin in hydro-alcoholic solution. *Journal of the Brazilian Chemical Society*. 17: 139–148.
[8] Ngamchuea, K.; Tharat, B.; Hirunsit, P.; Suthirakun, S. 2020. Electrochemical oxidation of resorcinol: mechanistic insights from experimental and computational studies. *RSC Advances*. 10: 28454–28462.
[9] Ferreira, M.; Varela, H.; Torresi, R.M.; Tremiliosi-Filho, G. 2006. Electrode passivation caused by polymerization of different phenolic compounds. *Electrochimica Acta*. 52: 434–442.
[10] Doménech-Carbó, A.; Doménech-Carbó, M.T.; Saurí-Peris, M.C. 2005. Electrochemical identification of flavonoid dyes in work of art samples by abrasive voltammetry at paraffin-impregnated graphite electrodes. *Talanta*. 66: 769–782.
[11] Doménech-Carbó, A.; Doménech-Carbó, M.T.; Calisti, M.; Maiolo, V. 2010. Sequential identification of organic dyes using the voltammetry of microparticles approach. *Talanta*. 81: 404–411.
[12] Hendrickson, H.P.; Kaufman, A.D.; Lunte, G.E. 1994. Electrochemistry of catechol-containing flavonoids. *Journal of Pharmaceutical and Biomedical Analysis*. 12: 325–334.
[13] Ghica, M.-E.; Oliveira-Brett, A.M. 2005. Electrochemical oxidation of Rutin. *Electroanalysis*. 17: 313–318.
[14] Janeiro, P.; Oliveira-Brett, A.M. 2007. Redox Behavior of Anthocyanins Present in *Vitis vinifera* L. *Electroanalysis*. 19: 1779–1786.
[15] Ramesova, S.; Sokolova, R.; Tarabek, J.; Degano, I. 2013. The oxidation of luteolin, the natural flavonoid dye. *Electrochimica Acta*. 110: 646–654.
[16] Ryan, P.; Hynes, M.J. 2008. The kinetics and mechanisms of the reactions of iron(III) with quercetin and morin. *Journal of Inorganic Biochemistry*. 102: 127–136.
[17] Gomes, A., Fernandes, E., Garcia, M.B.Q., Silva, A.M.S., Pinto, D.C.G.A., Santos, C.M.M., Lima, J.L.F.C. 2008. Cyclic voltammetric analysis of 2-styrylchromones: Relationship with the antioxidant activity. *Bioorganic & Medicinal Chemistry*. 16: 7939–7943.
[18] Teixeira, J.; Gaspar, A.; Garrido, E.M.; Garrido, J.; Borges, F. 2013. Hydroxycinnamic acid antioxidants: An electrochemical overview. *BioMedical Research Internacional*. 2013: artic. 251754.
[19] Doménech-Carbó, A.; Doménech-Carbó, M.T.; Saurí-Peris, M.C.; Gimeno-Adelantado, J.V.; Bosch-Reig, F. 2003. Electrochemical identification of anthraquinone-based dyes in solid microsamples by square wave voltammetry using graphite/polyester composite electrodes. *Analytical and Bioanalytical Chemistry*. 375: 1169–1175.
[20] Doménech-Carbó, A.; Doménech-Carbó, M.T.; Calisti, M.; Maiolo, V. 2010. Identification of naphtoquinonic and anthraquinonic dyes via sequential potential steps applied to the voltammetry of microparticles methodology. *Journal of Solid State Electrochemistry*. 14: 465–477.
[21] Doménech-Carbó, A.; Valle-Algarra, F.M.; Doméech-Carbó, M.T.; Osete-Cortina, L.; Domine, M. (2013) 'Maya Chemistry' of Organic-inorganic Hybrid Materials: Isomerization, Cyclicization and Redox Tuning of Organic Dyes Attached to Porous Silicates. *RSC Advances*. 3: 20099–21005.
[22] González, F.J. 1998. Cyclic voltammetry of two analogue K-group vitamin compounds in dimethylsulfoxide. *Electroanalysis*. 10:638–642.
[23] Dharmaraj, K.; Román-Silva, J.I.; Kahlert, H.; Lendeckel, U.; Scholz, F. 2020. The acid–base and redox properties of menaquinone MK-4, MK-7, and MK-9 (vitamin K_2) in DMPC monolayers on mercury. *European Biophysics Journal*. 49: 279–288.
[24] Ziyatdinova, G.; Ziganshina, E.; Budnikov, H. 2013. Electrochemical reduction and quantification of menadione in sodium dodecyl sulfate micellar media. *Journal of Solid State Electrochemistry*. 17: 2679–2685.
[25] Vanholme, R.; Demedts, B.; Morreel, K.; Ralph, J.; Boerjan, W. 2010. Lignin biosynthesis and structure. *Plant Physiology*. 153: 895–905.
[26] Cottyn, B.; Rivard, M.; Majira, A.; Beauhaire, J.; Allais, F.; Martens, T.; Baumberger, S.; Ducrot, P.-H. 2015. Comparative electrochemical study on monolignols and dimers relevant for the comprehension of the lignification process. *Phytochemistry Letters*. 13: 280–285.
[27] Doménech-Carbó, A.; Doménech-Carbó, M.T.; Ferragud-Adam, X.; Ortiz-Miranda, A.S.; Montoya, N.; Pasíes-Oviedo, T.; Pieiró-Ronda, M.A.; Vives-Ferrándiz, J.; Carrión, Y. 2017. Identification of vegetal species in wooden objects using *in situ* microextraction-assisted voltammetry of microparticles. *Analytical Methods*. 9: 2041–2048.

[28] Di Turo, F.; Mai, C.; Haba, A.; Doménech-Carbó, A. 2019. Discrimination of papers used in C&R by the means of the voltammetry of immobilized microparticles technique. *Analytical Methods*. 11: 4431–4439.
[29] Corduneanu, O.; Janeiro, P.; Oliveira-Brett, A.M. 2006. On the electrochemical oxidation of resveratrol. *Electroanalysis*. 18: 757–762.
[30] Manaia, M.A.N.; Diculescu, V.C.; Gil, E.S.; Oliveira-Brett, A.M. 2012. Guaicolic spices curcumin and capsaicin electrochemical oxidation behaviour at a glassy carbon electrode. *Journal of Electroanalytical Chemistry*.682: 83–89.
[31] Doménech-Carbó, A.; Doménech-Carbó, M.T.; Saurí-Peris, M.C.; Gimeno-Adelantado, J.V.; Bosch-Reig, F. 2005. Identification of curcuma and safflower dyes by voltammetry of microparticles using paraffin-impregnated graphite electrodes. *Microchimica Acta*. 152: 75–84.
[32] Lee, J.H.Q.; Lauw, S.J.L.; Webster, R.D. 2016. The Electrochemical Study of Vanillin in Acetonitrile. *Electrochimica Acta*. 211: 533–544.
[33] Deng, P.; Xu, Z.; Zeng, R.; Ding. C. 2015. Electrochemical behavior and voltammetric determination of vanillin based on an acetylene black paste electrode modified with graphene–polyvinylpyrrolidone composite film. *Food Chemistry*. 180: 156–163.
[34] Chen, L.; Chaisiwamongkhol, K.; Chen, Y.; Compton, R.G. 2019. Rapid Electrochemical Detection of Vanillin in Natural Vanilla. *Electroanalysis*. 31: 1067–1074.
[35] Rogers, R.N. 2005. Studies on the radiocarbon simple from the shroud of Turin. *Thermochimica Acta*. 425: 189–194.
[36] Doménech-Carbó, A.; Doménech-Carbó, M.T. 2023. *Electrochemistry for Cultural Heritage*. Monographs in Electrochemistry series, Scholz, F.; Peter, L., Eds. Spirnger, Berlin-Heidelberg.
[37] Pisoschi, A.M.; Pop, A.; Serban, A.I.; Fafaneate, C. 2014. Electrochemical methods for ascorbic acid determination. *Electrochimica Acta*. 121: 443–460.
[38] Osial, M.; Warczak, M.; Kulesza, P.J.; Kyrsiński, P.; Gniadek, M. 2020. Hybrid polyindole-gold nanobrush for electrochemical oxidation of ascorbic acid. *Journal of Electroanalytical Chemistry*. 877: artic. 114664.
[39] Rueda, M.; Aldaz, A.; Sanchez-Burgos, F. 1978. Oxidation of L-ascorbic acid on a gold electrode. *Electrochimica Acta*. 23: 419–424.
[40] Hu, I.; Kuwana, T. 1986. Oxidative mechanism of ascorbic acid at glassy carbon electrodes. *Analytical Chemistry*. 58: 3235–3239.
[41] Straub, O. 1987. in Pfander, H., Ed. *Key to Carotenoids*, 2nd ed., Birkhaüser, Basel.
[42] Liu, D.; Gao, Y.; Kispert, L.D. 2000. Electrochemical properties of natural carotenoids. *Journal of Electroanalytical Chemistry*. 488: 140–150.
[43] Grant, J.L.; Kramer, V.J.; Ding, R.; Kispert, L.D. 1988. Carotenoid Cation Radicals: Electrochemical, Optical, and EPR Study. *Journal of the American Chemical Society*. 110: 2151–2157.
[44] Čižmek, L.; Komorsky-Lovrić, Š. 2019. Study of Electrochemical Behaviour of Carotenoids in Aqueous Media. *Electroanalysis*. 31: 83–90.
[45] Hasharoni, K.; Levanon, H.; Tang, J.; Bowman, M.K.; Norris, J.R.; Gust, D.; Moore, T.A.; Moore, A.L. 1990. Singlet photochemistry in model photosynthesis: identification of charge separated intermediates by Fourier transform and CW-EPR spectroscopies. *Journal of the American Chemical Society*. 112: 6477–6481.
[46] Ion, A.; Partali, V.; Sliwka, H.-R.; Banica, F.G. 2002. Electrochemistry of a carotenoid self-assembled monolayer. *Electrochemistry Communications*. 4: 674–678.
[47] Tan, Y.S.; Urbancok, D.; Webster, R.D. 2014. Contrasting Voltammetric Behavior of Different Forms of Vitamin A in Aprotic Organic Solvents. *Journal of Physical Chemistry B*. 118: 8591–8600.
[48] Chan, Y.Y.; Yue, Y.; Webster, R.D. 2014. Voltammetric Studies on Vitamins D_2 and D_3 in Organic Solvents. *Electrochimica Acta*. 138: 400–409.
[49] Yao, W.W.; Peng, H.M.; Webster, R.D. 2009. Electrochemistry of α-Tocopherol (Vitamin E) and α-Tocopherol Quinone Films Deposited on Electrode Surfaces in the Presence and Absence of Lipid Multilayers. *Journal of Physical Chemistry C*. 113: 21805–21814.
[50] Giacomelli, C.; Giacomelli, F.C.; Ortigara Alves, L.; Timbola, A.K.; Spinelli, A. 2004. Electrochemistry of Vitamin E Hydro-Alcoholic Solutions. *Journal of the Brazilian Chemical Society*. 15: 748–755.
[51] Scholz, F.; Schröder, U.; Gulabowski, R.; Doménech-Carbó, A. 2014. *Electrochemistry of Immobilized Particles and Droplets*, 2nd ed. Springer, Berlin-Heidelberg.

ADDITIONAL LITERATURE

Phenolic Compounds

Abdel-Hamid, R.; Newair, E.F. 2011. Electrochemical behaviour of antioxidants: I. Mechanistic study on electrochemical oxidation of gallic acid in aqueous solutions at glassy carbon electrode. *Journal of Electroanalytical Chemistry*. 657: 107–112.

Cuartero, M.; Ortuno, J.A.; Truchado, P.; Garcia, M.S.; Tomas-Barberan, F.A.; Albero, M.I. 2011. Voltammetric behaviour and square-wave voltammetric determination of the potent antioxidant and anticarcinogenic agent ellagic acid in foodstuffs. *Food Chemistry*. 128: 549–554.

Fernandes, I.P.G.; Oliveira, S.C.B.; Ghalkhani, M.; Shahrokhian, S.; Oliveira-Brett, A.M. 2012. Electrochemical oxidation mechanisms of the antioxidants daidzein and 7-hydroxy-4-chromone. *Electroanalysis*. 24: 618–626.

Gil, E.S.; Cout, R.O.; Couto, R.O. 2013. Flavonoid electrochemistry: A review on the electroanalytical applications. *Revista Brasileira de Farmacognosia*. 23: 542–558.

Janeiro, P.; Oliveira-Brett, A.M. 2004. Catechin electrochemical oxidation mechanisms. *Analytical Chimica Acta*. 518: 109–115.

Kilmartin, P.A.; Hsu, C.F. 2003. Characterisation of polyphenols in green, oolong, and black teas, and in coffee, using cyclic voltammetry. *Food Chemistry*. 82: 501–512.

Novak, I.; Seruga, M.; Komorsky-Lovrić, Š. 2009. Electrochemical characterization ofepigallocatechin gallate using square wave voltammetry. *Electroanalysis*. 21: 1019–1025.

Novak, I.; Seruga, M.; Komorsky-Lovrić, Š. 2009. Square-wave and cyclic voltammetry of epicatechin gallate on glassy carbon electrode. *Journal of Electroanalytical Chemistry*. 631: 71–75.

Oliveira-Brett, A.M.; Ghica, M.-E. 2003. Electrochemical oxidation of quercetin. *Electroanalysis*. 15: 1745–1750.

Oliveira-Neto, J.R.; Rezende, S.G.; deFatima Reis, C.; Benjamin, S.R.; Rocha, M.L.; Gil, E.S. 2016. Electrochemical behavior and determination of major phenolic antioxidants in selected coffee samples. *Food Chemistry*. 190: 506–512.

Popa, O.M.; Diculescu, V.C. 2013. Electrochemical behaviour of isoflavones genistein and biochanin a at a glassy carbon electrode. *Electroanalysis*. 25: 108–115.

Yang, L.; Yang, T.; Li, G.; Ma, J. X.; Yu, Y. 2022. Electrochemical Studies of Polyphenols, Anthocyanins, and Flavonoids Extracted from Blueberry Fruit. *International Journal of Electrochemical Science*. 17: artic. 22038.

Zhou, A.; Kikandi, S.; Sadik, O.A. 2007. Electrochemical degradation of quercetin: Isolation and structural elucidation of the degradation products. *Electrochemistry Communications*. 9: 2246–2255.

Quinones

Alonso, L.; Palmero, S.; Muñoz, E.; Sanllorente, S.; García-García, M.A. 2000. Electrochemical behavior of menadione on glassy carbon rotating disk electrode (RDE). *Electroanalysis*. 12: 757–762.

Cannes, C.; Kanoufi, F.; Bard, A.J. 2002. Cyclic voltammetric and scanning electrochemical microscopic study of menadione permeability through a self-assembled monolayer on a gold electrode. *Langmuir*. 18: 8134–8141.

Marchal, D.; Boireau, W.; Laval, J.M.; Moiroux, J.; Bourdillon, C. 1997. An electrochemical approach of the redox behavior of water insoluble ubiquinones or plastoquinones incorporated in supported phospholipids layers. *Biophysical Journal*. 72: 2679–2687.

Prince, R.C.; Dutton, P.L.; Bruce, J.M. 1983. Electrochemistry of ubiquinones menaquinones and plastoquinones I aprotic solvents. *FEBS Letters*. 160: 273–276.

Revenga, J.; Rodríguez, F.; Tijero, J. 1994. Study of the redox behavior of anthraquinone in aqueous medium. *Journal of the Electrochemical Society*. 141: 330–333.

Ascorbic Acid, Carotenoids, Tocopherols, Calciferols, and Retinol

Böhmdorfer, S.; Rosenau, T. 2021. Oxidation with a "Stopover" – Stable Zwitterions as Intermediates in the Oxidation of **α**-Tocopherol (Vitamin E) Model Compounds to their Corresponding *ortho*-Quinone Methides. *ChemistryOpen*. 10: 421–429.

Bradshaw, M.P.; Prenzler, P.D.; Scollary, G.R. 2002. Square-wave voltammetric determination of hydrogen peroxide generated from the oxidation of ascorbic acid in a model wine base. *Electroanalysis*. 14: 546–550.

Ferreira, V.C.; Solla-Gullón, J.; Aldaz, A.; Silva, F.; Abrantes, L.M. 2011. Progress in the understanding surface structure and surfactant influence on the electrocalatylic activity of gold nanoparticles. *Electrochimica Acta*. 56: 9568–9574.

Heinze, J.; Tschuncky, P.; Smie, A. 1998. The oligomeric approach - the electrochemistry of conducting polymers in the light of recent research. *Journal of Solid State Electrochemistry*. 2: 102–109.

Kalakodimi, R.P.; Nookala, M. 2012. Electrooxidation of ascorbic acid on a Polyanilinedeposited nickel electrode: surface modification of a non-platinum metal for an electrooxidative analysis. *Analytical Chemistry*. 74: 5531–5537.

Peng, H.M.; Webster, R.D. 2008. Investigation into Phenoxonium Cations Produced during the Electrochemical Oxidation of Chroman-6-ol and Dihydrobenzofuran-5-ol Substituted Compounds. *Journal of Organic Chemistry*. 73: 2169–2175.

Webster, R.D. 2022. Electrochemical and Spectroscopic Characterization of Oxidized Intermediate Forms of Vitamin E. *Molecules*. 27: artic. 6194.

Ziyatdinova, G.; Morozov, M.; Budnikov, H. 2012. MWNT-modified electrodes for voltammetric determination of lipophilic vitamins. *Journal of Solid State Electrochemistry*. 16: 2441–2447.

Solid-State Electrochemistrry

Bosch-Reig, F.; Doménech-Carbó, A.; Doménech-Carbó, M.T.; Gimeno-Adelantado, J.V. 2007. H-Point Standard Addition Method Applied to Voltammetry of Microparticles. Quantitation of Dyes in Pictorial Samples. *Electroanalysis*. 19: 1575–1584.

Chen, A.; Rogers, E.I.; Compton, R.G. 2010. Abrasive stripping voltammetric studies of lignin and lignin model compounds. *Electroanalysis*. 22: 1037–1044.

Čižmek, L.; Komorsky-Lovrić, Š. 2019. Study of Electrochemical Behaviour of Carotenoids in Aqueous Media. *Electroanalysis*. 31: 83–90.

Doménech-Carbó, A.; Doménech-Carbó, M.T. 2006. Chronoamperometric study of proton transfer/electron transfer in solid state electrochemistry of organic dyes. *Journal of Solid State Electrochemistry*. 10: 949–958.

Grygar, T.; Kučková, Š.; Hradil, D.; Hradilová, D. 2003. Electrochemical analysis of natural solid organic dyes and pigments. *Journal of Solid State Electrochemistry*. 7: 706–713.

Janeiro, P.; Oliverira-Brett, A.M. 2005. Solid State Electrochemical Oxidation Mechanisms of Morin in Aqueous Media. *Electroanalysis*. 17: 733–738.

Martini, N.; de Carvalho, L.M.; Blasco-Blasco, A.; Doménech-Carbó, A. 2015. Screening and authentication of herbal formulations based on microextraction-assisted voltammetry of microparticles. *Analytical Methods*. 7: 5740–5747.

6 Electrochemistry of Vegetal Components II

6.1 INTRODUCTION

Nitrogen plays an essential role in plants where it participates in a wide variety of compounds. This element is incorporated from soils where nitrogen is present as inorganic forms, such as nitrate and ammonium salts, and as organic compounds, such as urea, amino acids, and short peptides [1, 2]. From a physiological perspective, nitrogen uptake is directly related to plasma membrane activity involving transport, receptor, and transceptor proteins in the root cells [3]. The incorporation of nitrogen from soil is, however, a complex process involving a variety of factors [4].

Because of a lack of space, the rich electrochemistry of inorganic nitrogen will not be treated here. In the following, we will study the electrochemistry of selected series of nitrogen compounds in plants having in common the presence of at least one nitrogen atom in an amine, pyridine, pyrrole, group. In this chapter, we will treat the electrochemistry of several groups, including amines, several vitamins, and alkaloids, which will be grouped according to their chemical structures. The electrochemistry of other essential nitrogen-containing components of plants, namely, chlorophyll, nicotinamide-adenine dinucleotides, and deoxirribonucleic acids will be also treated in this chapter.

The biogenic amines are related structurally and metabolically to the alkaloids. Alkaloids are produced by plants, where they play a frequently unknown biochemical role. Other compounds involved in metabolic cycles include terpenoids, steroids, and a series of nitrogen-based compounds. The latter have at least one heterocyclic N atom, and in general, they do not have non-heterocyclic N-containing groups, but the separation between amines and alkaloids is frequently quite arbitrary. Their beneficial or negative effects on human health have motivated their extensive study and application and, hence, the analysis of their electrochemical behavior. The chemical structures of nitrogen-based alkaloids are extremely variable. Electrochemical studies interest for analytical purposes but also for their capability to provide information on biosynthetic pathways and pharmacological applications.

6.2 AMINES

Amines are endogenous constituents of plants, where they are involved in different physiological processes, such as fruit development and senescence, and their metabolism appears to be coordinated with the cellular cycle [5]. Biogenic amines are synthesized from amino acids by decarboxylation or through aldehyde transamination and are mainly catabolized by oxidative deamination via diamine- and polyamine oxidases. The amines present in plant tissues can be divided into aliphatic monoamines, aliphatic di- and polyamines, and aromatic amines. The electrochemistry of several amines – namely, putrescine (diamine), spermidine, and spermine (polyamines) – has received considerable attention as acting as markers for the deterioration of food products [6].

The oxidation of amines in organic solvents has been extensively studied. In general, the process starts with a one-electron transfer yielding a radical cation. Depending on the amine structure, an enormous variety of subsequent processes can occur. Among others, they involve attack by nucleophiles and polymerization reactions [7].

One representative example of amine electrochemistry is provided by atropine, a tropane alkaloid. Figure 6.1 shows a series of cyclic and differential pulse voltammograms recorded at a boron-doped diamond electrode for a solution of atropine in a strongly acidic medium [8]. In this case, an apparently irreversible oxidation peak is recorded in strongly acidic aqueous solutions. The

DOI: 10.1201/9781032697826-6

oxidation process involves the formation of an iminium intermediate followed by a reaction with water with the concomitant loss of two protons and two electrons and the release of HCHO [8]. The same pathway has been described for nicotine oxidation, schematized in Figure 6.2.

Data in Figure 6.1 compares the voltammetry of the $Fe(CN)_6^{3-}/Fe(CN)_6^{4-}$ couple, a well-known reversible system with the voltammetry of atropine in an acidic medium. Here a boron-doped diamond (BDD) working electrode was used after the application of different potential inputs. One can see that the voltammetric response of the $Fe(CN)_6^{3-}/Fe(CN)_6^{4-}$ couple remains essentially insensitive to the electrode activation. However, the position and intensity of the voltammetric peaks for atropine oxidation are slightly dependent on the application of different electrochemical inputs to activate the working electrode. Since, as previously noted, most electrode processes are quite sensitive to the working electrode, the mechanical (polishing, ultrasonication, etc.), chemical, and electrochemical activation of solid electrodes can influence significantly the voltammetric response. In the case of carbon electrodes, electrochemical treatments can produce different oxygen functionalities, resulting in the modification of the electrode behavior [9].

Two other families of alkaloids, pyrrolidines and pyrrolizidines, show amine-centered electrochemistry. The structural core of these compounds includes the five-membered pyrrolidine ring eventually accompanied by a pyridine unit. This group includes, as a representative example, nicotine (3-(1-methyl-2-pyrrolidinyl) pyridine). In aqueous media at biological pHs, nicotine displays a unique, apparently irreversible oxidation wave near 1.0 V vs. SCE. Although this process could involve the pyridine ring, it has been proposed that the oxidation occurs at the pyrrolidine ring with the release of CH_3OH [10, 11], as is also schematized in Figure 6.2.

Figure 6.3 shows the square wave voltammetric response of nicotine at pencil graphite electrodes at three different pHs [11]. An apparently irreversible oxidation peak whose position and intensity are pH-dependent, as illustrated in Figure 6.4.

For reversible, proton-assisted electrochemical processes linear variations of the formal potential (or peak potentials) on the pH are expected. The slope of these representations at 298 K should be $59q/n$ mV decade^{-1} where n is the number of electrons and q is the number of protons involved in the electrochemical process. In general, this relationship applies to the rate-determining step under conditions of reversibility ($59q/\alpha n$ for non-reversible processes, α being the electron transfer coefficient whose meaning can be revised in [12]). This implies that, often, proton-assisted electrochemical processes can display pH-independent peak potentials.

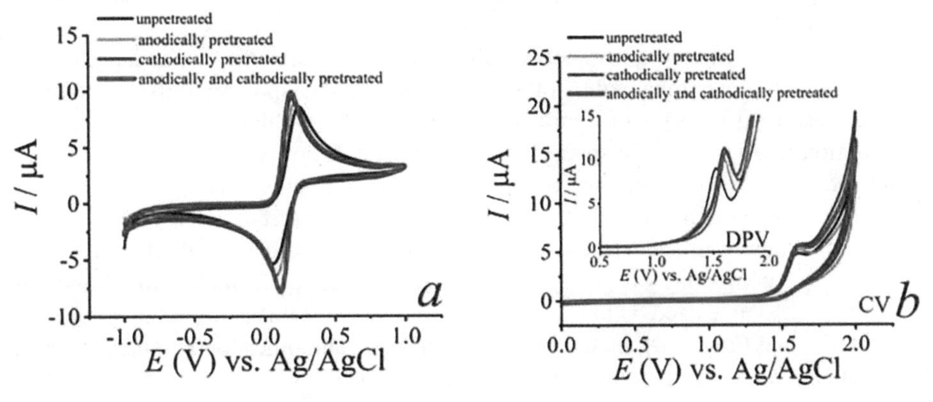

FIGURE 6.1 Cyclic voltammograms and differential pulse voltammograms (inset) at the BDD electrode of (a) 2.5 mM $Fe(CN)_6^{3-}/Fe(CN)_6^{4-}$ in 0.1 M KCl and (b) 0.1 mM atropine in 2 M $HClO_4$ after the application of different electrochemical inputs to activate the working electrode at the scan rate of 50 mV s^{-1}. (Adapted from ref. [8], with permission. Copyright Elsevier 2022.).

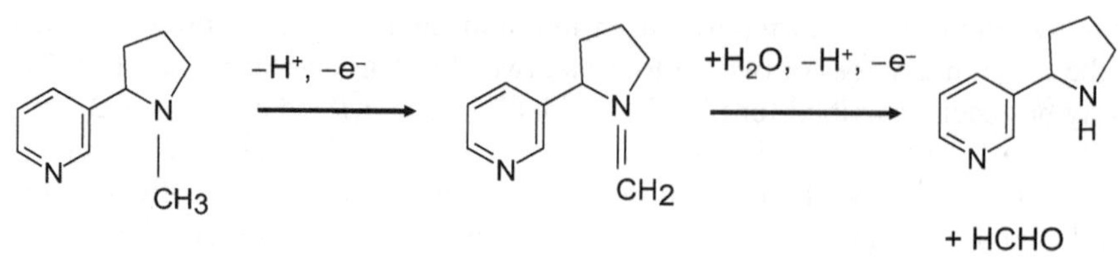

FIGURE 6.2 Electrochemical oxidation of nicotine.

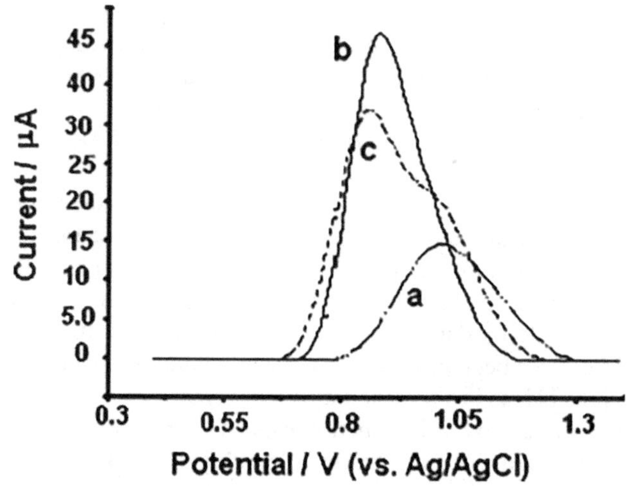

FIGURE 6.3 Square wave voltammograms at graphite pencil electrode of 0.5 mM nicotine solutions in phosphate buffer at pH values of 6 (a), 7 (b), and 8 (c). Step potential 8 mV; pulse amplitude 25 mV; frequency 25 Hz. (Reproduced from ref. [11], with permission. Copyright Elsevier 2009.).

6.3 INDOLIC COMPOUNDS

Most plant tissues contain compounds with indole structural units. The biochemical synthesis of most components originates from indole-3-acetic acid (IAA) that results in conjugated forms via amide linkage to amino acids, as well as esters or acyl alkyl acetal, with simple sugars or cyclitols. Among other indole derivatives, tryptophan, serotonin, and melatonin are important biochemically active molecules and several alkaloids.

The electrochemistry of indole in aqueous solution is characterized by an apparently irreversible oxidation peak which is accompanied by other secondary signals upon repetitive voltammetry [13]. This has been described in terms of a multistep pathway where indole experiences an initial one-electron oxidation to give a radical species. This process is followed/accompanied by the nucleophilic attack of water to yield reactive oxyindole and/or quinone imine species, which undergo subsequent electrochemical processes centered in the phenolic motif.

The essential trends of this electrochemistry are common to other indole derivatives with substituents in C3. One compound of widely studied electrochemistry is melatonin. Although mainly known as a hormone synthesized by the pineal parenchymal cells, this compound has been identified in plants, bacteria, and fungi [14], including *Vitis vinifera* L. and other food plants [15]. In an aqueous solution, melatonin displays a main oxidation wave at ca. 0.8 V vs. Ag/AgCl at neutral pH. As can be seen in Figure 6.5 [16], this is accompanied by two other oxidation waves in the initial anodic scan. In the subsequent cathodic scan, a reduction peak appears coupled with an additional anodic peak in the second forward scan.

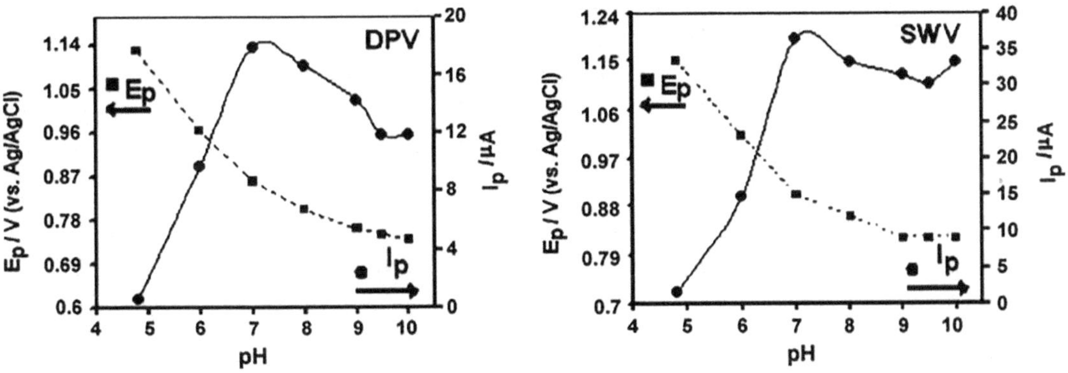

FIGURE 6.4 Variation of the peak potential and peak current on the pH for nicotine solutions in conditions such as in Figure 6.3. (Reproduced from ref. [11], with permission. Copyright Elsevier 2009.).

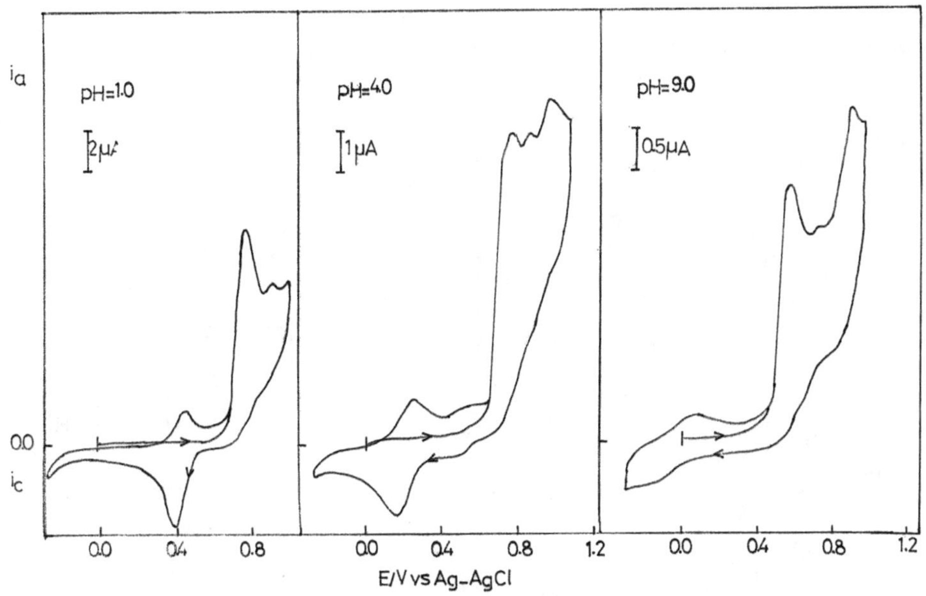

FIGURE 6.5 Cyclic voltammetry at carbon paste electrode of melatonin in aqueous solution (concentration 4×10^{-5} M) at different pHs. Potential scan rate 50 mV s^{-1}. (Reproduced from ref. [16], with permission. Copyright Elsevier 1998.).

In aprotic media, two apparently irreversible oxidation waves are recorded (Figure 6.6). This voltammetry can be interpreted in terms of an initial one-electron oxidation forming a radical cation, followed by the coupling of this radical with melatonin to yield a dimeric radical cation that is oxidized at more anodic potential. Schemes summarizing the electrochemistry of indolic compounds in protic and aprotic solvents are presented in Figures 6.7 and 6.8, respectively.

The biosynthesis of melatonin starts from tryptophan. The electrochemistry of this compound and serotonin, also involved in melatonin synthesis [17], has been studied in detail [18, 19]. The electrochemistry of indolic compounds at electrodes modified with redox mediators [20] and the solid-state electrochemistry after attachment to silica hosts of different porosity [21] have also been studied. Figure 6.9 permits the comparison of the voltammetric response of microparticulate films of different indolic compounds deposited onto glassy carbon electrode in contact with the aqueous

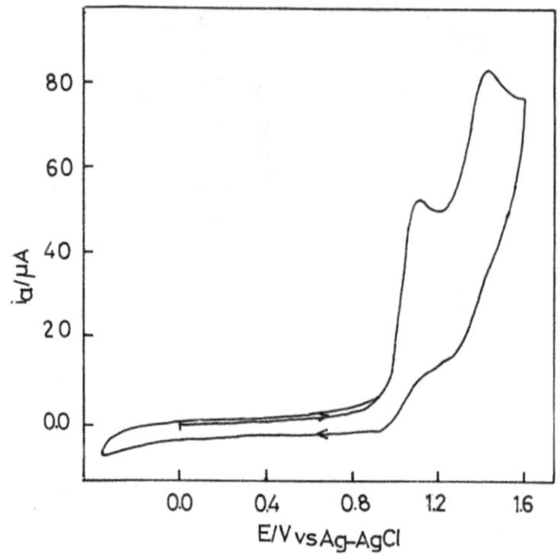

FIGURE 6.6 Cyclic voltammetry of melatonin in 0.05 M Bu_4NPF_6/MeCN solution. Potential scan rate 50 mV s^{-1}. (Reproduced from ref. [16], with permission. Copyright Elsevier 1998.).

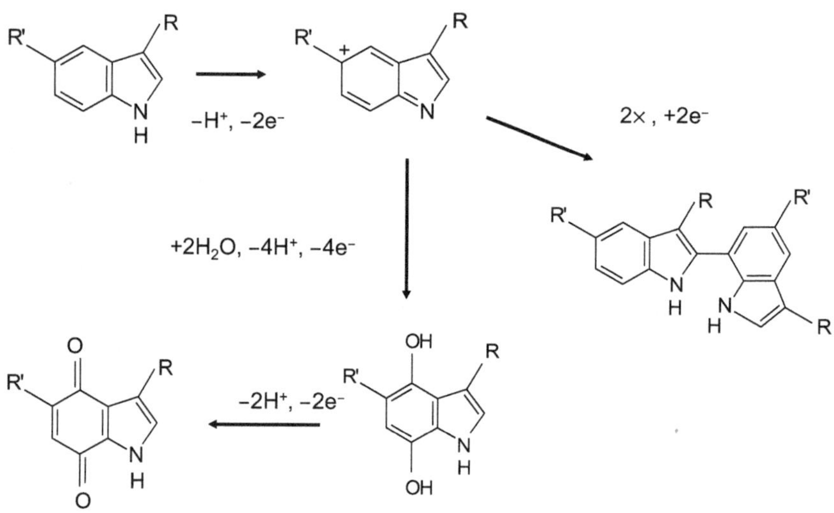

FIGURE 6.7 Simplified scheme for the electrochemical oxidation of C3-substituted indolic compounds in protic solvents.

acetate buffer [22]. In all cases, a main apparently irreversible oxidation peak appears that can be attributed, tentatively, to the ultimate generation of a quinone condensing the reaction scheme in Figure 6.8.

The application of this solid-state methodology has been used to test changes in mutant strains of *Saccharomyces* with deleted genes of the tryptophan metabolism, revealing significant differences in their voltammetric profiles [22].

Among other compounds directly related to indolic families, indigoids have received attention as a result of their use as pigments. Indigotin (or indigo) and indirubin are the more representative components of this group. They are synthesized in plants from 3-hydroxyl glucopyranosides (indican and isatan), which in turn can be seen as L-tryptophan derivatives [23]. Indole results from the action of tryptophanase on L-tryptophan, and the subsequent action of oxygenase yields indoxyl. Then the oxidase-assisted dimerization provides the indigoids.

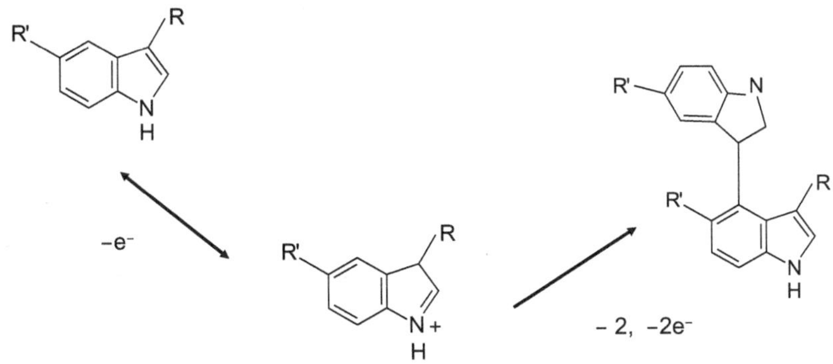

FIGURE 6.8 Simplified scheme for the electrochemical oxidation of C3-substituted indolic compounds in aprotic solvents.

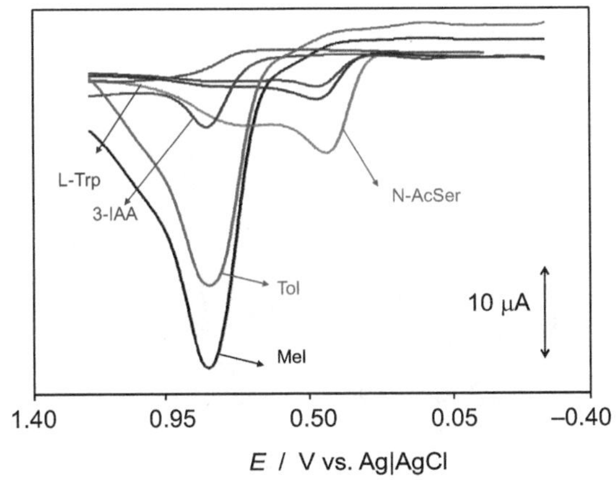

FIGURE 6.9 Square wave voltammograms of films of melatonin, serotonin, 5-hydroxitryptamin, N-acetylserotonin, tryptophol, 3-indoleacetic acid, and L-tryptophan on GCE in contact with 0.25 M HAc/NaAc aqueous buffer, pH 4.75. Potential scan initiated at –0.25 V in the positive direction, potential step increment 4 mV, square wave amplitude 25 mV, frequency 5 Hz. (Adapted from ref. [22], with permission. Copyright Springer 2017. Note that the axes are opposite to the IUPAC convention.).

Indigotin displays a well-defined electrochemistry in aprotic solvents, but it is notably influenced by traces of proton-donor species. Figure 6.10 shows the cyclic voltammetry of this compound in a dimethylsulfoxide (DMSO) solution [24]. The voltammograms show two successive one-electron reduction steps at midpeak potentials of −0.73 and −1.26 V vs. SCE. Both processes exhibit the essential characteristics of reversible, diffusion-controlled electrochemical processes and can be described in terms of the generation of an anion radical and a dianion.

In the positive-going potential scan, a two-electron oxidation process appears at 0.64 V, resulting in a more complex series of subsequent reduction signals. This process can be interpreted as the two-proton, two-electron oxidation to dehydroindigo [24]. The voltammetry of indigo in the presence of water traces is complex and offers the possibility of isomerization processes [25]. Given the extremely low solubility of indigo in water, most research has been developed using solid-state techniques, in particular concerning the so-called Maya blue problem [26]. Figure 6.11 shows the square wave voltammogram recorded from a microparticulate deposit of an indigo specimen prepared from leaves of *Indigofera tinctoria* mimicking traditional Mayan procedures [27]. The solid particles were attached to a GCE and put in contact with an aqueous acetate buffer. The voltammogram shows two main indigo-centered peaks (I and II) accompanied by a series of minor signals

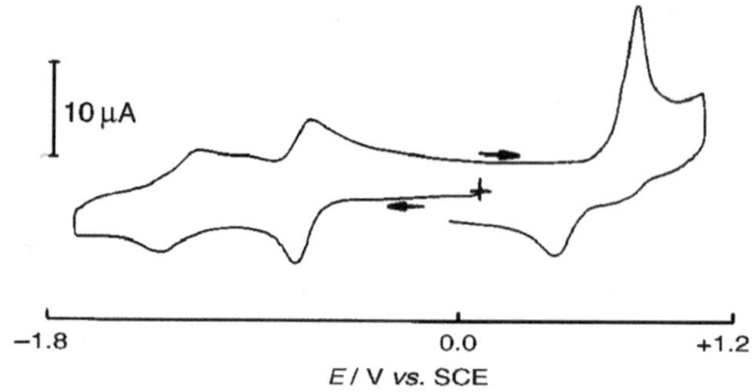

FIGURE 6.10 Cyclic voltammogram at GCE of 0.5 mM indigo solution in 0.10 M Bu_4NPF_6/DMSO. Potential scan initiated at 0.0 V vs. SCE in the negative direction. Potential scan rate 200 mV s^{-1}. (Adapted from ref. [24], with permission. Copyright the Royal Society of Chemistry 1997.).

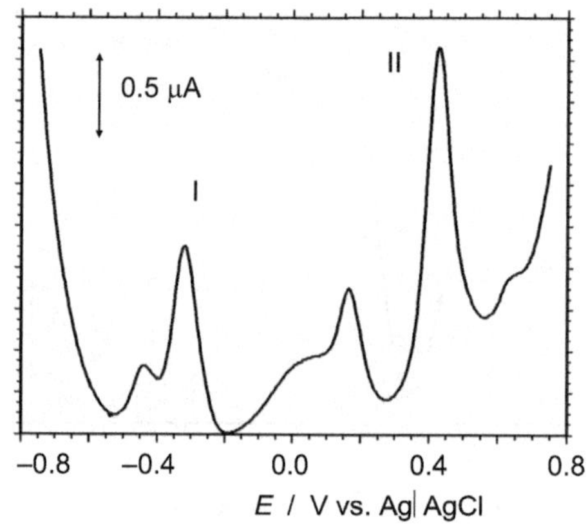

FIGURE 6.11 Square wave voltammogram of a microparticulate deposit of "natural" indigo attached to GCE in contact with 0.25 M HAc/NaAc aqueous solution at pH 4.75. The solid was prepared from Indigofera tinctoria leaves, mimicking traditional Mayan procedures described in [27]. Potential scan initiated at −0.75 V in the positive direction; potential step increment 4 mV; square wave amplitude 25 mV; frequency 5 Hz.

attributable to other indigoids. The indigo signals correspond to the reduction of indigo to leucoindigo (process I) and the oxidation of indigo to dehydroindigo (process II), both occurring as solid-state processes. Here, the issue of electrons is necessarily coupled, by reasons of charge conservation, with the release of protons from the solid lattice to the electrolyte solution. Similarly, the entrance of electrons is coupled with the ingress of protons into the solid. This electrochemistry can be expanded to include reactions with oxygen and other components such as isatin involved in biochemical cycles. A scheme of such processes appears in Figure 6.12.

6.4 PYRIDINES AND QUINOLINES

The structural core of these compounds consists of pyridine, quinoline, or isoquinoline units. In the first group are included alkaloids (nicotine) and nicotinic acid (pyridine 3-carboxylic acid, vitamin B_3) as representative examples. Quinoline and isoquinoline alkaloids include compounds such as quinine and papaverine, respectively. Quinine exhibits a complex electrochemistry, characterized by multiple reduction processes. In buffered aqueous solutions, two reduction waves appear at mercury electrodes, corresponding to one-electron, one-proton processes where protonation occurs on the nitrogen of the quinoline moiety [28, 29], but these processes are obscured by adsorption. At BDD

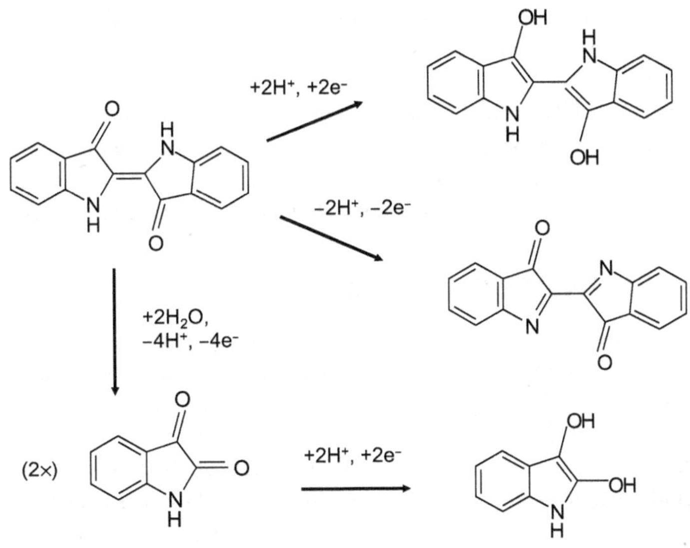

FIGURE 6.12 Scheme for the solid-state electrochemistry of indigo in contact with aqueous electrolyte solutions.

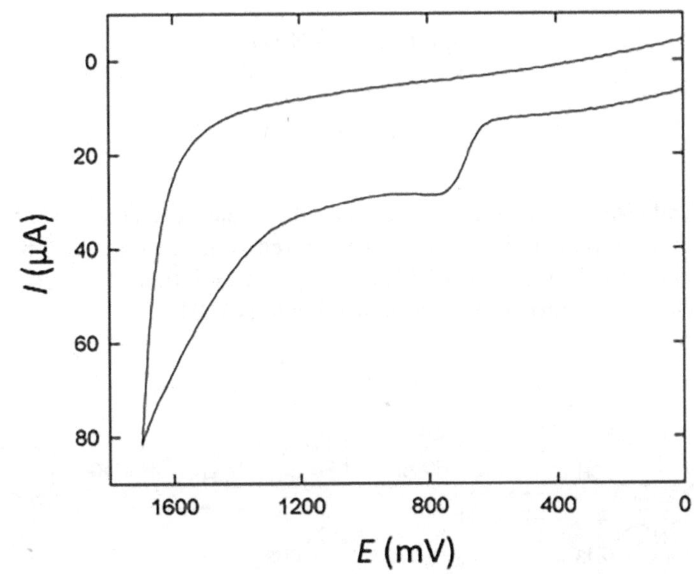

FIGURE 6.13 Cyclic voltammogram at glassy carbon electrode of a 1 mM solution of pyridoxine in phosphate buffer 0.1 M at pH = 7. Potential scan rate 100 mV s^{-1}. Note that the axes are opposite to the IUPAC convention. (Reproduced from ref. [31], with permission. Copyright Elsevier 2020.).

electrodes, independent oxidation and reduction processes appear [30], but no conclusive mechanistic considerations have been reported.

In the case of pyridoxine (one of the components of the vitamin B_6 group), another pyridine-based compound containing phenol units, different electrochemical pathways has been described. Figure 6.13 shows a representative example of cyclic voltammetry of pyridoxine at a glassy carbon electrode at neutral pH [31]. The voltammetric response is dependent on the pH, and the material and state of the working electrode being adsorption processes on gold is particularly important [32]. Of course, the electrochemistry in aprotic solvents exhibits different features [33]. Figure 6.14 depicts proposed schemes for the oxidation of pyridoxine in neutral and alkaline aqueous solutions [31, 34].

There is an enormous variety of nitrogenated compounds in plants, including polycyclic systems. Among other individual examples are riboflavin (vitamin B_2), folic acid (vitamin B_9), and biotin (vitamin B_7), whose electrochemistry has been widely studied [35]. Among other families of alkaloids, benzophenanthridines [36] and naphtyridones have been electrochemically characterized [37].

6.5 HYDROGEN TRANSFER COENZYMES

These important components of living organisms can be seen as having pyridine plus adenine (nicotinamide-adenine dinucleotide, NAD^+, and nicotinamide-adenine dinucleotide phosphate, $NADPD^+$), or flavin (riboflavin 5'-phosphate or flavin mononucleotide, FMN) structural units. These compounds regulate plant cellular redox homeostasis and signaling via the generation and scavenging of reactive oxygen species, as well as participate in other functions such as transcriptional regulation and microtubule alterations, among others [38, 39]. Although structurally similar, they are members of two independent redox systems where NAD^+ acts preferentially as an oxidizing agent for the removal of hydrogen from substrates. In turn, $NADP^+$ is employed in its reduced form, NADPH, as the reducing agent for biosynthetic purposes. The electrochemistry of NAD^+ and $NADP^+$ has been widely studied. The electrochemical reduction of these compounds occurs in non-aqueous media as a one-electron process followed by dimerization [40, 41].

$$NAD^+ + e^- \rightarrow NAD\bullet \tag{6.1}$$

$$2NAD\bullet \rightarrow (NAD)_2 \tag{6.2}$$

Notice that NAD^+ and $NADP^+$ are formulated as cationic species, even though at biological pHs, the net charge of the entire molecule is negative. The aforementioned processes involve the nicotinamide ring as schematized in Figure 6.15 [42]. In aqueous solution, the reduction of NAD^+ and $NADP^+$ involves a second electron transfer step leading to NADH.

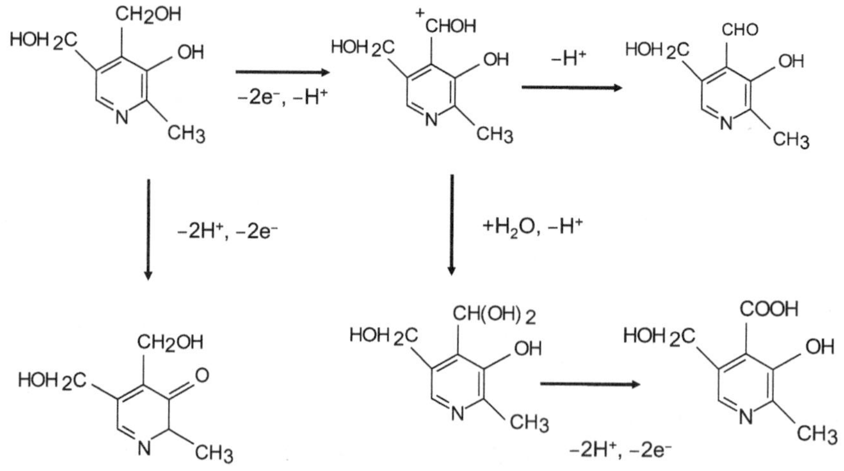

FIGURE 6.14 Electrochemical oxidation processes proposed for pyridoxine in aqueous media.

$$NAD\bullet + H^{+} + e^{-} \rightarrow NADH \tag{6.3}$$

The electrochemical pathway is complicated by adsorption processes and the acid-assisted disproportionation of the intermediate species into $NADP^+$ and NADH. Notice that the dimeric species are not enzymatically active.

Figure 6.16 shows the cyclic voltammogram at GCE of a $NADP^+$ solution in a neutral medium [43]. The cathodic wave A corresponds to the reduction process followed by dimerization described by Eqs. (6.1) and (6.2), whereas the shoulder B can be attributed to the second reductive step represented by Eq. (6.3). The process C can be assigned to the oxidation of the dimmer to $NADP^+$. Finally, the process D to the NADPH oxidation is mediated by a layer of $NADP^+$ adsorbed onto the electrode surface, but complications are associated with disproportionation/regeneration pathways [44].

Figure 6.16 also reflects the modulation of the electrochemical response due to the coordination by macrocyclic receptors. In this case, the macrocyclic polyamine 1,4,7,10,13,16,19-heptaazacyclohenicosane ([21]aneN_7) produces the slight positive shift of the reduction signals A and B, and the negative shift of the oxidation signal D while the dimmer oxidation peak C vanishes. These features can be considered as consistent with molecular dynamics calculations showing that NAD^+ and $NADP^+$ conformations are different in their complexed and free forms and that there are differences in their respective arrangement in the complexes (see Figure 6.17). The electrostatic interaction between phosphate groups and positively charged nitrogens of the macrocyclic receptor plays an important role in the stabilization of the complexes.

The aforementioned electrochemistry becomes significantly modified when the solid components (sodium salts) are embedded into carbon paste electrodes. As illustrated in Figure 6.18 [43], dimer formation and adsorption complications are minimized. Apparently, the 'direct' NAD^+/NADH interconversion (both reductive and oxidative) is favored. However, it is unclear if a 'true' solid-state electrochemistry is involved.

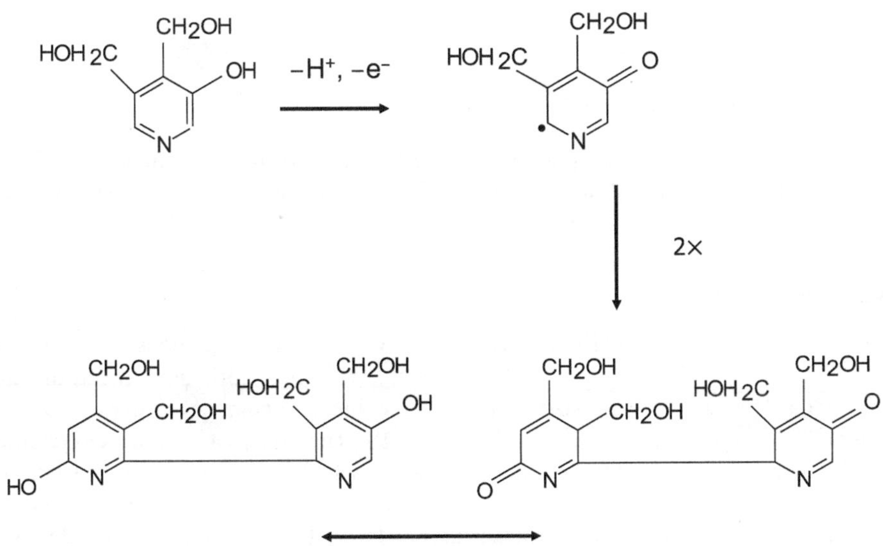

FIGURE 6.15 Scheme for the main processes involved in the NAD^+/NADH electrochemistry. R designates the chain of phosphate groups.

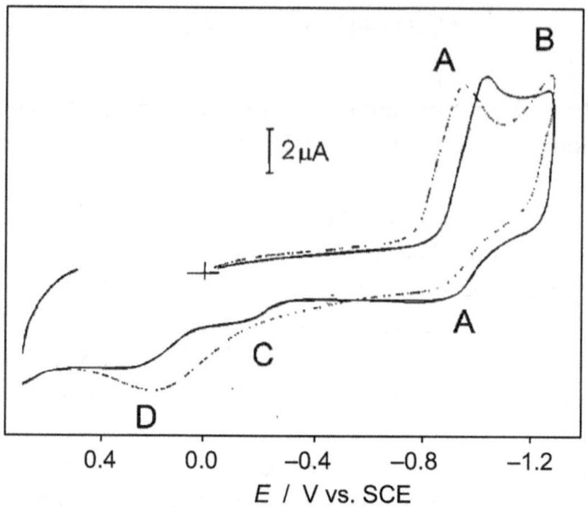

FIGURE 6.16 Cyclic voltammograms at GCE of 1.95 mM $NADP^+$ (continuous line) and 1.95 mM $NADP^+$ plus 3.80 mM [21]AneN_7 (dotted line) in 0.15 M $NaClO_4$ aqueous solution: Potential scan rate 100 mV s^{-1}. Note that the axes are opposite to the IUPAC convention. (Adapted from ref. [43], with permission. Copyright the Royal Society of Chemistry 1999.).

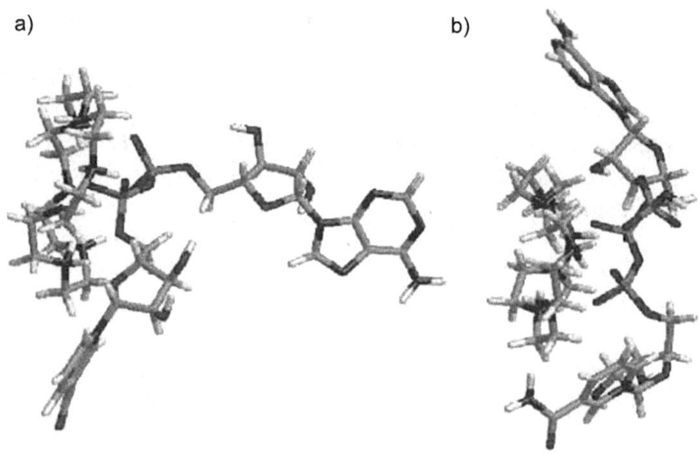

FIGURE 6.17 Lowest energy structures of the complexes between (a) NAD^+ and (b) $NADP^+$ with [21]aneN_7 from 500 ps dynamic simulations. (Adapted from ref. [43], with permission. Copyright the Royal Society of Chemistry 1999.).

6.6 NUCLEIC ACIDS

Deoxyribonucleic acid (DNA) and ribonucleic acid (RNA) are polynucleotides formally constituted by a base, a sugar (either d-ribose or D-2-deoxyribose), and phosphoric acid linked through phosphodiester bonds. The bases can be purines (purine, adenine, and guanine) or pyrimidines (cytosine and thymine (DNA) or uracil (RNA); see Figure 6.19). The study of their electrochemistry has increased exponentially in the last decades associated with DNA sequencing and labeling [45]. From the earlier studies on mercury electrodes, a variety of working electrodes have been used to characterize the electrochemistry of free bases and different DNA (single-strand, ssDNA, double-strand, dsDNA, G-quadruplex, etc.) types. This electrochemistry is dominated by the oxidation of adenine and guanine units and the reduction of adenine, guanine, and cytosine.

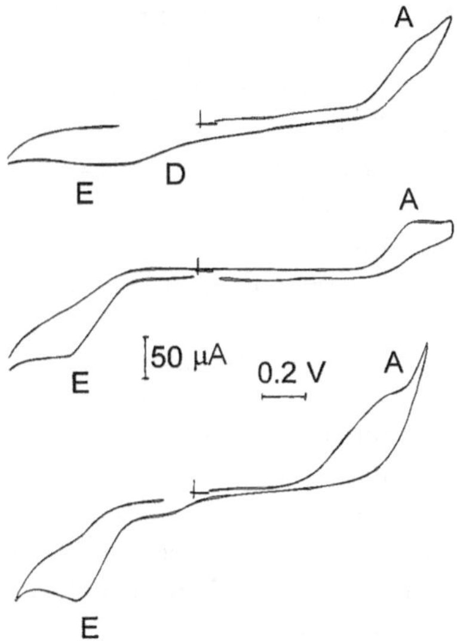

FIGURE 6.18 Cyclic voltammograms of carbon paste electrodes including (a) NAD^+, (b) NADH, and (c) NAD^+ plus [21]AneN_7 in contact with 0.15 M $NaClO_4$. Potential scan rate 100 mV s^{-1}. Note that the axes are opposite to the IUPAC convention. (Reproduced from ref. [43], with permission. Copyright the Royal Society of Chemistry 1999.).

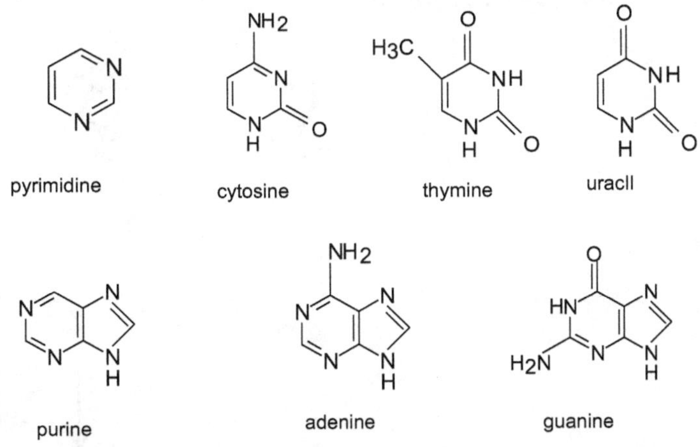

FIGURE 6.19 Chemical formulas of purine and pyrimidine bases.

At mercury electrodes, the reduction of adenine, guanine, and cytosine in neutral aqueous media occurs ca. −1.5 V vs. Ag/AgCl. In turn, guanine, thymine, adenine, and cytosine experience one-electron oxidation at carbon electrodes at potentials between 1.0 and 1.5 V vs. Ag/AgCl. The reduction processes involve the N7 and C8 positions of the imidazole ring whereas the oxidative processes involve $4H^+$, $4e^-$, and $6H^+$, $6e^-$ processes, but this electrochemistry is sensitive to the material forming the working electrode [46]. The electrochemistry of adenine-based compounds will also be treated in Chapter 8.

The direct electrochemistry of nucleic acids in solution is complemented with those performed after immobilization on electrode surfaces via adsorption or covalent binding and those involving the labeling with different redox indicators [47] or label-free hybridization [48]. A basic analytical target is discrimination between dsDNA and other nucleic acid forms. It is pertinent to note that the direct electrochemistry of DNA involves the tunneling of electrons from/to the electrode. Then, it is conceivable that only the electroactive units located near the biopolymer surface can participate in electrochemical processes. This means that electroactive units will not be easily accessible for DNA double helix [49]. Redox probes can be electroactive species such as $Fe(CN)_6^{3-}$ or $Ru(NH_3)_6^{3+}$, which interact electrostatically with the dsDNA backbone. In the case of a positively charged indicator, it is attracted by the dsDNA backbone forming an adduct (duplex) species. As a result, the voltammetric or amperometric signal of the redox probe is intensified. In the case of a negatively charged probe, there is repulsion with the dsDNA backbone, and the signal of the redox indicator is depleted [46].

This strategy can also be used with electroactive intercalators or species displaying groove binding or covalent binding with dsDNA. Among a plethora of compounds, ferrocene derivatives have been extensively studied [50]. The reversible electrochemistry of ferrocene units favors the electrochemical characterization of nucleic acids.

Another widely extended strategy consists of the immobilization of nucleic acids onto electrode surfaces. This was initiated by the assembly of close-packed DNA duplexes onto gold electrodes via alkylthiol linkers, producing an efficient reduction of methylene blue [51]. Figure 6.20 depicts a label-free hybridization scheme consisting of the direct oxidation of guanine bases after the hybridization with an inosine-modified capture probe [52]. Other interesting strategy involves guanine oxidation catalyzed by a redox mediator [53].

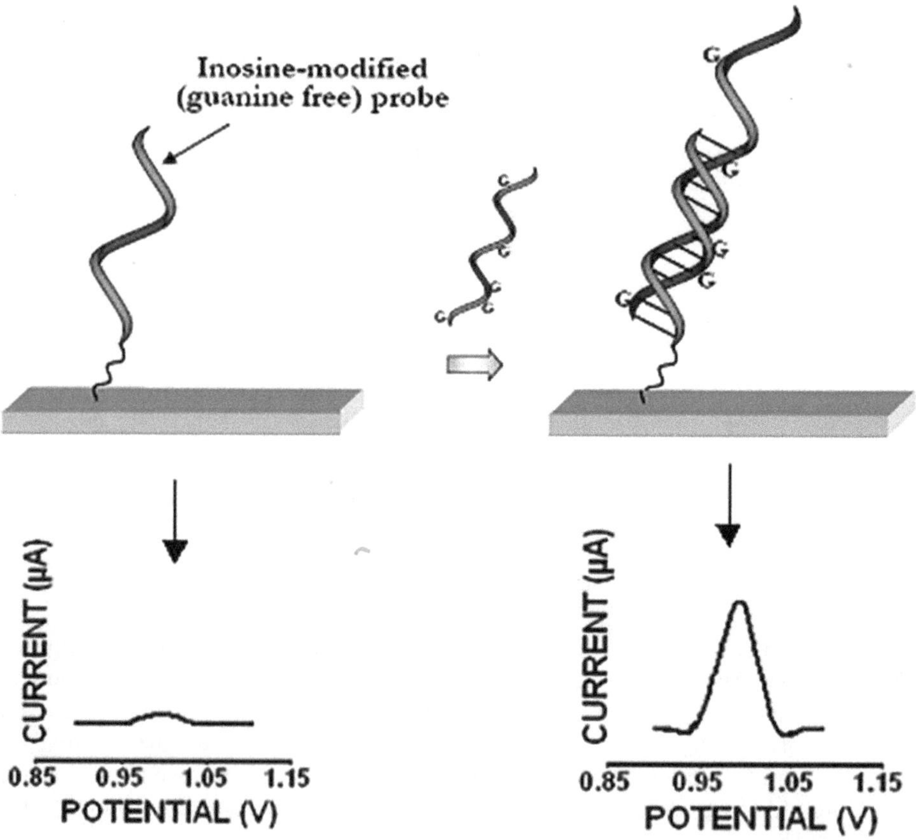

FIGURE 6.20 Overview of label-free dsDNA hybridization scheme via the direct oxidation of target DNA guanine bases after the hybridization with an inosine-modified probe. (Adapted from ref. [52], with permission. Copyright the American Chemical Society 2009.).

The immobilization of nucleic acids onto electrode surfaces involves an interesting aspect, the existence of fast electronic communication over long distances. This involves the π-electron system of stacked-based. This has been interpreted in terms of [46]: (i) a fast charge transport by long-distance tunneling through the DNA bases ("superexchange"), (ii) multistep random walk via short distance tunneling intervals linked by nucleotide sequences, and (iii) electron hopping between adjacent bases. However, despite intensive research, the mechanism of DNA-mediated electrochemistry remains not clearly understood [54]. Energy gaps arising for bridge-mediated superexchange and hole hopping, appear to be too high to facilitate electrochemical reactions. Resonance "flickering", or coherent tunneling along electric-field-induced delocalized DNA states has been recently discussed [55, 56], but it has also been proposed that these electrochemical processes are not DNA-mediated at all [57]. In this view, the origin of electrochemical signals of redox probes results from dynamic motions of individual DNA helices producing the direct contact of the probes with the electrode surface. A scheme of the two extreme situations is illustrated in Figure 6.21. Here, the DNA helices are linked to a gold surface by alkylthiol chains, and the redox probes are methylene blue molecules bound to the helix edge [54].

An alternative methodology for discriminating different DNA forms derives from solid-state electrochemistry, directly inspired by the voltammetry of immobilized particles methodology [58]. It consists of the record of the voltammetric response of a microparticulate deposit of an electroactive compound able to interact with the different DNA forms [59]. The interaction in the solution phase is evidenced in Figure 6.22 where the cyclic voltammograms of a DMSO solution of canthin-6-one, a natural β–carboline alkaloid isolated from *Zanthoxylum chiloperone* Mart ex Engl. var. *Angustifolium*, and other plants of the Rutaceae and Simaroubaceae families, are depicted [59]. The voltammogram shows a series of cathodic peaks that can be attributed to the reduction of the naphthyridone motif, yielding successively the anion radical and the dianion accompanied by other secondary reactions. In the presence of either dsDNA or ssDNA in the solution, the aforementioned voltammetry becomes modified significantly, thus denoting the existence of DNA-canthin-6-one interactions.

These interactions also occur when the naphthyridone is anchored to the electrode surface, forming a microparticulate film put in contact with an aqueous electrolyte. Now (see Figure 6.23), the voltammetric response can be described in terms of two to proton-assisted solid-state reduction processes as schematized in Figure 6.24 [59].

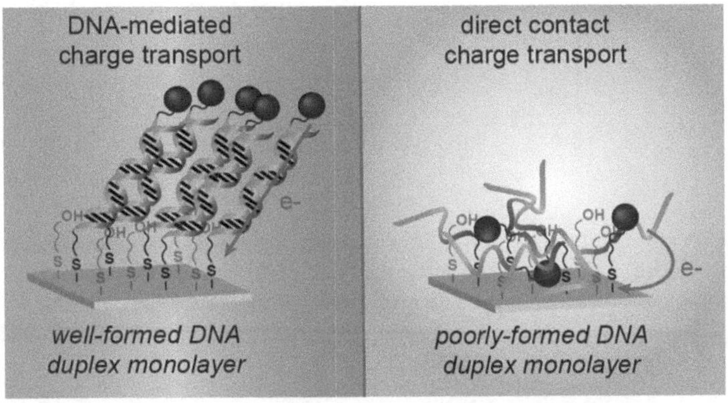

FIGURE 6.21 Schemes for the extreme cases of charge transport between a redox probe such as methylene blue (spheres) and a solid electrode mediated by DNA helices immobilized onto the electrode surface through alkylthiol units. (Adapted from ref. [54], with permission. Copyright the American Chemical Society 2021.).

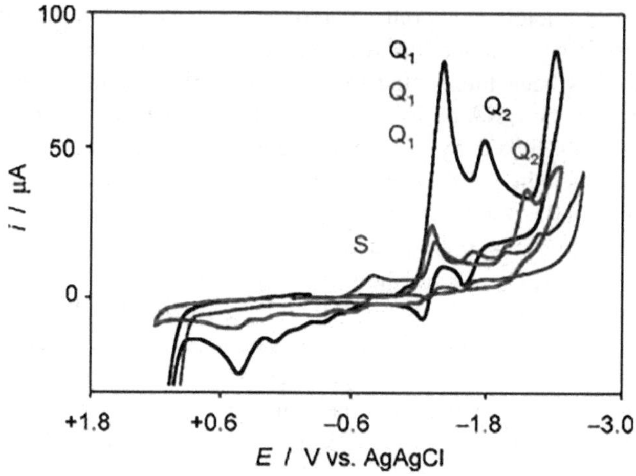

FIGURE 6.22 Cyclic voltammograms at GCE of 1.0 mM solution of canthin-6-one alkaloid in 0.10 M Bu_4NPF_6/DMSO (black line) after adding ssDNA (pale grey line) and dsDNA (dark grey). Potential scan rate 50 mV s^{-1}. Note that the axes are opposite to the IUPAC convention.

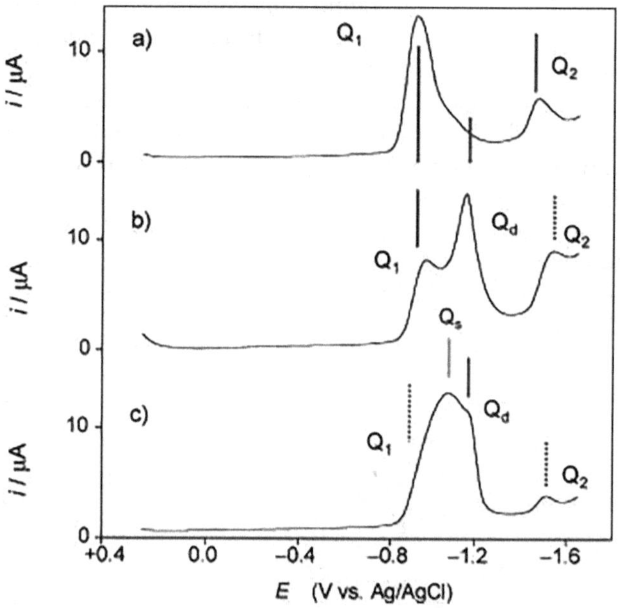

FIGURE 6.23 Square wave voltammograms of microparticulate deposits of canthin-6-one on GCE in contact with 0.10 M PBS, pH 7.4, (a) before and (b, c) after the addition of 0.03 mM dsDNA (b) and 0.03 mM ssDNA (c). Potential scan initiated at 0.25 V in the negative direction; potential step increment 4 mV; square wave amplitude 25 mV; frequency 5 Hz. (Adapted from ref. [59], with permission. Note that the axes are opposite to the IUPAC convention. Copyright Elsevier 2014.).

As shown in Figure 6.23, in the presence of ssDNA and dsDNA in the solution, the voltammetric pattern changes as a result of DNA-naphthyridone duplex formation. This results in the appearance of a significant peak splitting in the first cathodic signal, more intense for dsDNA than for ssDNA. As a result, characteristic voltammetric profiles were recorded for different DNA forms, including G-quadruplex DNA and nucleosomal DNA [59].

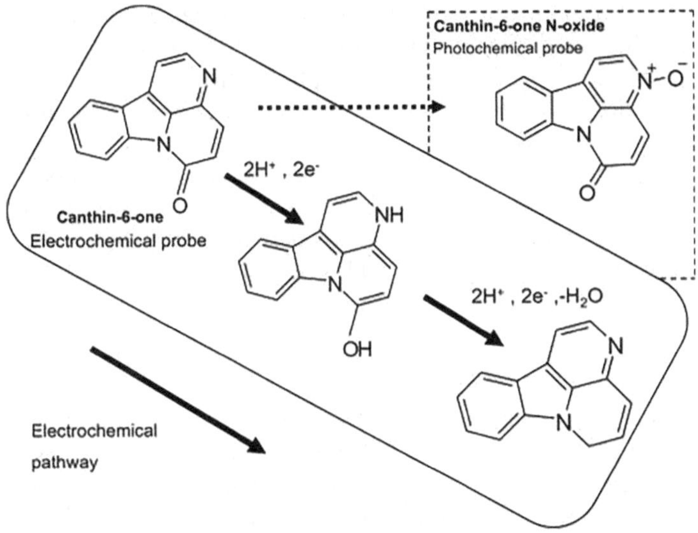

FIGURE 6.24 Main electrochemical processes involved in canthin-6-one reduction.

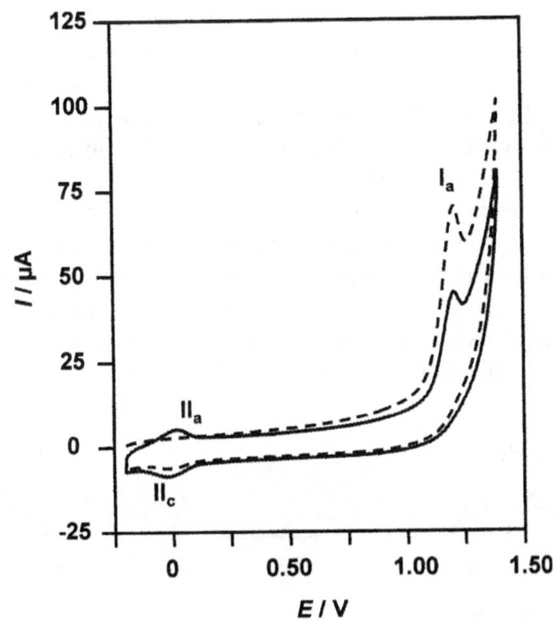

FIGURE 6.25 Cyclic voltammogram at pyrolitic graphite electrode of a 0.50 mM ATP solution in 0.1 M aqueous phosphate solution at pH 10.0. First (dotted line) and second (continuous line) potential scans. Potential scan rate 50 mV s^{-1}. (Reproduced from ref. [60], with permission. Copyright Elsevier 2001.).

6.7 ADENOSINE PHOSPHATES

The electrochemistry of adenosine phosphates is similar to that previously described for adenine nucleosides. Figure 6.25 shows the cyclic voltammetry of the pyrolitic graphite electrode of an alkaline solution of adenosine triphosphate (ATP) [60]. The initial positive-going potential scan yields an irreversible oxidation wave of ca. 1.2 V vs. Ag/AgCl (I_a). This process is followed, in the second and subsequent scans, by a weak, essentially reversible couple (II_c/II_a), whose symmetric profile and zero peak-to-peak potential separation characterize an adsorption-controlled behavior. The main oxidation process of adenosine phosphates can be described in terms of a cascade mechanism where the first step is a two-proton, two-electron oxidation to the 2-hydroxy derivative, further oxidized to the corresponding 2,8-dihydroxy adenosine derivative competing with dimerization reactions [61]. Figure 6.26 shows one possible scheme for processes I and II in Figure 6.25 [60].

This complex electrochemistry can involve a variety of catalytic processes promoted electrochemically. For example, it has been described that the oxidized forms of ATP adsorbed on carbon electrodes act catalytically on the electrochemical oxidation of NADH [60]. Here, the catalyst species are generated electrochemically as species adsorbed on the electrode surface.

6.8 AMINO ACIDS

Amino acids are essential biocomponents because of their role as building blocks and depart stage for most synthetic routes in plants and play active roles in plant growth and defense responses [62]. They also serve as precursors for the synthesis of primary and secondary metabolites. Amino acids are important in human nutrition; they are essential dietary components and sources of nutraceutical compounds. As it is well-known, there are 21 proteinogenic amino acids, 9 of which cannot be synthesized in animals, and several others are synthesized in quantities below metabolic needs.

Several common amino acids – namely, tyrosine, tryptophan, cysteine, histidine, and methionine – are electroactive. There is general agreement that electroactivity resides in the side chain of these amino acids. The voltammetric signals in aqueous solution at biological pHs fall in general between 0.0 and 1.0 V vs. Ag/AgCl and the discrimination of individual components in mixtures of amino acids is difficult. The carboxyl units, although potentially oxidizable at highly anodic potentials, can be seen usually as electrochemically silent.

Figure 6.27 depicts the overall oxidation processes for several amino acids. Cysteine oxidation proceeds via one-electron transfer to the thiol group followed by dimerization whereas in the case of methionine, the oxidation of the thioether unit results in the formation of sulfate derivative. The oxidation of histidine and tyramine proceeds via degradative processes releasing NH_3, CO_2, and other products. Tryptophan is oxidized through the indole function. In general, the electrochemical response of the of D- and L- enantiomers is essentially identical.

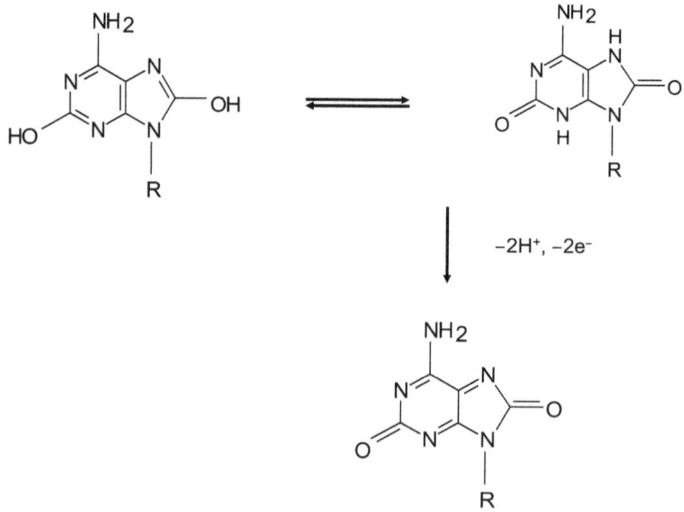

FIGURE 6.26 Scheme for the electrochemistry of ATP.

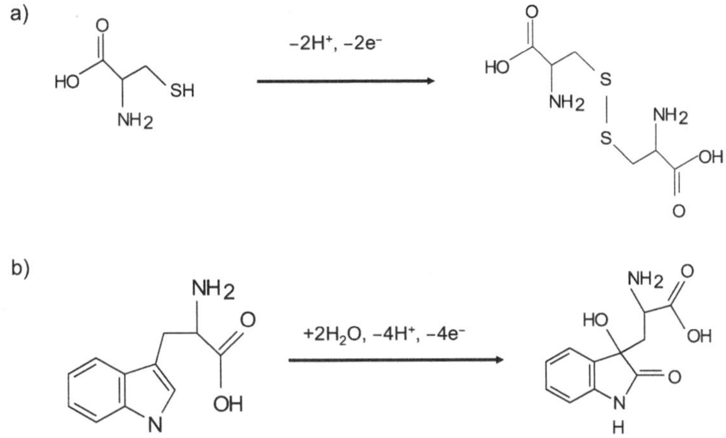

FIGURE 6.27 Electrochemical oxidation processes of (a) cysteine and (b) tryptophan.

REFERENCES

[1] Andrews, M.; Raven, J.A.; Lea, P.J. 2013. Do Plants Need Nitrate? The Mechanisms by Which Nitrogen Form Affects Plants: Do Plants Need Nitrate? *Annals of Applied Biology*. 163: 174–199.

[2] Xu, G.; Fan, X.; Miller, A.J. 2012. Plant Nitrogen Assimilation and Use Efficiency. *Annual Review of Plant Biology*. 63: 153–182.

[3] Muratore, C.; Espen, L.; Prinsi, B. 2021. Nitrogen Uptake in Plants: The PlasmaMembrane Root Transport Systems from a Physiological and Proteomic Perspective. *Plants*. 10: artic. 681.

[4] Owen, A.G.; Jones, D.L. 2001. Competition for Amino Acids between Wheat Roots and Rhizosphere Microorganisms and the Role of Amino Acids in Plant N Acquisition. *Soil Biology and Biochemistry*. 33: 651–657.

[5] Smith, T.A. 1985. Polyamines. *Annual Reviews in Plant Physiology*. 36: 117–143.

[6] Esti, M.; Volpe, G.; Massignan, L.; Compagnone, D.; La Notte, E.; Palleschi, G. 1998. Determination of Amines in Fresh and Modified Atmosphere Packaged Fruits Using Electrochemical Biosensors. *Journal of Agricultural and Food Chemistry*. 46: 4233–4237.

[7] Hammerich, O.; Speiser, R. Eds. 2019. *Organic Electrochemistry*, 5th ed. CRC Press, Boca Raton.

[8] Dushna, O.; Dubenska, L.; Vojs, M.; Marton, M.; Patsay, I.; Ivakh, S.; Plotyeya, S. 2022. Highly Sensitive Determination of Atropine in Pharmaceuticals, Biological Fluids and Beverage on Planar Electrochemical Cell with Working Boron-Doped Diamond Electrode. *Electrochimica Acta*. 432: artic. 141182.

[9] Chen, P.; McCreery, R.L. 1996. Control of Electron Transfer Kinetics at Glassy Carbon. Electrodes by Specific Surface Modification. *Analytical Chemistry*. 68: 3958–3965.

[10] Suffredini, H.B.; Santos, M.C.; De Souza, D.; Codognoto, L.; Homem-de-Mello, P.; Honorio, K.M.; Da Silva, A.B.F.; Machado, S.A.S.; Avaca, L.A. 2005. Electrochemical Behavior of Nicotine Studied by Voltammetric Techniques at Boron-Doped Diamond Electrodes. *Analytical Letters*. 38: 1587–1599.

[11] Levent, A.; Yardim, Y.; Senturk, Z. 2009. Voltammetric behavior of nicotine at pencil graphite electrode and its enhancement determination in the presence of anionic surfactant. *Electrochimica Acta*. 55: 190–195.

[12] Guidelli, R.; Compton, R.G.; Feliu, J.M.; Gileadi, E.; Lipkowski, J.; Schmickler, W.; Trasatti, S. 2014. Defining the transfer coefficient in electrochemistry: An assessment (IUPAC Technical Report). *Pure and Applied Chemistry*. 86: 245–258.

[13] Enache, T.A.; Oliveira-Brett, A.M. 2011. Pathways of electrochemical oxidation of indolic compounds. *Electroanalysis*. 23: 1337–1344.

[14] Tan, D.X.; Manchester, L.C.; Terron, M.P.; Flores, L.J.; Reiter, R.J. 2007. One molecule, many derivatives: a never-ending interaction of melatonin with reactive oxygen and reactive nitrogen species?. *Journal of Pineal Research*. 42:2 8–42.

[15] Iriti, M.; Varoni, E.M. 2015. Melatonin in Mediterranean diet, a new perspective. *Journal of Science of Food and Agriculture*. 95: 2355–2359.

[16] Radi, A.; Bekhiet, G.E. 1998. Voltammetry of melatonin at carbon electrodes and determination in capsules. *Bioelectrochemistry and Bioenergetics*. 45: 275–279.

[17] Feng X, Wang M, Zhao Y et al 2014. Melatonin from different fruit sources, functional roles, and analytical methods. *Trends in Food Science &Technology*. 37:21–31.

[18] Wrona, M.Z.; Dryhurst, G. 1989. Electrochemical oxidation of 5-hydroxy-tryptamine in acidic aqueous solution. *Journal of Organic Chemistry*. 54: 2718–2721.

[19] Liu, H.; Chen, Y.; Liu, Y.; Yang, Z. 2013. A sensitive sensor for determination of l-tryptophan based on gold nanoparticles/poly(alizarin red S)-modified glassy carbon electrode. *Journal of Solid State Electrochemistry*. 17: 2623–2631.

[20] Caro, C.A.; Lillo, L.; Valenzuela, F.J.; Cabello, G.; Lang, E.; Vallejos, D.; Castillo, C. 2016. Oxidation of melatonin on a glassy carbon electrode modified with metallic glucosamines. Synthesis and characterization. *Journal of Solid State Electrochemistry*. 20: 993–1000.

[21] Cohen, J.L.; Cox, J.A. 2004. Role of pore size on the electrochemical oxidation of hydroxytryptophan in a silica sol–gel matrix. *Journal of Solid State Electrochemistry*. 8: 886–891.

[22] Muñiz-Calvo, S.; Guillamón, J.M.; Domínguez, I.; Doménech-Carbó, A. 2017. Detecting and monitoring the production of melatonin and other related indole compounds in different *Saccharomyces* strains by solid-state electrochemical techniques. *Food Analytical Methods*. 10: 1408–1418.

[23] Choi, K.-Y. 2020. A review of recent progress in the synthesis of bio-indigoids and their biologically assisted end-use applications. *Dyes and Pigments*. 181: artic. 108570.

[24] Bond, A.M.; Marken, F.; Hill, E.; Compton, R.G.; Hügel, H. 1997. The electrochemical reduction of indigo dissolved in organic solvents and as a solid mechanically attached to a basal plane pyrolytic graphite electrode immersed in aqueous electrolyte solution. *Journal of the Chemical Society Perkin Transactions 2*. 1735–1742.

[25] Doménech-Carbó, A.; Costero, A.; Gil, S.; Montoya, N.; López-Lorente, A.; Sáez, J.A.; Arroyo, P.; Doménech-Carbó, M.T. 2021. Isomerization and Redox Tuning: Reorganizing the Maya Blue Puzzle from Synthetic, Spectral and Electrochemical issues. *Journal of Physical Chemistry C*. 125: 26188–26200.

[26] Doménech-Carbó, A. 2021. *Electrochemistry of Porous Materials*, 2nd ed. Taylor & Francis, Boca Raton.

[27] Doménech-Carbó, A.; Doménech-Carbó, M.T.; Vázquez de Agredos-Pascual, M.L. 2007. Electrochemical monitoring Maya Blue preparation from Maya's ancient procedures. *Journal of Solid State Electrochemistry*. 11: 1335–1346.

[28] Cover, R.E.; Folliard, J.T. 1971. The elucidation of electrochemical mechanisms using the vibrating dropping mercury electrode: The non-catalytic waves of quinine, quinoline and 3-aminoquinoline. *Journal of Electroanalytical Chemistry*. 30: 143–159.

[29] Dar, R.A.; Brahman, P.K.; Tiwari, S.; Pitre, K.S. 2012. Electrochemical studies of quinine in surfactant media using hanging mercury drop electrode: A cyclic voltammetric study. *Colloids and Surfaces B: Biointerfaces*. 98: 72–79.

[30] Dushna, O.; Dubenska, L.; Marton, M.; Hatala, M.; Vojs, M. 2023. Sensitive and selective voltammetric method for determination of quinoline alkaloid, quinine in soft drinks and urine by applying a boron-doped diamond electrode. *Microchemical Journal*. 191: artic. 108839.

[31] Gonzalez-Rodriguez, J.; Sevilla, J.M.; Pineda, T.; Blazquez, M.; López-Guerrero, M.M. 2020. A study on the electrooxidation of vitamin B_6 compounds on glassy carbon and polycrystalline gold electrodes. *Journal of Electroanalytical Chemistry*. 877: artic. 114525.

[32] Cao, Z.J.; Xie, Q.J.; Li, M.; Yao, S.Z. 2004. Simultaneous EQCM and fluorescence detection of adsorption/desorption and oxidation for pyridoxol in aqueous KOH on a gold electrode. *Journal of Electroanalytical Chemistry*. 568: 343–351.

[33] Lee, J.H.Q.; Yue, Y.; Ganguly, R.; Webster, R.D. 2015. Electrochemical Study of Pyridoxine (Vitamin B_6) in Acetonitrile. *ChemElectroChem*. 2: 412–420.

[34] Desai, P.B.; Kotkar, R.M.; Srivastava, A.K. 2008. Electrochemical behaviour of pyridoxine hydrochloride (vitamin B_6) at carbon paste electrode modified with crown ethers. *Journal of Solid State Electrochemstry*. 12: 1067–1075.

[35] Lovander, M.D.; Lyon, J.D.; Parr IV, D.L.; Wang, J.; Parke, B.; Leddy, J. 2018. Electrochemical properties of 13 vitamins: a critical review and assessment. *Journal of the Electrochemical Society*. 165: G18–G49.

[36] Hrbáč, J.; Broncová, G.; Matějka, P.; Bystroň, T.; Vrba, J.; Volf, R.; Vičar, J. 2005. Electrochemistry of Benzophenanthridine Alkaloids. Formation and Characterization of Redox Active Films from Products of Sanguinarine and Chelerythrine Oxidation. *Electroanalysis*. 23: 2175–2181.

[37] Cebrián-Torrejón, G.; Doménech-Carbó, A.; Scotti, M.T.; Fournet, A.; Figadère, B.; Poupon, E. 2015. Experimental and theoretical study of possible correlation between the electrochemistry of canthin-6-one and the anti-proliferative activity against human cancer stem cells. *Journal of Molecular Structure*. 1102: 242–246.

[38] Hashida, S.-N.; Takhashi, H.; Uchimiya, H. 2009. The role of NAD biosynthesis in plant development and stress responses. *Annals of Botany*. 103: 819–824.

[39] Smith, E.N.; Schwarzländer, M.; Ratcliffe, R.G.; Kruger, N.J. 2021. Shining a light on NAD- and NADP-based metabolism in plants. *Trends in Plant Science*. 26: 1072–1086.

[40] Santhanam, K.S.V.; Elving, P.J. 1973. Electrochemical Redox Pattern for Nicotinamide Species in Nonaqueous Media. *Journal of the American Chemical Society*. 95: 5482–5490.

[41] Braun, R.D.; Santhanam, K.S.V.; Elving, P.J. 1975. Electrochemical Oxidation in Aqueous and Nonaqueous Media of Dihydropyridine Nucleotides NMNH, NADH, and NADPH. *Journal of the American Chemical Society*. 97: 2591–2598.

[42] Ali, I.; Soomro, B.; Omanovic, S. 2011. Electrochemical regeneration of NADH on a glassy carbon electrode surface: The influence of electrolysis potential. *Electrochemistry Communications*. 13: 562–565.

[43] Doménech-Carbó, A.; García-España, E.; Ramírez, J.A.; Celda, B.; Martínez, M.C.; Monleón, D.; Tejero, R.; Bencini, A.; Bianchi, A. 1999. A thermodynamic, electrochemical and molecular dynamics study on NAD and NADP recognition by 1,4,7,10,13,16,19-heptaazacyclohenicosane ([21]aneN_7). *Journal of the Chemical Society Perkin Trnasactions 2*. 23–32.

[44] Hapiot, P.; Moiroux, J.; Savéant, J.-M. 1990. Electrochemistry of NADH/NAD^+ analogs. A detailed mechanistic kinetic and thermodynamic analysis of the 10-methylacridan/10-methylacridinium couple in acetonitrile. *Journal of the American Chemical Society*. 112: 1337–1343.

[45] Paleček, E.; Bartošík, M. 2012. Electrochemistry of Nucleic Acids. *Chemical Reviews*. 112: 3427–3481.

[46] Ferapontova, E.E. 2018. DNA Electrochemstry and Electrochemical Sensors for Nucleic Acids. *Annual Reports of Analytical Chemistry*. 11: 197–218.

[47] Vrábel, M.; Horáková, P.; Pivonková, H.; Kalachova, L.; Cernocká, H.; Cahová, H.; Pohl, R.; Sebest, P.; Havran, L.; Ocek, M.; Fojta, M. 2009. Base-Modified DNA Labeled by $[Ru(bpy)_3]^{2+}$ and $[Os(bpy)_3]^{2+}$ Complexes: Construction by Polymerase Incorporation of Modified Nucleoside Triphosphates, Electrochemical and Luminescent Properties, and Applications. *Chemistry A European Journal*. 15: 1144–1154.

[48] Wang, J.; Li, S.; Zhang, Y. 2010. A sensitive DNA biosensor fabricated from gold nanoparticles, carbon nanotubes, and zinc oxide nanowires on a glassy carbon electrode. *Electrochimica Acta*. 55: 4436–4440.

[49] Suprun, E.V. 2021. Direct electrochemistry of proteins and nucleic acids: The focus on 3D structure. *Electrochemistry Communications*. 125: artic. 106983.

[50] Luo, X.T.; Lee, T.M.H.; Hsing, I.M. 2008. Immobilization-Free Sequence-Specific Electrochemical Detection of DNA Using Ferrocene-Labeled Peptide Nucleic Acid. *Analytical Chemistry*. 80: 7341–7346.

[51] Kelley, S.O.; Barton, J.K.; Jackson, N.M.; Hill, M.G. 1997. Electrochemistry of methylene blue bound to a DNA-modified electrode. *Bioconjugate Chemistry*. 8: 31–37.

[52] Lusi, E.A.; Passamano, M.; Guarascio, P.; Scarpa, A.; Schiavo, L. 2009. Innovative Electrochemical Approach for an Early Detection of microRNAs. *Analytical Chemistry*. 81: 2819–2822.

[53] Armistead, P.M.; Thorp, H.H. 2000. Modification of Indium Tin Oxide Electrodes with Nucleic Acids: Detection of Attomole Quantities of Immobilized DNA by Electrocatalysis. *Analytical Chemistry*. 72: 3764–3770.

[54] Nano, A.; Furst, A.L.; Hill, M.G.; Barton, J.K. 2021. DNA Electrochemistry: Charge-Transport Pathways through DNA Films on Gold. *Journal of the American Chemical Society*. 143: 11361–11640.

[55] Zhang, Y.; Liu, C.; Balaeff, A.; Skourtis, S.S.; Beratan, D.N. 2014. Biological charge transfer via flickering resonance. *Proceedings of the National Academy of Sciences USA*. 111: 10049–10054.

[56] Michaeli, K.; Beratan, D.N.; Waldeck, D.H.; Naaman, R. 2019. Voltage-induced long-range coherent electron transfer through organic molecules. *Proceedings of the National Academy of Sciences USA*. 116: 5931–5936.

[57] Dauphin-Ducharme, P.; Arroyo-Currás, N.; Plaxco, K.W. 2019. High-Precision Electrochemical Measurements of the Guanine-, Mismatch-, and Length-Dependence of Electron Transfer from Electrode-Bound DNA Are Consistent with a Contact-Mediated Mechanism. *Journal of the American Chemical Society*. 141: 1304–1311. See *corrigendum* in *Journal of the American Chemical Society*. 2019, 141: 8000.
[58] Scholz, F.; Schröder, U.; Gulabowski, R.; Doménech-Carbó, A. 2014. *Electrochemistry of Immobilized Particles and Droplets*, 2nd ed. Springer, Berlin-Heidelberg.
[59] Doménech-Carbó, A.; Cebrián-Torrejón, G.; de Miguel, L.; Tordera, V.; Rodrigues-Furtado, D.; Assad-Kahn, S.; Fournet, A.; Figadère, B.; Vázquez-Manrique, R.P.; Poupon, E. 2014. dsDNA, ssDNA, G-quadruplex DNA, and nucleosomal DNA electrochemical screening using canthin-6-one alkaloid-modified electrodes. *Electrochimica Acta*. 115: 546–552.
[60] de Los Santos Álvarez, N.; Muñiz Ortea, P.; Montes Pañeda, A.; Lobo Castañón, M.J.; Miranda Ordieres, A.J.; Tuñón Blanco, P. 2001. A comparative study of different adenine derivatives for the electrocatalytic oxidation of β-nicotinamide adenine dinucleotide. *Journal of Electroanalytical Chemistry*. 502: 109–117.
[61] Goyal, R.N.; Sangal, A. 2003. Electrochemical oxidation of adenosine monophosphate at a pyrolytic graphite electrode. *Journal of Electroanalytical Chemistry*. 557: 147–155.
[62] Trovato, M.; Funck, D.; Forlani, G.; Okumoto, S.; Amir, R. 2021. Editorial: Amino Acids in Plants: Regulation and Functions in Development and Stress Defense. *Frontiers in Plant Science*. 12: artic. 772810.

ADDITIONAL LITERATURE

Biochemistry and Molecular Biology

Hooykaas, P.J.J.; Hall, M.A.; Libbenga, K.R. Eds. 1999. *Biochemistry and Molecular Biology of Plant Hormones*. Elsevier, Amsterdam.
Metzler, D.E. 1977. *Biochemistry. The Chemical Reactions of Living Cells*. Academic Pres, New York.
Tan, D.X.; Manchester, L.C.; Helton, P.; Reiter, R.J. 2007. Phytoremediative capacity of plants enriched with melatonin. *Plant Signaling & Behaviour*. 2: 514–516.

Amines, Indolic Compounds Pyridines and Quinolines

Becker, J.Y.; Miller, L.L.; Stermitz, F.R. 1976. Electrode processes in the oxidation of tetrahydroisoquinoline derivatives. *Journal of Electroanalytical Chemistry*. 68: 181–191.
Capucciati, A.; Cacciatore, L.; Protti, S.; Profumo, A.; Merli, D. 2022. Electrochemical analysis and characterization of psychoactive substances glaucine and tetrahydropalmatine. *Journal of Electroanalytical Chemistry* 907: artic.116032.
Doménech-Carbó, A.; Doménech-Carbó, M.T.; Vázquez de Agredos-Pascual, M.L. 2006. Dehydroindigo: a New Piece into the Maya Blue Puzzle from the Voltammetry of Microparticles Approach. *Journal of Physical Chemistry B*. 110: 6027–6039.
Doménech-Carbó, A.; Doménech-Carbó, M.T.; Vázquez de Agredos-Pascual, M.L. 2007. Chemometric Study of Maya Blue from the Voltammetry of Microparticles Approach. *Analytical Chemistry*. 79: 2812–2821.
Doménech-Carbó, A.; Holmwood, S.; Di Turo, F.; Montoya, N.; Valle-Algarra, F.M.; Edwards, H.G.M.; Doménech-Carbó, M.T. 2019. On the Composition and Color of Maya Blue: A Reexamination of Literature Data Based on the Dehydroindigo Model. *Journal of Physical Chemistry C*. 123: 770–782.
García-Parrilla, M.C.; Cantos, E.; Troncoso, A.M. 2009. Analysis of melatonin in foods. *Journal of Food Composition and Analysis*. 22: 177–183.
Karaaslan, C.; Suzen, S. 2011. Electrochemical Behavior of Biologically Important Indole Derivatives. *International Journal of Electrochemistry*. 2011: artic. 154804.
Laube, B.L.; Asirvatham, M.R.; Mann, C.K. 1977. Electrochemical oxidation of tropanes. *Journal of Organic Chemistry*. 42: 670–674.
Marken, F.; Compton, R.G.; Goeting, C.H.; Foord, J.S.; Bull, S.D.; Davies, S.G. 2001. Fast electrochemical triple-interface processes at boron-doped diamond electrodes. *Journal of Solid State Electrochemistry*. 5: 88–93.
Mruthunjaya, A.K.V.; Torriero, A.A.J. 2023. Mechanistic Aspects of the Electrochemical Oxidation of Aliphatic Amines and Aniline Derivatives. *Molecules*. 28: artic. 471.

Nelsen, S.F.; Hintz, P.J. 1972. Electrochemical Oxidation of Tertiary Amines. The Effect of Structure upon Reversibility. *Journal of the American Chemical Society*. 94: 207114–207117.

Olmos, J.M.; Pereira, C.M. 2021. Ion transfer electrochemistry of the alkaloids berberine and palmatine: Sensing and physicochemical characterization. *Journal of Electroanalytical Chemistry*. 895: artic. 115506.

Russell, G.A.; Kaupp, G. 1969. Oxdation of Carbanions IV. Oxidation of Indoxyl to Indigo in Basic Solution. *Journal of the American Chemical Society*. 91: 3851–3859.

Süzen, S.; Demírcígíl, B.T.; Buyukbingol, E.; Özkan, S.A. 2003. Electroanalytical evaluation and determination of 5-(3-indolyl)-2-thiohydantoin derivatives by voltammetric studies: possible relevance to in vitro metabolism. *New Journal of Chemistry*. 27: 1007–1011.

Tan, L.; Xie, Q.J.; Yao, S.Z. 2004. Electrochemical and Spectroelectrochemical studies on pyridoxine hydrochloride using a poly(methylene blue) modified electrode. *Electroanalysis*. 16: 1592–1597.

Wang, X.; Yang, N.; Wan, Q. 2006. Cyclic voltammetric response of nicotinic acid and nicotinamide on a polycrystalline gold electrode. *Electrochimica Acta*. 52: 361–368.

Wrona, M.Z.; Dryhurst, G. 1987. Oxidation Chemistry of 5-Hydroxytryptamine. 1. Mechanism and Products Formed at Micromolar Concentrations. *Journal of Organic Chemistry*. 52: 2817–2825.

Wrona, M.Z.; Lemordant, D.; Lin, L.; Blank, C.L.; Dryhurst, G. 1986. Oxidation of 5-hydroxytryptamine and 5,7-dihydroxytryptamine. A new oxidation pathway and formation of a novel neurotoxin. *Journal of Medicinal Chemistry*. 29: 499–505.

Zeng, C.C.; Liu, F.J.; Ping, D.W.; Hu, L.M.; Cai, Y.L.; Zhong, R.G. 2009. One-pot electrochemical synthesis of fused indole derivatives containing active hydroxyl groups in aqueous medium. *Journal of Organic Chemistry*. 74: 6386–6389.

Hydrogen Transfer Coenzymes

Aamer, E.; Thöming, J.; Baune, M.; Reimer, N.; Dringen, R.; Romero, M.; Bösing, I. 2022. Influence of electrode potential, pH and NAD^+ concentration on the electrochemical NADH regeneration. *Scientific Reports*. 12: artic. 16380.

Ali, I.; Gill, A.; Omanovic, S. 2012. Direct electrochemical regeneration of the enzymatic cofactor 1, 4-NADH employing nano-patterned glassy carbon/Pt and glassy carbon/Ni electrodes. *Chemical Engineering Journal*. 188: 173–180.

Damian, A.; Omanovic, S. 2006. Electrochemical reduction of NAD^+ on a polycrystalline gold electrode. *Journal of Molecular Catalysis A*. 253: 222–233.

Gakière, B.; Hao, J.; de Bont, L.; Pétriacq, P.; Nunes-Nesi, A.; Fernie, A.R. 2018. NAD^+ Biosynthesis and Signaling in Plants. *Critical Reviews in Plant Science*. 37: 1–49.

Hashida, S.-N.; Takahashi, H.; Uchimiya, H. 2009. The role of NAD biosynthesis in plant development and stress responses. *Annals of Botany*. 103, 819–824.

Jaegfeldt, H. 1981. A study of the products formed in the electrochemical reduction of nicotinamide-adenine-dinucleotide. *Journal of Electroanalytical Chemistry*. 128: 355–370.

Le, T.X.H.; Etienne, M.; Lapicque, F.; Hehn, A.; Vilà, N.; Walcarius, A. 2020. Local removal of oxygen for $NAD(P)^+$ detection in aerated solutions. *Electrochimica Acta*. 353: artic. 136546.

Nucleic Acids

Batchelor-McAuley, C.; Wildgoose, G.G.; Compton, R.G. 2009. The physicochemical aspects of DNA sensing using electrochemical methods. *Biosensors and Bioelectronics*. 24: 3183–3190.

Boussicault F, Robert M. 2008. Electron transfer in DNA and DNA-related biological processes. Electrochemical insights. *Chemical Reviews*. 108: 2622–2645.

Ferapontova, E.E.; Domínguez, E. 2003. Direct electrochemical oxidation of DNA on polycrystalline gold electrodes. *Electroanalysis*. 15: 629–634.

Inclán, M.;Albelda, M.T.; Frías, J.C.; Blasco, S.; Verdejo, B.; Serena, C.; Salat-Canela, C.; Díaz, M.L.; García-España, A.; García-España, E. 2012. Modulation of DNA Binding by Reversible Metal-Controlled Molecular Reorganizations of Scorpiand-like Ligands. *Journal of the American Chemical Society*. 134: 9644–9656.

Jortner, J.; Bixon, M.; Voityuk, A.A.; Rosch, N. 2002. Superexchange Mediated Charge Hopping in DNA. *Journal of Physical Chemistry A* 106: 7599–7606.

Kanvah, S.; Joseph, J.; Schuster, G.B.; Barnett, R.N.; Cleveland, C.L.; Landman, U. 2010. Oxidation of DNA: Damage of Nucleobases. *Accounts of Chemical Research*. 43: 280–287.

Kékedy-Nagy, L.; Ferapontova, E.E. 2018. Sequence-Specific Electron Transfer Mediated by DNA Duplexes Attached to Gold through the Alkanethiol Linker. *Journal of Physical Chemistry B*. 122: 10077–10085.

Kelley, S.O.; Jackson, N.M.; Hill, M.G.; Barton, J.K. 1999. Long Range Electron Transfer Through DNA Films. *Angewandte Chemie International Edition*. 38: 941–945.

Labuda, J.; Oliveira-Brett, A.M.; Evtugyn, G.; Fojta, M.; Mascini, M.; Ozsoz, M.; Palchetti, I.; Paleček, E.; Wang, J. 2010. Electrochemical nucleic acid-based biosensors: Concepts, terms, and methodology (IUPAC Technical Report). *Pure and Applied Chemistry*. 82: 1161–1187.

Schuster, G.B. 2000. Long-range charge transfer in DNA: transient structural distortions control the distance dependence. *Accounts of Chemical Research*. 33: 253–260.

Amino Acids

Deo, R.P.; Lawrence, N.S.; Wang, J. 2004. Electrochemical detection of amino acids at carbon nanotube and nickel-carbon nanotube modified electrodes. *Analyst*. 129: 1076–1081.

Galili, G.; Amir, A.; Fernie, A. 2016. The regulation of essential amino acid synthesis and accumulation in plants. *Annual Reviews in Plant Biology*. 67: 53–78.

Moulaee, K.; Neri, G. 2021. Electrochemical Amino Acid Sensing: A Review on Challenges and Achievements. *Biosensors*. 11: artic. 502.

Nguyen, N.T.; Wrona, M.Z.; Dryhurst, G. 1986. Electrochemical oxidation of tryptophan. *Journal of Electroanalytical Chemistry*. 199: 101–126..

Wu, J.; Ran, P.; Zhu, S.; Mo, F.; Wang, C.; Fu, Y. 2019. A highly sensitive electrochemiluminescence sensor for the detection of L-cysteine based on the rhombus-shaped rubrene microsheets and platinum nanoparticles. *Sensors and Actuators l*. 278: 97–102.

7 Antioxidant Capacity

7.1 INTRODUCTION

The term "antioxidant capacity" refers to the ability of a given system (individual component or a natural product) to react with oxidizing agents. This is a rather vague definition because reactivity involves thermochemical and kinetic aspects associated with concrete reaction conditions. The antioxidant capacity (or antioxidant potential, or antioxidant power) is used in the context of food chemistry and pharmaceutical chemistry with a more restricted meaning: It refers to the reactivity with oxidants involved in redox biochemical processes. Most of these processes (energy production, cellular growth, phagocytosis, etc.) involve oxidizing free radicals so that the antioxidant capacity is to a great extent evaluated as radical quenching/scavenging capacity.

An antioxidant compound can be defined as when present in low concentration relative to the substrate that will undergo oxidation, it delays, retards, or prevents its oxidation mediated by free radicals (or its autooxidation) making stable the radical resulting from the scavenging process [1].

The interest in the estimate of the antioxidant capacity of plant components derives from the wide spectrum of their biological activity. These have anti-inflammatory, anti-atherosclerosis, anti-cancer, and anti-aging, properties [1]. In plants, the major antioxidants are polyphenolic compounds (phenolic acids and ester derivatives, flavonoids and other benzopyran derivatives, quinones, chalcones, lignins, tannins, lignans, stilbenes, and curcuminoids, among others), ascorbic acid, terpenes, and tocopherols. Only the group of polyphenols includes more than 10,000 compounds [2].

In this context, the concept of antioxidant capacity is applied not only to individual compounds but also to foods, beverages, and parts of plants (mainly leaves but also seeds, buds, flowers, stems, roots, rhizomes, barks, etc.), as well as products (decoctions, infusions, tinctures, and essential oils) derived from the same [3]. To evaluate the total antioxidant capacity (TAC) or, equivalently, the total antioxidant potential (TAP) [4], chemical, biochemical, and electrochemical tests have been proposed [5]. When considering complex systems such as fruits and plant extracts, several remarks have to be accounted for:

a) Of course, this TAC is test-sensitive, i.e., each test defines its scale [6].
b) The measured TAC is also dependent on the conditions (solvent, pH) in which the assay is performed.
c) The composition profile of plants is affected by the chemotypes of plant species and varieties, as well as local variations in the cultivar and harvesting conditions, water uptake, seasonal climate variations, environmental physicochemical conditions, gathering, and storage conditions, among other factors.
d) Due to synergistic (positive or negative) effects between the individual compounds, the TAC, defined as the measure of the overall antioxidant activity of all antioxidants in each sample, generally will differ from the sum of the TACs of the separate components [4].

7.2 CHEMICAL AND ELECTROCHEMICAL ASSAYS

Evaluation of the antioxidant properties of plants is usually performed by assays based on the reactions of antioxidants with radicals or the reduction of metal ions. In general, chemical assays require complex conditioning steps frequently involving extraction with organic solvents. The reactions with the test reagents are usually monitored by spectrophotometry and fluorimetry. In the case of complex systems such as juices and plant extracts, estimates of the TAC can be made by determining the antioxidant capacity of the individual antioxidant constituents, previously separated usually using liquid chromatography. The most frequent chemical tests [3, 5] are summarized in Table 7.1.

DOI: 10.1201/9781032697826-7

Given the intrinsic redox nature of the antioxidant capacity, electrochemical methods aimed at evaluating the TAC have received considerable attention. The electrochemical methods can be grouped as methods based on the electrochemical monitoring of the following processes [6–8]:

- The direct electrochemical oxidation of plant components
- The reaction with oxidants and oxidants generated electrochemically
- The reactions with radical scavenger reagents and radical scavengers generated electrochemically
- The reactions with radical scavenger reagents immobilized onto electrode surfaces
- The interaction of plant components with ROS and ROS generated electrochemically

These methods involve different electrochemical techniques, mainly coulometry, potentiometry, voltammetry, amperometry, and chrono methods. The majority are applied following the common guidelines of conventional, solution-phase electrochemistry, but solid-state electrochemistry strategies can also be applied [6]. It must be noted, however, that the electrochemical response of redox probes may be partially inhibited by the presence of elicitors, plant hormones, metabolic inhibitors, and electron transport inhibitors in the samples [9].

In general, the estimated TAC must be referred to a reference antioxidant component. In the case of coulometric methods, the "absolute" concentration of antioxidants is determined, and there is no need for a standard.

7.3 METHODS BASED ON THE DIRECT MEASUREMENT OF OXIDATION POTENTIALS

These methods are based on the record of the voltammetric response of the electroactive antioxidant components. This is a simple and reproducible method typically performed by applying cyclic voltammetry and pulsed techniques, frequently square wave voltammetry and

TABLE 7.1
Main Chemical Assays Used to Evaluate the Antioxidant Capacity of Medicinal Plants

Essay	Description
Trolox equivalence antioxidant capacity (TEAC)	Reaction with 2,20-azino-bis (3-ethylbenzothiazoline-6-sulfonic acid) (ABTS·) radical
2,2-diphenyl-1-picrylhydrazyl (DPPH) assay	Reaction with DPPH• radical
Oxygen radical absorbance capacity assay (ORAC)	Inhibition of peroxyl-radical-induced oxidation
Total radical trapping antioxidant potential (TRAP)	Suppression of the oxidation of 2,20-azobis-2-methyl-propanimidamide, dihydrochloride (AAPH), or 2,20-azobis(2-amidinopropane) dihydrochloride (ABAP)
Inhibition of low-density lipoprotein (LDL) oxidation	Effect of the antioxidant on the peroxidation of linoleic acid or LDL induced by Cu(II) or AAPH
Total phenolic contents by reaction with Folin–Ciocalteu reagent	Reduction of molybdotungsto-phosphoric heteropolyanion to molybdenum-tungsten blue
Inhibition of β-carotene bleaching	Spectrophotometric monitoring of β-carotene emulsion with linoleic acid
Inhibition of lipid peroxidation using thiobarbituric acid-reactive substances (TBARS)	Reaction with thiobarbituric acid (TBA)
Ferric-reducing antioxidant power (FRAP)	Reduction of the Fe(III) complex with 2,4,6-tripyridyl-S-triazine
Cupric reducing antioxidant capacity (CUPRAC)	Reduction of the Cu(II) complex with neocuproine

differential pulse voltammetry. A variety of electrode materials and electrolytes can be used. Figure 7.1 shows the cyclic voltammograms recorded for solutions of trolox, caffeic acid, gallic acid, and ascorbic acid in 0.15 M phosphate buffer aqueous solution at pH 3.5 [10]. One can see that these compounds display well-defined oxidation peaks in the potential range between 0.0 and 1.0 V vs. Ag/AgCl.

Figure 7.2 shows the differential pulse voltammograms recorded at a sonogel-carbon electrode for gallic acid, caffeic acid, ferulic acid, and syringic acid, (+)-catechin, and rutin in Britton-Robinson buffer solution at pH 2 [11]. Each compound produces a characteristic voltammetric profile, but the position and intensity of the peaks, in general, depend on the electrode material, concentration of the electroactive species, pH of the solution, and electrochemical conditions (potential scan rate in linear potential scan voltammetry). This means that strict uniform experimental conditions have to be ensured.

Data in Figures 7.1 and 7.2 suggest that the voltammetry of plant extracts, juices, etc., will be complex. These are multicomponent systems and the voltammetric response in general shows strongly overlapped peaks. This can be seen in Figure 7.3, where the differential pulse voltammogram obtained for a sample of defatted walnut kernel extract in Britton-Robison solution (pH 2.0) is depicted.

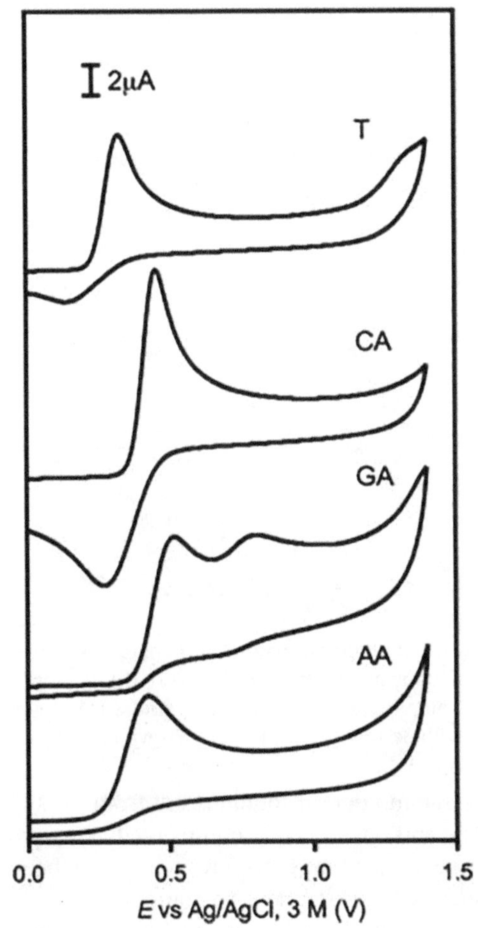

FIGURE 7.1 Cyclic voltammograms at glassy carbon electrode of 0.50 mM solutions of trolox (T), caffeic acid (CA), gallic acid (GA), and ascorbic acid (AA) in 0.15 M phosphate buffer solution at pH 3.5. Potential scan rate 100 mV s^{-1}. (Reproduced from ref. [10], with permission. Copyright Elsevier 2014.).

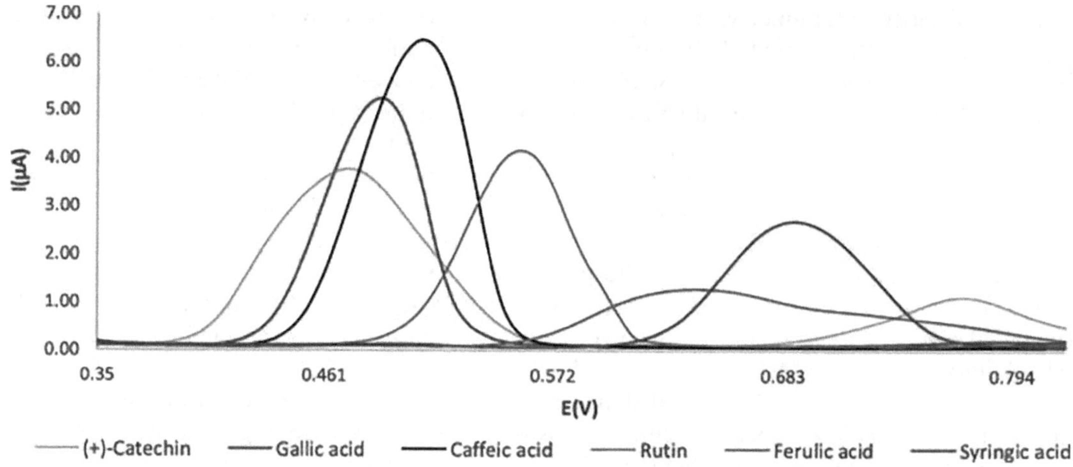

FIGURE 7.2 Differential pulse voltammograms at a sonogel-carbon electrode for gallic acid, caffeic acid, ferulic acid, syringic acid, (+)-catechin, and rutin in Britton-Robinson buffer solution at pH 2. Modulation amplitude 25 mV; step potential 8 mV; interval time 0.4 s; modulation time 0.05 s. (Adapted from ref. [11], with permission. Copyright Elsevier 2022.).

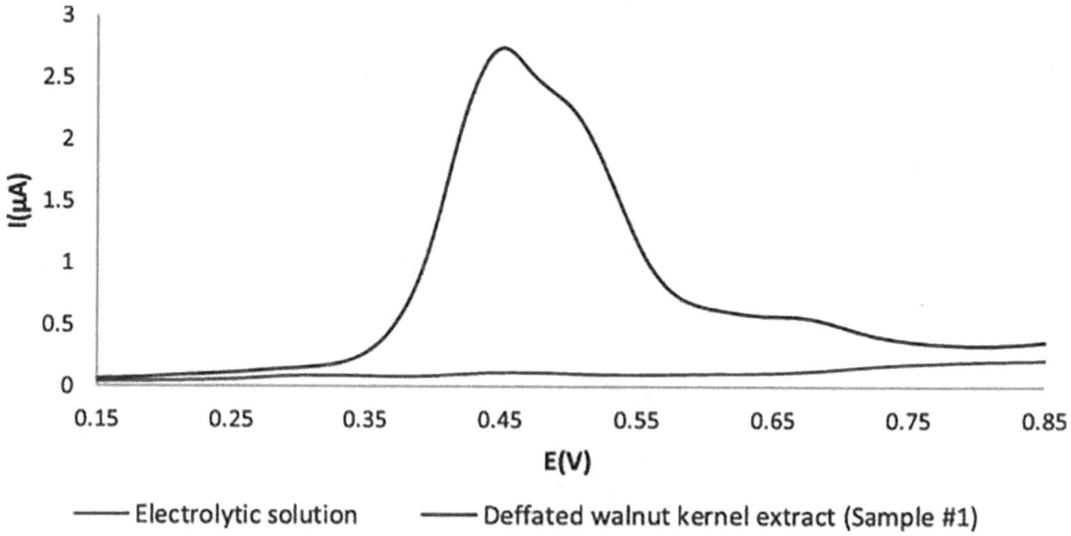

FIGURE 7.3 Differential pulse voltammogram recorded at sonogel-carbon electrode of 500 μL of defatted walnut kernel extract in Britton-Robinson solution at pH 2.0 (red line) superimposed to the response of the electrolytic solution. Modulation amplitude 25 mV; step potential 8 mV; interval time 0.4 s; modulation time 0.05 s. (Adapted from ref. [11], with permission. Copyright Elsevier 2022.).

In principle, the resulting voltammogram should result from the superposition of the voltammograms of the individual components. However, as commented in Chapter 4, there is another problem to be accounted for: In multicomponent systems, cross-reactions between the intermediates (typically semiquinone radical intermediates in phenolic compounds) and the parent compounds could distort the voltammetric response respect to the expected if all compounds oxidized independently.

The simplest evaluation of TAC from voltammetric data is based on the measurement of the peak potentials [12]. Under ideal, reversible conditions, these have thermochemical meaning so that the lower the oxidation potential, the more oxidizable the component. A more elaborated approach is the so-called electrochemical index (EI), defined for the determination of the antioxidant capacity of polyphenols in foodstuffs [13]. The idea is to balance the oxidation peak potential (E_{pa}) and the peak current (i_{pa}) recorded under fixed conditions. The anodic peak potential is representative of the electron donor ability of the compound whereas the peak current is representative of the concentration of the electroactive species. The EI for a series of N compounds displaying separated –and strictly independent – voltammetric peaks is defined as

$$EI = \frac{i_{pa1}}{E_{pa1}} + \frac{i_{pa2}}{E_{pa2}} + \ldots \frac{i_{paN}}{E_{paN}} \tag{7.1}$$

Given the limitations previously exposed, the EI has to be seen as an operational parameter to be used with caution. It is pertinent to note that the peak potentials recorded in voltammetric experiments do not necessarily possess thermodynamic significance. This requires electrochemical reversibility, a condition that frequently does not apply.

Different methods based on the measurement of electrode potentials (including the use of an auxiliary redox probe [14]), as well as a combination of peak currents and electrode potentials [15], have been proposed. Other electrochemical techniques such as amperometry [16] and chronocoulometry [17] have been reported within this context. An increasing strategy is to enhance sensitivity and selectivity using electrodes modified with different materials to study the voltammetric response of selected, individual components [18]

Apart from the aforementioned methods based on solution-phase electrochemistry, methods based on the voltammetry of immobilized microparticles methodology to dry extracts of fruits and vegetables [19], and potentiometric methods applied to plant microsuspensions have been proposed [20].

7.4 METHODS BASED ON THE REACTION WITH OXIDANTS

These involve monitoring the reactivity of plant extracts and cell cultures with redox probes such as $Fe(CN)_6^{3-}/Fe(CN)_6^{4-}$ and 2,4-quinone/2,4-hydroquinone couples [21]. One possible approach consists of the potentiometric measurement of the change of the platinum electrode potential in the ferricyanide/ferrocyanide system upon the addition of the plant extract generally in phosphate buffer at pH 7.0 or 7.4. The antioxidant capacity is expressed in terms of the antioxidant concentration, c_{AO}, given by

$$c_{AO} = q\frac{c_{III} - ac_{II}}{1+a}. \tag{7.2}$$

where c_{III} and c_{II} are the concentrations of $Fe(CN)_6^{3-}$ and $Fe(CN)_6^{4-}$ in the solution, q is the dilution factor (usually $q = 1$ for individual compounds and ranging between 10 and 50 for plant extracts) and a is a coefficient determined from the electrode potentials measured before (E_1) and after (E_2) the reaction,

$$a = \frac{c_{III}}{c_{II}}\exp\frac{F(E_2 - E_1)}{RT}. \tag{7.3}$$

Another possibility is the measurement of the decrease in peak current values for the $Fe(CN)_6^{3-}/Fe(CN)_6^{4-}$ couple upon additions of known concentrations of the antioxidant. Figure 7.4 depicts a series of cyclic voltammograms recorded on a $K_3Fe(CN)_6$ solution in citrate phosphate buffer at pH 6.0 before and after the addition of increasing concentrations of epigallocatechin gallate, a flavonoid [22]. The peak currents of the $Fe(CN)_6^{3-}/Fe(CN)_6^{4-}$ couple diminish after each addition of flavonoid, leading to a linear dependence between the absolute value of the variation of peak current with the concentration of the antioxidant.

Another example of this methodology is the electrochemical version of the CUPRAC assay [23]. The Cu(II)-neocuproine complex, $[Cu(Nc)_2]^{2+}$, is reversibly reduced to its Cu(I) analog and $[Cu(Nc)_2]^+$ in an aqueous ammonium acetate buffer solution. The Cu(II) complex oxidizes the antioxidants resulting in the enhancement of the anodic signal as can be seen in Figure 7.5.

The increase in the peak current for the oxidation peak, i_{pa}, becomes proportional to the concentration of antioxidants, as can be seen in calibration plots in Figure 7.6. The slope of the linear peak current vs. concentration plots differs from the different antioxidants. This has been interpreted as representative of the different stoichiometries of the Cu(II)-antioxidant reaction [23]. Ultimately, they could be taken as a measure of the TAC.

The oxidizing agents can also be generated electrochemically, and their reaction with the problem components can be monitored via coulometric or potentiometric titrations using, among other species, bromine and triiodide ions. The electrochemical generation of Br_2 [24] from KBr solutions and $Fe(CN)_6^{3-}$ from $K_4Fe(CN)_6$ solutions have been studied. In this second case, the determined TAC is termed “ferric-reducing powder”. In the case of bromine, reactions of electrophilic substitution in aromatic systems and electrophilic addition to multiple CC bonds can occur. As a result, the antioxidant capacity includes not only the reactivity of ascorbic acid and phenolic antioxidants but also the reactivity of terpenes and tocopherols.

7.5 METHODS BASED ON THE REACTION WITH RADICAL SCAVENGERS

These are based on the record of the modifications produced in the electrochemical response of the antioxidant components and/or the radical scavenger as a result of their mutual interaction. Paradigmatic cases are DPPH and TEMPO (2,2,6,6-tetramethylpiperidinyl-1-oxy). DPPH experiences a one-electron reduction to the radical DPPH•, which can react with the antioxidant components of the sample under study [25]. Potentiometric, voltammetric, and amperometric monitoring

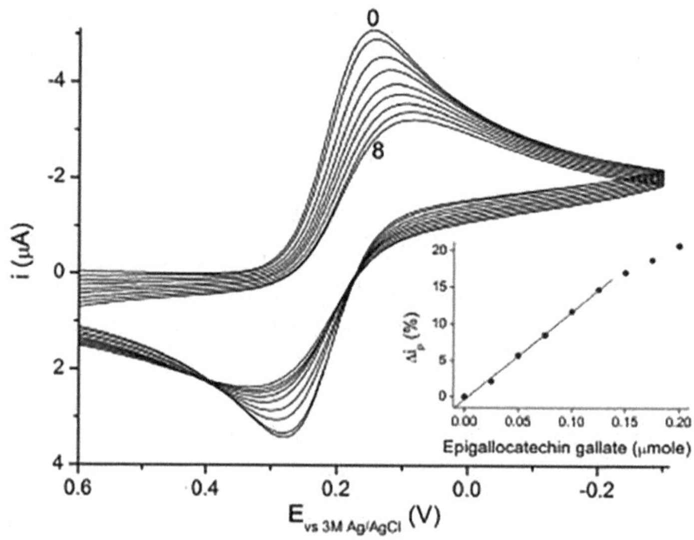

FIGURE 7.4 Cyclic voltammograms of 5 mL of 1.0 mM $K_3Fe(CN)_6$ solution in citrate phosphate buffer at pH 6.0 recorded before (0) and during titration with eight aliquots (1–8) of 12.5 μL of 2.0 mM epigallocatechin gallate solution. Inset: plot of the absolute value of the peak current increment vs. the amount of epigallocatechin gallate added. **(Reproduced from ref. [22], with permission. Copyright Elsevier 2020.).**

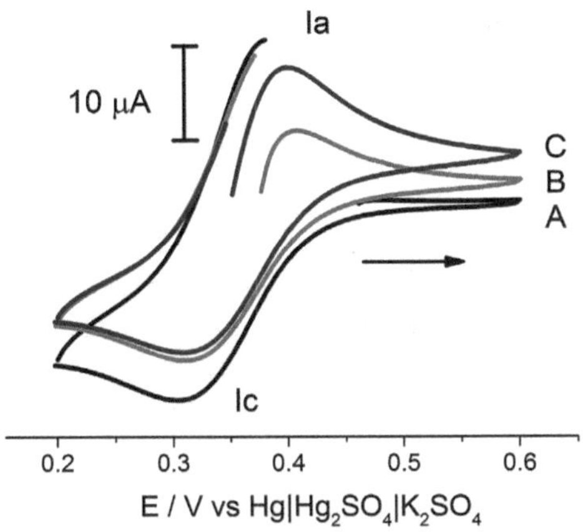

FIGURE 7.5 Cyclic voltammograms at GCE of 3.0 mM Cu(II) plus 6.0 mM Nc solution in ammonium acetate buffer (pH 7), upon adding different concentrations of Trolox: (A) 0; (B) 0.363, (C) 0.666 mM. The arrow indicates the initial direction of the potential scan, beginning from the open circuit potential (OCP). Potential scan rate 100 mV s^{-1}. **(Adapted from ref. [23], with permission.).**

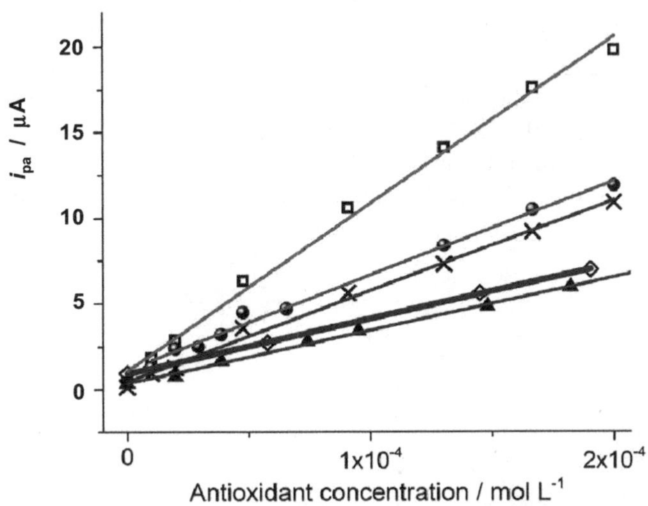

FIGURE 7.6 Calibration plots between anodic peak current (i_{pa}) and antioxidant concentration from cyclic voltammetric data of $[Cu(Nc)_2]^{2+}$ solutions in conditions such as in Figure 7.4. Circles: Trolox, solid circles: (+)-Catechin, squares: Quercetin, X: Gallic acid, solid triangles: AA. (Adapted from ref. [23], with permission. Copyright Elsevier 2014.).

of the reaction can be carried out [26, 27]. In the first case, it is convenient to add solutions with different concentrations of the problem system and obtain the corresponding linear (electrode potential) vs. (ln(concentration)) plots. Ideally, this representation can be fitted to a Nernstian plot of slope 59 mV decade^{-1}. Then the TAC is expressed in terms of the concentration of DPPH•, c_{DPPH}, which reacts as

$$c_{DPPH} = \frac{c_{DPPH}\left(1-k\right)-kc_{DPPH}}{1-k}, \tag{7.4}$$

where k is a function of the equilibrium potentials measured before (E_1) and after (E_2) each addition and b is the slope of the Nernstian plot:

$$k = \exp\left[\frac{E_1 - E_2}{b}\right]. \tag{7.5}$$

The radical scavenger can be electrochemically generated or regenerated in situ: In the second case, there is the possibility of automated amperometric titrations [27]. The case of DPPH• has been widely studied. In acidic media, the radical species undergoes a one-electron reversible reduction to DPPH⁻, as illustrated in Figure 7.7 [28]. In the presence of an antioxidant A-H, the voltammetric profile changes due to the reaction:

$$\text{DPPH}\bullet + \text{A}-\text{H} \rightarrow \text{DPPH}-\text{H} + \text{A}\bullet. \tag{7.6}$$

The parent A-H species can be regenerated electrochemically through the process:

$$\text{A}\bullet + \text{H}^+ + \text{e}^- \rightarrow \text{A}-\text{H}. \tag{7.7}$$

As a result, both the voltammetric profile associated with the DPPH- and A-H-localized processes become modified. A typical example is illustrated in Figure 7.7. Here, the DPPH•/DPPH⁻ appears as a pair of peaks at ca. 0.35 V, while the voltammetry of piperine is more complex. It consists of three anodic waves attributable to the formation of a piperine diol, subsequently oxidized to a dione compound, and then oxidized with the formation of two carboxylic acids. For our purposes, the relevant point to emphasize is that, in the DPPH• plus piperine mixture, the intensity of the piperine oxidation signals is diminished, thus denoting the advance of the reaction with DPPH•. Quantitative estimates of the TAC can be made based on measurements of the decrease in peak currents.

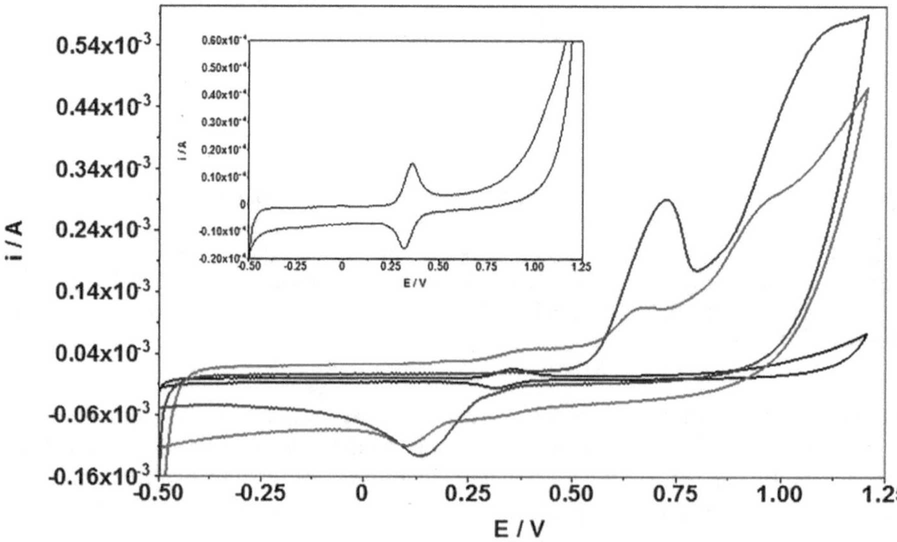

FIGURE 7.7 Cyclic voltammograms at screen printed electrode of 0.5 mM DPPH•, 0.5 mM piperine solutions in 0.1 M $HClO_4$, and a 0.5 mM DPPH• + 0.5 mM piperine solution in the same medium after 45 min incubation. Inset: cyclic voltammogram of 0.5 mM DPPH• in 0.1 M $HClO_4$. (Adapted from ref. [28], with permission. Copyright Elsevier 2021.).

Another possibility for estimating the TAC of plant extracts is the immobilization of the radical scavenger reagents onto electrode surfaces. This approach involves more or less complex procedures of electrode modification. Methods based on the immobilization of DPPH• [29] and galvinoxyl [30] radicals have been recently devised. As methods in solution, the essential idea is to study the variations in the voltammetric response of the radical scavenger and/or the antioxidant component.

7.6 METHODS BASED ON THE INTERACTION WITH ROS

As studied in Chapter 5, the interaction with ROS plays an essential role in plant biochemistry. In particular, most of the antioxidant activity of plant components refers to their interaction with ROS. The most studied components, phenolic compounds, donate hydrogen atoms from phenoxyl groups to radicals, inactivating their oxidative activity. Different alternative mechanisms have been proposed, hydrogen atom transfer (HAT), sequential proton loss electron transfer (SPLET), and electron transfer followed by proton transfer (SET-PT) [31–33]. About TAC estimates, this interaction can be monitored electrochemically using different strategies.

The possibly simplest approach exploits the well-known electrochemistry of dissolved oxygen in aprotic solvents [34, 35]. Here O_2 experiences a reversible one-electron reduction to radical anion superoxide [36],

$$O_2 + e^- \rightarrow O_2\bullet^-, \tag{7.8}$$

where both O_2 and $O_2\bullet^-$ are solvated species. Just the solvation of the superoxide radical anion determines its stability at the time scale of electrochemical measurements. Figure 7.8a shows a typical cyclic voltammogram recorded at GCE in an air-saturated DMSO solution. Here, the process described by Eq. (7.8) appears as a typical diffusion-controlled reversible couple at a midpeak potential of −0.9 V vs. Ag/AgCl. In the presence of proton donors, such as acetate buffer at pH 4.75 (see Figure 7.8b), the anodic peak vanishes, thus denoting that the superoxide radical anion electrochemically generated reacts rapidly relative to the time scale of the experiment. The simplest approach considers two reaction pathways. In the HAT pathway, this process leads to the formation of hydroperoxide radicals,

$$O_2\bullet^- + HA \rightarrow HO_2\bullet + A^-, \tag{7.9}$$

while in the radical transfer pathway, a hydroperoxide anion is formed:

$$O_2\bullet^- + HA \rightarrow HO_2^- + A\bullet. \tag{7.10}$$

In both cases, the result of the superoxide-antioxidant interaction is the depletion of the oxidation peak of the $O_2/O_2\bullet^-$ couple. This effect can be seen in Figure 7.9, corresponding to the voltammetric record of successive additions of 4-chlorophenol, *tert*-butylhydroquinone, catechol, and quercetin to an air-saturated N,N'-dimethylformaamide (DMF) solution [35].

Figure 7.10 shows a two-dimensional diagram containing the values of the ratio between the peak currents for O_2 reduction in the presence of antioxidant (i_{pc}^{AO}) and in its absence (i_{pc}°), and the logarithm of the antioxidant concentration necessary to decrease 10% of the initial anodic peak current ($c_{10\%}$) [35]. The diagram, recorded from cyclic voltammograms, such as depicted in Figure 7.9, clearly discriminates the antioxidant activity of different compounds.

The radical hydroperoxide can be partially adsorbed onto the electrode surface and disproportionates into hydrogen peroxide and oxygen,

$$2HO_2\bullet \rightarrow H_2O_2 + O_2, \tag{7.11}$$

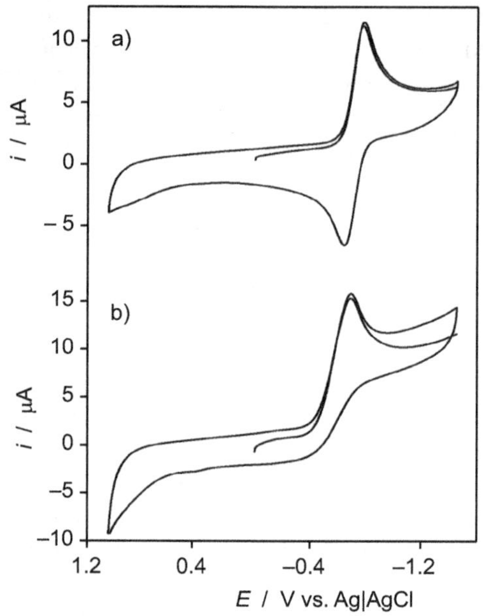

FIGURE 7.8 Cyclic voltammograms at GCE of an air-saturated 0.10 M Bu_4NPF_6/DMSO solution before (a) and after (b) the addition of HAc/NaAc aqueous buffer at pH 4.75 reaching a concentration 0.01 M. Potential scan initiated at 0.0 V vs. Ag/AgCl in the negative direction; potential scan rate 50 mV s^{-1}. Note that the axes are opposite to the IUPAC convention.

competing with the electrochemical reduction process at more negative potentials,

$$HO_2\bullet(ads) + HA + e^- \rightarrow H_2O_2 + A^-, \quad (7.12)$$

occurring in excess over proton donors [37]. Notice that the disproportionation process described by Eq. (7.11) generates O_2, i.e., can be seen as a catalytic regeneration of O_2, resulting in an increase of the peak current for O_2 reduction. In the presence of water traces in the aprotic solvents, a second reduction wave for O_2 is recorded at highly cathodic potentials, being attributed [38] to the reduction of the hydrogen-bonded complex of superoxide with water or another weak acid:

$$O_2\bullet^- + H_2O + e^- \rightarrow HO_2^- + OH^-. \quad (7.12)$$

For weak acids such as water and alcohols, the overall decomposition of $O_2\bullet^-$ is further complicated by the base-induced decomposition of H_2O_2 in aprotic solvents generating hydroxyl radical [37, 39]:

$$H_2O_2 + A^- \rightarrow HO_2^- + HA, \quad (7.13)$$

$$HO_2^- + H_2O_2 \rightarrow O_2\bullet^- + OH\bullet + H_2O. \quad (7.14)$$

In this complex scenario, the relevant point to emphasize is that most antioxidant compounds in plants exert the aforementioned effect on the voltammetric response of O_2. Then their antioxidant capacity can be evaluated in terms of the decrease in the anodic signal of the $O_2/O_2\bullet^-$ couple. Among other parameters [7], an antioxidant coefficient K_{AO} has been defined as [7]

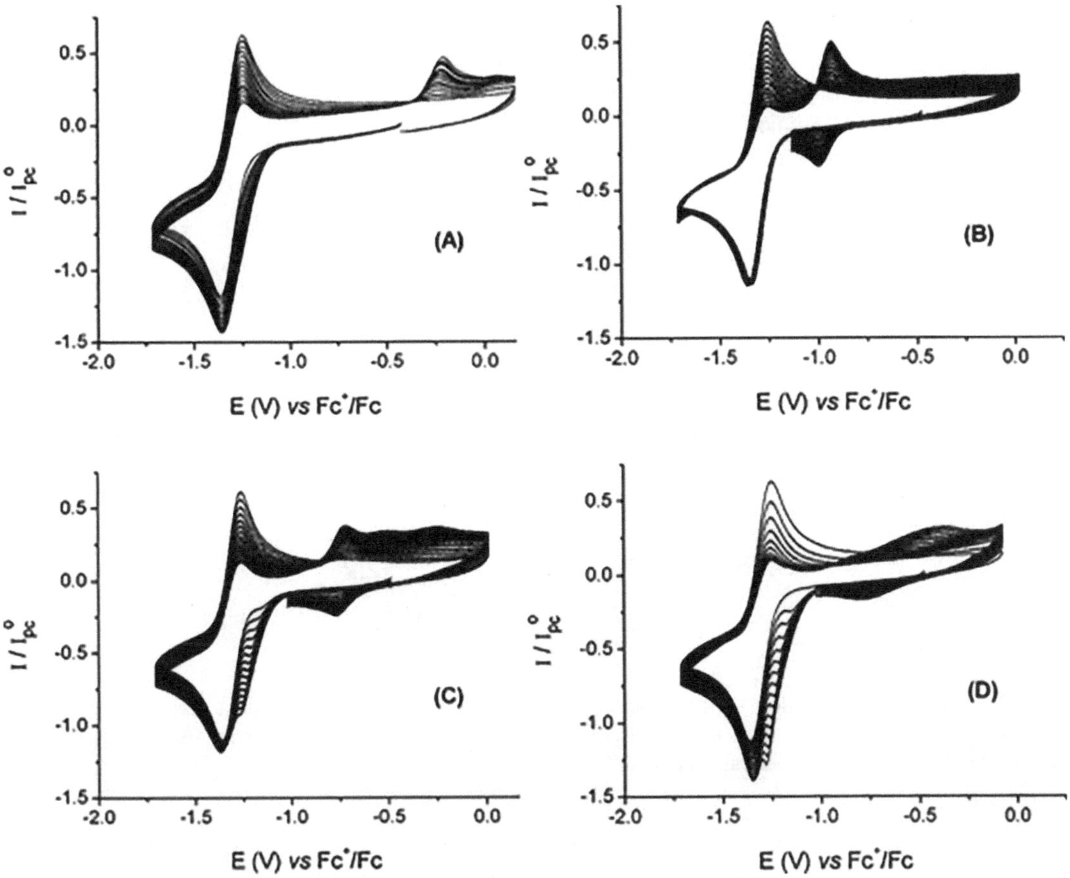

FIGURE 7.9 Cyclic voltammograms at GCE of the $O_2/O_2^{\bullet-}$ system in the presence of increasing concentrations of phenolic compounds in 0.10 M Bu_4NPF_6/DMF. 10^{-4} M successive additions of (a) 4-chlorophenol, (b) ***tert***-butylhydroquinone, (c) catechol, and (d) quercetin. Potential scan rate 100 mV s^{-1}. (Reproduced from ref. [35], with permission. Copyright the American Chemical Society 2010.).

$$K_{AO} = \frac{\Delta i_p}{\left(i_p^{\,lim} - i_p^{\,res}\right)\Delta c_{AO}}. \tag{7.15}$$

In this equation, Δi_p represents the increase in the peak current for O_2 reduction, i_p^{lim} the limiting peak current in the absence of antioxidants, and i_p^{res} the residual current at the same potential in the absence of O_2. Finally, Δc_{AO} denotes the variation in the concentration of the antioxidant.

In an aqueous solution, the reaction with hydroxyl ions generated chemically has been used to evaluate TAC [40, 41]. The hydroxyl radicals are produced via the Fenton reaction:

$$Fe^{2+} + H_2O_2 \rightarrow Fe^{3+} + HO\bullet + OH^-. \tag{7.16}$$

A competing strategy is in general used. It consists of the immobilization of an electrochemically active species (DNA) or an electrochemically silent probe (hexanethiol self-assembled monolayers), which are also sensitive to the HO• attack. In the first case, the voltammetric signal of the surface-attached probe (typically guanine oxidation of DNA probes) is diminished in the presence of the

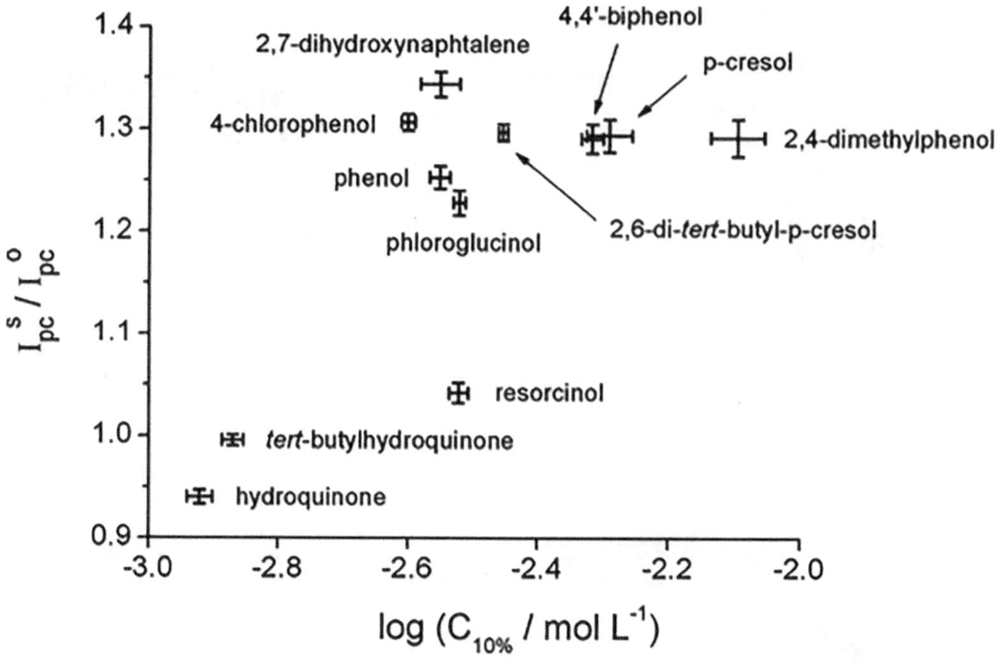

FIGURE 7.10 Representation of the values of the i_{pc}^{AO}/i_{pc}^{o} ratio vs. log$c_{10\%}$ for a series of antioxidant compounds using cyclic voltammetric data such as depicted in Figure 7.8. (Adapted from ref. [35], with permission. Copyright the American Chemical Society 2010.).

Fenton reagent. If an antioxidant is subsequently added, the signal of the surface-attached probe is enhanced because of the reaction of the antioxidant with hydroxyl radicals. In the case of hexanethiol self-assembled monolayers, a second redox probe, $[Ru(NH_3)_6]^{3+}$, is added. This complex displays a reversible, well-defined voltammetric response whose intensity increases in the presence of hydroxyl radicals. This is due to the reaction of the hexanethiol units with HO•. Then the addition of an antioxidant decreases the $[Ru(NH_3)_6]^{3+}$ signals as a result of the antioxidant-hydroxyl radical interaction. The antioxidant capacity is expressed in these cases as percentages of relative inhibition. The drawback of these methods is the possibility of side reactions by effect of the direct reaction of H_2O_2 with the antioxidants and/or the complexation of transition metal ions by the antioxidants. The latter effect can be prevented by adding a strong complexing agent (ethylenediaminetetraacetic acid, EDTA) to the cleavage solution.

A plausible strategy in aqueous media is the electrochemical generation of ROS, including hydroxyl radicals [10], superoxide [42], and peroxyl radicals [43]. The possibility of electrochemical ROS generation also refers to studies dealing with environmental chemistry [44–46]. This strategy has also been applied through the contact probe voltammetry technique for *in situ* monitoring of the reactivity of phenolic tomato compounds with ROS [47] and studying the ROS reactivity of plant extracts using solid-state approaches [48]. Figure 7.11 shows the cyclic voltammetric response, after semi-derivative convolution, of microparticulate deposits prepared from ethanolic extracts of Ceylan uva black tea leaves on GCE using air-saturated phosphate buffer at pH 7.0 as the electrolyte [48]. The voltammograms show a prominent cathodic signal at −0.85 V corresponding to the reduction of dissolved oxygen (C_{ox}) accompanied by a series of anodic signals in the 0.0 to 1.0 V potential region assigned to the oxidation of polyphenolic compounds (lignins, flavonoids, etc.) labeled as A_{AO}. The profile of this region changes depending on the previous electrochemical generation of ROS in the process C_{ox} (see Chapter 5).

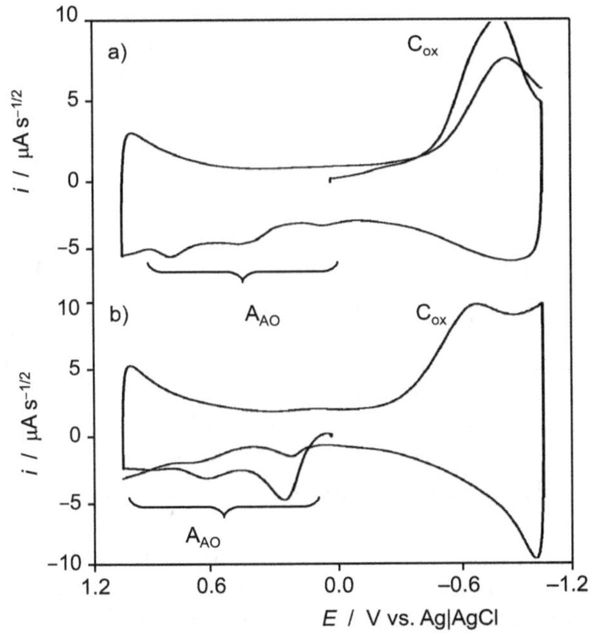

FIGURE 7.11 Cyclic voltammograms of microparticulate deposits on GCE of ethanolic extracts of Ceylan uva black tea leaves in contact with air-saturated 0.10 M aqueous potassium phosphate buffer at pH 7.0. Potential scan initiated at 0.0 V (a) in the negative and (b) in the positive direction. Potential scan rate 100 mV s^{-1}. Semi-derivative convolution was applied to current/potential curves to increase peak resolution. Note that the axes are opposite to the IUPAC convention. (Adapted from ref. [48], with permission. Copyright Elsevier 2015.).

In the initial negative-going potential voltammograms (Figure 7.11a), the C_{ox} process precedes the appearance of A_{AO} signals, consisting of oxidation peaks at 0.05, 0.35, 0.45, and 0.80 V. If the scan is initiated at 0.0 V in the positive direction (Figure 7.11b), there is no initial ROS generation, and the A_{AO} signals differ from the aforementioned. Here, anodic peaks at 0.25, 0.60, and 0.90 V are recorded. The height of such peaks decreases in the second anodic scan, recorded after the ROS generation in the previous C_{ox} process. The essential idea is that the interaction of the parent antioxidant components with the different ROS species generated electrochemically leads to new species, several being electroactive.

This type of voltammetric feature is potentially useful to evaluate antioxidant capacities using the ratio between the intensity of peaks C_{ox} and/or A_{AO} in the first and second potential scans [47, 48]. The drawback is the sensitivity of ROS generation to the electrochemical conditions (potential range, potential scan rate) and, hence, of the A_{AO} profile to these conditions. Ultimately, this type of experiment could lead to select conditions for the generation of a given ROS and estimate the antioxidant capacity of plant extracts relative to this ROS.

7.7 INDIVIDUAL ANTIOXIDANT COMPONENTS OF PLANTS

Another form of study of the antioxidant properties of plants is the analysis of individual components. This matter will be treated more extensively in Chapter 10. As far as the estimate of the antioxidant capacity of plants, foods, beverages, etc., is concerned, the essential idea is to express such capacity in terms of the concentration of the majority component or a particular component of interest – for instance, catechin in tea. This demands high sensitivity and high selectivity in the analytical detection/determination so that considerable research is being developed based on more or less complex procedures of electrode modification. These include irreversible adsorption, covalent binding, inclusion in and/or functionalization of polymer films, embedding into carbon paste, deposition via drop casting, screen printing, sol-gel technologies, formation of inclusion complexes, and molecularly imprinted polymeric materials, among others. Electrode modification includes an enormous variety of materials, from metal nanoparticles and different metal and metal compound nanoarchitectures to carbon nanotubes, graphene and porous carbons, silica, silicates and aluminosilicates, layered double hydroxides, natural and synthetic polymers, among others, including DNA and aptamers, enzymes and other biocomponents [49].

The most widely studied components are ascorbic acid, baicalein, chlorogenic acid, eugenol, ferulic acid, gallic acid, naringenin, quercetin, rutin, and vanillin, among others. The voltammetric analysis is a technique largely predominating, but potentiometric, amperometric, and impedimetric sensing are also under intensive research. The analytical approach can involve (i) the record of the direct electrochemical response of the electroactive analyte or (ii) the monitoring of an auxiliary electrochemical process. In this second case, the analyte modified the auxiliary electrochemical process via catalysis, complexation, etc., or blocks intensively the response of the auxiliary electroactive species.

Another approach is the determination of a family of components. Reported methods include the voltammetric determination of total phenolic components [50] and other parameters [51]. For practical purposes, the electrochemical methods have to accomplish not only the usual analytical demands, accuracy, precision, repeatability, reproducibility, sensitivity, and selectivity, but also those concerning robustness, traceability, as well as the conditions associated with environmental safety. In general, the analysis of individual components in vegetal samples offers difficulty derived from the multicomponent nature of such samples, i.e., the possibility of serious interference and matrix effects.

REFERENCES

[1] Halliwell, B.; Gutteridge, J. M. C. 2015. *Free Radicals in Biology and Medicine*, 4th ed., Oxford University Press, Oxford.

[2] Chiorcea-Paquim, A.-M.; Enache, T.A.; De Souza Gil, E.; Oliveira-Brett, A.M. 2020. Natural phenolic antioxidants electrochemistry: Towards a new food science methodology. *Comprehensive Reviews in Food Science and Food Safety*. 19: 1680–1726.

[3] Ziyatdinova, G.; Kalmykova, A. 2023. Electrochemical Characterization of the Antioxidant Properties of Medicinal Plants and Products: A Review. *Molecules*. 28: artic. 2308.

[4] Glod, K.; Czaspi, G.A.; Haddad, P.R. 2000. Application of high-performance liquid chromatography to the investigation of free radical reactions in biological systems, *Trends in Analytical Chemistry*. 19: 492–497.

[5] Barros, L.; Cabrita, L.; Vilas Boas, M.; Carvalho, A.M.; Ferreira, I.C.F.R. 2011. Chemical, biochemical and electrochemical assays to evaluate phytochemicals and antioxidant activity of wild plants. *Food Chemistry*.127: 1600–1608.

[6] Doménech-Carbó, A. 2021. Electrochemistry of plants: Basic theoretical research and applications in plant science. *Journal of Solid State Electrochemistry*. 25: 2747–2757.

[7] Haque, M.A.; Morozova, K.; Ferrentino, G.; Scampicchio, M. 2021 Electrochemical methods to evaluate the antioxidant activity and capacity of foods: A review. *Electroanalysis*. 33: 1419–1435.

[8] Ziyatdinova, G.; Kalmykova, A. 2023. Electrochemical Characterization of the Antioxidant Properties of Medicinal Plants and Products: A Review. *Molecules*. 28: artic. 2308.

[9] Otte, O.; Barz, W. 1996 The elicitor-induced oxidative burst in cultured chickpea cells drives the rapid insolubilization of two cell wall structural proteins. *Planta*. 200: 238–246.

[10] Oliveira, R.; Geraldo, D.; Bento, F. 2014 Radical scavenging activity of antioxidants evaluated by means of electrogenerated HO radical. *Talanta*. 129: 320–327.

[11] Ruiz-Caro, P.; Espada-Bellido, E.; García-Guzmán, J.J.; Bellido-Milla, 2022 An electrochemical alternative for evaluating the antioxidant capacity in walnut kernel extracts. *Food Chemistry*. 323: artic. 133417.

[12] Yang, B.; Kotani, A.; Arai, K.; Kusu, F. 2001 Estimation of the antioxidant activities of flavonoids from their oxidation potentials. *Analytical Sciences*. 17: 599–604.

[13] Blasco, A.J.; González, M.C.; Escarpa, A. 2004 Electrochemical approach for discriminating and measuring predominant flavonoids and phenolic acids using differential pulse voltammetry: Towards an electrochemical index of natural antioxidants. *Analytica Chimica Acta*. 511: 71–81.

[14] Bordonaba, J.G.; Terry, L.A. 2012 Electrochemical behaviour of polyphenol rich fruit juices using disposable screen-printed carbon electrodes: Towards a rapid sensor for antioxidant capacity and individual antioxidants. *Talanta*. 90: 38–45.

[15] Glod, B.K.; Kiersztyn, I.; Piszcz, P. 2014 Total antioxidant potential assay with cyclic voltammetry and/or differential pulse voltammetry measurements. *Journal of Electroanalytical Chemistry*. 719: 24–29.

[16] Cosio, M.S.; Buratti, S.; Mannino, S.; Benedetti, S. 2006 Use of an electrochemical method to evaluate the antioxidant activity of herbal extracts from the Labiatae family. *Food Chemistry*. 97: 725–731.

[17] Ziyatdinova, G.; Kozlova, E.; Morozova, E.; Budnikov, H. 2018 Chronocoulometric method for the evaluation of antioxidant capacity of medicinal plant tinctures. *Analytical Methods*. 10: 4995–5003.

[18] Hoyos-Arbeláez, J.; Vázquez, M.; Contreras-Calderón, J. 2017 Electrochemical methods as a tool for determining the antioxidant capacity of food and beverages: A review. *Food Chemistry*. 221: 1371–1381.

[19] Komorsky-Lovrić, Š.; Novak, I. 2009 Estimation of antioxidative properties of tea leaves by abrasive stripping electrochemistry using paraffin-impregnated graphite electrode. *Collection of Czechoslovak Chemical Communications*. 74: 1467–1475.

[20] Brainina, K.; Stozhko, N.; Bukharinova, M.; Khamzina, E.; Vidrevich, M. 2019 Potentiometric method of plant microsuspensions antioxidant activity determination. *Food Chemistry*. 278: 653–658.

[21] Kinpara, T.; Murakami, Y.; Yokoyama, K.; Tamiya, E. 2000 Electrochemical Monitoring of Plant Stress Responses. *Electroanalysis*. 13: 451–456.

[22] Pastor, F.T.; Šegan, D.M.; Gorjanovic, S.Ž.; Kalušević, A.M.; Sužnjvić, D.Ž. 2020 Development of voltammetric methods for antioxidant activity determination based on Fe(III) reduction. *Microchemical Journal*. 155: artic. 104721.

[23] Cárdenas, A.; Gómez, M.; Frontana, C. 2014 Relationship between the chemical structures of antioxidants and the differences in their Cupric Ion Reducing Antioxidant Capacity (CUPRAC) by electrochemical methods. *Journal of Electroanalytical Chemistry*. 729: 116–120.

[24] Abdullin, I.F.; Turova, E.N.; Gaisina, G.K.; Budnikov, G.K. 2002 Use of electrogenerated bromine for estimating the total antioxidant capacity of plant raw materials and plant-based medicinal preparations. *Journal of Analytical Chemistry*. 57: 557–560.

[25] Solon, E.; Bard, A.J. 1964 The electrochemistry of diphenylpicrylhydrazyl radical. *Journal of the American Chemical Society*. 86: 1926–1928.

[26] Gerasimova, E.; Gazizullina, E.; Kolbaczkaya, S.; Ivanova, A. 2022 The novel potentiometric approach to antioxidant capacity assay based on the reaction with stable radical 2,20-diphenyl-1-picrylhydrazyl. *Antioxidants*. 11: artic. 1974.

[27] Oliveira, G.K.F.; Tormin, T.F.; Sousa, R.M.F.; de Oliveira, A.; de Morais, S.A.L.; Richter, E.M.; Munoz, R.A.A. 2016 Batch-injection analysis with amperometric detection of the DPPH radical for evaluation of antioxidant capacity. *Food Chemistry*. 192: 691–697.

[28] Carp, O.E.; Moraru, A.; Pinteala, M.; Arvinte, A. 2021 Electrochemical behaviour of piperine. Comparison with control antioxidants. *Food Chemistry*. 339: artic. 128110.

[29] Ziyatdinova, G.; Snegureva, Y.; Budnikov, H. 2017 Novel approach for the voltammetric evaluation of antioxidant activity using DPPH•-modified electrode. *Electrochimica Acta*. 247: 97–106.

[30] Ziyatdinova, G.; Zelenova, Y.; Budnikov, H. 2020 Novel modified electrode with immobilized galvinoxyl radical for the voltammetric determination of antioxidant activity. *Jorunal of Electroanalytical Chemistry*. 856: artic. 113677.

[31] Litwinienko, G.; Ingold, K.U. 2004 Abnormal Solvent Effects on Hydrogen Atom Abstractions. 2. Resolution of the Curcumin Antioxidant Controversy. The Role of Sequential Proton Loss Electron Transfer. *Journal of Organic Chemistry*. 69: 5888–5896.

[32] Litwinienko, G.; Ingold, K. U. 2007. Solvent effects on the rates and mechanisms of reaction of phenols with free radicals. *Accounts of Chemical Research*, *40*, 222–230.

[33] Galano, A.; Raúl Alvarez-Idaboy, J. 2019 Computational strategies for predicting free radical scavengers' protection against oxidative stress: where are we and what might follow? *International Journal of Quantum Chemistry*. 119: artic. e25665

[34] Le Bourvellec, C.; Hauchard, D.; Darchen, A.; Burgot, J.-L.; Abasq, M.-L. 2008 Validation of a new method using the reactivity of electrogenerated superoxide radical in the antioxidant capacity determination of flavonoids. *Talanta*. 75: 1098–1103.
[35] René, A.; Abasq, M.-L.; Hauchard, D.; Hapiot, P. 2010 How Do Phenolic Compounds React toward Superoxide Ion? A Simple Electrochemical Method for Evaluating Antioxidant Capacity. *Analytical Chemistry*. 82: 8703–8710.
[36] Sawyer, D.T.; Chiericato, G.; Angelis, C.T.; Nanni, E.J.; Tsuchiya, T. 1982 Effects of Media and Electrode Materials on the Electrochemical Reduction of Dioxygen. *Analytcal Chemistry*. 54: 1720–1724.
[37] Cofré, P.; Sawyer, D.T. 1986 Electrochemical Reduction of Dioxygen to Perohydroxyl (HO_2•) in Aprotic Solvents That Contain Bronsted Acids. *Analytcal Chemistry*. 58: 1057–1062.
[38] Singh, P.S.; Evans, D.H. 2006 Study of the Electrochemical Reduction of Dioxygen in Acetonitrile in the Presence of Weak Acids. *Journal of Physical Chemistry B*.110: 637–644.
[39] Chin, D.-H.; Chiericato, G.; Nanni, E.J.; Sawyer, D.T. 1982 Proton-Induced Disproportionation of Superoxide Ion in Aprotic Media. *Journal of the American Chemical Society*. 104: 1296–1299.
[40] Labuda, J.; Bučková, M.; Heilerová, L.; Čaniová-Žiaková, A.; Brandšteterová, E.; Mattusch, J.; Wennrich, R. 2002 Detection of antioxidative activity of plant extracts at the DNA-modified screen-printed electrode. *Sensors*. 2: 1–10.
[41] Hilgemann, M.; Scholz, F.; Kahlert, H.; de Carvalho, L.M.; da Rosa, M.B.; Lindequist, U.; Wurster, M.; do Nascimento, P.C.; Bohrer, D. 2010 Electrochemical assay to quantify the hydroxyl radical scavenging activity of medicinal plant extracts. *Electroanalysis*. 22: 406–412.
[42] Juárez-Gómez, J.; Ramírez-Silva, M.T.; Guzmán-Hernández, D.S.; Romero-Romo, M.; Palomar-Pardave, M. 2020 Novel electrochemical method to evaluate the antioxidant capacity of infusions and beverages, based on in situ formation of free superoxide radicals. *Food Chemistry*. 332: artic. 127209.
[43] Ivanova, A.V.; Gerasimova, E.L.; Gazizullina, E.R. 2019 New antiradical capacity assay with the use potentiometric method. *Analytica Chimica Acta*. 1046, 69–76.
[44] Marsellis, B.; Garcia-Gomez, J.; Michaud, P.A.; Rodrigo, M.A.; Comninellis, C. 2003 Electrogeneration of hydroxyl radicals on boron-doped diamond electrodes. *Journal of the Electrochemical Society*. 150: D79–D83.
[45] Enache, T.A.; Chiorcea-Paquim, A.M.; Fatibello-Filho, O.; Oliveira-Brett, A.M. 2009 Hydroxyl radicals electrochemically generated in situ on a boron doped diamond electrode. *Electrochemistry Communications*. 11: 1342–1345.
[46] Kapałka, A.; Foti, G.; Comninellis, C. 2009 The importance of electrode material in environmental electrochemistry: Formation and reactivity of free hydroxyl radicals on boron-doped diamond electrodes. *Electrochimica Acta*. 54: 2018–2023.
[47] Doménech-Carbó, A.; Gavara, R.; Hernández, P.; Domínguez, I. 2015 Contact probe voltammetry for *in situ* monitoring of the reactivity of phenolic tomato (*Solanum lycopersicum* L.) compounds with ROS. *Talanta*. 144: 1207–1215.
[48] Domínguez, I.; Doménech-Carbó, A. 2015 Screening and authentication of tea varieties based on microextraction-assisted voltammetry of microparticles. *Sensors and Actuators B*. 210: 491–499.
[49] Doménech-Carbó, A. 2021 *Electrochemistry of Porous Materials*, 2nd edit. CRC Press, Boca Raton.
[50] Piovesan, J.V.; Jost, C.L.; Spinelli, A. 2015 Electroanalytical determination of total phenolic compounds by square-wave voltammetry using a poly(vinylpyrrolidone)-modified carbon-paste electrode. *Sensors and Actuators B*. 216: 192–197.
[51] Koc, T.B.; Savan, E.K.; Karabulut, I. 2022 Electrochemical determination of the antioxidant capacity, total phenolics, and ascorbic acid in fruit and vegetables by differential pulse voltammetry (DPV) with a *p*-toluene sulfonic acid modified glassy carbon electrode (TSA/GCE). *Analytical Letters*. 56: 881–894.

ADDITIONAL LITERATURE

Abdel-Hamid, R.; Newair, E.F. 2011. Electrochemical behavior of antioxidants: I. Mechanistic study on electrochemical oxidation of gallic acid in aqueous solutions at glassy-carbon electrode. *Journal of Electroanalytical Chemistry*. 657: 107–112.
Apak, R.; Gorinstein, S.; Böhm, V.; Schaich, K.M.; Özyürek, M.; Güçlü, K. 2013 Methods of measurement and evaluation of natural antioxidant capacity/activity (IUPAC Technical Report). *Pure and Applied Chemistry*. 85: 957–998.

Ayaz, S.; Üzer, A.; Dilgin, Y.; Apak, R. 2022 A novel flow injection amperometric method for sensitive determination of total antioxidant capacity at cupric-neocuproine complex modified MWCNT glassy carbon electrode. *Microchimica Acta*. 189: artic. 167.

Barroso, A.F.; Ramalhosa, M.J.; Alves, A.C.; Dias, A.; Soares, C.M.D.; Oliva-Teles, M.T.; Delerue-Matos, C. 2016 Total antioxidant capacity of plant infusions: Assessment using electrochemical DNA-based biosensor and spectrophotometric methods. *Food Control*. 68: 153–161.

Brainina, K.; Stozhko, N.; Bukharinova, M.; Khamzina, E.; Vidrevich, M. 2019 Potentiometric method of plant microsuspensions antioxidant activity determination. *Food Chemistry*. 278: 653–658.

Ferreira, R. de Q.; Avaca, L.A. 2008. Electrochemical determination of the antioxidant capacity: The ceric reducing/antioxidant capacity (CRAC) assay. *Electroanalysis*. 20: 1323–1329.

Glod, B.K.; Kiersztyn, I.; Piszcz, P. 2014 Total antioxidant potential assay with cyclic voltammetry and/or differential pulse voltammetry measurements. *Journal of Electroanalytical Chemistry*. 719: 24–29.

Hashkavayi, A.B.; Hashemnia, S.; Osfouri, S. 2020 Investigations of antioxidant potential and protective effect of Acanthophora algae on DNA damage: An electrochemical approach. *Microchemical Journal*. 159: artic. 105455.

Intarakamhang, S.; Schulte, A. 2012 Automated electrochemical free radical scavenger screening in dietary samples. *Analytical Chemistry*. 84: 6767–6774.

Ivanova, A.V.; Gerasimova, E.L.; Gazizullina, E.R. 2020 An integrated approach to the investigation of antioxidant properties by potentiometry. *Analytica Chimica Acta*. 1111: 83–91.

Kilmartin, P.A.; Zou, H.; Waterhouse, A.L. 2001 A cyclic voltammetry method suitablefor characterizing antioxidant properties of wine and wine phenolics. *Journal of Agricultural and Food Chemistry*. 49: 1957–1965.

Komorsky-Lovrić, Š.; Novak, I. 2011 Abrasive stripping square-wave voltammetry of blackberry, raspberry, strawberry, pomegranate, and sweet and blue potatoes. *Journal of Food Science*. 76: C916–C920.

Korotkova, E.I.; Voronova, O.A.; Dorozhko, E.V. 2012 Study of antioxidant properties of flavonoids by voltammetry. *Journal of Solid State Electrochemistry*. 16: 2435–2440.

Lauw, S.J.L.; Chiang, Z.; Lee, J.H.Q.; Webster, R.D. 2018 Comparing the Relative Reactivities of Structurally Varied Alcohols toward Electrochemically Generated Superoxide. *ChemElectroChem*. 5: 643–650.

Mello, L.D.; Hernandez, S.; Marrazza, G.; Mascini, M.; Kubota, L.T. 2006 Investigations of the antioxidant properties of plant extracts using a DNA-electrochemical biosensor. *Biosensors and Bioelectronics*. 21: 1374–1382.

Milardović, S.; Iveković, D.; Grabarić, B.S. 2006 A novel amperometric method for antioxidant activity determination using DPPH free radical. *Bioelectrochemistry*. 68: 175–180.

Newair, E.F.; Abdel-Hamid, R.; Kilmartin, P.A. 2017 Electrochemical determination of the antioxidant activity in Echinacea Purpurea roots using square wave voltammetry. *Electroanalysis*. 29: 1131–1140.

Oliveira, R.; Bento, F.; Sella, C.; Thouin, L.; Amatore, C. 2013 Direct electroanalytical method for alternative assessment of global antioxidant capacity using microchannel electrodes. *Analytical Chemistry*. 85: 9057–9063.

Piovesan, J.V.; Jost, C.L.; Spinelli, A. 2015 Electroanalytical determination of total phenolic compounds by square-wave voltammetry using a poly(vinylpyrrolidone)-modified carbon-paste electrode. *Sensors and Actuators B*. 216: 192–197.

Villela Maciel, J.; Almeida Silva, T.; Dias, D.; Fatibello-Filho, O. 2018 Electroanalytical determination of eugenol in clove oil by voltammetry of immobilized microdroplets. *Journal of Solid State Electrochemistry*. 22: 2277–2285.

8 Electrochemistry of Proteins, Enzymes, and Photosynthesis

8.1 INTRODUCTION

Proteins and enzymes are essential components of living organisms. Direct electron transfer processes between redox-active proteins and electrodes have been studied upon immobilization of the biopolymer onto electrode surfaces, frequently using a redox mediator. In the case of enzymes, there is the possibility of recording the electrochemical response of the redox cofactor comprising the active site in the absence of a substrate and/or observing the catalytic response in the presence of the substrate [1].

Although electron transfer processes between proteins, enzymes, and electrodes differ from the conditions in which redox transformations occur in vivo, the study of the electrochemistry of such biopolymers can be helpful for the interpretation of redox pathways in the biological context [2].

Among the plethora of biochemical processes occurring in plants, photosynthesis plays a crucial role. Photosynthesis can be seen as the process through which plants, algae, and cyanobacteria convert solar energy into stored energy in the form of reduced carbon. The process starts with the absorption of light by the reaction centers, which contain chlorophyll pigments. In eukaryotic photosynthetic cells, these are located on the thylakoid, the internal membrane structure of chloroplasts. Photosynthesis involves electron transport through several membrane protein complexes. In oxygenic phostosynthesis, the photosystem II (PS II) P680 complex becomes excited and activates the electron transport chain to photosystem I (PS I), an integral protein complex in which antenna chlorophylls drive photon absorption and energy transfer to a special chlorophyll A dimer (P700). The electron chain finally reduces $NADP^+$ to NADPH, yielding ATP used by the organisms as biological energy currency [3].

The photosynthetic process can be compared with a photocatalytic process involving an electron transfer between an electrode and solution-phase species, thus motivating an increasing interest in the design of photoelectrochemical cells [4]. The incorporation of biological photosystems into (bio)hybrid photovoltaic devices will be treated in Chapter 14. In this chapter, we will treat the electrochemical studies aimed to describe the photosynthetic pathways where the chlorophylls play a dual role (i) as light collectors and antenna excitons in the photosynthetic complexes [5] and (ii) as electron carriers in the reaction centers, promoting the electron transport across the photosynthetic membrane [6].

The energy metabolism of cells involves the adenylate system where adenosine triphosphate (ATP) serves as a carrier of chemical energy. This molecule is synthesized in the photosynthetic membranes of chloroplasts and can be hydrolyzed to adenosine diphosphate (ADP) and adenosine monophosphate (AMP). ATP plays an essential role in most biological processes. These include cellular metabolism, DNA biosynthesis, replication and transcription, and enzymatic activity [7]. ATP also participates in the photosynthetic reduction of CO_2.

8.2 ELECTROCHEMISTRY OF PROTEINS

The first reports on protein electrochemistry refer to classical polarographic studies, further expanded via immobilization of the proteins in solid electrodes [8] (see Additional literature). Currently, it is recognized that proteins possess a variety of potential electroactive centers, including aminoacid residues and DNA and RNA nucleobase residues. On solid electrodes, proteins yield electrochemical signals associated with the oxidation of their amino acid residues and/or reduction and oxidation

DOI: 10.1201/9781032697826-8

of their prostatic groups such as heme or flavin. However, despite intensive research, there is a lack of detailed mechanistic studies characterizing the electrochemical processes and the relationship between protein structure and reactivity [2].

Figure 8.1 shows a schematic representation of the electron relay between cytochrome c and a solid electrode [9]. This illustrates the common problem to interpret the electrochemistry of proteins and enzymes: The direct electron exchange between the electrode and redox-active centers requires that they must be positioned close enough to the protein surface to allow electron tunneling [10].

The voltammetry of proteins has been studied preferentially upon attachment to solid electrodes using, among other linkers, long-chain alkane thiols. Since most proteins tend to aggregate or denature on the electrode, thereby blocking electron transfer, amphiphilic tensioactives and/or promoters such as 4,4'-dipyridyl are frequently used. Even in the case of soluble proteins, electron transfer mediators/promoters are needed to amplify the intensity of voltammetric responses [11]. This can be seen in Figure 8.2 where the cyclic voltammograms of a cytochrome c solution at biological pH at a membrane glassy carbon electrode without and with coverage of a mercury film are shown [12]. The voltammogram shows a reversible one-electron pair corresponding to the Fe(III)/Fe(II) redox couple.

In the case of multi-heme proteins, different voltammetric signals can be registered. Figure 8.3a shows the cyclic voltammogram recorded in a solution of cythochrome c_3 extracted from *Desulfovibrio norvegicum* Genthner et al. The voltammogram shows two well-defined couples whose peak currents are pH-dependent [12]. The difference between such peak currents remains essentially constant (see Figure 8.3b) thus suggesting that two different redox centers behaving independently are involved.

The essential idea for interpreting protein electrochemistry is that only the surface located functional groups of monomeric units of proteins and other biopolymers can participate in electrochemical processes. Such units should be inaccessible, for example, in protein/peptide aggregates or the DNA double helix [2].

8.3 ELECTROCHEMISTRY OF ENZYMES

As previously indicated, the electrochemistry of enzymes has characteristics similar to that of proteins but is enriched by the possibility of monitoring the interaction with either electroactive or non-electroactive substrates [1]. The electrochemistry of metalloenzymes is dominated by the response of the metal center(s). In the case of superoxide dismutases (SODs), the enzymes can be divided into four groups: copper/zinc-containing (Cu/Zn-SOD), manganese-containing (Mn-SOD), iron-containing (Fe-SOD), and nickel-containing (Ni-SOD) [13]. Figure 8.4 shows cyclic voltammograms of solutions of Fe-SOD and Mn-SOD extracted from *Escherichia coli* (Migula) Castellani and Calmers and Cu/Zn-SOD from bovine erythrocyte in PBS at 3-mercaptopropionic acid-modified gold electrodes [14]. The voltammograms show one couple at potentials between 0.14 and 0.23 V

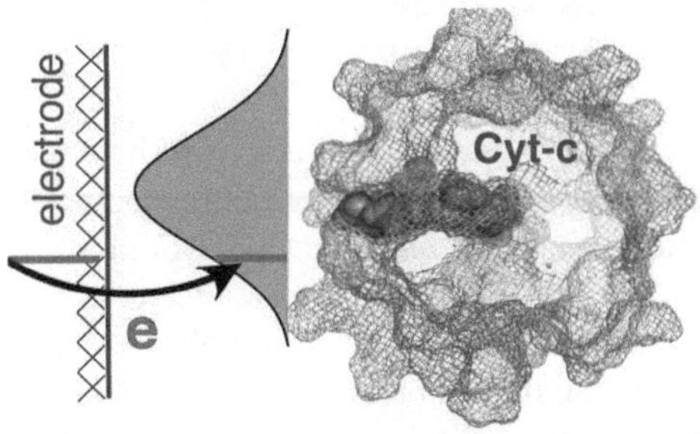

FIGURE 8.1 Schematic of the electron transfer between cytochrome c and a solid electrode. (Adapted from ref. [9], with permission. Copyright the American Chemical Society 2017.).

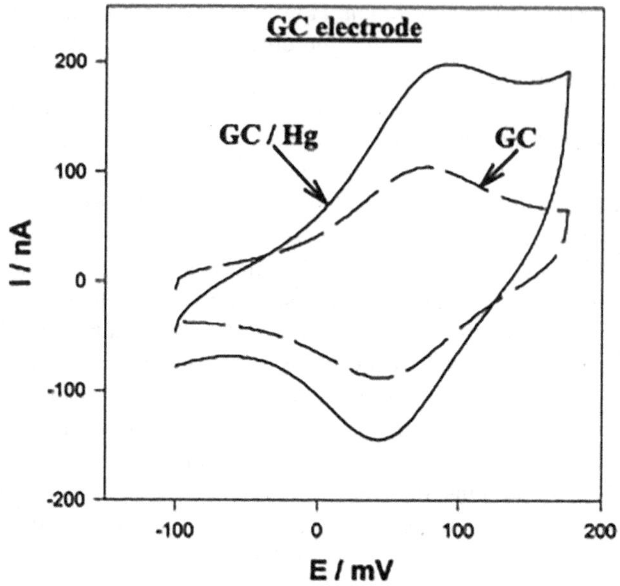

FIGURE 8.2 Cyclic voltammograms at modified GCEs of 137 μM cytochrome ***c*** in 10 mM TRIS chloride aqueous buffer at pH 7.6. Voltammgrams at membrane GCE without (dotted line) and with (continuous line) mercury film coating are superimposed. Potential scan rate, 20 mV s^{-1}. (Reproduced from ref. [12], with permission. Copyright Elsevier 2000.).

vs. Ag/AgCl, which in all cases is pH-sensitive. The electrochemical process is favored by the promoter, 3-mercaptopropionic acid, and involves the respective metal centers [14].

An example of the electrochemical monitoring of enzyme-substrate interactions is illustrated in Figure 8.5. Here, the enzymes were immobilized on the 3-mercaptopropionic acid-modified gold electrode and submitted to inputs of superoxide radical anion [14]. This was generated by the addition of aliquots of xanthine to air-saturated PBS containing xanthin oxidase. As can be seen in this figure, the addition of the superoxide substrate produces an increase in the peak currents of both reduction and oxidation processes. These are accompanied by a drastic increase of the current at the extreme cathodic potentials associated with the reduction of O_2 and/or H_2O_2. The increase of the intensity of SOD-centered oxidation and reduction processes can be interpreted as due to the bifunctional electrocatalytic ability of the enzymes. They catalyze both the reduction of superoxide to H_2O_2 and the oxidation of superoxide to O_2. This kind of study offers a solid basis for the detection of species involved in enzymatic catalysis and gain information on mechanistic pathways operating *in vivo.*

8.4 ELECTROCHEMISTRY OF CHLOROPHYLLS

8.4.1 Electrochemistry in Solution Phase

The chlorophylls are a group of green pigments characterized by a tetrapyrrolic structural unit. The major light-adsorbing chlorophylls are chlorophyll *a* (Chl a) and chlorophyll *b* (Chl b) [15]. The structures of Chl a and Chl b are illustrated in Figure 8.6. Both have a porphyrin ring coordinating with a central magnesium ion, only differing by the presence of a methyl (Chl b) or a formyl substituent in position 7. The chelation of Mg(II) facilitates light absorption, and the long hydrocarbon chain accompanying the tetrapyrrolic unit imparts hydrophobicity.

The direct electrochemistry of chlorophylls in an aqueous solution is also illustrated in Figure 8.6, where the cyclic voltammograms at the glassy carbon electrode of solutions of Chl a and Chl

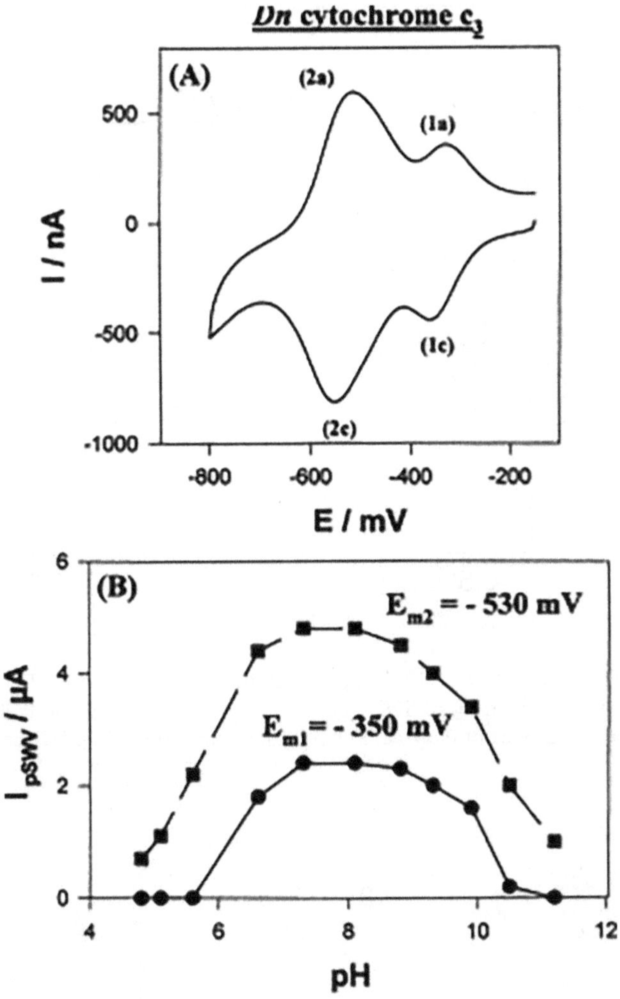

FIGURE 8.3 (a) Cyclic voltammogram at membrane-modified pyrolitic graphite electrode of 50 μM cythochrome c_3 extracted from *Desulfovibrio norvegicum* in 10 mM TRIS chloride plus 10 mM sodium borate buffer solution at pH 7.6. Potential scan rate, 20 mV s^{-1}. (b) pH-dependence of peak currents recorded for the above voltammetric signals from square wave voltammograms. (Reproduced from ref. [12], with permission. Copyright Elsevier 2000.).

b in PBS are shown [16]. The voltammograms show an apparently irreversible oxidation peak at 0.54 V vs. SCE for Chl a and at 0.62 V vs. SCE for Chl b. The electrochemical oxidation processes are mediated by adsorption, as denoted by the linear variation of the peak current with potential scan rate.

The oxidation of Chl b occurs at a slightly more positive potential than the oxidation of Chl a. This feature can be attributed to the major electron-withdrawing effect of the –CHO group (Chl b) relative to the $-CH_3$ group (Chl a), resulting in a reduction of the electron density of the chlorophyll π-conjugated system [16].

In aprotic media, Chl a is reversibly oxidized to a stable radical cation while the reduction occurs via two one-electron processes yielding a radical anion and a dianion, successively. These processes are complicated by side reactions, in particular, those involving proton donors [17]. Figure 8.7 shows the cyclic voltammetry of Chl in DMF solution at Pt electrode [17].

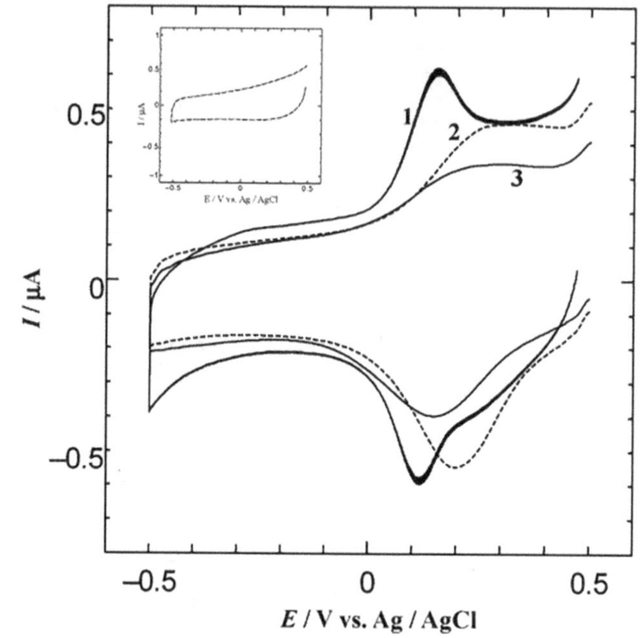

FIGURE 8.4 Cyclic voltammograms at 3-mercaptopropionic acid-modified gold electrodes in 5 mM PBS solution at pH 7.0 containing 0.32 mM Fe-SOD (1), 0.40 mM Mn-SOD (2), and 0.20 mM Cu/Zn-SOD (3). Inset: voltammogram in the absence of enzyme. Potential scan rate: 100 mV s^{-1}. (Reproduced from ref. [14], with permission.).

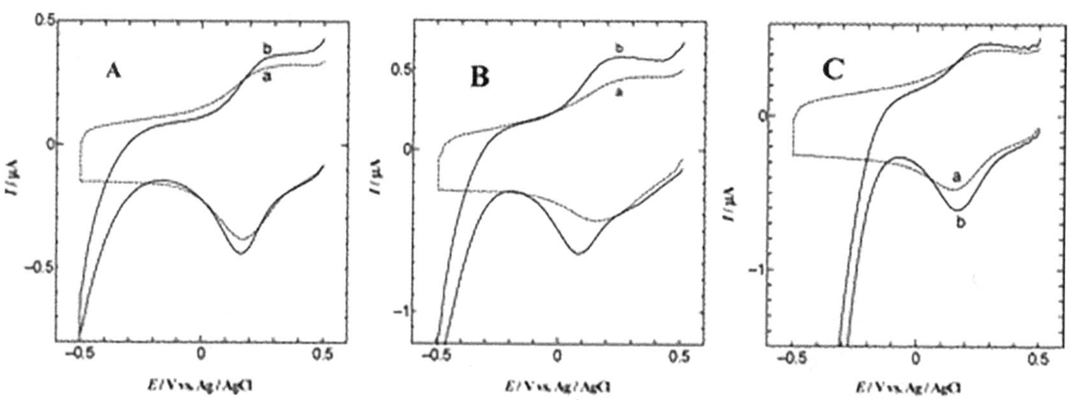

FIGURE 8.5 Cyclic voltammograms at SOD-3-mercaptopropionic acid-modified gold electrodes in 25 mM PBS solution at pH 7.5 recorded in the absence (dotted lines) and presence (solid lines) of 1.8 μM min^{-1} O_2^-. (a) Cu,Zn-SOD, (b) Fe-SOD, /MPA/Au, (c) Mn-SOD. Potential scan rate: 100 mV s^{-1}. (Reproduced from ref. [14], with permission. Copyright the American Chemical Society 2004.).

8.4.2 Adsorptive and Solid-State Electrochemistry of Chlorophylls

The low solubility of chlorophylls in water has motivated their study as adsorbates and microcrystalline films on different electrode substrates. Examples of adsorptive electrochemistry are chlorophylls adsorbed on mercury [18–20] and on hydroxyl-terminated hexanethiol-modified gold electrodes [21]. Figure 8.8 shows the cyclic voltammetry at the mercury electrode of Chl a dissolved in water containing a small mole fraction of an organic solvent (DMF) [18]. There is a small but

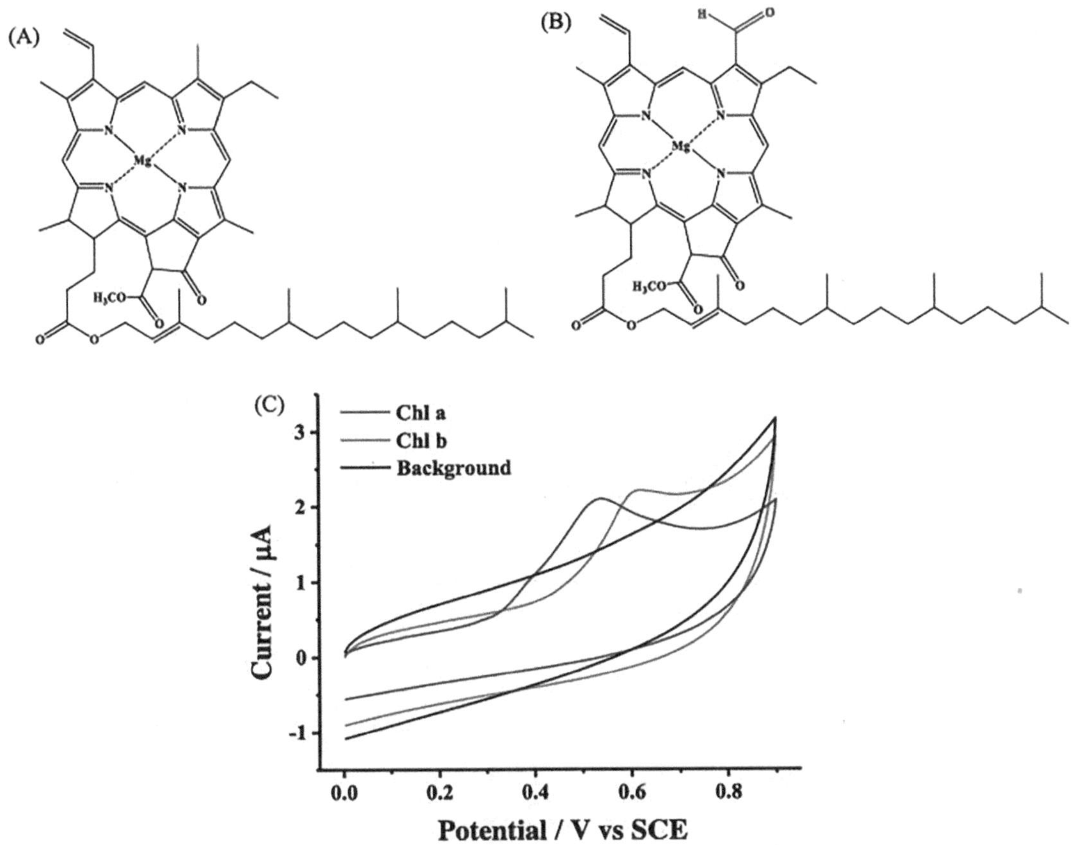

FIGURE 8.6 Chemical structures of Chl a (a) and Chl b (b). Cyclic voltammetry at glassy carbon electrode of 23 µM solutions of Chl a and Chl b in PBS at pH 7.4 accompanied by the background voltammogram. Potential scan rate 100 mV s^{-1}. (Reproduced from ref. [16], with permission. Copyright Elsevier 2020.).

consistent difference between the voltammograms recorded immediately after chlorophyll addition and after 1 min of contact with the mercury electrode. This has been interpreted in terms of the progressive aggregation of chlorophyll, also detected by spectral changes in the solution [18]. The peak at −0.66 V was attributed to the reduction of the C=O group, which precedes the formation of radical anion and dianion [19]. This signal is enhanced when chlorophyll aggregation is produced, a feature associated with a structural reorganization in which the carbonyl groups are oriented toward the external bulk water solution, thus favoring the adsorption on the mercury surface.

The electrochemistry of microparticulate deposits of chlorophylls is illustrated in Figure 8.9. Here, four successive cycles of potential were applied to a microparticulate film of Chl a deposited on an indium-doped tin oxide electrode (ITO) in contact with 0.1 M $LiClO_4$ aqueous solution [22]. In the initial anodic scan, the voltammogram shows an oxidation peak at 0.97 V vs. SCE preceded by shoulders at 0.57 and 0.77 V. In the reverse scan, two or three overlapping peaks between 0.5 and −0.5 V are recorded. In successive potential cycles, the anodic region exhibits two peaks at 0.77 and 1.0 V while in the cathodic region, the signals tend to concentrate in the more positive wave.

To interpret this electrochemistry, it has to be taken into account that solid-state oxidation/reduction processes must retain the electroneutrality of the solid film. Accordingly, and assuming that in the studied range of potentials Chl a experiences one-electron oxidation, the main anodic process can be formulated as

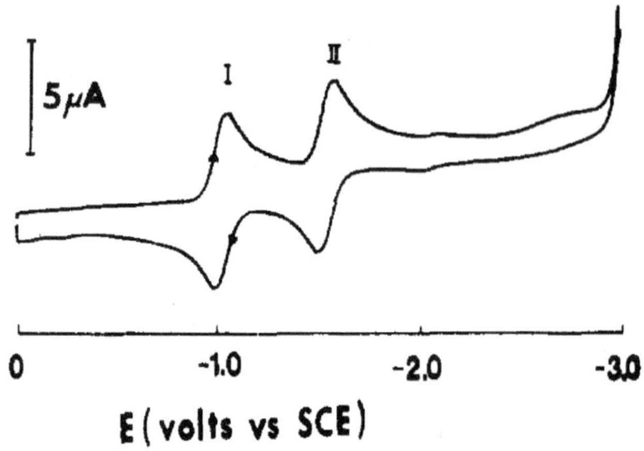

FIGURE 8.7 Cyclic voltammogram at Pt electrode of 0.42 mM Chl a solution in 0.1 M TBAP/DMF. Potential scan rate 200 mV s^{-1}. (Adapted from ref. [17], with permission. Copyright the American Chemical Society 1977.).

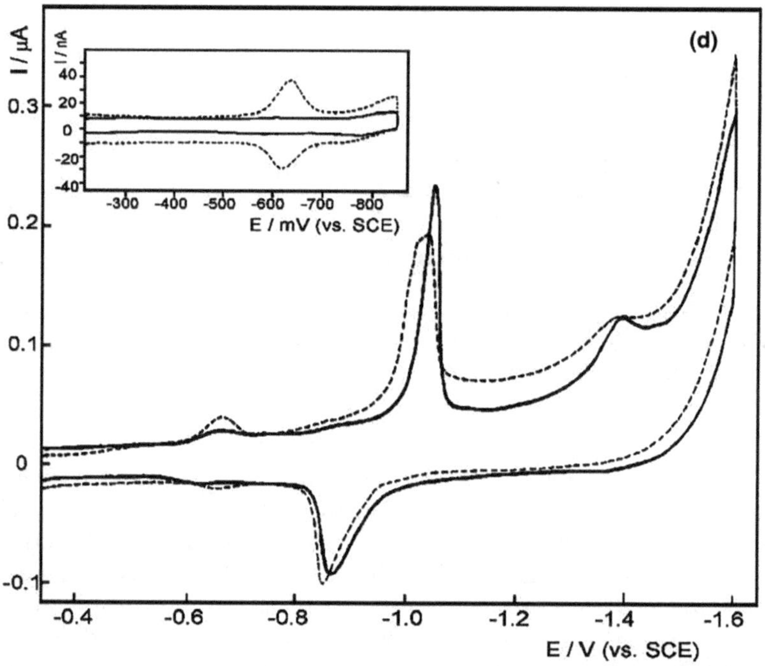

FIGURE 8.8 Cyclic voltammograms at Hg electrode of a 10^{-5} M Chl a solution in water containing a 0.0002-mole fraction of DMF (continuous line) and after 1 min of adsorption (dotted line) Inset: voltammogram recorded in a restricted potential range. (Reproduced from ref. [18], with permission. Note that the axes are opposite to the IUPAC convention. Copyright Elsevier 2004.).

$$\{\text{Chl a}\}_{\text{solid}} + \text{ClO}_4{}^-{}_{\text{aq}} \rightarrow \left\{(\text{Chl a})^{\bullet+} \cdots \text{ClO}_4{}^-\right\}_{\text{solid}} + e^- \tag{8.1}$$

The main cathodic signal will correspond to the inverse of the aforementioned process. The large anodic-to-cathodic peak potential separation, typical of solid-state processes with miscibility gaps [23], can be assigned to the existence of significant structural rearrangements in the film or

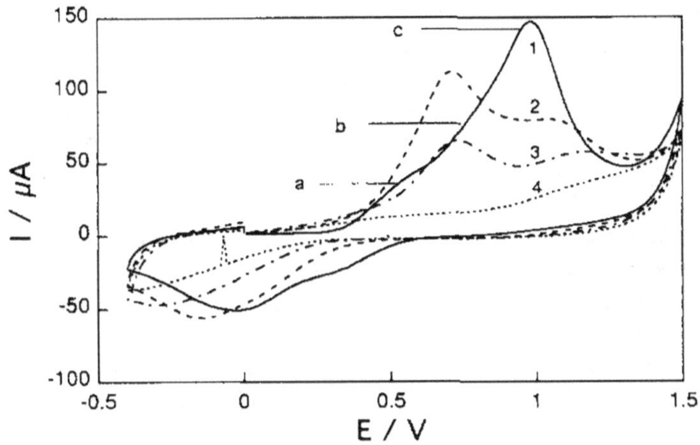

FIGURE 8.9 Repetitive cyclic voltammetry (four potential scans) of a microparticulate deposit of Chl a on ITO electrode in contact with 0.1 M $LiClO_4$ aqueous solution. The film thickness was estimated as 150 nm. Potential scan initiated at 0.0 V vs. SCE in the positive direction; potential scan rate is 100 mV s^{-1}. (Reproduced from ref. [22], with permission. Copyright Elsevier 1995.).

conformational changes in the chlorophyll [18]. These factors could determine the superposition of different oxidation and reduction signals. As described in the solid-state electrochemistry of malachite green [24], this effect can also result from the superposition of cation-assisted and anion-assisted processes, as well as from the superposition of different kinetic routes, as described for ion-insertion electrochemistry [25]. If sufficiently cathodic potentials were attained, the cation-assisted solid-state reduction of chlorophyll,

$$\{\text{Chl a}\}_{\text{solid}} + \text{Li}^+_{\text{aq}} + \text{e}^- \rightarrow \left\{(\text{Chl a})^{\bullet-} \cdots \text{Li}^+\right\}_{\text{solid}}, \tag{8.2}$$

may occur.

8.4.3 Photoelectrochemistry of Chlorophylls

In the solution phase, the spectrum of Chl a shows absorption maxima at 428 and 660 nm. These bands are accompanied by bands in the 700–740 nm region when oligomeric chlorophyll forms exist [26]. The fluorescence spectrum shows a main band at 680 nm with a shoulder at about 730 nm [17].

Photocurrents on a variety of substrates covered by monolayers of chlorophylls have been observed upon excitation with red light. Figure 8.10 shows the light-on and light-off currents recorded upon illuminating with the red light of a Chl a coating on mercury in contact with 0.1 M KCl aqueous solution at pH 8.5 [20]. Here, the chlorophyll coating was obtained by a soaking procedure, and the photocurrent response was significantly dependent on the applied potential. If a potential of −0.160 V is applied, the light-on current is negative, i.e., corresponding to a flow of negative charges from the electrode toward the solution, while the subsequent light-off current decays exponentially to zero. Repeating the experiment with the application of more negative potentials results in an increase of the negative light-on current until a maximum is reached when the applied potential is −0.460 V. At more negative potentials, the light-on current decreases, and, at a potential of −0.760 V, the light-on photocurrent becomes positive.

This complex behavior has been interpreted in terms of the sum of two contributions [20]. The negative light-on photocurrents can be associated with a reduction process (conceivably water reduction to hydrogen) mediated by photoexcited chlorophyll molecules. The positive light-on photocurrents correspond to an oxidation process (conceivably water oxidation to oxygen). The photocurrents are notably influenced by several factors, such as the solution pH, excitation wavelength, and coating thickness, among others, being important in the degree of aggregation of the chlorophyll.

The aforementioned behavior refers to the photoelectrochemistry of semiconductors, which are summarized in Figure 8.11. In this figure, the redox potentials of the O_2/H_2O and H_2O/H_2 couples at pH 8.5 (respectively, 0.73 and −0.50 V vs. SHE) are superimposed to the potentials of the (Chl/$Chl^{\bullet-}$) and ($Chl^{\bullet+}$/Chl) couples (respectively, −0.94 and 0.80 vs. SHE). The corresponding redox potentials of the couples for the singlet and triplet excited states of Chl are shown in Figure 8.11. These were calculated by taking singlet and triplet excitation energies of 1.85 and 1.33 eV [27]. In this figure, the electronic energy scale is referred to as the vacuum energy scale by subtracting the usual conversion factor of 4.4 eV [28].

The electrogenerated chemiluminiscence of Chl a can be stimulated by the presence of coreactants. An illustrative case is that of tri-n-propyl amine (TPrA) in acetonitrile. Figure 8.12a shows the cyclic voltammogram at GCE of a Chl a solution in MeCN. The voltammogram at GCE shows two one-electron couples at 0.76 and 1.0 V vs. Ag/AgCl corresponding to the stepwise oxidation of chlorophyll to the radical cation and the dication. Figure 8.12b depicts the electrochemiluminiscence-potential (ECL-E) curves in the presence and in absence of a large excess of TPrA. In the absence of TPrA, the ECL current is weak, but it becomes drastically enhanced in the presence of the coreactant. To test the influence of the different chlorophyll oxidized forms in the chemiluminiscence, ECL measurements were performed (Figure 8.12c), reaching potentials before the first chlorophyll oxidation signal, intermediate between the first and the second oxidation waves, and past the second signal. The intermediate potential is close to that corresponding to the one-electron oxidation of TPrA to its corresponding radical cation [30].

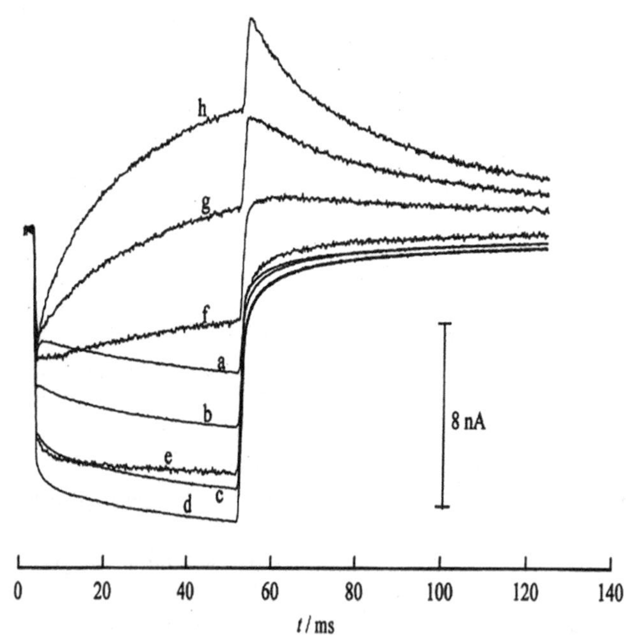

FIGURE 8.10 Light-on and light-off currents recorded upon illuminating a mercury drop coated by Chl a in contact with 0.1 M KCl aqueous solution at pH 8.5 after illumination with red light for 50 ms at potentials of (a) −0.160, (b), −0.260, (c) −0.360, (d) −0.460, (e) −0.560, (f) −0.660, (g) −0.760, and (h) −0.860 V. The measurements were performed starting from the most positive potential and waiting for the stabilization of the background current before illuminating the drop at each potential. (Reproduced from ref. [20], with permission. Copyright Elsevier 2003.).

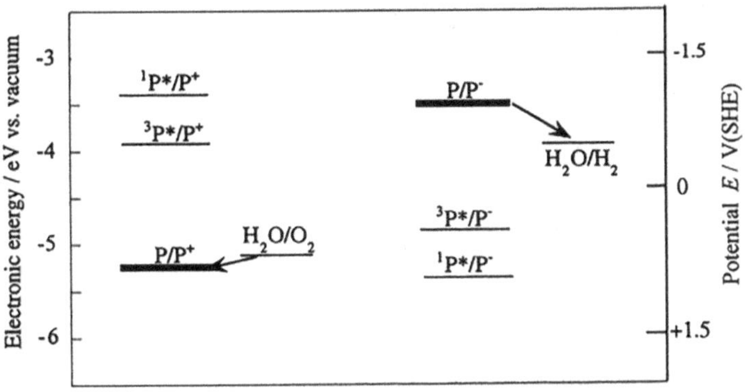

FIGURE 8.11 Potential energy diagram showing the redox processes proposed to describe the cathodic and anodic photocurrents in Figure 8.10. Chl a is denoted by P for simplicity. The asterisks mark excited states. (Reproduced from ref. [20], with permission. Copyright Elsevier 2003.).

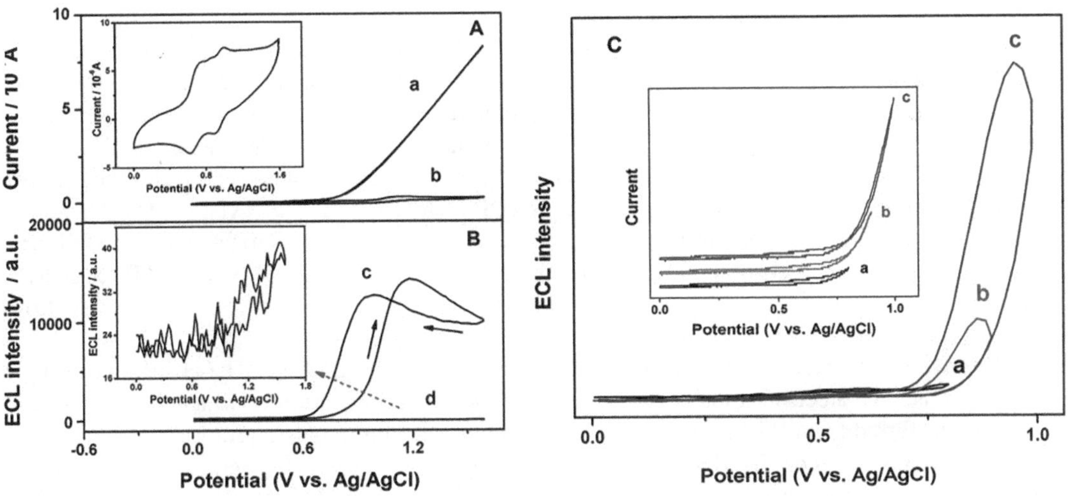

FIGURE 8.12 (a) Cyclic voltammogram at GCE of 5 μM Chl a in 0.025 M 1-butyl-3-methylimidazolium hexafluorophosphate solution in MeCN. (b) ECL-potential curves in the presence (a, c) and in the absence (b, d) of 0.05M TPrA. Potential scan rate 200 mV s^{-1}. (c) ECL-potential curve when the electrode was cycled between 0.0 and 0.8 V (a), 0.0 and 0.9 V (b), and 0.0 and 1.0 V (c). The inset in (a) shows the cyclic voltammogram of 10^{-4} M Chl a in MeCN at 0.050 mV s^{-1}. The inset in (b) shows the enlarged view of curve d. The inset in (c) shows the voltammetric curves at the above three potential windows. (Adapted from ref. [29], with permission. Copyright Elsevier 2015.).

These features have been interpreted as indicative of the preferential role of the (Chl $^{\bullet+}$/Chl) couple in the electrochemimluminiscent process. One possible simplified reaction scheme can be summarized as follows (the asterisk denotes excited states).

$$(\text{Chl a}) \rightarrow (\text{Chl a})^{\bullet+} + e^{-} \tag{8.3}$$

$$\text{TPrA} \rightarrow \text{TprA}^{\bullet+} + e^{-} \tag{8.4}$$

$$TprA^{\bullet +} \rightarrow TprA^{\bullet} + H^{+} \quad (8.5)$$

$$(Chl\,a)^{\bullet +} + TprA^{\bullet} \rightarrow (Chl\,a)^{*} + TprA^{ox} \quad (8.6)$$

$$(Chl\,a)^{*} \rightarrow (Chl\,a) + h\upsilon \quad (8.7)$$

The ECL emission spectrum shows a maximum at 670 nm, coincident with that obtained by photoexcitation with light at 420 nm [29]. This suggests that the excited states generated electrochemically are the same as those generated photochemically. The possibility of using chlorophylls as environmentally friendly ECL luminophores is being under intensive research.

8.5 NATURAL ELECTRON TRANSPORT IN THE THYLAKOID MEMBRANE

In higher plants, the photosynthesis takes place in chlorophyll-protein (CP) complexes localized in thylakoids, the inner membranes of chloroplasts [31]. These complexes, of trimer structural units, absorb light transferring energy to photosystem II (PSII) reaction centers [32, 33]. The monomer units comprise a polypeptide of roughly 230 amino acid residues, which include three transmembrane α-helices and short amphipathic helices. Each monomer binds 6 molecules of Chl b, 8 molecules of Chl a, and 4 molecules of carotenoid. The light-excited photosystem II activates the electron transport chain to photosystem I leading to the $NADP^{+}$/NADPH and ADP/ATP cycles. The entire process involves a series of electron transfer processes with different coenzymes and metalloproteins (plastoquinone, cytochrome b6f, plastocyanin, ferredoxin, and ferredoxin-NADP reductase). Figure 8.13 shows a scheme of a thylakoid membrane with a flux diagram of the flow of electrons and protons [34].

These involve two electron transport chains schematized in Figure 8.13. The absorption of light by PSII initiates the electron transfer chain, resulting in the net oxidation of water to O_2. The electrons are transferred from PSII to plastoquinone which is reduced to plastoquinol. This binds to the cytochrome b6f and is reoxidized to plastoquinone whereas protons are released into the lumen. The electrons are transferred from cytochrome b6ft to a copper-containing protein, plastocyanin. The reduced form of the same transfers electrons to the PSI, which in turn transfers electrons to ferredoxin. Now, two electron transfer pathways can operate. In the cyclic pathway, the reduced form of ferredoxin reduces $NADP^{+}$ to NADPH aided by the enzyme ferredoxin-$NADP^{+}$ reductase. This second pathway increases the production of NADH while the cyclic pathway increases the production of ATP [34].

From the earlier work on the bacterial photosynthetic reaction center (bRC) [35], the direct electrochemistry of soluble proteins and enzymes has been studied during the last decades. Figure 8.14 shows a scheme of the cascade mechanisms operating in photosynthetic reaction centers PSI and bRC [36]. The symbols in this figure and the approximate values of the midpeak potentials of the redox centers involved in PSI, PSII, and bRC photosynthetic pathways are indicated in Table 8.1 [36, 37].

It is worth noting that, in general, proteins provide a nonpolar environment for electron transfer, making difficult the access of water molecules to the active sites. However, the accessibility of redox-active sites is favored by protein hydration, causing ionization of the surface groups in turn favoring the solubility in water [9].

The voltammetry at electrodes modified with intact thylakoids has been extensively studied concerning photocurrent production [34, 38, 39]. The voltammetric responses involve transmembrane

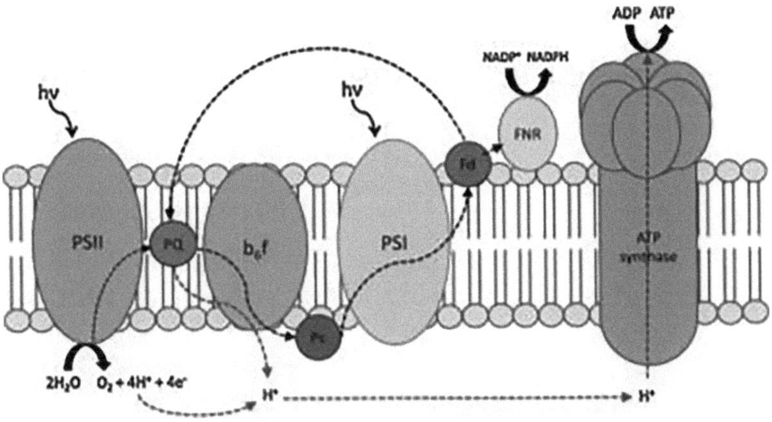

FIGURE 8.13 Schematic representation of thylakoid membrane showing electron flow (black arrows) and proton flow (grey arrows). (Adapted from ref. [34], with permission. Copyright Elsevier 2014.).

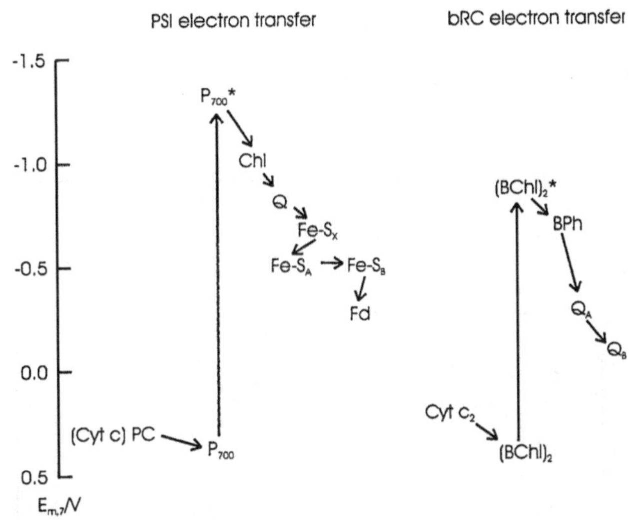

FIGURE 8.14 Cascade electron transfer pathways of PSI and bRC photosystems scaled according to approximate midpeak potentials vs. SHE. The meaning of symbols is listed in Table 8.1. (Reproduced from ref. [36], with permission. Copyright Elsevier 2001.).

electrochemistry pathways [40] that will be treated in Chapter 12. In most cases, the observed voltammetric responses result from the superposition of the responses of individual components. Figure 8.15 shows the cyclic voltammetry of thylakoid-modified carbon electrodes immersed in citrate buffer at pH 5.5 in the presence of $NADP^+$ and ADP. The former promotes the noncyclic electron transport pathway, while the second promotes the cyclic pathway.

The voltammograms in Figure 8.15 show two oxidation peaks at 0.2 and 0.4 V vs. Ag/AgCl and one reduction peak at –0.35 V, which can be assigned to the Fe-S protein in the cytochrome b6f complex and plastocyanin, respectively [41, 42]. The signals at 0.65 and 0.80 V are assigned to the background, probably due to oxygen functionalities in the carbon surface. The removal of ferredoxin-$NADP^+$ reductase [43], results in the appearance of a reversible-like couple at a midpeak potential of –0.05 V, attributable to plastoquinone [44]. Hen $NADP^+$ is added to the electrolyte solution, the noncyclic pathway is activated, and only a weak plastocyanin peak at 0.4 V is recorded. This signal

TABLE 8.1
Approximate Midpeak Potentials (E_{mp}, V vs. SHE) of the Redox Centers in the Photosystems PSI and bRC

Systems PSI and PSII	E_{mp}	System bRC	E_{mp}
Cytochrome c (Cyt c)	0.25	Cytochrome c_2 (Cyt c_2)	0.34
Cytochrome f	0.35		
Cytochrome b_6	0.05		
Plastocyanin (PC)	0.37	BacteriochlorophyllII dimmer $(BChl)_2$	0.45
Chlorophyll (P_{700})	0.49	Bacetriopheophytin (BPh)	−0.60
Chlorophyll (P_{680})	0.9		
Chlorophyll (Chl)	−1.0	Quinone A (Q_A)	−0.05
Plastoquinone	0.0		
Phylloquinone (Q)	−0.80	Quinone B (Q_B)	0.02
F_4S_4 clusters (Fe_X)	−0.70		
(Fe_A)	−0.53		
(Fe_B)	−0.58		
Ferredoxin (F_d)	−0.42		

Adapted from refs. [36, 37]. Copyright Elsevier 2001, 2014.

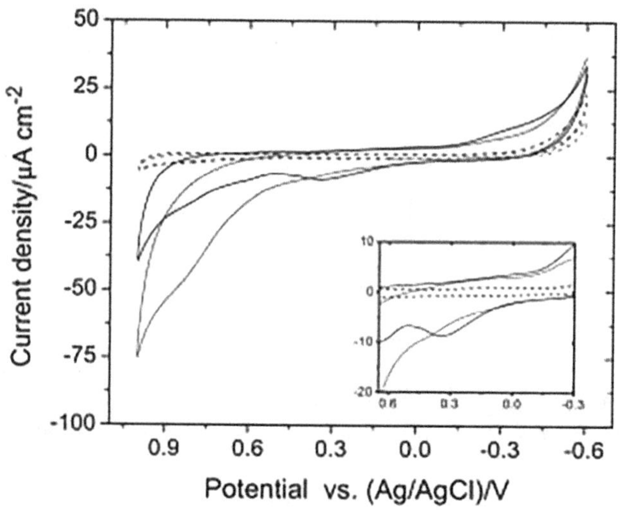

FIGURE 8.15 Cyclic voltammetry of thylakoid-modified carbon electrodes in contact with 0.1 M citrate buffer, pH 5.5 (continuous lines), and blank voltammogram with electrodes containing no thylakoids (dashed line). The voltammograms in the presence of 2 mM $NADP^+$ (grey line) and 2 mM ADP (black line) are superimposed. CVs were collected in the presence of light. Potential scan rate 50 mV s^{-1}. Adapted from ref. [34], with permission.
(Copyright Elsevier 2014. Note that the axes are opposite to the IUPAC convention.).

becomes enhanced upon illumination. When ADP (activating the cyclic pathway) is added to the solution, the plastocyanin signal is increased significantly. Removal, activation/deactivation of selected thylakoid suggested that photosystems I and II, plastoquinone, cytochrome b6f, and plastocyanin are involved in the direct electron transfer mechanism and both electron transport pathways in photosynthesis (cyclic and noncyclic) contribute to the photocurrent [34].

Attachment of photosystems to solid electrodes (either forming films or suspensions) has been also studied and the influence of redox mediators and stabilizers, inhibitors, and cross-linkers systematically treated [37, 45]. Figure 8.16 depicts a simplified scheme of the role of redox mediators in such systems exemplified in the case of phenazine methosulfate [37]. The redox mediator captures an electron from the PSI system of the thylakoid membrane, and the reduced mediator form diffuses toward the electrode surface where it is reoxidized. Then, the regenerated mediator diffuses back to the PSI to repeat the cycle. In experiments such as those schematized in Figure 8.16, the concentration of the redox mediator is crucial. At low mediator concentrations, the photocurrent increases with increasing mediator concentration until a maximum is reached. A subsequent increase in the mediator concentration produces a progressive decrease in the photocurrent. This effect can be interpreted as a result of the competition between the reduction and the oxidation processes [37].

Inhibitors produce the inactivation of electron transfer chains. Figure 8.17 shows the variation in the photocurrent recorded for a suspension of thylakoid pellets in the absence and the presence of an inhibitor, 3-(3',4'-dichlorophenyl)-1,1-dimethylurea [37]. One can see that the inhibitor produces a drastic decrease or disappearance of the photocurrent blocking the electron transport chain.

Interestingly, the photoelectrochemical activity of thylakoid membranes and other systems such as chloroplasts can be monitored using scanning electrochemical microscopy (SECM). Figure 8.18 shows a schematic representation of the operation mode in which the $Fe(CN)_6^{3-}/Fe(CN)_6^{4-}$ couple is used as the redox probe. The substrate thylakoid membrane is in contact with a solution of ferricyanide that is reduced to ferrocyanide on the substrate surface. In turn, ferrocyanide is collected at the electrode tip, applying a bias potential to promote its oxidation. Then the fluctuations in the tip current are representative of the electrochemical activity of the surface [46].

In these experiments, ferricyanide interacts with thylakoid membranes upon removing the chloroplast envelope [47]. In contact with the thylakoid membrane, the redox probe can act on the stromal side of PSII via light-driven electron transfer chains accompanied by oxygen evolution [48, 49]. This last process can be monitored electrochemically using Ag microelectrodes at the tip and applying a bias potential cathodic enough to promote the reduction of dissolved oxygen [47].

Figure 8.19 depicts the time variation of the tip current for ferrocyanide oxidation in contact with a thylakoid membrane submitted to illumination in the absence and presence of ferricyanide when the tip is placed near the thylakoid membrane and when placed in the bulk solution containing ferricyanide [47]. When the tip is placed in the bulk ferricyanide solution or near the thylakoid membrane in the absence of ferricyanide, the tip current does not reflect significant variations during illumination. In contrast, when the tip is close to the thylakoid membrane and ferricyanide is present in the solution, the current increases significantly upon illumination and decays when the illumination ceases.

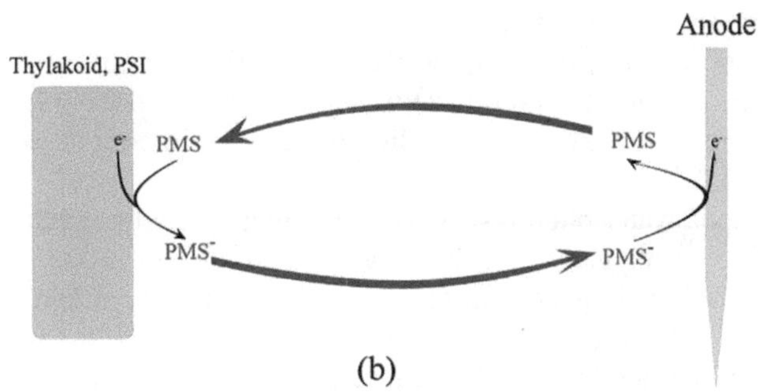

FIGURE 8.16 Scheme for the role of phenazine methosulfate mediator in thylakoid electrochemistry. (Adapted from ref. [37], with permission. Copyright Elsevier 2014.).

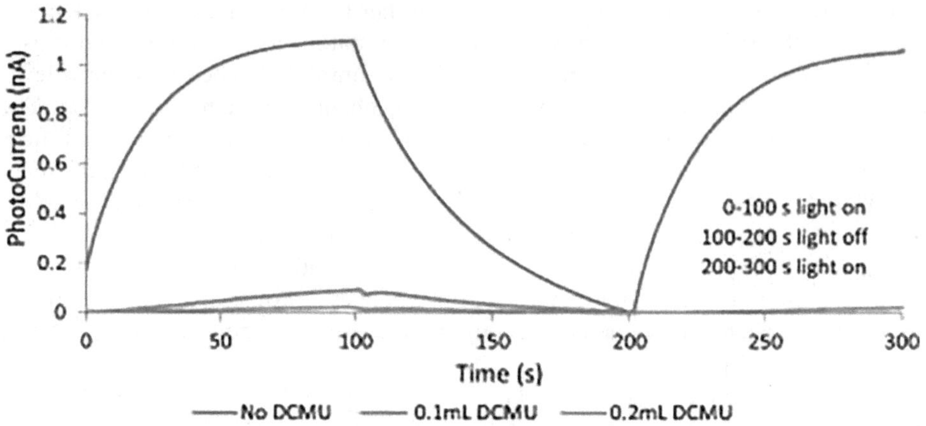

FIGURE 8.17 Photocurrents measured in the absence and the presence of 3-(3',4'-dichlorophenyl)-1,1-dimethylurea inhibitor (two successive additions) for a suspension of thylakoid pellets extracted from spinach in a 0.2 M sucrose, 3 mM $MgCl_2$, 10 mM KCl, and 20 mM Tricine/NaOH aqueous solution buffered at pH 7.5. (Adapted from ref. [37], with permission. Copyright Elsevier 2014.).

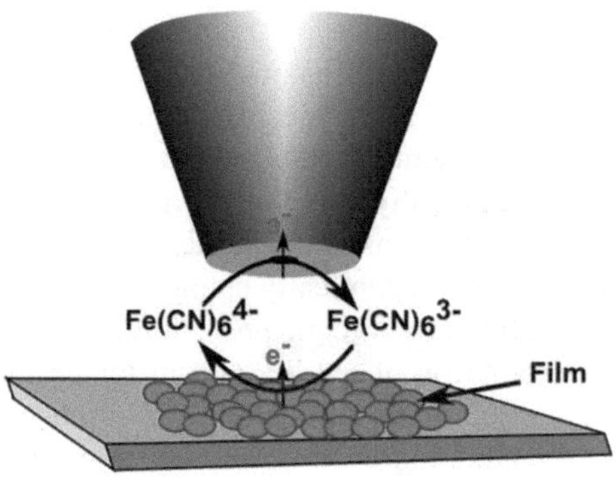

FIGURE 8.18 Scheme for the SECM operation mode to study photosynthetic electron flows in thylakoid membranes using $Fe(CN)_6^{3-}$/$Fe(CN)_6^{4-}$ as the redox probe. (Adapted from ref. [46], with permission. Copyright the American Chemical Society 2013.).

The experiments in Figure 8.19 illustrate the capability of SECM for detecting the photoelectrochemical activity of thylakoid and chloroplast membranes. The monitoring of the production of oxygen by such membranes upon illumination can be seen in Figure 8.20. Here, current/time curves recorded for an Ag tip held at a potential promoting the reduction of generated oxygen are recorded when the electrode was placed near chloroplast and thylakoid films. Light inputs of different intensities were applied to the film in contact with different buffered solutions at pH 8.0 [47].

Oxygen is produced via water photolysis during the light reactions at PSII. Since the electrolyte solutions contain dissolved oxygen, there is a background current in the dark that must be subtracted. When the chloroplast films are illuminated, a sharp increase in the tip current is observed (Figure 8.20a). This spike denotes a drastic increase of oxygen flux at the tip electrode and decays rapidly after the peak reaches a lower steady-state value. When the illumination is turned off, a small negative transient is recorded. Both positive and negative transients are enhanced on increasing intensity of illumination.

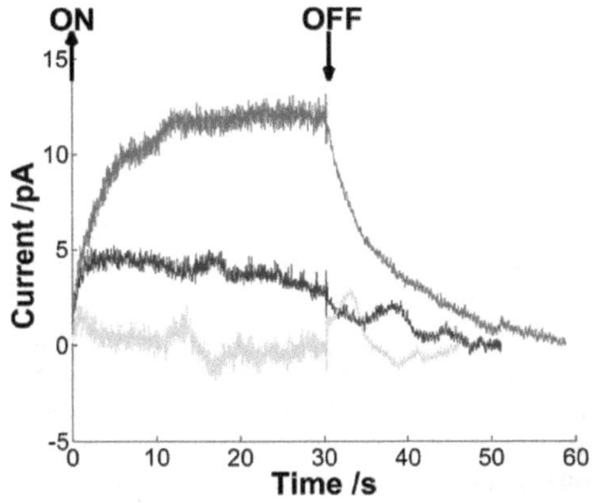

FIGURE 8.19 Time variation of the tip current for ferrocyanide oxidation in HEPES-magnesium buffer at pH 8.0 during illumination for 30 s in contact with a thylakoid membrane. The tip was a 25 μm diameter Pt disk electrode held at 0.2 V vs. SCE located 12.5 μm above the thylakoid membrane. Curves were recorded in the absence (middle) and presence (upper of 3.3 mM ferricyanide and placing the tip electrode in bulk solution with 3.3 mM ferricyanide (lower). (Adapted from ref. [46], with permission. Copyright the American Chemical Society 2013.).

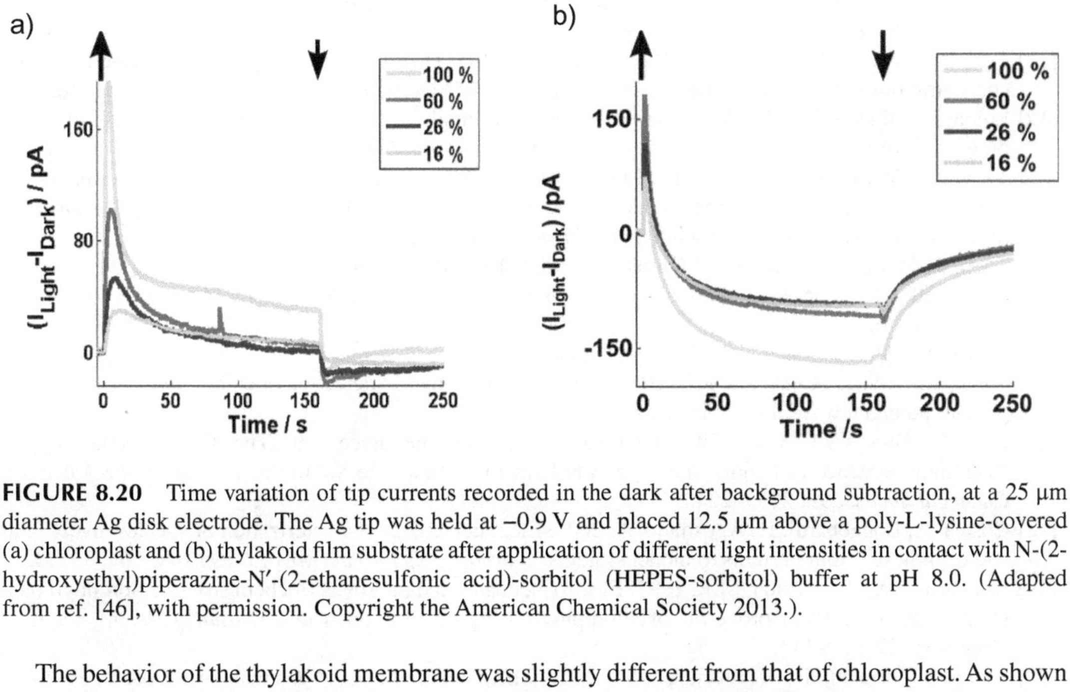

FIGURE 8.20 Time variation of tip currents recorded in the dark after background subtraction, at a 25 μm diameter Ag disk electrode. The Ag tip was held at −0.9 V and placed 12.5 μm above a poly-L-lysine-covered (a) chloroplast and (b) thylakoid film substrate after application of different light intensities in contact with N-(2-hydroxyethyl)piperazine-N′-(2-ethanesulfonic acid)-sorbitol (HEPES-sorbitol) buffer at pH 8.0. (Adapted from ref. [46], with permission. Copyright the American Chemical Society 2013.).

The behavior of the thylakoid membrane was slightly different from that of chloroplast. As shown in Figure 8.20b, upon illumination a sharp initial photocurrent spike was also obtained. This was followed by a rapid decay until a steady state. In contrast with chloroplast membranes, when the illumination was interrupted, a small increase in the oxygen flux appeared. This difference can be attributed to the fact that in thylakoid membranes, although containing the light-driven components of photosynthesis, several important terminal acceptors of electrons (NADP in particular) are absent. Then, the electron transport chain, and hence the final oxygen production, differs from that in chloroplasts [47]. These results denote the capability of electrochemical techniques for the detection of subtle changes in the reactivity of biochemical systems.

REFERENCES

[1] Ferapontova, E.E.; Shleev, S.; Ruzgas, T.; Stoica, L.; Christenson, A.; Tkac, J.; Yaropolov, A.I.; Gorton, L. 2005. *Perspectives in Bioanalysis*. Elsevier, Amsterdam, pp. 59–81.

[2] Suprun, E.V. 2021. Direct electrochemistry of proteins and nucleic acids: The focus on 3D structure. *Electrochemistry Communications*. 125: artic. 106983.

[3] Hall, D.O. Rao, K.K. 1999. *Photosynthesis*. Cambridge University Press, Cambridge.

[4] Blankenship, R.E.; Tiede, D.M.; Barber, J.; Brudvig, G.W.; Fleming, G.; Ghirardi, M.; Gunner, M.R.; Junge, W.; Kramer, D.M.; Melis, A.; Moore, T.A.; Moser, C.C.; Nocera, D.G.; Nozik, A.J.; Ort, D.R.; Parson, W.W.; Prince, R.C.; Sayre, R.T. 2011. Comparing Photosynthetic and Photovoltaic Efficiencies and Recognizing the Potential for Improvement. *Science*. 332: 805–809.

[5] Haran, G.; Wynne, K.; Moser, C.C.; Dutton, P.L.; Hochstrasser, R.M. 1996. Level Mixing and Energy Redistribution in Bacterial Photosynthetic Reaction Centers. *Journal of Physical Chemistry*. 100: 5562–5569.

[6] Kimaier, C.; Holten, D. 1993. In Deisenhofer, J.; Norris, J.R., Eds. *The Photosynthetic Reaction Center*, vol. 2, Academic Press, San Diego, p. 49.

[7] Dennis, P.B.; Jaeschke, A.; Saitoh, M.; Fowler, B.; Kozma, S.C.; Thomas, G. 2016. Mammalian TOR: A homeostatic ATP sensor. *Science*. 294: 1102–1105.

[8] Blandford, C.F. 2013. The Birth of protein electrochemistry. *Chemical Communications*. 49: 11130–11132.

[9] Seyedi, S.S.; Waskasi, M.M.; Matyushov, D.V. 2017. Theory and Electrochemistry of Cytochrome c. *Journal of Physical Chemistry B*. 121: 4958–4967.

[10] Armstrong, F.A.; Cheng, B.; Herold, R.A.; Megarity, C.F.; Siritanaratkul, B. 2023. From Protein Film Electrochemistry to Nanoconfined Enzyme Cascades and the Electrochemical Leaf. *Chemical Reviews*. 123: 5421–5458.

[11] Ugo, P.; Pepe, N.; Moretto, L.M.; Battagliarin, M. 2003. Direct voltammetry of cytochrome c at trace concentrations with nanoelectrode ensembles. *Journal of Electroanalytical Chemistry*. 560: 51–58.

[12] Lojou, E.; Bianco, P. 2000. Membrane electrodes can modulate the electrochemical response of redox proteins — direct electrochemistry of cytochrome c. *Journal of Electroanalytical Chemistry*. 485: 71–80.

[13] Ye, Q.; Li, W.; Wang, Z.; Zhang, L.; Tan, X.; Tian, Y. 2014. Direct electrochemistry of superoxide dismutases (Mn-, Fe-, and Ni-) from human pathogen Clostridium difficile: Toward application to superoxide biosensor. *Journal of Electroanalytical Chemistry*. 729: 21–26.

[14] Tian, Y.; Mao, L.; Okajima, T.; Ohsaka, T. 2004. Electrochemistry and Electrocatalytic Activities of Superoxide Dismutases at Gold Electrodes Modified with a Self-Assembled Monolayer. *Analytical Chemistry*. 76: 4162–4168.

[15] Scheer, H. 1991. *Chlorophylls*. CRC Press, Boca Raton.

[16] Wang, Y.; Chen, L.; Compton, R.G. 2020. Selective voltammetric detection of chlorophylls using a semicircular potential wave. *Food Chemistry*. 323: artic. 126844.

[17] Saji, T.; Bard, A.J. 1977. Electrogenerated Chemiluminescence. 29. The Electrochemistry and Chemiluminescence of Chlorophyll a in N,N-Dimethylformamide Solutions. *Journal of the American Chemical Society*. 99: 2235–2240.

[18] Cosma, P.; Longobardi, F.; Agostiano, A. 2004. Electrochemical characterization of species involved in photosynthesis: from proteins to model systems. *Journal of Electroanalytical Chemistry*. 564: 35–43.

[19] Khanova, L.A.; Tarasevich, M.R. 1987. Electrochemistry and photoelectrochemistry of chlorophyll on a metallic electrode: Part I. properties of chlorophyll adsorption films on metals. *Journal of Electroanalytical Chemistry*. 227: 99–114.

[20] Tadini Buonisegni, F.; Becucci, L.; Moncelli, M.R.; Guidelli, R.; Agostiano, A.; Cosma, P. 2003. Electrochemical and photoelectrochemical behavior of chlorophyll a films adsorbed on mercury. *Journal of Electroanalytical Chemistry*. 550–551: 229–240.

[21] Ciobanu, M.; Kincaid, H.A.; Lo, V.; Dukes, A.D.; Kane Jennings, G.; Cliffel, D.E. 2007. Electrochemistry and photoelectrochemistry of photosystem I adsorbed on hydroxyl-terminated monolayers. *Journal of Electroanalytical Chemistry*. 599: 72–78.

[22] Suponeva, E.P.; Hotchandani, S.; Kisselev, B.A.; Arbour, C. 1995. Cyclic voltammetric and spectrophotometric study of microcrystalline chlorophyll a electrodeposited on the ITO optically transparent electrode. *Journal of Electroanalytical Chemistry*. 387: 123–126.

[23] Lovrić, M.; Hermes, M.; Scholz, F. 2000. Solid state electrochemical reactions in systems with miscibility gaps. *Journal of Solid State Electrochemistry*. 4: 394–401.

[24] Doménech-Carbó, A.; Dias, D.; Doménech-Carbó, M.T. 2020. Cation and anion electrochemically assisted solid-state transformations of malachite gren. *PhysChemChemPhys*. 22: 1502–1510.

[25] Doménech-Carbó, A.; López, S.; Muñoz, E. 2024. logistic Approximation to Ion-Intercalation Electrochemistry: Application to Potassium-Copper(II) Hexacyanoferrate in DMSO Electrolytes. *Jurnal of Physical Chemistry C*. 128: 1487–1495.

[26] Agostiano, A. 1994. Chlorophyll a in aqueous organic solvents: Spectroscopic and electrochemical investigation of the factors effecting molecular aggregation and photoreactivity. *Trends in Photochemistry & Photobiology*. 3: 371–382.

[27] Seely, G.R. 1978. The energetics of electron-transfer reactions of chlorophyll and other compounds. *Photochemistry and Photobiology*. 27: 639–654.

[28] Trasatti, S. 1986. The Absolute Electrode Potential: An Explanatory Note. *Pure and Applied Chemistry*. 27: 955–966.

[29] Li, X.; Qin, X.; Zheng, H.; Yuan, H.; Guo, Y.; Xiao, D. 2015. Highly efficient electrogenerated chemiluminescence of natural chlorophyll a. *Electrochemistry Communications*. 61: 66–69.

[30] Lai, RY.; Bard, A.J. 2003. Electrogenerated chemiluminescence. 70. The application of ECL to determine electrode potentials of tri-n-propylamine, its radical cation, and intermediate free radical in MeCN/benzene solutions. *Journal of Physical Chemistry A*. 107: 3335–3340.

[31] Rochaix, J.D. 2011. Regulation of photosynthetic electron transport. *Biochimica Biophysica Acta*. 1807: 878–886.

[32] Barros, T.; Kuhlbrandt, W. 2009. Crystallisation, structure and function of plant light harvesting complex II. *Biochimica Biophysica Acta*. 1787: 753–772.

[33] Kouřil, R.; Dekker, J.P.; Boekema, E.J. 2012. Supramolecular organization of photosystemII in green plants. *Biochimica Biophysica Acta*. 1817: 2–12.

[34] Rasmussen, M.; Minteer, S.D. 2014. Investigating the mechanism of thylakoid direct electron transfer for photocurrent generation. *Electrochimica Acta*. 126: 68–73.

[35] Kong, J.; Lu, Z.; Lvov, Y.M.; Desamero, R.Z.B.; Frank, H.A.; Rusling, J.F. 1998. Direct electrochemistry of cofactor redox sites in a bacterial photosynthetic reaction center protein. *Journal of the American Chemical Society*. 120: 7371–7372.

[36] Kievit, O.; Brudvig, G.W. 2001. Direct electrochemistry of photosystem I. *Journal of Electroanalytical Chemistry*. 497: 139–149.

[37] Yu, Y.; Zuo, F.; Li, C.-Z. 2014. Mediator-assisted photocurrent extraction from the thylakoids. *Electrochimica Acta*. 144: 263–267.

[38] Kanso, H.; Pankratova, G.; Bollella, P.; Leech, D.; Hernandez, D.; Gorton, L. 2018. Sunlight photocurrent generation from thylakoid membranes on gold nanoparticle modified screen-printed electrodes. *Journal of Electroanalytical Chemistry*. 816: 259–264.

[39] Piotrowska, P.; Łazicka, M.; Palińska-Saadi, A.; Paterczyk, B.; Kowalewska, L.; Grzyb, J.; Maj-Źurawska, M.; Garstka, M. 2019. Electrochemical characterization of LHCII on graphite electrodes –Potential-dependent photoactivation and arrangement of complexes. *Bioelectrochemistry*. 127: 37–48.

[40] Hammarström, L.; Almgren, M.; Lind, J.; Merényi, G.; Norrby, T.; Åkermark, B. 1993. Mechanisms of Transmembrane Electron Transfer: Diffusion of Uncharged Redox Forms of Viologen, 4,4'-Bipyridine, and Nicotinamide with Long Alkyl Chains. *Journal of Physical Chemistry*. 97: 10083–10091.

[41] Joliot, P.; Joliot, A. 1986. Mechanism of proton-pumping in the cytochrome b/f complex. *Photosynthesis Research*. 9: 113–124.

[42] McLeod, D.D.N.; Freeman, H.C.; Harvey, I.; Lay, P.A.; Bond, A.M. 1996. Voltammetry of plastocyanin at a graphite electrode: effects of structure, charge, and electrolyte. *Inorganic Chemistry*. 35: 7156–7165.

[43] Matthijs, H.C.; Coughlan, S.J.; Hind, G. 1986. Removal of ferredoxin:$NADP^+$ oxidoreductase from thylakoid membranes, rebinding to depleted membranes, and identification of the binding site. *Journal of Biological Chemistry*. 261: 12154–12158.

[44] Marchal, D.; Boireau, W.; Laval, J.M.; Moiroux, J.; Bourdillon, C. 1997. An electrochemical approach of the redox behavior of water insoluble ubiquinones or plastoquinones incorporated in supported phospholipid layers. *Biophysical Journal*. 72: 2679–2687.

[45] Yang, S.; Robinson, M.T.; Mwambutsa, F.; Cliffel, D.E.; Jennings, G.K. 2016. Effect of Cross-linking on the Performance and Stability of Photocatalytic Photosystem I Films. *Electrochimica Acta*. 222: 926–932.

[46] McKelvey, K.; Martin, S.; Robinson, C.; Unwin, P.R. 2013. Quantitative Local Photosynthetic Flux Measurements at Isolated Chloroplasts and Thylakoid Membranes Using Scanning Electrochemical Microscopy (SECM). *Journal of Physical Chemistry B*. 117: 7878–7888.

[47] Nakatani, H.Y.; Barber, J. 1977. An Improved Method for Isolating Chloroplasts Retaining their Outer Membranes. *Biochimica Biophysica Acta*. 461: 510–512.

[48] Banaszak, J.; Barr, R.; Crane, F.L. 1976. Evidence for Multiple Sites of Ferricyanide Reduction in Chloroplasts. *Journal of Bioeneretics and Biomembranes*. 8: 83–92.

[49] Barr, R.; Crane, F.L. 1981. Ferricyanide Reduction in Photosystem II of Spinach Chloroplasts. *Plant Physiology*. 67: 1190–1194.

ADDITIONAL LITERATURE

Protein Electrochemistry

Betso, S.R.; Klapper, M.H.; Anderson, L.B. 1972. Electrochemical studies of heme proteins. Coulometric, polarographic and combined spectroelectrochemical methods for reduction of the heme prosthetic group in cytochrome c. *Journal of the American Chemical Society*. 94: 8197–8204.

Eddowes, M.J.; Hill, H.A.O. 1977. Novel method for the investigation of the electrochemistry of metalloproteins: cytochrome c. *Journal of the Chemical Society Chemical Communications*. 771–772.

Herles, F.; Vancura, A. 1932. Research on the cause of the characteristic "wave" on the polarographic curve of human serum. *Bulletin International de l'Academie des Sciences de Boheme*. 33: 119–120.

Heyrovsky, J.; Babicka, J. 1930. Polarographic studies with the dropping mercury cathode. Part XIII – the effect of albumins. *Collection of Czechoslovak Chemical Communications*. 2: 370–378.

Leger, C.; Elliott, S.J.; Hoke, K.R.; Jeuken, L.J.C.; Jones, A.K.; Armstrong, F.A. 2003. Enzyme electrokinetics: Using protein film voltammetry to investigate redox enzymes and their mechanisms. *Biochemistry*. 42: 8653–8662.

Malfoy, B.; Reynaud, J.A. 1980. Electrochemical investigations of amino acids at solid electrodes. Part II. Amino acids containing no sulfur atoms: trypthophan, tyrosine, histidine and derivatives. *Journal of Electroanalytical Chemistry*. 114: 213–223.

Palecek, E. 1958. Oszillographische polarographie der nucleinsäuren und ihrer bestandteile. *Die Naturwissenschaften*. 45: 186–187.

Reynaud, J.A.; Malfoy, B.; Canesson, P. 1980. Electrochemical investigations of amino acids at solid electrodes. Part I. Sulfur components: cystine, cysteine, methionine. *Journal of Electroanalytical Chemistry*. 114: 195–211.

Yeh, P.; Kuwana, T. 1977. Reversible electrode reaction of cytochrome c. *Chemical Letters*. 1145–1148.

Chlorophyll and Photosynthesis

Essmann, V.; Zhao, F.; Hartmann, V.; Nowaczyk, M.M.; Schuhmann, W.; Conzuelo, F. 2017. In operando investigation of electrical coupling of photosystem 1 and photosystem 2 by means of bipolar electrochemistry. *Analytical Chemistry*. 89: 7160–7165.

Hamidi, H.; Hasan, K.; Emek, S.C.; Dilgin, Y.; Åkerlund, H.-E.;Albertsson, P.-Å.; Leech, D.; Gorton, L. 2015. Photocurrent generation from thylakoid membranes on osmium-redox-polymer-modified electrodes. *ChemSusChem*. 8: 990–993.

Hasan, K.; Dilgin, Y.; Emek, S.C.; Tavahodi, M.; Åkerlund, H.-E.; Albertsson, P.-Å.; Gorton, L. 2014. Photoelectrochemical communication between thylakoid membranes and gold electrodes through different quinone derivatives. *ChemElectroChem*. 1: 131–139.

Hasan, K.; Grippo, V.; Sperling, E.; Packer, M.A.; Leech, D.; Gorton, L. 2017. Evaluation of photocurrent generation from different photosynthetic organisms. *ChemElectroChem*. 4: 412–417.

Hill, H.A.O.; Walton, N.J.; Whitford, D. 1985. The coupling of heterogeneous electron transfer to photosystem 1. *Journal of Electroanalytical Chemistry*. 187: 109–119.

Iwaki, M.; Kumazaki, S.; Yoshihara, K.; Erabi, T.; Itoh, S. 1996. ΔG^0 Dependence of the Electron Transfer Rate in the Photosynthetic Reaction Center of Plant Photosystem I: Natural Optimization of Reaction between Chlorophyll *a* (A0) and Quinone. *Journal of Physical Chemistry*. 100: 10802–10809.

Kuroiwa, Y.; Kato, Y.; Watanabe, T. 2009. Negative shift of chlorophyll a oxidation potential by aggregation in acetonitrile/ionic liquid mixed solvents. *Journal of Photochemistry and Photobiology A*. 202: 191–195.

Luz, R.A.S.; Crespilho, F.N. 2016. Gold nanoparticle-mediated electron transfer of cytochrome c on a self-assembled surface. *RSC Advances*. 6: 62585–62593.

Miao, W.; Choi, J.-P.; Bard, A.J. 2002. Electrogenerated chemiluminescence 69: The tris(2,2′-bipyridine) ruthenium(II), (Ru(bpy)32+)/Tri-n-propylamine (TPrA) system revisited a new route involving TPrA•+ cation radicals. *Journal of the American Chemical Society*. 124: 14478–14485.

Shipman, L.L.; Cotton, T.M.; Norris, J.R.; Katz, J.J. 1976. An análisis of the visible absorption spectrum of chlorophyll a monomer, dimer, and oligomers in solution.*Journal of the American Chemical Society*. 98: 8222–8230.

Wilson, D.F.; Dutton, M.; Erecińska, M.; Lindsay, J.G.; Soto, N. 1972. Mitochondrial electron transport and energy conservation. *Accounts of Chemical Research*. 5: 234–241.

Solid-State Electrochemistry

Gavilán-Arriazu, E.M.; Mercer, M.P.; Pinto, O.A.; Oviedo, O.A.; Barraco, D.E.; Hoster, H.E.; Leiva, E.P.M. 2020. Numerical simulations of cyclic voltammetry for lithium-ion intercalation in nanosized systems: finiteness of diffusion versus electrode kinetics. *Journal of Solid State Electrochemistry*. 24: 3279–3287.

Gavilán-Arriazu, E.M.; Mercer, M.P.; Barraco, D.E.; Hoster, H.E.; Leiva, E.P.M. 2022. Voltammetric behaviour of LMO at the nanoscale: a map of reversibility and diffusional limitations. *ChemPhysChem* 23: 1–12.

Malaie, K.; Scholz, F.; Scröder, U. 2023. A Thermodynamic Model for the Insertion Electrochemistry of Battery Cathodes. *ChemElectroChem*. 3: artic. 202201118.

Nafady, A.; Bond, A.M.; Qu, V.; Martin, L.L. 2013. Kinetic and thermodynamic interplay of cation ingress and egress at a TCNQ-modified electrode in contact with aqueous electrolyte mixtures containing Co(II) and Ni(II) cations. *Journal of Solid State Electrochemistry*. 17: 1609–1620.

Scholz, F.; Lovrić, M.; Stojek, Z. 1997. The role of redox mixed phases $\{ox_x(C_n red)_{1-x}\}$ in solid state electrochemical reactions and the effect of micibility gaps in voltammetry. *Journal of Solid State Electrochemistry*. 1: 134–142.

Scanning Electrochemical Microscopy

Abucayon, E.; Ke, N.; Cornut, R.; Patelunas, A.; Miller, D.; Nishiguchi, M.K.; Zoski, C.G. 2014. Investigating Catalase Activity Through Hydrogen Peroxide Decomposition by Bacteria Biofilms in Real Time Using Scanning Electrochemical Microscopy. *Analytical Chemistry*. 86: 498–505.

Bard, A.J.; Fan, F.R.F.; Kwak, J.; Lev, O. 1989. Scanning electrochemical microscopy: introduction and principles. *Analytical Chemistry*. 61: 132–138.

Bard, A.J.; Mirkin, M.V., Eds. 2001. *Electrochemical Microscopy*, Taylor & Francis, Boca Raton.

Beaulieu, I.; Kuss, S.; Mauzeroll, J.; Geissler, M. 2011. Biological Scanning Electrochemical Microscopy and Its Application to Live Cell Studies. *Analytical Chemistry*. 83: 1485–1492.

Edwards, M.A.; Martin, S.; Whitworth, A.L.; Macpherson, J.V.; Unwin, P.R. 2006. Scanning electrochemical microscopy: Principles and applications to biophysical systems. *Physiological Measurement*. 27: R63–R108.

Huang, L.; Li, Z.; Lou, Y.; Cao, F.; Zhang, D.; Li, X. 2018. Recent Advances in Scanning Electrochemical Microscopy for Biological Applications. *Materials*. 11: artic. 1389.

Lee, C.; Kwak, J.; Bard, A. J. 1990. Application of Scanning Electrochemical Microscopy to Biological Samples. *Proceedings of the National Academy of Sciences USA*. 87: 1740–1743.

Polcari, D.; Dauphin-Ducharme, P.; Mauzeroll, J. 2016. Scanning Electrochemical Microscopy: A Comprehensive Review of Experimental Parameters from 1989 to 2015. *Chemical Reviews*. 116: 13234–13278.

9 Electrochemistry of Reactive Oxygen Species (ROS)

9.1 INTRODUCTION

The term "reactive oxygen species" (ROS) is applied to a group of chemical species derived from molecular oxygen via redox reactions or electronic excitation [1]. This collective term encompasses different molecules and free radicals and can be extended to N- and S- reactive species, in particular, nitrogen oxide (NO•), nitrogen dioxide (NO_2•), and peroxynitrite/peroxonitrous acid ($ONOO^-$/ONOOH) [2–4]. The main ROS are summarized in Table 9.1 [1].

ROS possess vastly different chemical properties and biological functions. In the biological domain, they range from signaling to causing cell damage. In physiology, H_2O_2 and O_2•$^-$ are recognized as the major ROS in the redox regulation of cellular activities [3, 4]. These ROS act as redox signaling and are formed by mitochondrial NADH-dependent systems, extramitochondrial NADPH-dependent systems, and several oxidases. In normally metabolizing cells, their concentrations are in the nanomolar range for H_2O_2 and the picomolar range for O_2•$^-$. The dismutation of this species (spontaneously or catalyzed by superoxide dismutases) leads to H_2O_2 and O_2, thus making O_2•$^-$ a major source of H_2O_2 [1].

H_2O_2 signaling occurs mainly through reversible oxidation of thiolate units of specific cysteine-containing proteins to sulfenic acid and subsequent redox cascade mechanisms [2]. Extracellular signaling is expected to differ from intracellular one's due to the higher oxidant concentrations in the extracellular fluids, typically at the micromolar level for H_2O_2 [5]. Signaling by O_2•$^-$ is not well understood. It probably proceeds by affecting the redox state of transition metal ion complexes in proteins [6] and can be modulated by superoxide dismutases [7].

H_2O_2 can be transported through water channels in the cellular membrane thus contributing to the maintenance of steady-state concentration gradients [8] and the integration and transmission of signals between the extracellular environment and the intracellular space. Other redox-active species are maintained at steady-state levels, defining, generically, the "redox tone" of the cell. ROS participate in a complex set of signaling pathways affecting protein function, enzyme activity, gene transcription, and membrane and genome integrity [1].

The "ordinary" cellular state of low-level H_2O_2 maintenance and its associated signaling is termed "oxidative eustress". Higher, supraphysiological H_2O_2 concentrations (approximately up to 100 nM) lead to the unspecific damage of proteins and other biomolecules, causing pathological cellular states. This situation is called "oxidative distress". Figure 9.1 shows a schematic representation of cellular responses associated with different intra- and extracellular levels of H_2O_2 [1].

The interest in electrochemistry in ROS studies is reinforced by the need for analytical tools for the real-time and site-specific detection and quantification of individual ROS in living organisms [2].

9.2 ROLE OF ROS IN PLANT BIOLOGY

Superoxide anion radical and hydrogen peroxide are generated by several endogenous enzymatic and non-enzymatic processes. The major cellular site of the generation of these ROS is the extracellular face of the plasma membrane [2, 3], a process that involves NADPH oxidases [9]. Other important sources of ROS are the mitochondrial electron transport chain, which primarily produces

DOI: 10.1201/9781032697826-9

TABLE 9.1
Main Molecular and Free Radical ROS [1]

Molecular ROS	Free Radical ROS
Hydrogen peroxide (H_2O_2)	Superoxide anion radical ($O_2{\bullet}^-$)
Singlet molecular oxygen (electronically excited O_2) ($^1O_2$)	Hydroxyl radical ($HO\bullet$)
Ozone (O_3)	Hydroperoxyl radical ($HO_2\bullet$)
Organic hydroperoxides (ROOH)	Alkoxyl radicals ($RO\bullet$)
Electronically excited carbonyl (R-C-O)	Peroxyl radicals ($ROO\bullet$)
Hypochlorous and hypobromous acids (HOCl, HOBr)	Carbonate radical anion ($CO_3{\bullet}^-$)

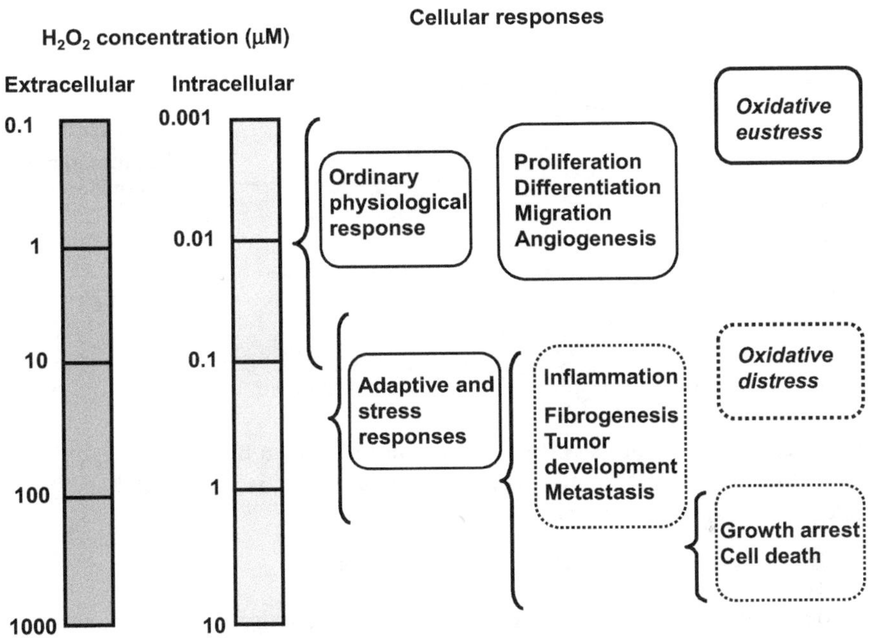

FIGURE 9.1 Cellular responses associated with different intra- and extracellular levels of H_2O_2. Notice that concentrations are expressed in logarithmic scale. (From data in ref. [1].).

superoxide, which is further converted to H_2O_2 and other oxidants [10], and the lumen of the endoplasmic reticulum Here, H_2O_2 is produced during disulfide bond formation in protein folding [2].

The contribution of the different sources to the cellular pool of ROS depends on the context and cellular metabolic state. ROS are not only generated by intracellular sources but also by environmental exposure, the so-called exposome [11]. This includes molecular factors (nutrients, pollutants), physicochemical stressors (radiation, mechanical damage), and even psychological stressors [1].

ROS signaling impacts a multitude of cellular processes and enables adaptation to the environment. Their transmission involves enzyme cascades and other targets which in turn induce changes in gene expression patterns [12]. The reaction of $O_2{\bullet}^-$ with iron-sulfur cluster proteins (Fe–S), and H_2O_2 with thiolate cysteine residues on redox-sensitive proteins, induces redox signaling and a series of physiological processes [2]. As schematized in Figure 9.2,

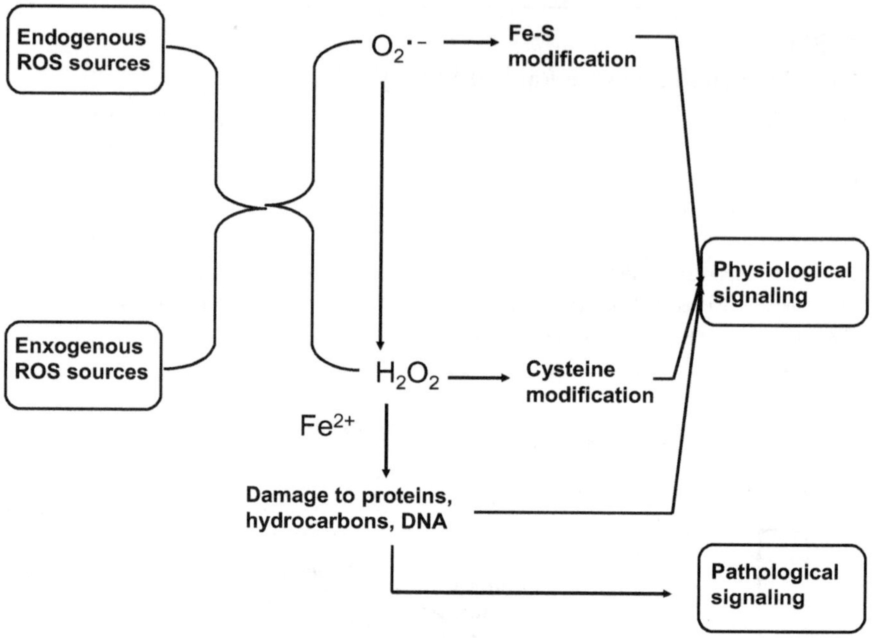

FIGURE 9.2 Simplified scheme for the generation of superoxide and hydrogen peroxide ROS and redox signaling in living cells. (From data in ref. [2].).

such processes lead to "ordinary" physiological signaling, and, eventually to dysfunction and disease.

Control of excess amounts of ROS is carried out in plants by a battery of enzymes starting with the superoxide dismutase (SOD) family. These are metalloproteins which catalyze the dismutation of superoxide radical anion into oxygen and hydrogen peroxide.

$$2O_2\bullet^- + 2H^+ \rightarrow O_2 + H_2O_2 \tag{9.1}$$

SODs reside in the chloroplasts, mitochondria, cytosols, peroxisomes, and apoplasts. Depending on the metal center, three isoenzymes have been identified in plants, having different structures and functions: Mn-SOD, Cu/Zn-SOD, and Fe-SOD, Cu/Zn-SOD being the most abundant [13, 14]. These SOD enzymes operate in conjunction with H_2O_2 – removing enzymes, such as catalases and glutathione (GSH) peroxidases.

Hydroxyl is a highly reactive radical. Its short half-life in aqueous solution (less than 1 ns) implies that when produced in vivo, it reacts rapidly close to its site of formation. The hydroxyl radical can be generated through different mechanisms, including ionizing radiations, and photolytic decomposition of alkylhydroperoxides, among others. The hydroxyl radical can be formed *in vivo* from superoxide and hydrogen peroxide through the Haber-Weiss reaction, catalyzed by metal ions (iron ions in particular) bound to biomolecules [15].

$$O_2\bullet^- + H_2O_2 \rightarrow O_2 + HO\bullet + OH^- \tag{9.2}$$

This reaction proceeds in two steps.

$$O_2\bullet^- + Fe(III) \rightarrow O_2 + Fe(II) \tag{9.3}$$

$$Fe(II) + H_2O_2 \rightarrow Fe(II) + HO\bullet + OH^- \quad (9.4)$$

This second step corresponds to the Fenton reaction that can be accompanied by other secondary processes [15].

$$HO\bullet + H_2O_2 \rightarrow H_2O + O_2\bullet^- + H^+ \quad (9.5)$$

$$HO\bullet + Fe(II) \rightarrow Fe(III) + OH^- \quad (9.6)$$

This defines a complex scenario in which electrochemical techniques offer interesting capabilities because of the electroactive character of most biological compounds and ROS. The scope of such techniques is enhanced by the possibility of generating ROS electrochemically.

9.3 OXYGEN ELECTROCHEMISTRY

In principle, ROS can be generated electrochemically by reducing O_2. In aqueous media, the obvious reduction products are H_2O_2 and H_2O. The processes of interest and their respective standard electrode potentials (V vs. SHE) are [15]

$$O_2(g) + 4H^+(aq) + 4e^- \rightarrow 2H_2O(l); E^\circ = 1.229\,V, \quad (9.7)$$

$$O_2(g) + 2H^+(aq) + 2e^- \rightarrow H_2O_2(aq); E^\circ = 0.695\,V, \quad (9.8)$$

$$H_2O_2(aq) + 2H^+(aq) + 2e^- \rightarrow 2H_2O(l); E^\circ = 1.763\,V. \quad (9.9)$$

The electrochemical processes, however, involve multi-step mechanisms and, depending on the experimental conditions, complications due to adsorption/desorption processes and coupled chemical reactions such as disproportionations. Figure 9.3 shows the cyclic voltammogram recorded at a GCE in an acidic air-saturated aqueous solution (0.50 M H_2SO_4). The voltammogram shows a well-defined cathodic peak at –0.65 V vs. Ag/AgCl with no anodic counterpart. This voltammetric signal

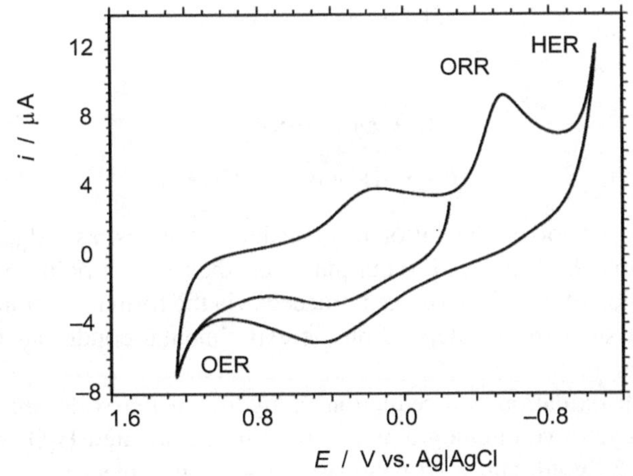

FIGURE 9.3 Cyclic voltammogram at GCE of an air-saturated 0.50 M H_2SO_4 aqueous solution. Potential scan initiated at 0.0 V vs. Ag/AgCl in the negative direction. Potential scan rate 50 mV s^{-1}. Note that the axes are opposite to the IUPAC convention.

corresponds to the oxygen reduction reaction (ORR) that, depending on the experimental conditions can be represented by Eqs. (97) or (9.8). At more negative potentials, the hydrogen evolution reaction (HER) appears as a rapidly increasing current. At potentials ca. 1.2 V, the oxygen evolution reaction (OER) produces the corresponding anodic current. Interestingly, at potentials between 0.2 and 0.5 V, an additional redox couple is recorded. This corresponds to the oxidation/reduction of oxygen functionalities generated in the glassy carbon surface [16, 17].

Three basic reductive pathways have been proposed for the electrochemical oxygen reduction process (ORR) at metallic electrodes where adsorption processes play a significant role [18]:

a) Dissociation:

$$O_2 \rightarrow 2O^{\bullet} \xrightarrow{2H^+,\,2e^-} 2HO^{\bullet} \xrightarrow{2H^+,\,2e^-} 2H_2O \tag{9.10}$$

b) Association:

$$O_2 \xrightarrow{H^+,\,e^-} HO_2^{\bullet} \rightarrow O^{\bullet} + HO^{\bullet} \xrightarrow{H^+,\,e^-} O^{\bullet} + H_2O \tag{9.11}$$

c) Peroxo pathway:

$$O_2 \xrightarrow{H^+,\,e^-} HO_2^{\bullet} \xrightarrow{H^+,\,e^-} H_2O_2 \rightarrow 2HO^{\bullet} \xrightarrow{2H^+,\,2e^-} 2H_2O \tag{9.12}$$

The association mechanism occurring at carbon electrodes in neutral and alkaline media can be described in terms of a sequence of steps [19] initiated by the one-electron generation of the radical anion superoxide:

$$O_2 + e^- \rightarrow O_2 \bullet^- . \tag{9.13}$$

This species rapidly protonates:

$$O_2 \bullet^- + H_2O \rightarrow HO_2 \bullet + OH^- . \tag{9.14}$$

The resulting hydroperoxyl radical experiences two competing processes: disproportionation,

$$2HO_2\bullet \rightarrow O_2 + H_2O_2, \tag{9.15}$$

and electron transfer,

$$HO_2 \bullet + e^- \rightarrow HO_2^-, \tag{9.16}$$

$$HO_2 \bullet + H^+ + e^- \rightarrow 2HO\bullet . \tag{9.17}$$

The cascade pathway involves several other plausible processes, as judged by thermochemical data summarized in Table 9.2 [20–22]. For our purposes, the relevant point to emphasize is that the electrochemical reduction of dissolved oxygen proceeds via the formation of intermediate ROS. The type and proportion of such species depend on the experimental conditions (material of the electrode, pH, potential scan rate, etc.).

The electrochemical oxidation and reduction of hydrogen peroxide are kinetically hindered. Figure 9.4 shows the cyclic voltammograms recorded in a deaerated H_2O_2 solution in phosphate buffer at pH 7.4 at GCE, gold, and gold nanoparticles-modified electrodes submitted to previous pretreatment by potential cycling [22].

One can see in this figure that only in the presence of nanoparticulate gold is a prominent H_2O_2 oxidation wave recorded for pretreated electrodes. This effect, which was also observed for H_2O_2 reduction, was attributed to the catalytic action of a film of gold oxides during the electrochemical

TABLE 9.2
Standard Electrode Potentials for Selected Processes Involved in the Aqueous Electrochemistry of Water and Oxygen with O_2 At 1 Atm [15, 20–22]

Process	E° (V vs. SHE)
One-electron reduction	
O_2 (g) + e^- → $O_2^{\bullet -}$ (aq)	−0.33
$1/2O_2$ (g) + H_2O (aq) + e^- → ½ H_2O_2 (aq) + OH^- (aq)	−0.13
½ O_2 (g) + ½ H_2O + e^- → ½ OH^- (aq) + ½ $HO_2^\bullet$ (aq)	−0.06
O_2 (aq) + H^+ (aq) + e^- → $HO_2^\bullet$ (aq)	−0.125
H^+ (aq) + e^- → ½ H_2 (g)	0.00
H^+ (aq) + O_2 (g) + e^- → $HO_2^\bullet$ (ads)	0.12
H^+ (aq) + e^- → $H^\bullet$ (ads)	0.32
$O_2^{\bullet -}$ (aq) + H_2O + e^- → HO_2^- (aq) + OH^-(aq)	0.20
¼ O_2 (g) + ½ H_2O + e^- → OH^- (aq)	0.42
$HO_2^\bullet$ (aq) + e^- → HO_2^- (aq)	0.74
H_2O_2 (aq) + H^+ (aq) + e^- → $HO^\bullet$ (aq) + H_2O	0.79
$O_2^{\bullet -}$ (aq) + $2H^+$ (aq) + e^- → H_2O_2 (aq)	1.35
$HO^\bullet$ (aq) + H^+ (aq) + e^- → H_2O (l)	2.72
Two-electron reduction	
O_2 (g) + $2H^+$ (aq) + $2e^-$ → H_2O_2 (aq)	0.71
H_2O_2 (aq) + $2H^+$ (aq) + $2e^-$ → H_2O	1.76
Four-electron reduction	
O_2 (g) + $4H^+$ (aq) + $4e^-$ → $2H_2O$	1.23

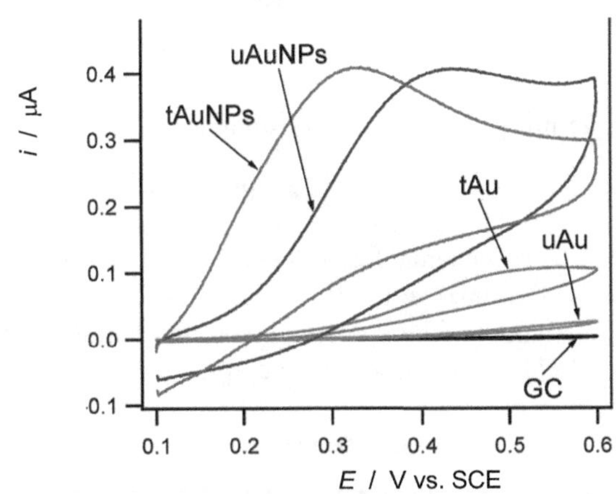

FIGURE 9.4 Cyclic voltammograms of 1.0 mM H_2O_2 solution in deaerated 0.05 M phosphate buffer at pH 7.4 at different working electrodes; glassy carbon (GC), gold, and gold coated by gold nanoparticles optionally submitted to pretreatment by potential cycling. Potential scan rate 50 mV s^{-1}. (Adapted from ref. [22], with permission. Copyright the American Society of Chemistry 2016.).

pretreatment, as supported by experiments at platinum electrodes [23]. The overall oxidation process can be represented as

$$H_2O_2 \rightarrow O_2 + 2H^+ + 2e^-, \tag{9.18}$$

where the electrochemical pathway involves the generation of hydroxyl radicals as intermediates, as evidenced by voltammetric records in the presence of radical-trap reagents [23]. In turn, the electrochemical reduction of hydrogen peroxide has been described in terms of an adsorptive step,

$$H_2O_2\,(aq) \rightarrow 2HO\bullet(ads), \tag{9.19}$$

followed by a one-electron electron transfer [25, 26],

$$HO\bullet(ads) + e^- \rightarrow OH^-\,(aq). \tag{9.20}$$

The net cathodic reaction is

$$H_2O_2\,(aq) + 2e^- \rightarrow 2OH^-\,(aq). \tag{9.21}$$

The catalytic effect exerted by a variety of electrode modifiers on this process (typically, Prussian blue [24, 25]) is exploited for analytical purposes in the biochemical context. The process OER also follows a pH- and electrode-dependent cascade mechanism where intermediate ROS can be generated.

9.4 ELECTROCHEMICAL GENERATION OF ROS

There are numerous mechanisms of chemical and photochemical generation of ROS. Among an enormous variety of possibilities, autooxidation of organic compounds (RH) such as fatty acids, oils, and lipids occurs as free radical chain reactions initiated by the photochemical generation of radicals R•. The following step is constituted by a chain of propagation reactions [26]:

$$R\bullet + O_2 \rightarrow ROO\bullet, \tag{9.22}$$

$$RH + ROO\bullet \rightarrow ROOH + R\bullet, \tag{9.23}$$

ending with a third step of termination reactions forming nonradical products:

$$ROO\bullet + ROO\bullet \rightarrow \text{products}. \tag{9.24}$$

Interestingly, ROS can also be generated electrochemically. Several electrochemical processes described in the precedent section involve the formation of different ROS, such as highly reactive hydroxyl radicals in particular. These can be generated electrochemically by direct water oxidation at potentials of ca. 1.2 V vs. Ag/AgCl at boron-doped diamond electrodes in alkaline media [27–30]:

$$H_2O \rightarrow HO\bullet + H^+ + e^-. \tag{9.26}$$

Figure 9.5 shows a series of differential pulse voltammograms recorded at the boron-doped diamond electrode (BDD) of several alkaline solutions [28]. The signal of interest, attributed to the generation of hydroxyl radicals via Eq. (9.18), appears (signal P2) as an inflection preceding the gross anodic current corresponding to water oxidation to O_2 (OER). In an ammonia buffer, a new intense signal (P3) was recorded and assigned to NH_3 oxidation [28]. The electrochemical generation of hydroxyl radicals at Pt [30] and Au [31] electrodes has also been reported. Hydroxyl generation at a variety of electrode materials is increasingly used for the electrochemical degradation of pollutants [32].

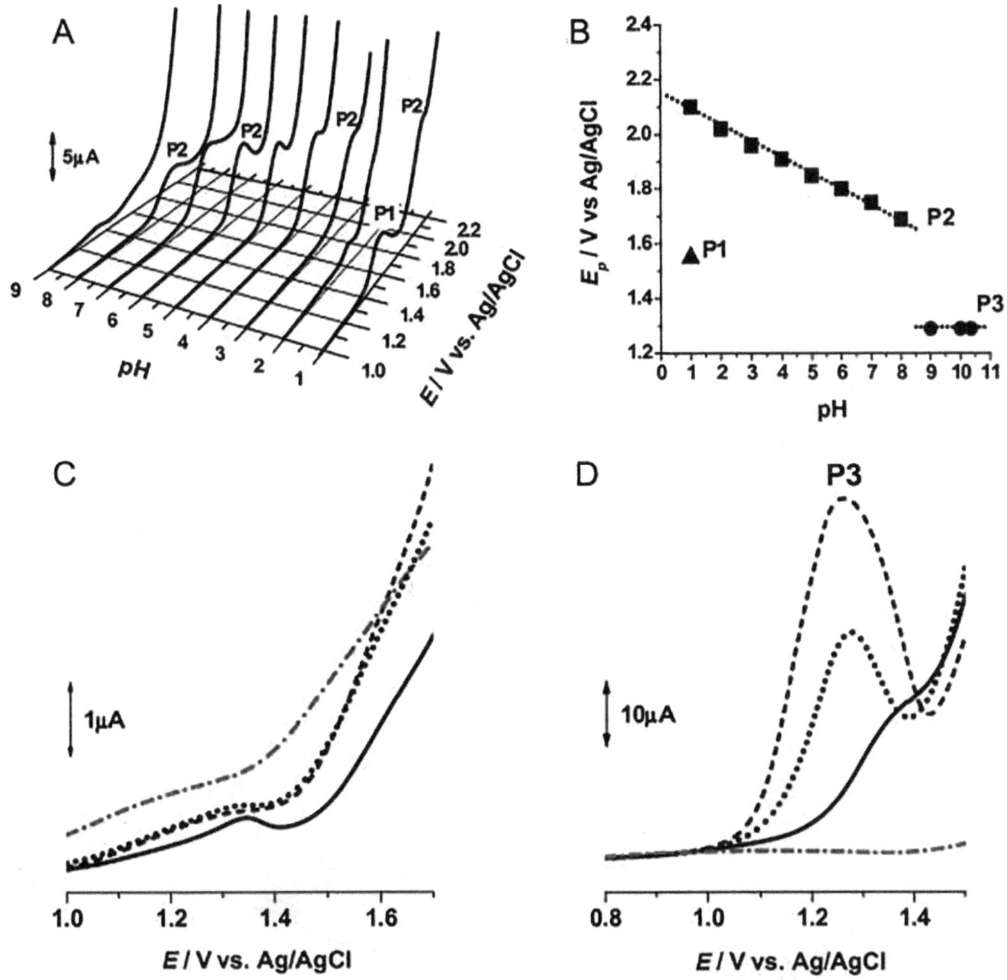

FIGURE 9.5 Differential pulse voltammograms at BDD electrode. A: 3D plot as a function of pH; B: peak potential vs. pH plots of peaks P1 (▲), P2 (■), and P3 (●)vs. pH; C: Voltammograms in $NaHCO_3/Na_2CO_3$ buffer at pH 9.0 (▬), 10.0 (●●●), 11.0 (▪▪▪), and NaOH pH 11.5 (•▪•▪•), D: Voltammograms in NH_3/NH_4^+ buffer at pH 8.9 (▬), 10.0 (●●●), 10.5 (▪▪▪), and NaOH pH 11.5 (•▪•▪•). Pulse amplitude 50 mV; pulse width 70 ms; potential scan rate 5 mV s^{-1}. (Reproduced from ref. [28], with permission. Copyright Elsevier 2009.).

It is worth noting that oxygen electrochemistry is strongly solvent-dependent. In aprotic solvents, O_2 is reversibly reduced to superoxide radical anion through an essentially reversible one-electron process, as described by Eq. (9.13) [33]. This can be seen in Figure 9.6 where the cyclic voltammetry at GCE of an air-saturated DMSO solution is depicted.

The generated ROS is stable at the time scale of conventional voltammetric experiments. However, it becomes sensitive to the presence of protonated species in the solution, resulting in the formation of radical hydroperoxide (see Figure 9.7). As a result, the reversible voltammetric pattern is replaced by an irreversible profile. This species rapidly disproportionates into O_2 and H_2O_2 in the presence of protic species (catechols, ascorbic acid, tocopherols) [34]. In the presence of electrophilic species, the current for oxygen reduction is enhanced as a result of their reaction with the nascent superoxide radical anion regenerating O_2 [34]. The reaction of superoxide radical anion with Brønsted acids can be represented as

$$O_2{\bullet}^{-} + HA \rightarrow HO_2{\bullet} + A^{-}, \tag{9.27}$$

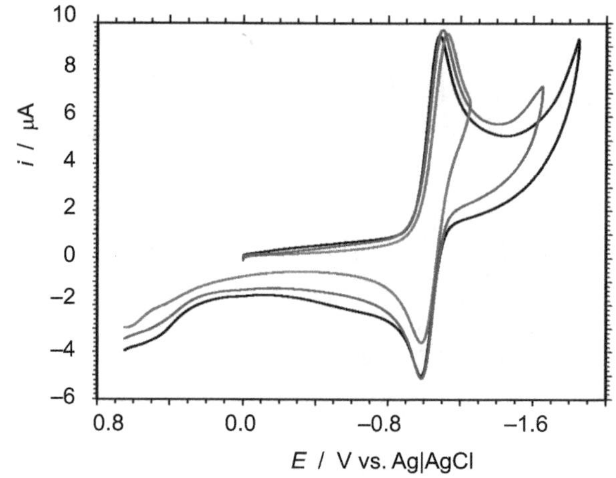

FIGURE 9.6 Cyclic voltammograms at GCE recorded at different potential ranges in an air-saturated 0.10 M Bu_4NPF_6/DMSO solution. A Pt pseudoreference electrode was used, with the potential scale being normalized by adding ferrocene as an internal standard and further converted into the Ag/AgCl scale. Potential scan initiated at 0.0 V vs. Ag/AgCl in the negative direction. Potential scan rate 50 mV s^{-1}. (Note that the axes are opposite to the IUPAC convention.).

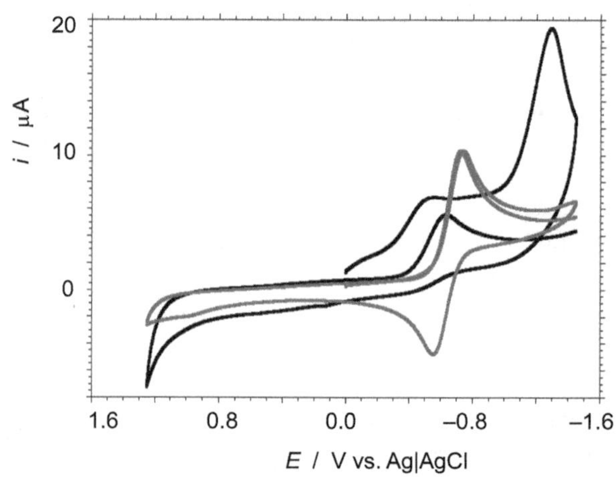

FIGURE 9.7 Cyclic voltammograms at GCE recorded in an air-saturated 0.10 M Bu_4NPF_6/DMSO solution before (gray line) and after (black line) the addition of 2.0 mM HAc/NaAc aqueous buffer at pH 4.75. A Pt pseudoreference electrode was used, with the potential scale being normalized by adding ferrocene as an internal standard and further converted into the Ag/AgCl scale. Potential scan initiated at 0.0 V vs. Ag/AgCl in the negative direction. Potential scan rate 50 mV s^{-1}. (Note that the axes are opposite to the IUPAC convention.).

$$2HO_2\bullet \rightarrow H_2O_2 + O_2. \quad (9.28)$$

The formation of H_2O_2 in the presence of water traces is denoted by the appearance of a weak anodic wave at ca. 0.4 V (see Figure 9.6). In the presence of relatively high proton concentrations, oxygen is directly reduced to hydroperoxide radical, lowering slightly the peak potential (see Figure 9.7).

$$O_2 + HA + e^- \rightarrow HO_2\bullet + A^- \quad (9.29)$$

This give rise to a second reduction signal at more negative potentials (see Figure 9.7) associated with the process.

$$HO_2 \bullet (ads) + HA + e^- \rightarrow H_2O_2 + A^- \quad (9.30)$$

The preceding processes are frequently mediated by the adsorption of the radical species on electrode surfaces [34].

9.5 INTERACTION OF ROS WITH PLANT COMPONENTS

There is an enormous variety of reactions between ROS and chemical components of plants. One of the most studied processes is the donation of hydrogen atoms from phenoxyl groups of polyphenolic compounds. Different alternative mechanisms have been proposed to describe such processes, whose result is ROS inactivation [35–37]. These could be applied, for example, to the reaction of phenols with superoxide radical anions [35]. In this context, hydrogen atom transfer (HAT) processes to phenolic species where a radical phenoxide is generated have been extensively studied [35–37]. Generically, these processes can be represented by the reaction of the phenolic substrate with a reactive radical R•.

$$R \bullet + ArOH \rightarrow RH + ArO \bullet \quad (9.31)$$

In the sequential electron-proton transfer mechanism (SEPT), an intermediate cationic species is formed.

$$R \bullet + ArOH \rightarrow R^- + ArOH^+ \quad (9.32)$$

$$R^- + ArOH^+ \rightarrow RH + ArO \bullet \quad (9.33)$$

In the proton loss electron transfer pathway (SPLET), protons are released in the first step,

$$ArOH \rightarrow ArO^- + H^+ \quad (9.34)$$

$$R \bullet + ArO^- + H^+ \rightarrow ArO \bullet + R^- + H^+ \quad (9.35)$$

$$ArO \bullet + R^- + H^+ \rightarrow RH + ArO \bullet \quad (9.36)$$

The reaction rates are influenced by the hydrogen-bond-accepting and ion-solvation abilities of solvents and by the electron affinities and energies of O-H bond dissociation of radicals. In turn, these properties are influenced by the ring substituents in the polyphenolic aromatic system [35–37].

The reactions between ROS and plant components can be monitored electrochemically, eventually aided by synthetic redox probes. Figure 9.8 shows the cyclic voltammograms recorded at GCE in contact with an acidic solution of methylene blue in N_2-saturated and O_2-saturated aqueous solutions [38]. In the O_2-containing solution, the voltammogram shows a cathodic peak at –0.70 V vs. Ag/AgCl corresponding to the ORR process. In these conditions, methylene blue remains electrochemically silent, as denoted by the absence of voltammetric signals in N_2-saturated solutions. In an O_2-saturated solution in the presence of methylene blue, the ORR signal becomes positively shifted and slightly diminished in height. These features can be interpreted in terms of the reaction of methylene blue with any oxygen species generated in O_2 reduction. Figure 9.9 shows the corresponding voltammetric response when 1,4-benzoquinone is added to the solution [38]. In the presence of the quinone, the ORR signal becomes, as before, positively shifted and slightly diminished. When methylene blue is added to the previous solution, the voltammetric peak is again enhanced

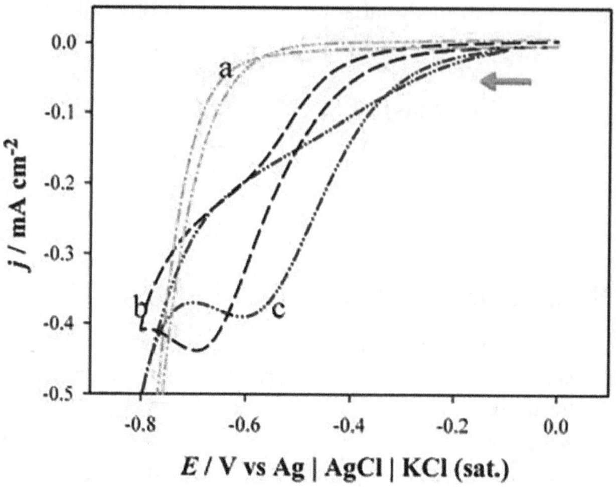

FIGURE 9.8 Cyclic voltammograms at GCE recorded in (a) N_2-saturated, 0.001 mM methylene blue; (b) O_2-saturated with methylene blue; and (c) O_2-saturated, 0.001 mM methylene blue solutions in 0.05 M H_2SO_4. Potential scan rate: 50 mV s^{-1}. (Reproduced from ref. [38], with permission. Copyright Elsevier 2020.).

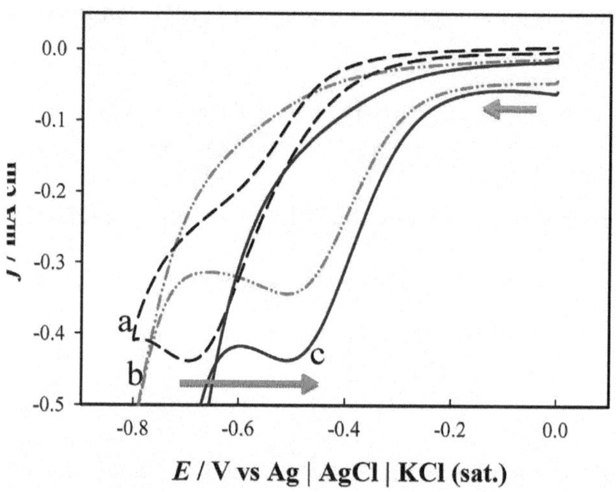

FIGURE 9.9 Cyclic voltammograms at GCE recorded in O_2-saturated 0.05 M H_2SO_4 aqueous solutions with (a) no methylene blue and no 1,4-benzoquinone, (b) 1 mM 1,4-benzoquinone, and (c) 0.001 mM methylene blue plus 1 mM 1,4-benzoquinone. Potential scan rate: 50 mV s^{-1}. (Reproduced from ref. [38], with permission. Copyright Elsevier 2020.).

without any additional peak potential shift. These features have been interpreted in terms of the reaction of the quinone with the nascent superoxide radical anion yielding hydrogen peroxide, which reacts with methylene blue [38].

Another electrochemical strategy for studying ROS reactivity is the attachment of the substrates to the electrode. An example is depicted in Figure 9.10 where the voltammetry of fluor-doped tin oxide electrodes (FTOs) modified with different flavonoids and their Fe(III) and Cu(II) complexes, aided with 3-aminopropyl)triethoxysilane as a coupling agent, are shown [39]. The voltammograms show the anodic peaks corresponding to the oxidation of the flavonoids. The intensity of such peaks

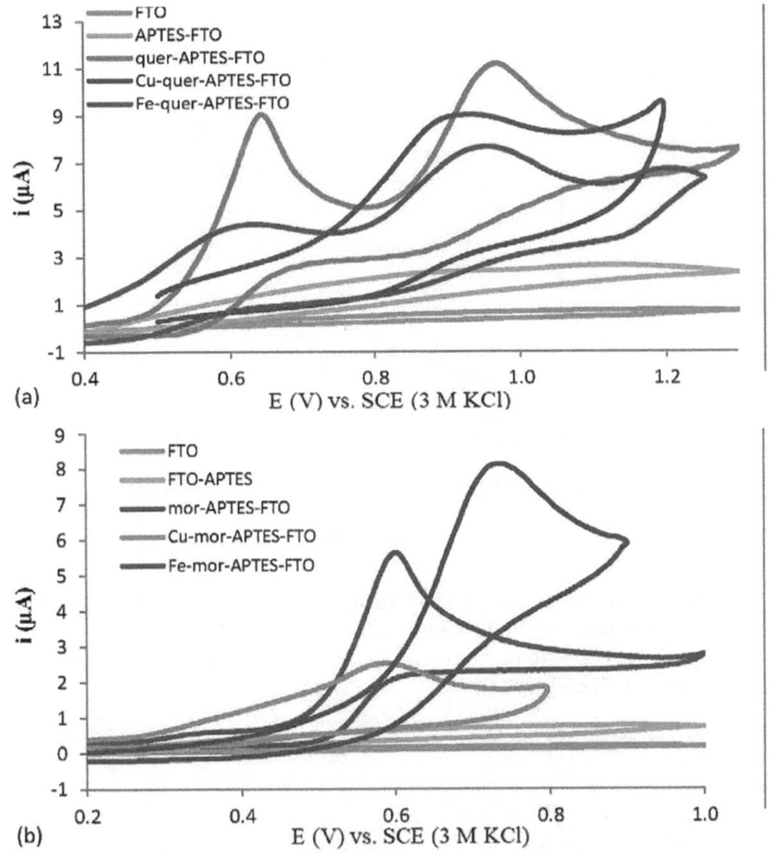

FIGURE 9.10 Cyclic voltammograms of unmodified and (3-aminopropyl) triethoxysilane-modified FTO electrodes in phosphate buffer saline (PBS) solution at pH 7.4 containing (a) quercetin (quer), (b) morin (mor), and their complexes with Fe(III) and Cu(II). Potential scan rate 100 mV s^{-1}. (Adapted from ref. [39], with permission. Copyright Elsevier 2017.).

decreases progressively upon successive additions of the hydroxyl-generating Fenton reagent, thus denoting the reaction of the flavonoids with the ROS and prompting an electrochemical method for hydroxyl quantification [39].

An example of electrochemical characterization of the interaction of ROS with plant components in a lipophilic environment is the study of chartreusin in DMSO solutions. This compound is an aromatic polyketide glycoside isolated from *Streptomyces chartreusis* Leach et al. that exhibits significant antibacterial and antitumoral chemotherapeutic properties [40, 41].

Figure 9.11 shows the cyclic voltammograms obtained at GCE in a chartreusin solution in DMSO containing a small proportion of water (1% *v*/*v*) [42]. On starting the potential scan at 0.0 V vs. Ag/AgCl in the negative direction, reduction peaks at −0.18 (C_1) and −0.75 V (C_2), accompanied by a shoulder at ca. −1.5 V (C_3). In the following positive-going potential scan, peaks C_1 and C_2 are coupled to weak oxidation counterparts at −0.58 (A_2) and 0.10 V (A_1). These are followed by oxidation peaks and beta-lapachione derivatives at 0.59 (A_3) and 0.90 V (A_4). At scan rates above 500 mV s^{-1}, peak A_4 retains its irreversible profile, while peak A_3 exhibits a reduction counterpart (C_3). In initial anodic scan voltammograms, the previous pattern becomes essentially unchanged with the exception of the vanishing of peak A_3.

This voltammetry has been interpreted based on the literature on quinone electrochemistry (see Chapter 4), and β-lapachone derivatives possessing two redox centers [43], and thermochemical calculations on elsamicin, a structurally similar bislactone aglycone (chartarin) [44], suggesting that

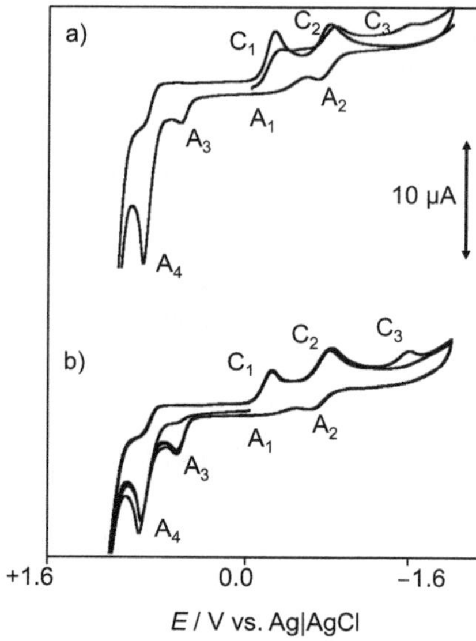

FIGURE 9.11 Cyclic voltammograms at GCE of a 0.03 mM chartreusin solution in 0.10 M Bu_4NPF_6 solution in DMSO containing 1% (v/v) water. (a) Initial cathodic scan and (b) initial anodic scan. Potential scan rate 50 mV s^{-1}. (Reproduced from ref. [42], with permission. Note that the axes are opposite to the IUPAC convention. Copyright the Royal Society of Chemistry 2017.).

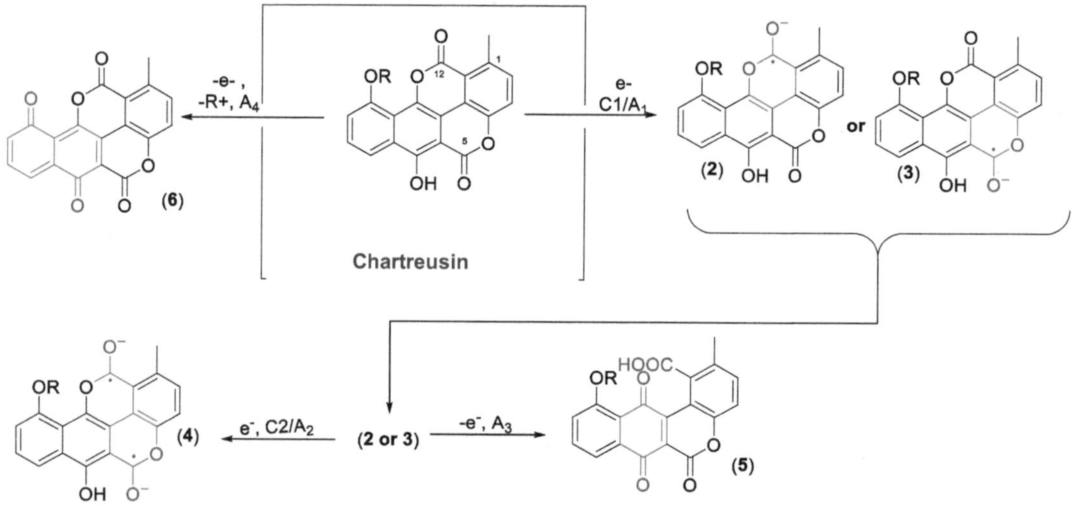

FIGURE 9.12 A possible scheme for the electrochemistry of chartreusin in DMSO solution. (Adapted from ref. [42], with permission. Copyright the Royal Society of Chemistry 2017.).

the most favored reductive pathway involves one of the oxo groups in either the C5 or the C12 positions. This results in successive one-electron transfer corresponding to peaks C_1 and C_2, followed by some structural reorganization giving rise to competing oxidative pathways. The parent chartreusin is directly oxidized (process A_4) via the –OH and –OR motifs. Figure 9.12 depicts a plausible scheme for such processes [42].

The aforementioned voltammetric response changes significantly in air-saturated solutions. As shown in Figure 9.13, the well-known reduction of dissolved oxygen at ca. –0.9 V (C_{ox}) [33] overlaps with C_1 and C_2 processes now accompanied by additional peaks (C_5 and C_6). The anodic region

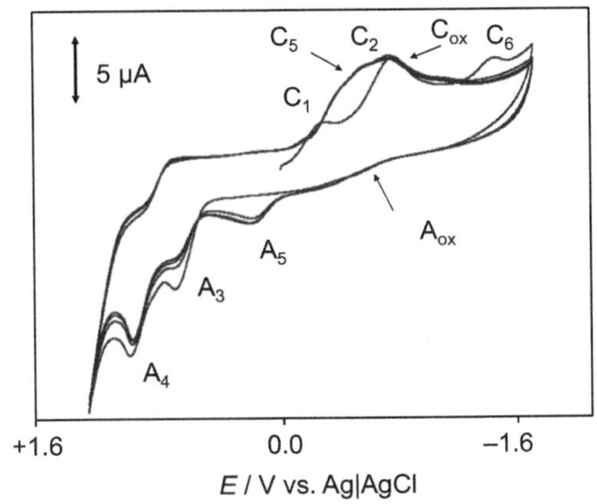

FIGURE 9.13 Cyclic voltammogram at GCE of air-saturated 0.03 mM chartreusin solution in 0.10 M Bu_4NPF_6/DMSO. Potential scan rate 50 mV s^{-1}. (Reproduced from ref. [42], with permission. Note that the axes are opposite to the IUPAC convention. Copyright the Royal Society of Chemistry 2017.).

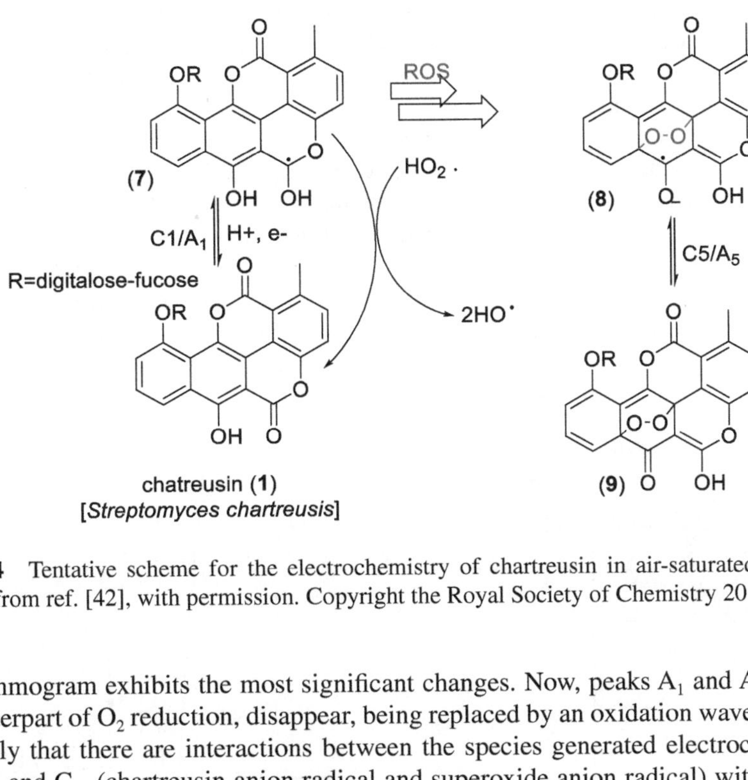

FIGURE 9.14 Tentative scheme for the electrochemistry of chartreusin in air-saturated DMSO solution. (Reproduced from ref. [42], with permission. Copyright the Royal Society of Chemistry 2017.).

of the voltammogram exhibits the most significant changes. Now, peaks A_1 and A_2, as well as the anodic counterpart of O_2 reduction, disappear, being replaced by an oxidation wave at 0.20 V. These features imply that there are interactions between the species generated electrochemically in the processes C_1 and C_{ox} (chartreusin anion radical and superoxide anion radical) with the parent species. A tentative pathway is schematized in Figure 9.14 [42].

9.6 ROS ELECTROCHEMISTRY IN PLANTS

The interaction of ROS with vegetal matter can also be studied *in situ*. Electrochemical studies directly conducted on berry fruits and lyophilized fruit samples provide information on the interaction of anthocyanins and anthocyianidins, principal antioxidant components of the fruits [45], with ROS [46, 47]. Anthocyanins and anthocyanidins form a sub-family of flavonoids (see Chapter 4) characterized by their high antioxidant capacity associated with the radical-scavenging ability of their components [48]. Production and accumulation of anthocyanins in plants can be triggered by a series of factors, including ROS. However, the effect of ROS in such processes and the role of anthocyanins and anthocyianidins in plant signaling pathways are not well understood [49].

Figure 9.15 shows the voltammetric response recorded at pH 7.4 on GCEs modified with microparticulate deposits resulting from the evaporation of ethanolic extracts of mature blueberry fruits [46]. Initiating the potential scan at potentials ca. −1.0 V vs. Ag/AgCl, both voltammograms in air-saturated and deaerated solutions show a series of overlapping anodic peaks (A) between −0.2 and 1.0 V. In air-saturated solutions, an additional, prominent peak at ca. −0.7 V (A*) is recorded. This signal disappears upon deoxygenation of the electrolyte solution but does not appear at unmodified GCEs in air-saturated and solutions of H_2O_2. These features denote that the signal A* results from the interaction between the plant components and oxygen species. Taking into account that the potentials at which A* appears are close to those of the ORR process, it is reasonable to assume that this signal reflects the presence of any species resulting from the interaction of ROS-plant components. Consistently, the peak A* diminishes when DMSO, a radical scavenger [50], is added to the electrolyte solution.

In turn, the A peaks can be associated with the oxidation of the most abundant anthocyanins in blueberry fruits [45]: pelargonidin-3-*O*-glucoside, peonidin-3-*O*-rutinoside, and delphinidin-3-*O*-rutinoside. This was confirmed by voltammetry of cyanidin, pelargonidin, and peonidin 3-glucoside standards in an aqueous solution under equivalent conditions [46, 47]. Similar electrochemistry was obtained on microparticulate deposits from fresh and lyophilized blueberry and raspberry fruits [47] and in situ on blueberry fruits using Pt microelectrodes inserted into the subepithelial region of mature fruits [46]. Interestingly, the square wave voltammetric profile was frequency dependent. As can be seen in Figure 9.16, the signal A* is accompanied by other intense signals at less negative potentials. At low frequencies, i.e., at relatively large experimentation times, the signal A* predominates. However, on increasing frequency (i.e., decreasing the experimentation time) the accompanying signals become progressively enhanced. These features can be interpreted by considering that, depending on the experimental conditions (electrode material, solution pH, time of electrogeneration), different ROS species in different proportions will be formed during the ORR. Consequently, different products of ROS reaction with anthocyanins and anthocyanidins will be formed at different reaction times.

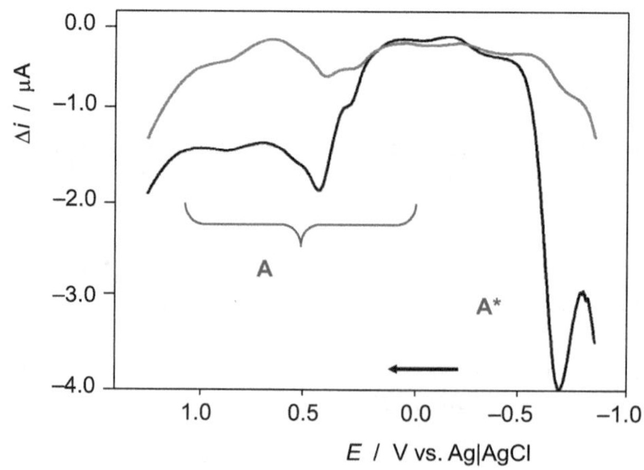

FIGURE 9.15 Square wave voltammograms at GCEs modified with microparticulate deposits of ethanolic extracts from blueberry fruits in contact with 0.10 M phosphate buffer at pH 7.4 when the potential scan was initiated at −1.05 V vs. Ag/AgCl in the positive direction. Black line: air-saturated; gray line: deoxygenated electrolyte solutions. Potential step increment 4 mV, square wave amplitude 25 mV, frequency 5 Hz. The arrow marks the direction of the potential scan. (Reproduced from ref. [46], with permission. Copyright Elsevier 2023.).

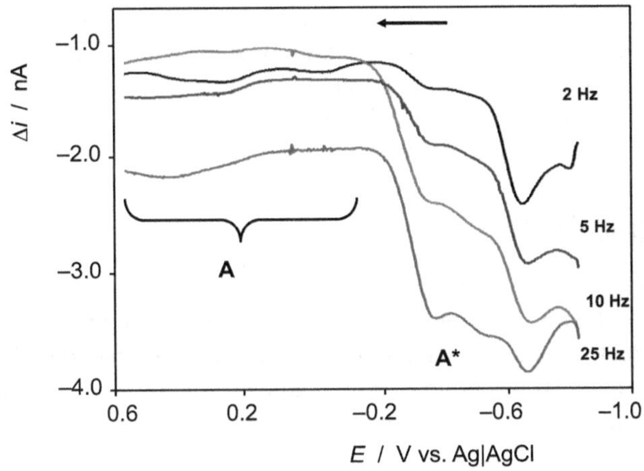

FIGURE 9.16 ***In situ*** square wave voltammograms recorded upon insertion of Pt microelectrode (diameter 25 μm) on blueberry fruits. The three-electrode arrangement was completed with Pt (auxiliary and pseudoreference) electrodes of 100 μm diameter. The electrodes were inserted perpendicularly to the fruit distributed in an area of ca. 1 cm^2 reaching a depth of 2 mm. Potential scan initiated at −0.85 V in the positive direction, potential step increment 4 mV, square wave amplitude 25 mV, and frequencies indicated in the figure. The arrow marks the direction of the potential scan. (Reproduced from ref. [46], with permission. Copyright Elsevier 2023.).

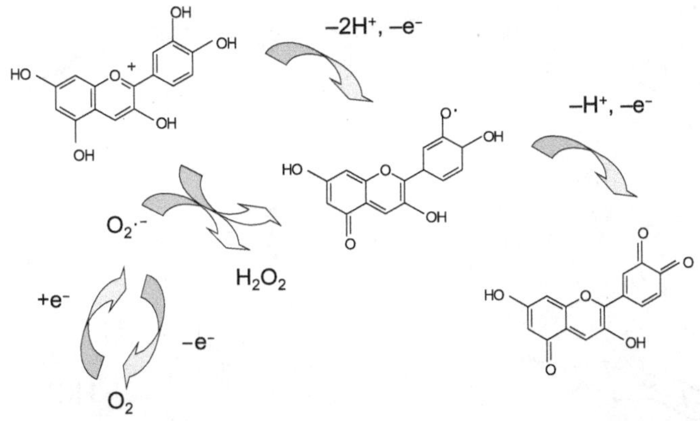

FIGURE 9.17 Scheme of the redox processes involved in electrochemistry of cyanidin in the presence of superoxide radical anion generated electrochemically. (Reproduced from ref. [47], with permission. Copyright Elsevier 2024.).

This voltammetric behavior cannot be explained in terms of the "ordinary" electrochemistry of anthocyanins and anthocyanidins consisting of A-signals appearing in the region of potentials between ca. 0.0 and ca. 1.2 V vs. Ag/AgCl [51–53]. This electrochemistry is dominated by the oxidation of *o*-catechol units to *o*-quinones in both aqueous and nonaqueous media. The interaction with ROS species generated electrochemically results in an increase of the peak currents through an electrocatalytic pathway. The simplest case, limited to the superoxide radical anion, is schematized in Figure 9.17 [47].

This electrochemical route can be complemented by others involving the opening of the C-ring [52, 53] and the formation of peroxo bridges [54, 55]. Nevertheless, the former process occurs at high anodic potentials, and it is conceivable that the second process involves potentials in the

A-region of potentials. Since the process A* takes place at very cathodic potentials, one can expect that it involves the generation of species with a high degree of electronic conjugation to provide a sufficiently high degree of electronic conjugation. A plausible scheme is hypothesized in Figure 9.18 for cyanidin 3-glucoside, where the release of the sugar unit is also included [47]. This release is consistent with data on the thermal degradation of anthocyanins [56].

9.7 ELECTROCHEMISTRY OF RNS

Reactive nitrogen species (RNS) exert a variety of functions in plants that are under intensive investigation, with NO being the most studied species. It is believed that NO synthesis proceeds via two main routes: (i) via reduction of nitrites and (ii) via oxidation of aminated molecules [57]. The focus is posed on the electrochemical detection of NO in aqueous media. This is usually achieved by promoting its oxidation process in a three-step pathway [58].

$$NO \rightarrow NO^+ + e^- \tag{9.37}$$

$$NO^+ + OH^- \rightarrow HNO_2 \tag{9.38}$$

$$HNO_2 + H_2O \rightarrow NO_3^- + 3H^+ + 2e^- \tag{9.39}$$

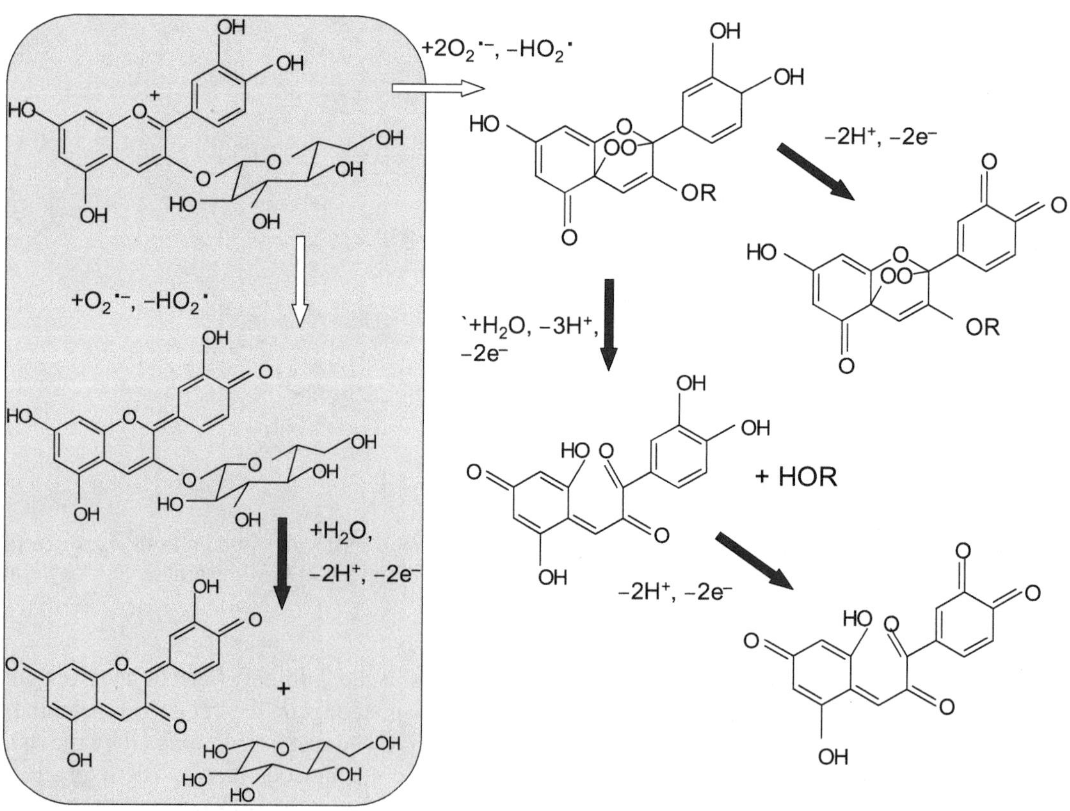

FIGURE 9.18 Scheme of possible redox pathways involved in the electrochemistry of cyanidin 3-glucoside in the presence of superoxide radical anion generated electrochemically. The gray zone marks the electrochemical route associated with signal A*. For simplicity, the glucoside unit has been labeled OR. The electrochemical processes are marked by dark gray arrows, while the chemical reactions are marked by white arrows. (Adapted from ref. [46], with permission. Copyright Elsevier 2024.).

A wide variety of electrochemical sensors have been reported, as well as for nitrite and nitrate, including *in vitro* and *in vivo* analysis. This is a research field under expansion directly dealing with plant physiology, including nitrogen uptake [59].

REFERENCES

[1] Sies, H.; Jones, D.P. 2020. Reactive oxygen species (ROS) as pleiotropic physiological signaling agents. *Nature Reviews Molecular Cell Biology*. 21: 363–383.

[2] Sies, H.; Belousov, V.V.; Chandel, N.S.; Davies, M.J.; Jones, D.P.; Mann, G.E.; Murphy, M.P.; Yamamoto, M.; Winterbourn, C. 2022. Defining roles of specific reactive oxygen species (ROS) in cell biology and physiology. *Nature Reviews Molecular Cell Biology*. 23: 499–515.

[3] Winterbourn, C.C. 2018. Biological production, detection, and fate of hydrogen peroxide. *Antioxidants & Redox Signaling*. 29: 541–551.

[4] Murphy, M.P.; Bayir, H.; Belousov, V.; Chang, C.J.; Davies, K.J.A.; Davies, M.J.; Dick, T.P.; Finkel, T.; Forman, H.J.; Hanssen-Heininger, Y.; Gems, D.; Kagan, V.E.; Lalyanaraman, B.; Larsson, N.-G.; Milne, G.L.; Nyström, T.; Poulsen, H.E.; Radi, R.; Van Remmen, H.; Schumacker, P.T.; Thornalley, P.J.; Toyokuni, S.; Winterbourn, C.C.; Halliwell, B. .2022. Guidelines for measuring reactive oxygen species and oxidative damage in cells and in vivo. *Nature Metabolism*. 4: 651–662.

[5] Forman, H.J.; Bernardo, A.; Davies, K.J.A. 2016. What is the concentration of hydrogen peroxide in blood and plasma? *Archives of Biochemistry and Biophysics*. 603: 48–53.

[6] Malard, E.; Valable, S.; Bernaudin, M.; Pérès, E.; Chatre, L. 2021. The reactive species interactome in the brain. *Antioxidants & Redox Signaling*. 35: 1176–1206.

[7] Wang, Y.; Branicky, R.; Noë, A.; Hekimi, S. 2018. Superoxide dismutases: dual roles in controlling ROS damage and regulating ROS signaling. *Journal of Cell Biology*. 217: 1915–1928.

[8] Henzler, T.; Steudle, E. 2000. Transport and metabolic degradation of hydrogen peroxide in Chara corallina: model calculations and measurements with the pressure probe suggest transport of H_2O_2 across water channels. *Journal of Expermiental Botany*. 51: 2053–2066.

[9] Schroeder, K. 2020. NADPH oxidases: current aspects and tools. *Redox Biology*. 34: artic. 101512.

[10] Zhang, Y.; Wong, H.S. 2021. Are mitochondria the main contributor of reactive oxygen species in cells? *Journal of Experimental Biology*. 224: artic. 221606.

[11] Niedzwiecki, M.M. et al. 2019. The exposome: molecules to populations. *Annual Review of Pharmacology and Toxicology*. 59: 107–127.

[12] Marinho, H. S.; Real, C.; Cyrne, L.; Soares, H.; Antunes, F. 2014. Hydrogen peroxide sensing, signaling and regulation of transcription factors. *Redox Biology*. 2: 535–562.

[13] Szőllősi, R. 2014. Superoxide Dismutase (SOD) and Abiotic Stress Tolerance in Plants: An Overview, in Ahmad, P. Edit. *Oxidative Damage to Plants*, Elsevier, Amsterdam.

[14] Leonowicz, G.; Trzebuniak, K.; Zimak-Peikarczyk, P.; Slesak, I.; Mysliwa-Kurdziel, B. 2018. The activity of superoxide dismutases (SODs) at the early stages of wheat deetiolation. *PLOS One*. 13: artic. e0194678.

[15] Valko, M.; Izakovic, M.; Mazur, M.; Rhodes, C.J.; Telser, J. 2004. Role of oxygen radicals in DNA damage and cancer incidence. *Molecular and Cellular Biochemistry*. 266, 37–56.

[16] Barisci, J.N.; Wallace, G.G.; Baughman, R.H. 2000. Electrochemical studies of single-wall carbon nanotubes in aqueous solutions. *Journal of Electroanalytical Chemistry*. 488: 92–98.

[17] Barisci, J.N.; Wallace, G.G.; Baughman, R.H. 2000. Electrochemical quartz crystal microbalance studies of single-wall carbon nanotubes in aqueous and non-aqueous solutions. *Electrochimica Acta*. 46: 509–517.

[18] Katsounaros, I.; Cherevko, S.; Zeradjanin, A.R.; Mayrhofer, K.J.J. 2014. Oxygen Electrochemistry as a Cornerstone for Sustainable Energy Conversion. *Angewandte Chemie International Edition*. 53: 102–121.

[19] Xu, J.; Huang, W.; McCreery, R.L. 1996. Isotope and surface preparation effects on alkaline dioxygen reduction at carbon electrodes. *Journal of Electroanalytical Chemistry*. 410: 235–242.

[20] Jinnouchi, R.; Anderson, A.B. 2008. Aqueous and Surface Redox Potentials from Self-Consistently Determined Gibbs Energies. *Journal of Physical Chemistry C*. 112: 8747–8750.

[21] Anderson, A.B.; Jinnouchi, R.; Uddin, J. 2013. Effective Reversible Potentials and Onset Potentials for O_2 Electroreduction on Transition Metal Electrodes: Theoretical Analysis. *Journal of Physical Chemistry C*. 117: 41–48.

[22] Shim, J.H.; Kim, J.; Lee, C.; Lee, Y. 2011. Electrocatalytic Activity of Gold and Gold Nanoparticles Improved by Electrochemical Pretreatment. *Journal of Physical Chemistry C*. 115: 305–309.

[23] Roberts, J.G.; Voinov, M.A.; Schmidt, A.C.; Smirnova, T.I.; Sombers, L.A. 2016. The Hydroxyl Radical is a Critical Intermediate in the Voltammetric Detection of Hydrogen Peroxide. *Journal of the American Chemical Society*. 138: 2516–2519.

[24] Karyakin, A.A.; Karyakina, E.E.; Gorton, L. 1999. On the mechanism of H_2O_2 reduction at Prussian Blue modified electrodes. *Electrochemistry Communications*. 1: 78–82.

[25] Araminaité, R.; Garjonyté, R.; Malinauskas, A. 2010. Electrocatalytic reduction of hydrogen peroxide at Prussian blue modified electrodes: a RDE study. *Journal of Solid State Electrochemistry*. 14: 149–155.

[26] Litwinienko, G.; Ingold, K.U. 2007. Solvent effects on the rates and mechanisms of reaction of phenols with free radicals. *Accounts of Chemical Research*. 40: 222–230.

[27] Marselli, B.; Garcia-Gomez, J.; Michaud, P.A.; Rodrigo, M.A.; Comninellis, C. 2003. Electrogeneration of hydroxyl radicals on boron-doped diamond electrodes. *Journal of the Electrochemical Society*. 150, D79–D83.

[28] Enache, T.A.; Chiorcea-Paquim, A.M.; Fatibello-Filho, O.; Oliveira-Brett, A.M. 2009. Hydroxyl radicals electrochemically generated in situ on a boron doped diamond electrode. *Electrochemistry Communications*. 11: 1342–1345.

[29] Kapałka, A.; Foti, G.; Comninellis, C. 2009. The importance of electrode material in environmental electrochemistry: Formation and reactivity of free hydroxyl radicals on boron-doped diamond electrodes. *Electrochimica Acta*. 54: 2018–2023.

[30] den Hartog, S.; Sammanipour, M.; Vincent Ching, H.Y.; Van Doorslaer, S.; Breugelmans, T.; Hubin, A.; Ustarroz, J. 2021. Reactive oxygen species formation at Pt nanoparticles revisited by electron paramagnetic resonance and electrochemical analysis. *Electrochemistry Communications*. 122: artic 106878.

[31] Kalaiyarasan, G.; Kumar, A.V.N.; Sivakumar, C.; Joseph, J. 2015. Electro-generated reactive oxygen species at Au surface as an indicator to explore glutathione redox chemistry and quantification. *Electrochemistry Communications*. 56: 29–33.

[32] Doménech-Carbó, A. 2021. *Electrochemistry of Porous Materials*, 2nd edit. Taylor & Francis, Boca Raton.

[33] Sawyer, D.T.; Chiericato, G.; Angelis, C.T.; Nanni, E.J.; Tsuchiya, T. 1982. Effects of Media and Electrode Materials on the Electrochemical Reduction of Dioxygen. *Analytical Chemistry*. 54: 1720–1724.

[34] Cofré, P.; Sawyer, D.T. 1986. Electrochemical Reduction of Dioxygen to Perohydroxyl (HO2·) in Aprotic Solvents That Contain Bronsted Acids. *Analytical Chemistry*. 58: 1057–1062.

[35] Litwinienko, G.; Ingold, K.U. 2003. Abnormal Solvent Effects on Hydrogen Atom Abstractions. 1. The Reactions of Phenols with 2,2-Diphenyl-1-picrylhydrazyl (dpph) in Alcohols. *Journal of Organic Chemistry*. 68: 3433–3438.

[36] Litwinienko, G.; Ingold, K.U. 2004. Abnormal Solvent Effects on Hydrogen Atom Abstractions. 2. Resolution of the Curcumin Antioxidant Controversy. The Role of Sequential Proton Loss Electron Transfer. *Journal of Organic Chemistry*. 69: 5888–5896.

[37] Litwinienko, G.; Ingold, K.U. 2005. Abnormal Solvent Effects on Hydrogen Atom Abstraction. 3. Novel Kinetics in Sequential Proton Loss Electron Transfer. *Journal of Organic Chemistry*. 70: 8982–8990.

[38] Hossain, M.S.; Mollah, M.Y.A.; Susan, M.A.B.H.; Islam, M.M. 2020. Role of in situ electrogenerated reactive oxygen species towards degradation of organic dye in aqueous solution. *Electrochimica Acta*. 344: artic. 136146.

[39] Jabeen, E.; Janjua, N.K.; Ahmed, S.; Domínguez-Álvarez, E.; Jacob, C. 2017. A selective and sensitive monitoring of the OH radical using flavonoid-modified electrodes. *Electrochimica Acta*. 258: 228–235.

[40] McGovern, J.P.; Neil, G.L.; Crampton, S.L.; Robinson, M.I.; Douros, J.D. 1977. Antitumor activity and preliminary drug disposition studies of chartreusin (NSC-5159). *Cancer Research*. 37: 1666–1672.

[41] Ueberschaar, N.; Xu, Z.; Scherlach, K.; Metsä-Ketelä, M.; Bretschneider, T.; Dahse, H.M.; Goerls, H.; Hertweck, C. 2013. Synthetic remodeling of the chartreusin pathway to tune antiproliferative and antibacterial activities. *Journal of the American Chemical Society*. 135: 17408–17416.

[42] Doménech-Carbó, A.; Cebrián-Torejón, G.; Montoya, N.; Ueberschaar, N.; Scotti, M.T.; Benfodda, Z.; Hetweck, C. 2017. Electrochemical monitoring of ROS generation by anticancer agents: the case of chartreusin. *RSC Advances*. 7: 45200–45210.

[43] Hernández, D.M.; de Moura, M.A.B.F.; Valencia, D.P.; González, F.J.; González, I.; de Abreu, F.C.; da Silva Junior, E.N.; Ferreira, V.F.; Ventura Pinto, A.; Goulart, M.P.F.; Frontana, C. 2008. Inner reorganization during the radical–biradical transition in a nor-β-lapachone derivative possessing two redox centers. *Organic & Biomolecular Chemistry*. 6: 3414–3420.
[44] Párraga, A.; Orozco, M.; Portugal, J. 1992. Experimental and modelling studies on the DNA cleavage by elsamicin A. *European Journal of Biochemistry*. 208: 227–233.
[45] Mustafa, A.M.; Angeloni, S.; Abouelenein, D.; Acquaticci, L.; Xiao, J.; Sagratini, G.; Maggi, F.; Vittori, S.; Caprioli, G. 2022. A new HPLC-MS/MS method for the simultaneous determination of 36 polyphenols in blueberry, strawberry and their commercial products and determination of antioxidant activity. *Food Chemistry*. 367: artic. 130743.
[46] Doménech-Carbó, A.; Domínguez, I. 2023. In situ electrochemical analysis of anthocyanin activation by ROS in blueberries. *Electrochemistry Communications*. 149: artic. 107468.
[47] Domínguez, I.; Romero, I.; Sánchez-Ballesta, M.T.; Escribano, M.I.; Merodio, C.; Doménech-Carbó, A. 2024. Electrochemistry of lyophilized blueberry and raspberry samples: ROS activation of the antioxidant ability of anthocyanins. *Food Chemistry*. 435: artic. 137649.
[48] Fan, L.; Lin, L.; Zhang, Y.; Li, S.; Tang, Z. 2023. Component characteristics and reactive oxygen species scavenging activity of anthocyanins from fruits of *Lonicera caerulea*. *Food Chemistry*. 403: artic. 134391.
[49] Xu, Z.; Mahmood, K.; Rothstein, S.J. 2017. ROS Induces Anthocyanin Production Via Late Biosynthetic Genes and Anthocyanin Deficiency Confers the Hypersensitivity to ROS-Generating Stresses in Arabidopsis. *Plant Cell Physiology*. 58: 1364–1377.
[50 Herscu-Kluska, R.; Masanwa, A.; Saphier, M.; Cohen, H.; Meverstein, D. 2008. Mechanism of the reaction of radicals with peroxides and dimethyl sulfoxide in aqueous solution. *Chem*. 14: 5880–5890.
[51] De Lima, A.A.; Sussuchi, E.M.; De Giovanni, W.F. 2007. Electrochemical and Antioxidant Properties of Anthocyanins and Anthocyanidins. *Croatica Chemica Acta*. 80: 29–34.
[52] Ramesova, S.; Sokolova, R.; Tarabek, J.; Degano, I. 2013. The oxidation of luteolin, the natural flavonoid dye. *Electrochimica Acta*. 110: 646–654.
[53] Chiorcea-Paquim, A.-M.; Enache, T.A.; De Souza Gil, E.; Oliveira-Brett, A.M. 2020. Natural phenolic antioxidants electrochemistry: Towards a new food science methodology. *Comprehensive Reviews in Food Chemistry and Food Safety*. 19: 1680–1726.
[54] Krishnamachari, V.; Levine, L.H.; Paré, P.W. 2002. Flavonoid oxidation by the radical generator AIBN: a unified mechanism for quercetin radical scavenging. *Journal of Agricultural and Food Chemistry*. 50: 4357–4363.
[55] Doménech-Carbó, A.; Gavara, R.; Hernández-Muñoz, P.; Domínguez, I. 2015. Contact probe voltammetry for in situ monitoring of the reactivity of phenolic tomato (solanum lycopersicum L.) compounds with ROS. *Talanta*. 144: 1207–1215.
[56] Sadilova, E.; Stintzing, F.C.; Carle, R. 2006. Thermal Degradation of Acylated and Nonacylated Anthocyanins. *Journal of Food Science*. 71: C504–C512.
[57] Astier, J.; Gross, I.; Durner, J. 2018. Nitric oxide production in plants: an update. *Journal of Experimental Botany*. 14: 3401–3411.
[58] Privett, B.J.; Shin, J.H.; Schoenfisch, M.H. 2010. Electrochemical nitric oxide sensors for physiological measurements. *Chemical Society Reviews*. 39: 1925–1935.
[59] Gupta, K.J. Edit. 2016. *Plant Nitric Oxide*. Springer Protocols. Springer, New York.

ADDITIONAL LITERATURE

Role of ROS in Plant Biology

Brookes, P.S.; Levonen, A.L.; Shiva, S.; Sarti, P.; Darley-Usmar, V.M. 2002. Mitochondria: Regulators of signal transduction by reactive oxygen and nitrogen species. *Free Radical Biology and Medicine*. 33: 755–764.
Guijarro Inclán, M.; Pitacrh-Jarque, J.; Doménech-Carbó, A.; Chicote, J.U.; Trefler, S.; García-España, E.; García-España, A.; Verdejo, B. 2017. Homo- and Heterobinuclear Cu^{2+} and Zn^{2+} Complexes of Ditopic Aza Scorpiand Ligands as Superoxide Dismutase Mimics. *Inorganic Chemistry*. 66: 13748–13768.
Martínez-Camarena, A.; Llinares, J.M.; Doménech-Carbó, A.; Alarcón, J.; García-España, E. 2019. A step forward in the development of superoxide dismutase mimetic nanozymes: the effect of the charge of the surface on antioxidant activity. *RSC Advances*. 9: 41549–41560.

Nordzieke, D.E.; Medraño-Fernandez, I. 2018. The plasma membrane: a platform for intra- and intercellular redox signaling. *Antioxidants*. 7: artic. 7110168.

Oxygen Electrochemistry

El-Deab, M.S.; Ohsaka, T. 2003. Quasi-reversible two-electron reduction of oxygen at gold electrodes modified with a self-assembled submonolayer of cysteine. *Electrochemistry Communications*. 5: 214–219.

ROS Reactivity

Arroyo-Currás, N.; Rosas-García, V.M.; Videa, M. 2016. Substituent Inductive Effects on the Electrochemical Oxidation of Flavonoids Studied by SquareWave Voltammetry and Ab Initio Calculations. *Molecules*. 21: artic. 1422.

Farhana, T.I.; Mollah, M.Y.A.; Susan, M.A.B.H.; M.M. Islam, Catalytic degradation of an organic dye through electroreduction of dioxygen in aqueous solution. *Electrochimica Acta*. 139: 244.

Galano, A.; Francisco-Marques, M. 2009. Reactions of OOH Radical with β-Carotene, Lycopene, and Torulene: Hydrogen Atom Transfer and Adduct Formation Mechanisms. *Journal of Physical Chemistry B*. 113: 11338–11345.

Galano, A.; Raúl Alvarez-Idaboy, J. 2019. Computational strategies for predicting free radical scavengers' protection against oxidative stress: where are we and what might follow? *International Journal of Quantum Chemistry*. 119: artic. e25665

Ingold, K.U. 1969. Peroxy Radicals. *Accounts of Chemical Research*. 2: 1–9.

Masek, A.; Chrzescijanska, R.; Zaborski, M. 2014. Electrooxidation of morin hydrate at a Pt electrode studied by cyclic voltammetry. *Food Chem*istry. 148: 18–23.

Kinetic Analysis

Inbaraj, J.J.; Gandhidasan, R.; Murugesan, R. 1999. Cytotoxicity and superoxide anion generation by some naturally occurring quinones. *Free Radical Biology and Medicine*. 26: 1072–1078.

Roginsky, V.A.; Pisarenko, L.M.; Bors, W.; Michel, C. 1999. The kinetics and thermodynamics of quinone–semiquinone–hydroquinone systems under physiological conditions. *Journal of the Chemical Society Perkin Transactions 2*. 871–876.

Wass, J.R.; Ahlberg, E.; Panas, I.; Schiffrin, D.J. 2005. Quantum chemical modeling of reduction of quinones. *Journal of Physical Chemistry A*. 110: 2005–2020.

Wraight, C.A. 2004. Proton and electron transfer in the acceptor quinone complex of photosynthetic reaction centers from Rhodobacter sphaeroides. *Frontiers in Biosciences*. 9: 309–337.

Electrochemistry of ROS and NRS

Astier, J.; Lindermayr, C. 2012. Nitric oxide-dependent posttranslational modification in plants: an update. *International Journal of Molecular Sciences*. 13: 15193–15208.

Ciszewski, A.; Milczarek, G. 2003. Electrochemical detection of nitric oxide using polymer modified electrodes. *Talanta*. 61: 11–26.

Zhao, S.; Zang, G.; Zhang, Y.; Liu, H.; Wang, N.; Cai, S.; Durkan, C.; Xie, G.; Wang, G. 2021. Recent advances of electrochemical sensors for detecting and monitoring ROS/RNS. *Biosensors and Bioelectronics*. 179: artic. 113052.

10 Electrochemical Analysis of Biochemical Processes in Plants

10.1 INTRODUCTION

There is an enormous variety of biochemical processes in plants. The current chapter is focused on those for which electrochemical methods of analysis and modeling have been proposed. Most processes, such as plant growth and plant stress, can be monitored electrochemically using the determination of the concentration (or their variations) of marker compounds [1]. As far as these methods endow within the frame of electroanalytical chemistry, they will be treated in Chapter 13. Here, the focus will be the processes where the electrochemistry is not specifically addressed to monitor a given plant component. Under this view, kinetic modeling of several plant processes has been proposed [2, 3]. This can be seen as a holistic approach whose implications for chemotaxonomy and phylogenetic analysis will be treated in Chapter 11.

10.2 OXYGEN GENERATION AT CHLOROPLASTS

A first example of how electrochemistry can be used for kinetic modeling of biochemical processes in plants is provided by oxygen generation at chloroplast films [2]. This formulation assumes that the rate of oxygen production depends on the concentration of oxidized plastoquinone form (N_{Ox}), which is reduced, and the concentration of the corresponding reduced form (N_{Rd}), which is reoxidized.

$$\text{Ox} \leftrightarrow \text{Rd} \tag{10.1}$$

Assuming that the direct (forward) and inverse (backward) processes (of rate constants k_f, k_b, respectively) occur under first-order kinetics and that the total number of plastoquinone remains constant (N_o), the reaction rate will be

$$\frac{dN_{Ox}}{dt} = -k_f N_{Ox} + k_b \left(N_o - N_{Ox}\right), \tag{10.2}$$

where t is the time passed after the illumination of the chloroplast. Integration of this equation leads to an estimate of the flux J of oxygen produced at the chloroplast film [2].

$$J = \frac{k_f N_o \left(k_b + k_f e^{-(k_f + k_b)t}\right)}{k_f + k_b} \tag{10.3}$$

This model can be tested using tip current measurements in SECM experiments described via finite element method simulations [2]. This assumes that the mass transport of oxygen between the tip and the substrate follows an asymmetric cylindrical diffusion described by Fick's second law. The diffusion coefficient of oxygen was taken as 9.1×10^{-6} cm^2 s^{-1}, and the bulk oxygen concentration was 0.4 mM [4]. For finite element method simulations it was assumed that (i) at the electrode surface, the oxygen concentration is zero because the applied potential reduces oxygen here at a

DOI: 10.1201/9781032697826-10

transport-limited rate; (ii) a non-normal, light-dependent flux of oxygen was established; and (iii) in the dark, there is no consumption of oxygen due to respiration or other processes.

After numerical simulations, a satisfactory agreement between experimental and simulated variations of tip currents with time recorded under different levels of illumination for a poly-L-lysine-covered chloroplast film substrate in contact with two different aqueous buffers at pH 8.0 (Figure 10.1) [2].

10.3 KINETIC ANALYSIS OF ROS GENERATION/REACTION

Different kinetic models have been proposed dealing with the generation of ROS and their interaction with the substrates. A first approach can be made based on thermochemical considerations for the superoxide production by semiquinone forms according to the cyclic scheme in Figure 10.2 [5].

The rate of superoxide production depends on the potential of the quinone/semiquinone (Q/SQ) couple, and the equilibrium constant (K_S) for the formation of the semiquinone from the quinone and the quinol (H_2Q). The former will determine the second-order rate constant for superoxide production, while the second will determine the total concentration of semiquinone when the equilibrium is reached.

$$K_S = \frac{[SQ]^2}{[Q][H_2Q]} \tag{10.4}$$

Introducing the standard potentials of the couples Q/H_2Q ($E°$(Q/H_2Q)) and (Q/Q•⁻) ($E°$(Q/Q•⁻)), and the acidity constant (K_a(SQ)) of the protonated semiquinone, HQ•, the following expression applies under standard conditions:

$$RT \ln K_S = -2FE^{\circ}\left(Q/H_2Q\right) + 2FE^{\circ}\left(Q/Q^{\bullet-}\right) - 0.059\left(pH - pK_a\left(SQ\right)\right). \tag{10.5}$$

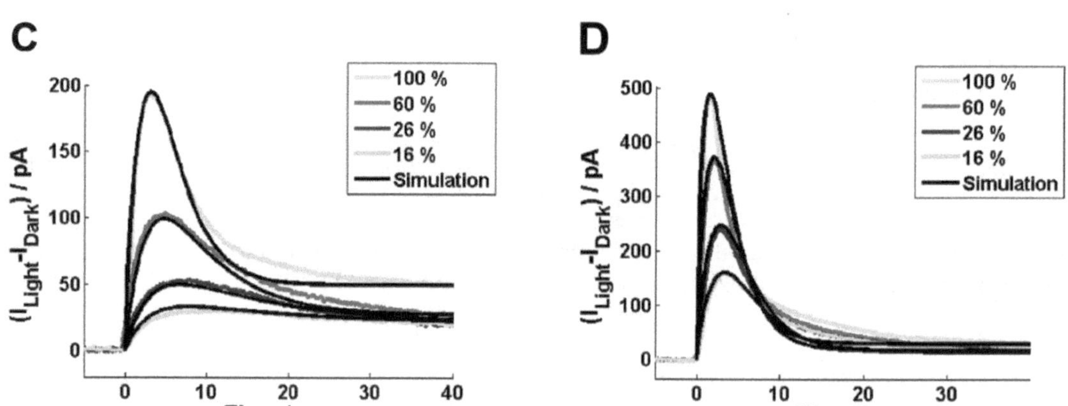

FIGURE 10.1 Experimental and simulated time variations of tip currents recorded at a 25 μm diameter Ag disk electrode. The Ag tip was held at −0.9 V and placed 12.5 μm above a poly-L-lysine-covered chloroplast film substrate after the application of different light intensities in contact with two different HEPES-sorbitol buffers at pH 8.0. The simulated curves were obtained based on Eq. (3) taking different values for N_o (around 3×10^{-11} mol cm^{-2}), k_f (around 0.5 s^{-1}), and k_b (around 0.02 s^{-1}). (Adapted from ref. [2], with permission. Copyright the American Chemical Society 2013.).

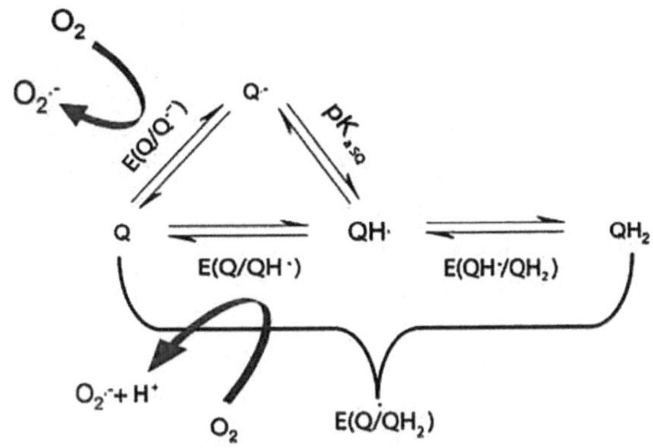

FIGURE 10.2 Thermochemical cycle for the superoxide production from O_2 and semiquinones. (Reproduced from ref. [5], with permission. Copyright Elsevier 2006.).

Apart from the difficulty of the evaluation of several thermochemical parameters in *in vivo* systems, the processes schematized in Figure 10.2 can be coupled to other processes so that equilibrium conditions do not hold.

A different kinetic analysis of particular interest for our purposes is based on the reaction between substrates and ROS generated electrochemically simultaneously to the OER process through [6]

$$H_2O \rightarrow HO\bullet + H^+ + e^-. \tag{10.6}$$

It was assumed that (i) the generated hydroxyl radicals remain strongly adsorbed on the working electrode (Pt), (ii) the substrates react with such radicals originating products that can react with the radicals, (iii) the substrates and their reaction products can be electroactive, and (iv) that all electrochemical and chemical reactions are slow relative to diffusion so that the concentration polarization can be neglected [6]. Then it is assumed that the substrate, S, reacts with nascent hydroxyl radicals yielding J products P_{SJ} as

$$S + n_S HO^{\bullet} \xrightarrow{k_S} P_{SJ}. \tag{10.7}$$

These products can also react with hydroxyl radicals.

$$P_{SJ} + n_{JK} HO^{\bullet} \xrightarrow{k_S} P_{JK}. \tag{10.8}$$

In the most general case, the previous reactions compete with the electrochemical oxidation of the substrate,

$$S \xrightarrow{k_{Se}} P_{SJ} + ne^-, \tag{10.9}$$

and the electrochemical oxidation of the hydroxyls,

$$2HO^{\bullet} \xrightarrow{k_{O2}} O_2 + 2H^+ + 2e^-. \tag{10.10}$$

In Eqs. (10.7) and (10.8), k_S and k_{SJ} represent the rate constants for the reactions of the original substrate and its products of reaction with hydroxyl radical. In Eqs. (10.9) and (10.10), k_{Se} and k_{O2}

represent the heterogeneous rate constants of the corresponding electron transfer processes. Since the formation of HO• is controlled by the current density j (A cm^{-2}) of the electrolysis, the rate of formation of hydroxyl radicals, v_{OH} (mol s^{-1} cm^{-2}), will be, $v_{OH} = j/F$. In turn, assuming that under the conditions of operation, oxygen is formed exclusively from the oxidation of hydroxyl radicals (and not by direct electrochemical oxidation of water), the formation rate, v_{O2}, will be $v_{O2} = k_{O2}\Gamma_{OH}$. Here, Γ_{OH} represents the density of hydroxyl radical at the electrode surface (mol cm^{-2}). If the system reaches a steady state, the formation rates of HO•, O_2, and the products of reactions ((10.8) and ((10.9) (v_{SJ}, v_{JK}) will satisfy the relationship.

$$v_{OH} = 2v_{O2} + n_{SJ}v_{SJ} + \sum n_{JK}v_{JK}. \tag{10.11}$$

Introducing the concentration of the S and P_J species, and taking $v_{SJ} = k_{SJ}\Gamma_{OH}[S]$, and $v_{JK} = k_{JK}\Gamma_{OH}[P_{SJ}]$, Eq. (10.11) leads to

$$v_{OH} = \frac{j}{F} = \left(2k_{O2} + n_{SJ}k_{SJ}\left[S\right] + \sum n_{JK}k_{JK}\left[P_J\right]\right)\Gamma_{OH}. \tag{10.12}$$

Combining the previous equations to eliminate Γ_{OH}, the rate of the reaction between the substrate and hydroxyl radicals (Eq. (9.40), $v_{S\text{-}OH}$, will be [6]

$$v_{S-OH} = \frac{j}{F}\left(\frac{k_{SJ}\left[S\right]}{2k_{O2} + n_{SJ}k_{SJ}\left[S\right] + \sum n_{JK}k_{JK}\left[P_J\right]}\right). \tag{10.13}$$

This treatment can reduce, under favorable conditions, to a pseudo-first-order decay of the concentration of substrate. Experimental data for the constant current oxidation at a Pt anode of ascorbic acid, caffeic acid, gallic acid, trolox in phosphate buffer at pH 3.5 agree with this expectance [6]. Pertinent data can be seen in Figure 10.3, where the logarithm of the ratio between the substrate concentration at a time t and the initial concentration of the substrate (determined from high-performance liquid chromatography, HPLC) is plotted vs. time. The treatment leading to Eq. ((10.13) should be refined to account for the effect of the electrochemical oxidation of hydroxyls (Eq. ((10.10)) and the direct oxidation of water to oxygen. Despite this complexity, lineal representations in Figure 10.3 fit satisfactorily to a pseudo-first-order decay of substrate concentration. Experiments such as those in Figure 10.3 conducted in conditions where the electrochemical oxidation of hydroxyls accompanies the direct water to O_2 oxidation led to rate constants higher than those carried out in potentiostatic electrolysis with no O_2 evolution [6].

10.4 REACTION KINETICS WITH ELECTROGENERATED ROS

As discussed in previous chapters, the electrochemical generation of ROS can provide valuable information on the oxidant capacity and other properties of vegetal matter. These properties can be studied in vivo or in solutions of the plant components after their extraction. In this second case, the sensitivity can be enhanced by adopting a solid-state strategy directly inspired by the voltammetry of immobilized particles methodology [7]: the record of the voltammetric response of the microparticulate deposits resulting from the evaporation of plant extracts in ethanol, acetone, chloroform, etc. [3].

In the previous treatment, it was assumed that hydroxyl radicals generated electrochemically were strongly adsorbed onto the electrode surface where the plant components and their reaction products react with them [6]. In this approach, the process of ROS generation occurs at potentials near to those where the OER process takes place, and the kinetic was monitored by determining the concentration of plant components through HPLC.

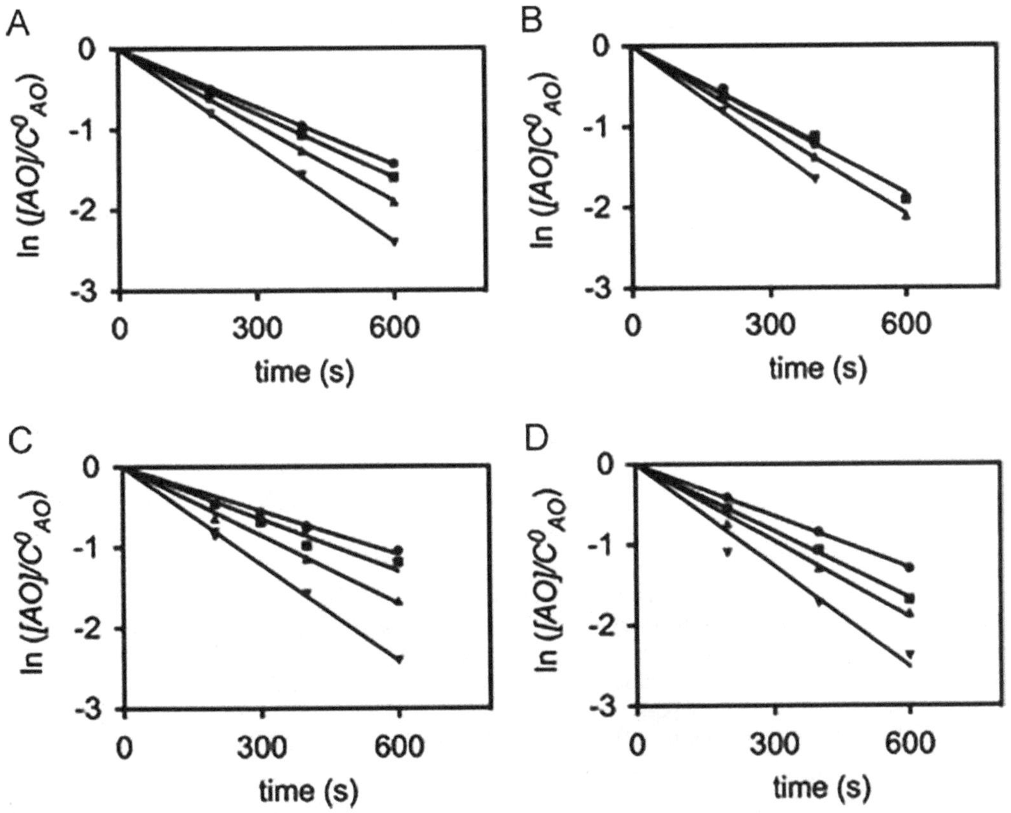

FIGURE 10.3 Decrease of the concentration of (a) ascorbic acid, (b) caffeic acid, (c) gallic acid, and (d) trolox (all 50 mM solutions in 0.15 M phosphate buffer at pH 3.5) determined by HPLC during the oxidation at a Pt anode at constant current densities of 50 (•), 268(■), 625(▲), and1250(▼) A m^{-2}. (Reproduced from ref. [6], with permission. Copyright Elsevier 2014.).

A different approach has been proposed based on the generation of ROS during the electrochemical reduction of dissolved oxygen [8]. As described in Chapter 9, the ORR process follows a multistep mechanism where superoxide radical ion, hydroperoxide radical, hydrogen peroxide, and hydroxyl radical can be generated. In air-saturated aqueous solutions in a wide range of pHs, the ORR process at carbon electrodes occurs at potentials around −0.7 V vs. Ag/AgCl. This process is separated in the potential scale from the majority of electrochemical processes affecting polyphenolic plant components. This can be seen in Figure 10.4 (see also Figure 7.11) where the cyclic voltammograms of microparticulate films on GCE resulting from ethanolic extracts of leaves of seedlings of two species of the tribe Inuleae in contact with acetate buffer are depicted [8]. When the potential cycle is initiated ca. 0.0 V vs. Ag/AgCl and scanned in the positive direction, a series of oxidation peaks (A_A) appear in the 0.0 to 0.9 V potential range. These are often accompanied, in the subsequent negative-going potential scan, by coupled cathodic signals (C_A), which precede the gross wave corresponding to the ORR process (C_{ox}). In the following (second) anodic scan, the voltammetric profile in the A_A region changes significantly. Depending on the nature of the vegetal sample and the electrochemical parameters (switching potential, potential scan rate), several peaks are diminished or even disappear while new oxidation peaks may appear.

These features, which have also been observed for different vegetal systems and flavonoids in solution [9], can be interpreted as the result of the interaction of electroactive plant components with the

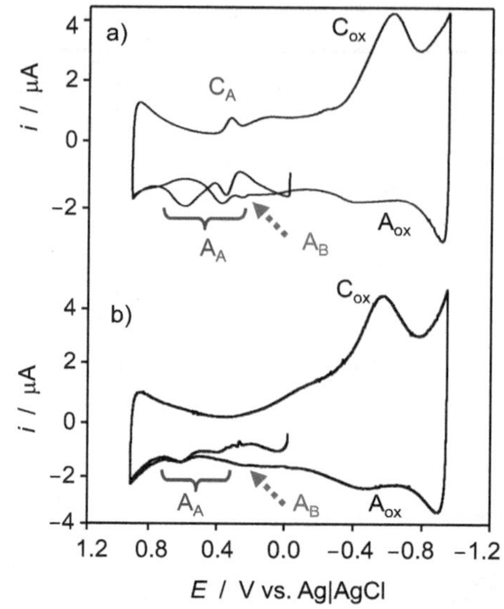

FIGURE 10.4 Cyclic voltammograms of microparticulate films on GCE resulting from ethanolic extracts of leaves of seedlings of two species of the tribe Inuleae in contact with air-saturated 0.25 M HAc/NaAc buffer at pH 4.75 after seven days of germination. (a) *Dittrichia viscosa* and (b) *Limbarda crithmoides*. Potential scan initiated at 0.0 V vs. Ag/AgCl in the positive direction. Potential scan rate 50 mV s^{-1}. Semi-derivative data convolution was carried out. Dotted lines: baselines employed for peak current measurements. (Reproduced from ref. [8], with permission. Copyright the Royal Society of Chemistry 2019. Note that the axes are opposite to the IUPAC convention.).

ROS generated during the previous reduction of dissolved oxygen. In the case of the tribe that includes the species in Figure 10.4, *Dittrichia viscosa* L. (Greuter) and *Limbarda crithmoides* L. (Dumort), compositional data from HPLC-MS confirm the presence of phenolic acids, flavonoids, and terpenoids [10, 11]. In the case of *D. viscosa*, different derivatives of caffeic acid and dicaffeoylquinic acid are particularly abundant components [12]. Accordingly, there are a significant number of components that are electroactive under the experimental conditions used. This interpretative scene is consistent with abundant literature on the electrochemistry of polyphenolic compounds (see Chapter 4) and studies on the chemical interaction of such compounds with ROS [13–15].

The electrochemical processes A_A and C_A for insoluble compounds can be described in terms of solid-state, proton-assisted electron transfer processes [7]. However, several other compounds can be more or less extensively dissolved during the voltammetric experiments. Then, solution-phase electrochemistry can also be involved. Reported data suggest that under a relatively wide range of experimental conditions, such electroactive species remain attached to or concentrated enough near the electrode surface to ensure a well-defined response at the time scale of the usual voltammetric experiments.

Accordingly, the kinetics of the reaction of plant components with ROS species electrochemically generated can be described assuming (i) the recorded voltammetry can be treated as occurring in the solution phase, (ii) the chemical reactions between electroactive components of plants and ROS can be treated in terms of homogeneous kinetics equations, (iii) all of the above processes are slow relative to diffusion of species in the electrolyte solution so that the concentration polarization can be neglected

To obtain kinetic equations, the time of reaction t can be varied by varying the potential scan rate of the voltammogram, v. Then an estimate of the reaction time can be made as the time interval between the appearance of the C_{ox} signal and the appearance of the signals A_A and/or A_B in the same voltammogram. If the peak potential separation (including direct and reverse sectors) between these signals is ΔE, the reaction time can be calculated from the relationship $t = \Delta E/v$. Of course, this is a rough approximation because the ROS species begin to be generated before reaching the peak potential of C_{ox} and, hence, the reaction with plant components.

Formally, two basic hypotheses will be made:

a) The vegetal sample attached to, or near to the electrode surface, contains a series of electroactive species, A, responsible for the signals A_A/C_A, each one containing a certain concentration c_A in the same.
b) b) Some of such species react with ROS electrochemically generated giving rise to new species some of which are electrochemically responsive yielding the A_B signals. As a result, the concentration of the parent vegetal species will decrease with time whereas the concentration of new species will increase.

By the first token, the kinetics of the reaction between A and the electrochemically generated ROS can be described in terms of a second-order rate equation:

$$\frac{dc_A}{dt} = -\sum k_A c_A c_{ROS}, \tag{10.14}$$

where k_A is the rate constant, and c_A, c_{ROS} are the concentrations of A and the ROS at the time t. The sum must be extended not only to all plant components reacting with ROS but also to all ROS generated electrochemically. For simplicity, it will be assumed that a unique plant component and a unique ROS species are involved whose initial concentrations are $c^o{}_A$, $c^o{}_{ROS}$. Notice that this approach implies that, once generated at the electrode surface, the ROS species solely react with the A component and do not experience side reactions. Then Eq. ((10.14) can be rewritten as

$$\frac{dx}{dt} = k_A\left(c^o{}_A - x\right)\left(c^o{}_{ROS} - x\right), \tag{10.15}$$

with x being the concentration of the species formed in the reaction between A and ROS. Integration of this equation yields

$$\ln\left(\frac{1 - x/c^o{}_A}{1 - x/c^o{}_{ROS}}\right) = \left(c^o{}_A - c^o{}_{ROS}\right)k_A t. \tag{10.16}$$

This treatment can be tested from voltammetric data assuming that the peak current of the process A in the J-scan, $i_A(J)$, is proportional to the concentration of A at the corresponding time. This is the customary relationship in diffusion-controlled processes that is converted into a proportionality, $i_A(J) = g_A c_{AJ}$, by introducing an electrochemical constant g_A, which only depends on the potential scan rate, electrolyte, etc. Qualitatively, experimental voltammetric data agree with this picture, showing variations of peak currents with reaction time as exemplified in Figure 10.5. Here, the variation of peak current ratios $i_A(1)/i_{ox}(1)$ and $i_B(2)/i_A(1)$ with reaction time in voltammograms of *Dittrichia viscosa* after seven days of germination are depicted. Interestingly, experimental data reveal a significant difference between the specimens developed in the absence of salinity and those developed in the presence of salinity in the substrate. It is worth noting that the use of microparticulate deposits implies that there is no accurate control of the net amount of plant extract deposited onto the base electrode surface. Accordingly, as is usual in the VIMP technique [7], peak current ratios rather than absolute values of peak currents have to be used for quantifications.

Taking voltammograms such as those shown in Figure 10.4, the peak current measured in the first scan (in which no ROS are generated) will be representative of the initial concentration of A. Then, the peak current recorded in the second anodic scan will be representative of the concentration of A at the time t. As previously noted, this is the time passed between the appearance of the peak C_{ox} and

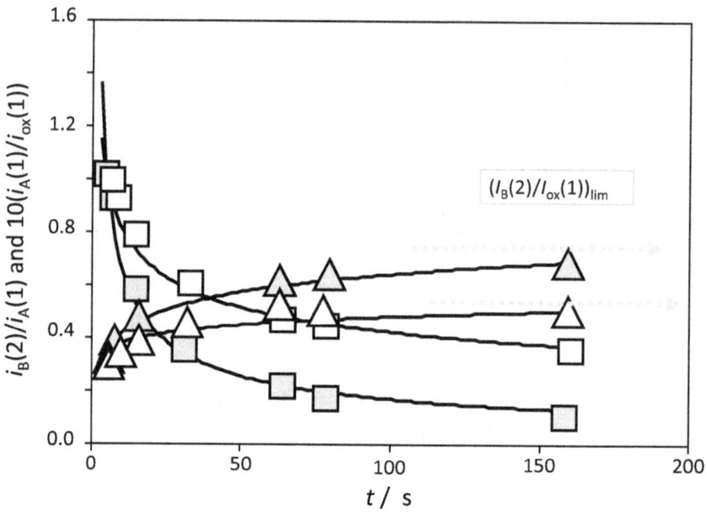

FIGURE 10.5 Variation of the peak current ratios $i_A(1)/i_{ox}(1)$ (squares) and $i_B(2)/i_A(1)$ (triangles) with time in voltammograms of *Dittrichia viscosa* after seven days of germination in the absence of salinity (gray figures) and in 100 mM NaCl (empty figures) in the substrate. From voltammetric data such as depicted in Figure 10.4 performed at different scan rates. (Adapted from ref. [8], with permission. Copyright the Royal Society of Chemistry 2019.).

the record of the signal AA in the second positive-going potential scan. Accordingly, the ratio x/c°_A can be calculated as

$$\frac{x}{c^{\circ}_A} = 1 - \frac{i_A(2)}{i_A(1)}. \tag{10.17}$$

The same criterion can be applied to the concentration of ROS, whose initial value will be proportional to the peak current for the process C_{ox} recorded in the first cathodic scan, $i_{ox}(1)$, i.e., $i_{ox}(1) = g_{ox}c^{\circ}_{ox}$. Accordingly, the x/c_{ROS} ratio will be

$$\frac{x}{c^{\circ}_{ROS}} = \frac{g_{ox}}{g_A}\left(\frac{i_A(1) - i_A(2)}{i_{ox}(1)}\right). \tag{10.18}$$

Eq. ((10.16) can be rewritten as

$$\ln\left(\frac{\dfrac{i_A(2)}{i_A(1)}}{1 - \left(\dfrac{g_{ox}}{g_A}\right)\left(\dfrac{i_A(1) - i_A(2)}{i_{ox}(1)}\right)}\right) = \ln I_A = \left(c^{\circ}_A - c^{\circ}_{ROS}\right)k_A t. \tag{10.19}$$

This equation predicts a linear variation of $\ln I_A$ on t that can be tested with experimental data providing that the electrochemical coefficient g is known. Notice that, in principle, the concentration of ROS generated in a voltammetric experiment should vary with the potential scan rate. Assuming that the rate of ROS formation is proportional to the current density [6], the large rate of ROS formation at high potential scan rates would be compensated by the lower time in which the potentials adequate for this generation operate. It appears conceivable that c°_{ROS} can be taken as approximately constant during voltammetric experiments performed in a relatively narrow interval of sweep rates.

Figure 10.6 depicts the plots of $\ln I_A$ vs. t, taking $g_{ox}/g_A = 1.0$ (a reasonable assumption for processes involving the same electron stoichiometry) for the reaction of the electroactive component(s) responsible for the peak at 0.25 V (A_A) measured in voltammograms at different scan rates for seedlings of *D. viscosa* and *L. crithmoides* growth in the absence and the presence of salinity in the culture substrate [8]. In all cases, $\ln I_A$ vs. t plots showed a satisfactory linearity with intercepts near to zero. The slope of the corresponding lines corresponds to the product $(c^\circ_{A\text{-}}c^\circ_{ROS})k_A$, which can be taken as an apparent rate constant, k_A^{app}. The values of k_A^{app} can be used as a quantitative marker to study the influence of salinity or other stress factors on the germination of plants [8].

An equivalent treatment can be made for the case of the component(s) responsible for the new voltammetric peaks appearing after ROS reaction (signals A_B in Figure 10.4). Assuming that an electroactive compound B is formed as a result of the reaction of a precursor compound P with ROS, i.e., P + ROS → B, the corresponding peak current for the B-localized process, $i_B(2)$, will increase with time. This peak current will be proportional to the concentration of B at a reaction time t, i.e., introducing the corresponding electrochemical constant, $i_B(2) = g_B c_B$. Experimental data in Figure 10.5 agree with this expectance and tend to a limiting value, $i_B^{lim}(2)$, that corresponds to (assuming the simplest reaction stoichiometry) $i_B^{lim} = g_B c^\circ_P$, c°_P being the initial concentration of the precursor compound P. Since, by the previously commented reasons, peak current ratios have to be used, this means that in VIMP-type experiments, the $i_B(2)/i_{ox}(1)$ ratio will tend (see Figure 10.5) to a limiting value, $(i_B(2)/i_{ox}(1))_{lim}$.

If the reaction of the precursor compound with ROS obeys the second-order kinetics described by Eqs. (10.14) to (10.16), the reaction advance can be evaluated from peak current measurements as,

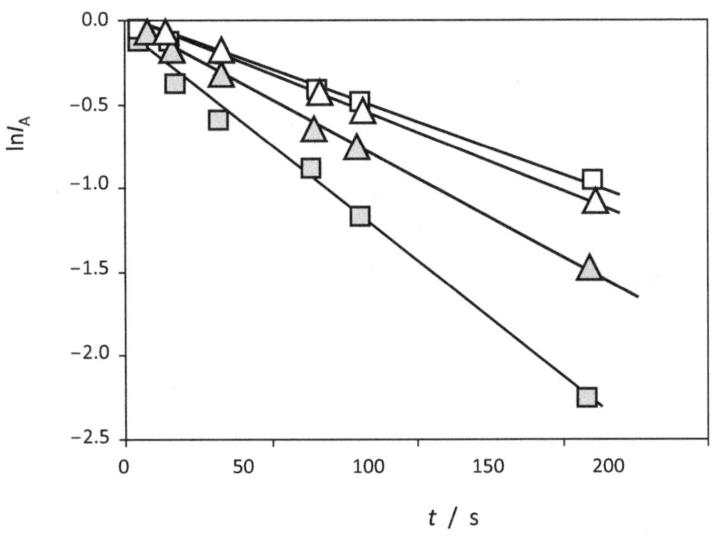

FIGURE 10.6 Plots of $\ln I_A$, calculated from Eq. ((10.19) taking $g_{ox}/g_A = 1.0$, vs. t for the reaction of the electroactive component(s) responsible for the peak at 0.25 V (A_A) in the voltammograms of extracts from ***Dittrichia viscosa*** (squares) and ***Limbarda crithmoides*** (triangles) seedlings growth in the absence of salinity (solid figures) and in 100 mM NaCl in the germination medium (empty figures). From voltammetric data such as depicted in Figure 10.4 performed at different scan rates. (Adapted from ref. [8], with permission. Copyright the Royal Society of Chemistry 2019.).

$$\frac{x}{c_P} = \frac{\frac{I_B(2)}{I_{ox}(1)}}{\left(\frac{I_B(2)}{I_{ox}(1)}\right)_{lim}} \tag{10.20}$$

Accordingly, the following relationship applies,

$$\ln\left(\frac{1-\frac{i_B(2)/i_{ox}(1)}{\left(i_B(2)/i_{ox}(1)\right)_{lim}}}{1-\left(\frac{g_{ox}}{g_B}\right)\left(\frac{i_B(2)}{i_{ox}(1)}\right)}\right) = \ln I_B = \left(c^o{}_P - c^o{}_{ROS}\right)k_B t \tag{10.21}$$

Figure 10.7 shows the plots of $\ln I_B$ vs. t, taking $g_{ox}/g_B = 1.0$ for extracts from *Dittrichia viscosa* seedlings growth with and without salinity, and *Inula helenium* [8]. Again, the expected linear relationship was consistent with experimental data. As before, apparent rate constants k_B^{app} (= ($c^o{}_{P-}$- $c^o{}_{ROS}$)k_B) can be calculated and used to compare the response to stress of different seedlings.

10.5 DYNAMIC PATHWAY OF PLANT COMPOUNDS

An aspect of particular interest in plant science is the elucidation of the factors influencing the composition of plant tissues, in particular, concerning their nutritional and pharmacological properties [16] and their relationship with plant physiology [17]. Antioxidants – namely, ascorbic acid and polyphenolic compounds – center a significant part of research in this line, directed to investigate

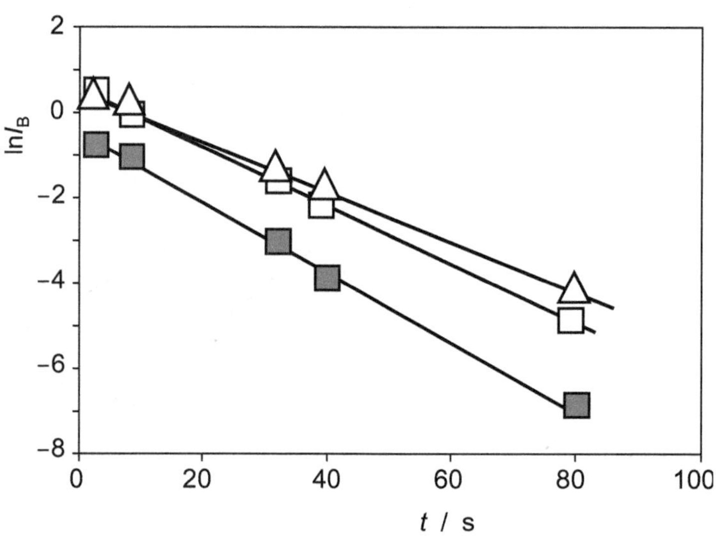

FIGURE 10.7 Plots of lnI_B, calculated from Eq. ((10.21) taking g_{ox}/g_B = 1.0, vs. t for the reaction of the electroactive component(s) responsible for the peak A_B at 0.10 V in the voltammograms of extracts from ***Dittrichia viscosa*** seedlings growth in the absence of salinity (gray squares) and in 100 mM NaCl in the germination medium (squares), and ***Inula helenium*** (triangles). From voltammetric data such as depicted in Figure 10.4 performed at different scan rates. (Adapted from ref. [8], with permission. Copyright the Royal Society of Chemistry 2019.).

their dynamics, i.e., the biosynthetic and regulatory pathways determining the presence of such components in plants at the genetic [18], epigenetic [19], and microbioma [20] levels.

Studies on plant component dynamics are made difficult by the large inter- and intraspecific variation of the content of antioxidants in the plants and their dependence on the maturity stage, and stress factors [21]. This difficulty is evident in fruit crops, where the location of the components within the fruit, time since harvesting conditions, and time of storage, among other factors, are influential [22]. In this context, there is growing interest in the development of *in vivo* or *in situ* electrochemical methods for monitoring the evolution of plant components [3, 9], for example, for monitoring H_2O_2 release [23]. A presentation in the electroanalytical context will be made in Chapter 13; here, we will focus on the treatment of electrochemical data for reactivity modeling.

The simplest experimental procedure for the *in situ* study of dynamic pathways consists of the insertion of electrodes into accessible regions of the plant, fruits in particular for obvious reasons [24]. Figure 10.8 shows the square wave voltammetric responses of six apple varieties immediately after the insertion of a set of microelectrodes [25]. At a graphite working microelectrode, the voltammograms show an intense anodic peak at 0.2 V vs. Ag/AgCl (I), followed by a broad peak centered at ca. 0.5 V (II). At potentials ca. 1.0 V, the initial portion of a third anodic wave with maximum intensity at 1.2 V (III) can be recorded. Depending on the apple variety, the aforementioned signals

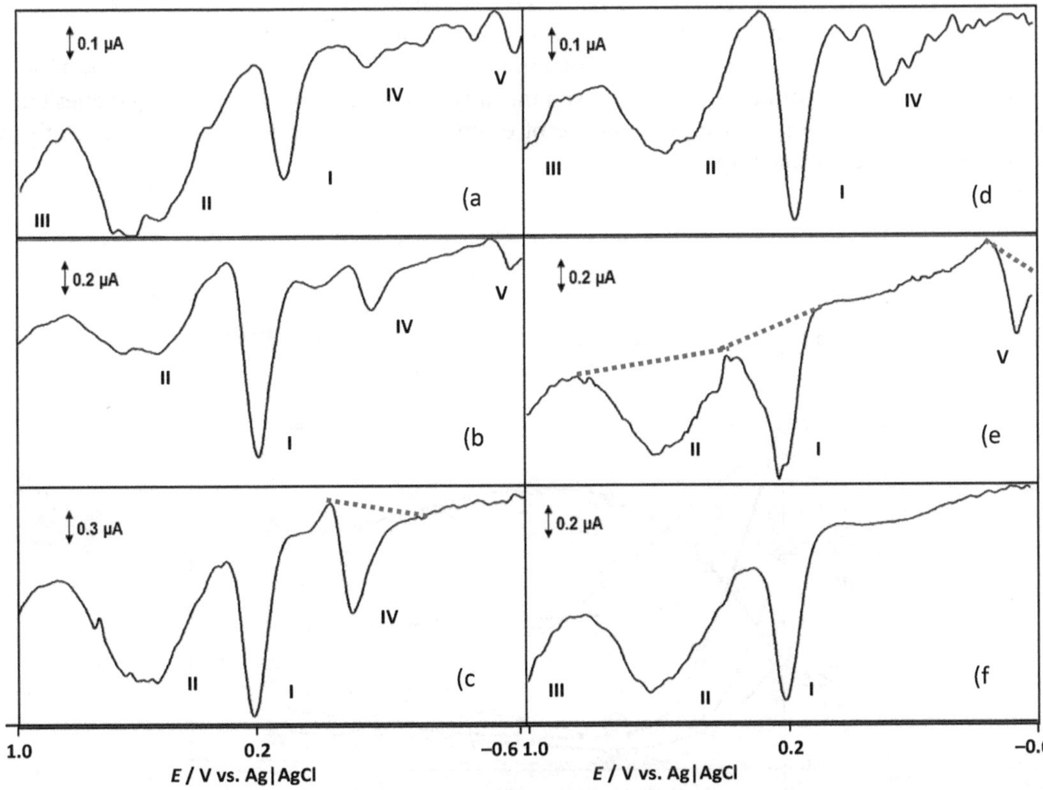

FIGURE 10.8 Square wave voltammograms at graphite microelectrode of (a) golden, (b) Granny Smith, (c) ***Reineta***, (d) red delicious, (e) Fuji, and (f) Braeburn apple specimens. Potential scan initiated at −0.80 V in the positive direction; potential step increment 4 mV, square wave amplitude 25 mV; frequency 10 Hz. Semi-derivative convolution was applied to the data. For simplicity, the graph is limited to the −0.6 to 1.0 V vs. Ag/AgCl potential range. (Adapted from ref. [25], with permission. Note that the axes are opposite to the IUPAC convention. Copyright Elsevier 2022.).

are accompanied by minor peaks between −0.5 and 0.0 V (IV, V). As judged by comparison with standards, peak I can be assigned to the oxidation of ascorbic acid while the other signals will correspond to different polyphenolic compounds. Among them, galactoside, glucoside, xyloside, arabinoside, and rhamnoside derivatives of epicatechin, chlorogenic acid, quercetin, are all abundant in apple fruits [26].

As described in Chapter 9 and the precedent section, the electrochemical generation of ROS can be used as a tool for studying the behavior of plant components. Under the experimental conditions of Figure 10.8, ROS can be generated in the vicinity of the graphite working microelectrode and react with the existing substrate compounds in an area close to the electrode. Accordingly, the voltammetric response of the electroactive compounds in apple fruit can be recorded either in the absence (initiating the potential scan at values less negative than −0.5 V vs. Ag/AgCl) or in the presence of ROS (now starting the scan at potentials more negative than −0.5 V vs. Ag/AgCl). As already noted, this second case will result in a variation in the intensity of the appearance of the voltammetric signals and/or the appearance of new peaks.

Figure 10.9 compares the variation of the peak current of the signal I, *i*(I), with the time after the cut of the fruit and immediate insertion of the three-electrode set, recorded in square wave voltammetry of six apple varieties [25]. Here, the potential scan was initiated at potentials in which there is no electrochemical ROS generation. Data for equivalent experiments when there is electrochemical generation of ROS are shown in Figure 10.10 [25]. Examination of Figure 10.9 reveals that, in the absence of electrochemical generation of ROS, there is a different time evolution of the local concentration of ascorbic acid (responsible for peak I) depending on the variety. As can be seen by comparing data in Figure 10.10, when the voltammetric record is accompanied by local ROS generation, the peak current/time curves change significantly. Now, the behavior of many varieties tends to be much more uniform than in the absence of electrochemically generated ROS. Qualitatively similar results were obtained for signal II [25].

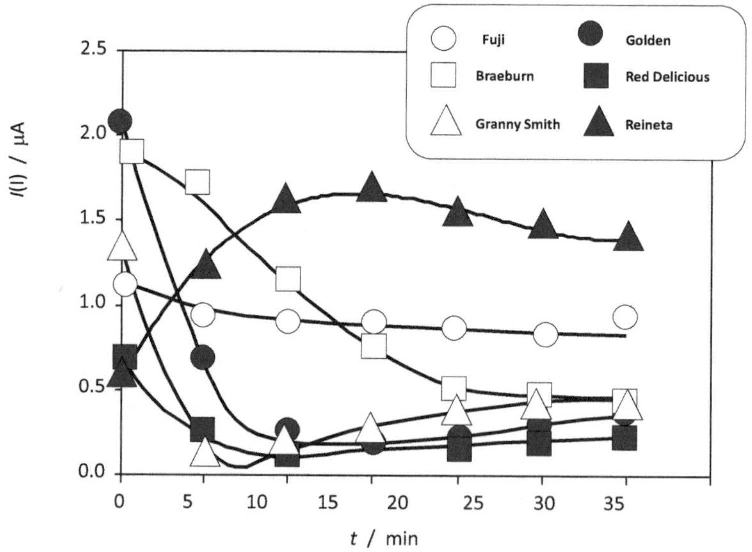

FIGURE 10.9 Time variation of *i*(I) in square wave voltammograms at graphite microelectrode (0.5 mm diameter) of different apple varieties. Potential scan initiated at 0.2 V in the positive direction, potential step increment 4 mV, square wave amplitude 25 mV, frequency 10 Hz. (Adapted from ref. [25], with permission. Copyright Elsevier 2022.).

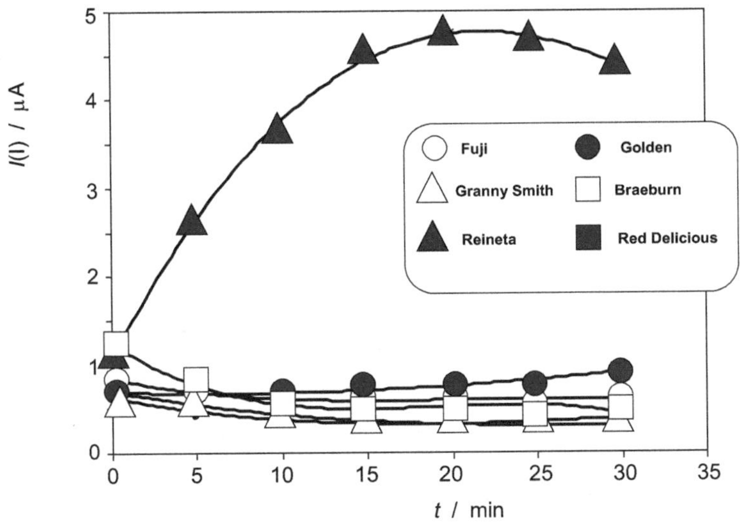

FIGURE 10.10 Variation of (a) I(I) with time in square wave voltammograms at graphite microelectrode (0.5 mm diameter) of different apple varieties. Potential scan initiated at –0.6 V in the positive direction, potential step increment 4 mV, square wave amplitude 25 mV, frequency 10 Hz. (Adapted from ref. [25], with permission. Copyright Elsevier 2022.).

To interpret the previous data, it is pertinent to underline that the content of ascorbic acid or any other species can vary notably among the fruits of different varieties or cultivars [27]. Consistently, the initial values of peak current show significant differences among the different varieties.

The obtained peak current/time curves can be modeled by considering that when cutting the fruit and inserting the electrodes, the electroactive compounds (responsible for signals I and II in Figure 10.8) are present in different concentrations, c°_{A}, in the fruit. The aforementioned operations produce the exposure of the fresh fruit to the atmosphere. The electroactive components (of an antioxidant nature) react more or less rapidly with oxygen that permeates the region where the working electrode has been inserted. Depending on the apple variety, the aforementioned process initially causes a decrease in the local concentration of the electroactive component, but it subsequently increases. A relatively simple kinetic model can be constructed assuming,

a) Under ordinary conditions, the component A is generated from a precursor P. When mechanical (or other) stress is produced, a second precursor Q activates the generation of the electroactive component.
b) Upon reaction with oxidants (oxygen in particular) A degrades to one (or more) compound(s) B.
c) The A-generation/degradation processes can be described in terms of the competence between two first-order consecutive reactions (P → A → B; Q → A → B).

Assuming that the pericarp behaves similarly to a homogeneous solution, the rate equations for the concentrations of P, Q, A, and B (c_P, c_Q, c_A, c_B, respectively) will be as follows.

$$\frac{dc_P}{dt} = -k_P c_P \tag{10.22}$$

$$\frac{dc_Q}{dt} = -k_Q c_Q \tag{10.23}$$

$$\frac{dc_A}{dt} = k_P c_P + k_Q c_Q - k_D c_A \tag{10.24}$$

$$\frac{dc_B}{dt} = k_D c_A \tag{10.25}$$

In Eqs. ((10.22)–(10.25), k_P, k_Q, k_D represent the respective rate constants. Assuming that the initial concentrations of A and their precursors are c°_P, c°_Q, c°_A, integration of the equations gives the following.

$$c_P = c_P^o e^{-k_P t} \tag{10.26}$$

$$c_Q = c_Q^o e^{-k_Q t} \tag{10.27}$$

Then,

$$\frac{dc_A}{dt} + k_D c_A = k_P c_P^o e^{-k_P t} + k_Q c_Q^o e^{-k_Q t}. \tag{10.28}$$

The multiplication of the two terms of the previous equation by e^{kDt}, and subsequent integration yields

$$c_A = c_A^o + \frac{k_P c_P^o}{k_D - k_P}\left[e^{-k_P t} - e^{-k_D t}\right] + \frac{k_Q c_Q^o}{k_D - k_Q}\left[e^{-k_Q t} - e^{-k_D t}\right] \tag{10.29}$$

Assuming, as usual, that the measured peak currents (i(A)) are proportional to the concentration of A, the previous equations can be rewritten in terms of peak currents by introducing an electrochemical constant, g_A. Accordingly, Eq. (10.29) leads to

$$i(A) = g_A c_A^o + g_A \frac{k_P c_P^o}{k_D - k_P}\left[e^{-k_P t} - e^{-k_D t}\right] + g_A \frac{k_Q c_Q^o}{k_D - k_Q}\left[e^{-k_Q t} - e^{-k_D t}\right], \tag{10.30}$$

Since the concentrations of A, P, and Q, are unknown, it is convenient to use the ratio between the peak current at a time t, i(A), and the current at time zero, i°(A), corresponding to

$$\frac{i(A)}{i^\circ(A)} = 1 + \frac{k_P\left(c_P^o / c^\circ{}_A\right)}{k_D - k_P}\left[e^{-k_P t} - e^{-k_D t}\right] + \frac{k_Q\left(c_Q^o / c^\circ{}_A\right)}{k_D - k_Q}\left[e^{-k_Q t} - e^{-k_D t}\right]. \tag{10.31}$$

Experimental data in Figures 10.9 and 10.10 can be fitted satisfactorily to theoretical curves using appropriate values for the kinetic parameters. Figure 10.11 illustrates several theoretical curves for different values of the rate constants and the parameters $R = c^\circ_P k_P / c^\circ_A (k_D\text{-}k_P)$, and $S = c^\circ_P k_P / c^\circ_A (k_D\text{-}k_P)$.

Data in Figure 10.9 can be analyzed by taking into account that the content of ascorbic acid (and other antioxidants) appears to be correlated with the oxidative status of the fruit, which in turn is determined by the activity of enzymes involved in the redox state and the H_2O_2 content [28]. Then the observed differences in the stress response cannot be attributed merely to differences in the pool of ascorbic acid (or other components).

The comparison between data in Figures 10.9 and 10.10 indicates that, although the response to stress of ascorbic acid and polyphenols in the presence of ROS electrochemically generated differs

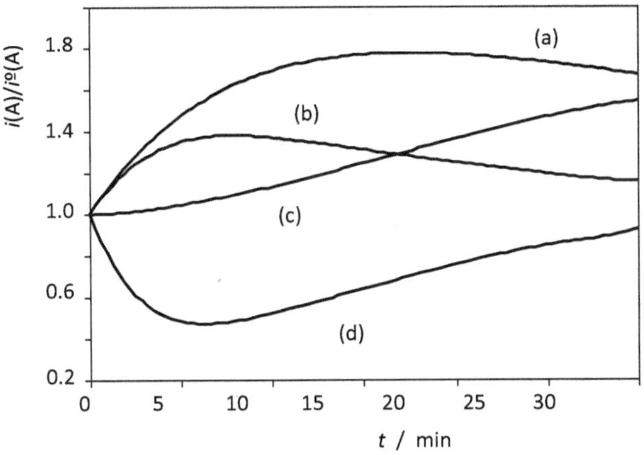

FIGURE 10.11 Theoretical $i(A)/i^{o}(A)$ vs. t curves according to Eq. (10.31) for different values of the rate constants and the parameters $R = c^{o}_{P}k_{P}/c^{o}_{A}(k_D - k_P)$, and $S = c^{o}_{P}k_{P}/c^{o}_{A}(k_D - k_P)$. (a) $k_P = 0.02$ min^{-1}, $k_D = 0.05$ min^{-1}, $k_Q = 0.10$ min^{-1}, $R = 1.0$, $S = 1.0$; (b) $k_P = 0.02$ min^{-1}, $k_D = 0.10$ min^{-1}, $k_Q = 0.20$ min^{-1}, $R = 0.2$, $S = 1.0$; (c) $k_P = 0.002$ min^{-1}, $k_D = 0.05$ min^{-1}, $k_Q = 0.10$ min^{-1}, $R = 0.2$, $S = 1.0$; (d) $k_P = 0.02$ min^{-1}, $k_D = 0.05$ min^{-1}, $k_Q =$ 0.30 min^{-1}, $R = 1.0$, $S = 1.0$.

from the response in its absence, the mechanistic pathway remains (qualitatively) apparently unaltered. This suggests that the ROS generated electrochemically adds to the "natural" pool of ROS in the fruit without (qualitative) alteration of the regeneration/degradation pathway [25].

10.6 PREDATION AND PLANT COMMUNICATION

Many plants respond to mechanical, chemical, biological, and external stress by employing different defensive strategies [29]. These can include, among others, the release of repelling compounds [30] and the attraction of parasitoids and predators of wounding insects [31]. Indeed, plants appear to distinguish between mechanical wounding and the attack of herbivores [32] and can induce defense responses in intact neighbor plants [33], i.e., there is chemical communication between plants and between plants and other organisms [34]. Plant communication involves volatile organic compounds (VOCs), alkaloids, terpenoids, cyanogenic glucosides, and phenolics.

This complex scenario involves molecular recognition events associated with the identification of the secondary metabolites acting as signaling and/or defensive compounds. The first strategy to study plant communication phenomena is the identification of VOCs (mainly, indole and methyl salicylate, cyclic and acyclic terpenoids, oximes, and nitriles). This analytical target is usually accomplished by chromatographic and multinuclear magnetic resonance techniques.

Electrochemical techniques can also be used in this context for studying trophic interactions [35] and evidencing plant communication [36]. The essential idea is that predation generates changes in the composition of vegetal matter because of the plant's defense response which can be detected electrochemically. This approach has been applied to detect the communication between plants of *Peperomia obtusifolia* A. Dietr., submitted to herbivory by the beetle *Monoplatus* sp. [36]. For this purpose, a series of specimens were divided into three groups confined in plastic boxes permeable to chemicals but not to insects. The herbivory group (Group I) was submitted to herbivory by beetles and allowed to communicate with an "internal control" group (II). A third group of plants, the 'external control' group (III), was isolated in a separate incubator. All three groups of plants remained for ten days under identical conditions of temperature, illumination, and humidity.

Figure 10.12 shows the cyclic voltammograms of extracts of *P. obtusifolia* leaves in contact with air-saturated PBS at pH 7.4 [36]. The external control group (Figure 10.12a) shows a cathodic signal ca. −0.40 V (C_1) preceding a second (C_2) strongly overlapped with the reduction of dissolved oxygen (C_{ox}). In the reverse scan, anodic signals at −0.10 (A_1), 0.20 (A_2), and 0.56 V (A_3) are recorded. This

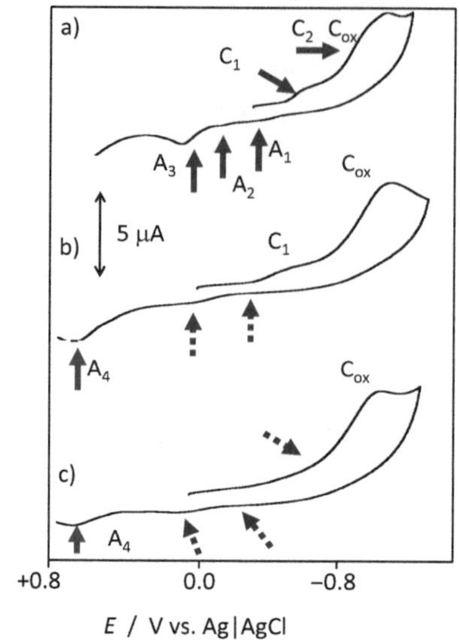

FIGURE 10.12 Cyclic voltammograms of microparticulate deposits on GCE resulting from the evaporation of ethanolic extracts of *P. obtusifolia* leaves, after 10 days of predation by *Monoplatus* sp., in contact with air-saturated 0.10 M PBS solution at pH 7.4. (a) External control, (b) herbivory group, (c) internal control plants. Potential scan initiated at 0.0 V vs. Ag/AgCl in the negative direction. Potential scan rate 50 mV s^{-1}. (Reproduced from ref. [36], with permission. Copyright the Royal Society of Chemistry 2015. Note that the axes are opposite to the IUPAC convention.).

voltammetric pattern becomes substantially changed in the herbivory group. In the voltammograms of this group (Figure 10.12b) of plant extracts, the peaks C_2 and A_1 vanish, and the peaks C_1, A_2, and A_3 decrease. This is followed by a new, intense oxidation wave at 0.85 V (A_4) and ill-defined anodic signals (A_5, A_6) between 0.0 and 0.4 V. The change in the voltammetric response can be attributed to the modification in the composition of plant secondary metabolites associated with predation/defense.

Remarkably, the voltammograms of the internal control group (Figure 10.12c), i.e., the plants potentially experiencing communication with predated plants, become intermediate between the aforementioned. Features characterizing the predation/response (in particular, the vanishing of peaks C_1 and A_1, and the appearance of peak A_4) also appear in the voltammograms of this group, in agreement with the hypothesis of the existence of plant communication. The aforementioned features can be seen in square wave voltammograms in Figure 10.13 [36], restricted to the region of potentials where the oxidation processes A_1-A_3 appear. Again, the voltammetric profile of the plants of the internal control group resembles that of the plants undergoing herbivory.

Figure 10.14 presents a schematic representation of how the processes of predation/defense/plant communication take place [36]. After receiving communication from the predated specimens, the plants of the internal control group become stressed and develop the chemical pattern corresponding to the response to herbivory.

Although there is a wide variety of plant components involved in the recorded electrochemistry, there is a possibility of attributing some of these oxidation processes to sesquiterpene compounds abundant in the essential oils from *P. obtusifolia*, α-eudesmol, guaiol, and δ-cadinene [37]. Figure 10.15 shows a tentative scheme for δ-cadinene-based processes [36]. This compound could produce oxidation signals corresponding to its electrochemical oxidation to 8-hydroxy-δ-cadinene. The same process can be produced biochemically under herbivory. The oxidized compound subsequently condensates to gossypol, a compound having *o*-catechol units that can be oxidized electrochemically to the corresponding *o*-quinones at potentials differing from the oxidation of the parent δ-cadinene.

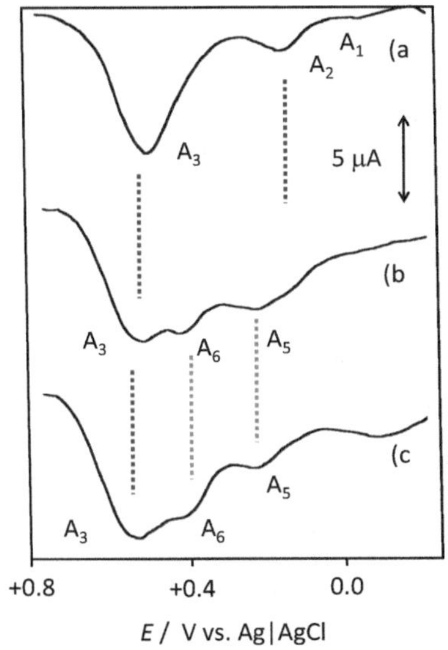

FIGURE 10.13 Square wave voltammograms, after semi-derivative convolution, of ethanolic leaf extracts from *P. obtusifolia* plants, after 30 days of predation by *Monoplatus* sp., in contact with air-saturated 0.10 M PBS solution at pH 7.4. (a) External control, (b) herbivory group, (c) internal control specimens. Potential scan initiated at −0.85 V in the positive direction, potential step increment 4 mV, square wave amplitude 25 mV, frequency 5 Hz. (Adapted from ref. [36], with permission. Copyright the Royal Society of Chemistry 2015. Note that the axes are opposite to the IUPAC convention.).

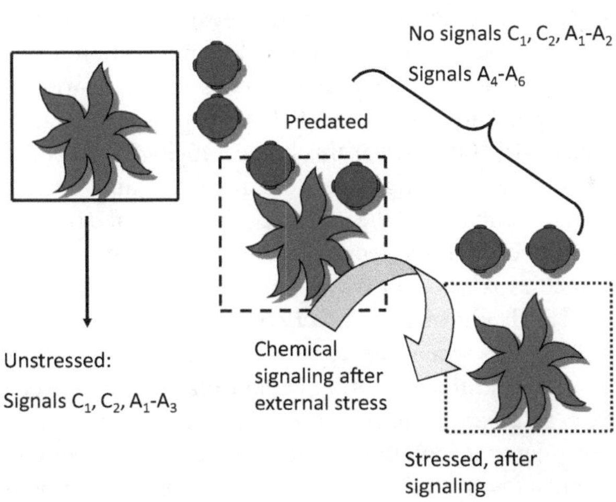

FIGURE 10.14 Schematic diagram illustrating the influence on the voltammetric response of the processes of plant external stress associated with predation and the corresponding defense and signaling/communication phenomena. (Adapted from ref. [36], with permission. Copyright the Royal Society of Chemistry 2015.).

It is worth noting that the expression of molecules with functions in plant immune systems and the production of secondary metabolites is determined by a variety of biotic and abiotic factors. Then, a complete description of plant defense responses involves different biochemical techniques of analysis. Due to its simplicity and versatility, electrochemical techniques can complement efficiently genetic and biochemical studies.

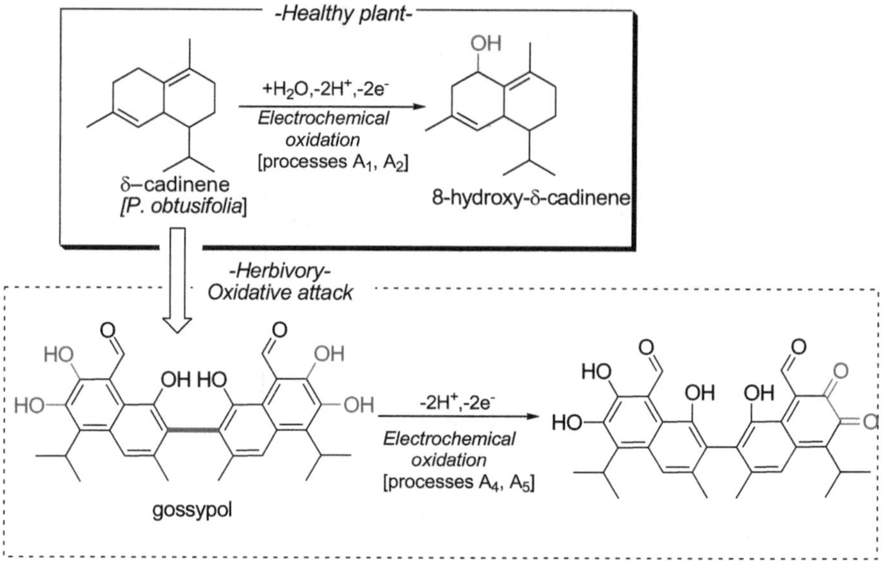

FIGURE 10.15 Possible chemical and electrochemical processes involved in δ-cadinene-based signaling. (Reproduced from ref. [36], with permission. Copyright the Royal Society of Chemistry 2015.).

10.7 INTERSECTING ELECTROCHEMISTRY AND PLANT SCIENCE

The sensitivity of electrochemical techniques to the composition of vegetal matter can be exploited for a variety of applications. These can include control of maturation and variety screening [24], harvesting conditions [38], and study of germination processes [8], as well as chemoecological and even forensic applications [39]. Directly related fields, such as soil analysis, are also accessible using solid-state techniques [40]. Of course, this same sensitivity implies that repeatability and reproducibility tests and accurate sampling have to be involved, and, eventually, multivariant chemometric techniques can aid in enhancing the analytical performance of electrochemical data.

One interesting application of electrochemical techniques is sex and age screening in *Ginkgo biloba* L. This dioecious species is the only living representative of the Ginkgoaceae family and is considered a living fossil [41]. The growth of *G. biloba* trees is relatively slow, especially between the ages of more than ca. 100 years when tree development remains approximately constant [42]. A study of ten *G. biloba* trees around the city of València (Spain) of ages between 30 and 150 years suggests that voltammetric data can be used for age and sex screening. Figure 10.16 shows the voltammetric responses of microparticulate deposits from ethanolic extracts of four specimens of diverse ages and sexes in contact with an acetate buffer [43]. The voltammograms exhibit two groups of overlapping peaks at ca. 0.25 (I) and 0.90 V (II), the second showing a more or less marked inflection at 0.70 V (III). The peak currents for the signals, i(J) (J = I, II, III), measured from the initial baseline centered ca. 0.0 V, can be used for discriminating sex and age. The first aspect is shown in Figure 10.17, where the i(II)/i(I) ratio is plotted vs. i(I) [43]. One can see that the experimental data points corresponding to replicate experiments are grouped in two tendency curves, allowing us to discriminate between female and male specimens. The variation of the i(III)/i(I) ratio on the age of the trees (known with a 10% uncertainty from documentation in botanic gardens) is shown in Figure 10.18. Now, no significant differences appear between male and female specimens, in both cases showing the i(III)/i(I) ratio, apparently, has a common tendency to increase with age.

This example and the previous list do not exhaust the fields where electrochemistry intersects or applies in the domain of plant science. The capabilities of electrochemistry to provide information on taxonomy and physiological and phyllogenetic aspects will be treated in the next chapter.

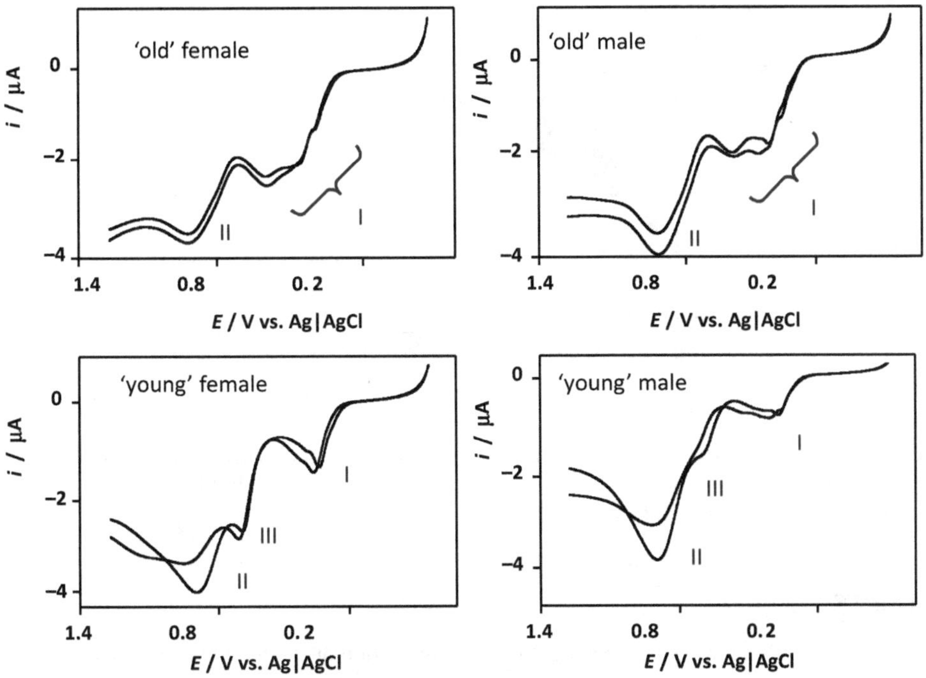

FIGURE 10.16 Linear potential scan voltammograms of microparticulate deposits on GCE resulting from the evaporation of ethanolic extracts of four specimens of *Ginkgo biloba* leaves in contact with 0.25 M HAc/NaAc solution at pH 4.75. Potential scan initiated at –0.50 V vs. Ag/AgCl. Potential scan rate 50 mV s^{-1}. Note that the axes are opposite to the IUPAC convention. (Reproduced from ref. [43], with permission. Copyright Elsevier 2024.).

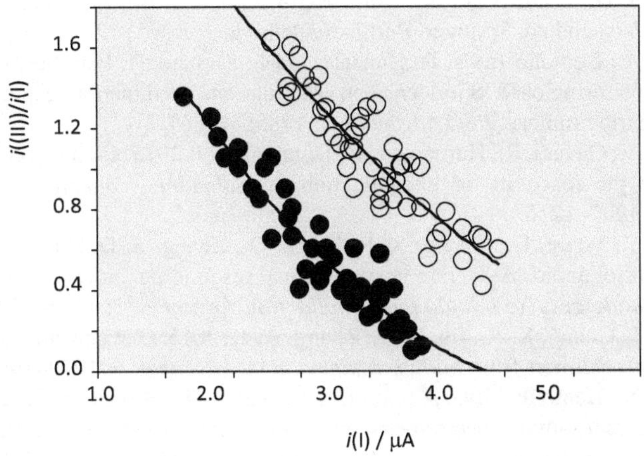

FIGURE 10.17 Plots of the ***i***(II)/***i***(I) ratio vs. ***i***(I) from linear potential scan voltammograms, such as in Figure 10.16 for female (solid circles) and male (circles) *G. Biloba* specimens of different ages. (Adapted from ref. [43], with permission. Copyright Elsevier 2024.).

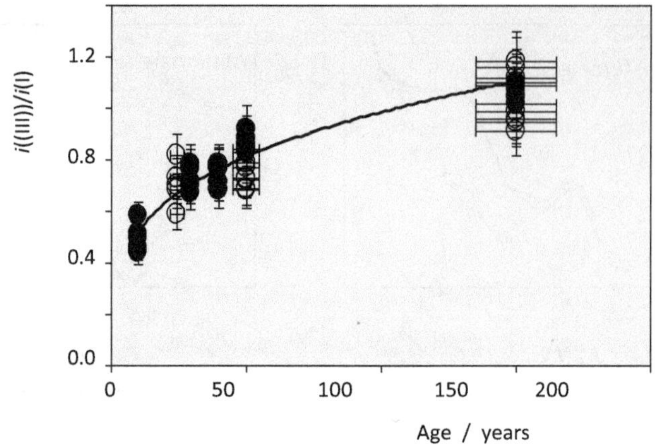

FIGURE 10.18 Variation of the i(III)/i(I) ratio on the age of female (solid circles) and male (circles) specimens of *G. Biloba*. From linear potential scan voltammograms, such as in Figure 10.16. (Adapted from ref. [43], with permission. Copyright Elsevier 2024.).

REFERENCES

[1] Li, Z.; Zhou, J.; Dong, T.; Xu, Y.; Shang, Y. 2021. Application of electrochemical methods for the detection of abiotic stress biomarkers in plants. *Biosensors & Bioelectronics*. 182: artic. 113105.

[2] McKelvey, K.; Martin, S.; Robinson, C.; Unwin, P.R. 2013. Quantitative Local Photosynthetic Flux Measurements at Isolated Chloroplasts and Thylakoid Membranes Using Scanning Electrochemical Microscopy (SECM). *Journal of Physical Chemistry B*. 117: 7878–7888.

[3] Doménech-Carbó, A.; Dias, D.; Donnici, M. 2021. In vivo electrochemical monitoring of signaling transduction of plant defense against stress in leaves of *Aloe vera* L. *Electroanalysis*. 33: 1024–1032.

[4] CRC Handbook of Chemistry and Physics, 90th ed.; Lide, D.R., Ed. 2009. CRC Press/Taylor & Francis: Boca Raton.

[5] Cape, J.L.; Bowman, M.K.; Kramer, D.M. 2006. Computation of the redox and protonation properties of quinones: Towards the prediction of redox cycling natural products. *Phytochemistry*. 67: 1781–1788.

[6] Oliveira, R.; Geraldo, D.; Bento, F. 2014. Radical scavenging activity of antioxidants evaluated by means of electrogenerated HO radical. *Talanta*. 129: 320–327.

[7] Scholz, F.; Schröder, U.; Gulabowski, R.; Doménech-Carbó, A. 2014. *Electrochemistry of Immobilized Particles and Droplets*, 2nd ed. Springer, Berlin-Heidelberg.

[8] Doménech-Carbó, A.; Cervelló-Bulls, P.; González, J.M.; Soriano, P.; Estrelles, E.; Montoya, N. 2019. Electrochemical monitoring of ROS influence on seedlings and germination response to salinity stress of three species of the tribe Inuleae. *RSC Advances*. 9: 17856–17867.

[9] Doménech-Carbó, A.; Gavara, R.; Hernández, P.; Domínguez, I. 2015. Contact probe voltammetry for *in situ* monitoring of the reactivity of phenolic tomato (*Solanum lycopersicum* L.) compounds with ROS. *Talanta*.144: 1207–1215.

[10] Qin, J.-J.; Huang, Y.; Wang, D.; Cheng, X.-R.; Zeng, Q.; Zhang, S.-D.;Hu, Z.-L.; Jin, H.-Z.; Zhang, W.-D. 2012. Lineariifolianoids A–D, rare unsymmetrical sesquiterpenoid dimers comprised of xanthane and guaiane framework units from *Inula lineariifolia*. *RSC Advances*. 3: 1307–1309.

[11] Zhang, X.-F.; Ren, J.; Cheng, X.-R.; Jin, H.-Z.; Zhang, W.-D. 2015. One new unusual sesterterpenoid and four new sesquiterpene dimers from *Inula britannica*. *RSC Advances*. 5: 1979–1982.

[12] Kheyar-Kraouche, N.; Bento da Silva, A.; Serra, A.T.; Bedjou, F.; Bronze, M.R. 2018. Characterization by liquid chromatography-mass spectrometry and antioxidant activity of an ethanolic extract of Inula viscosa leaves. Journal of *Pharmaceutical and Biomedical Analysis*. 156: 297–306.

[13] Krishnamachari, V.; Levine, L.H.; Paré, P.W. 2002. Flavonoid oxidation by the radical generator AIBN: a unified mechanism for quercetin radical scavenging. *Journal of Agricultural and Food Chemistry*. 50: 4357–4363.

[14] Zenkevich, I.G.; Eshchenko, A.Y.; Makarova, S.V.; Vitenberg, A.G.; Dobryakov, Y.G.; Utsal, V.A. 2007. Identification of the products of oxidation of quercetin by air oxygen at ambient temperature. *Molecules*. 12: 654–672.

[15] Gülçin, I. 2012. Antioxidant capacity of food constituents: an overview. *Archives of Toxicology*. 86: 345–391.

[16] Fang, T.; Zhen, Q.; Liao, L.; Owiti, A.; Zhao, L.; Korban, S.S.; Han, Y. 2017. Variation of ascorbic acid concentration in fruits of cultivated and wild apples. *Food Chemistry*. 225: 132–137.

[17] Mauro, M.A.; Dellarosa, N.; Tylewicz, U.; Tappi, S.; Laghi, L.; Rocculi, P.; Rosa, M.D. 2016. Calcium and ascorbic acid affect cellular structure and water mobility in apple tissue during osmotic dehydration in sucrose solutions. *Food Chemistry*. 195: 19–28.

[18] Davey, M.W. 2006. Genetic control of fruit vitamin C contents. *Plant Physiology*. 142: 343–351.

[19] Conklin, P.L. 2001. Recent advances in the role and biosynthesis of ascorbic acid in plants. *Plant, Cell and Environment*. 24: 383–394.

[20] Yashiro, E.; Spear, R.N.; McManus, P.S. 2011. Culture-dependent and culture-independent assessment of bacteria in the apple phyllosphere. *Journal of Applied Microbiology*. 110: 1284–1296.

[21] Fenech, M.; Amaya, I.; Valpuesta, V.; Botella, M.A. 2019. Vitamin C Content in Fruits: Biosynthesis and Regulation. *Frontiers in Plant Science*. 9: artic. 2006.

[22] Lu, L.; Zuo, W.; Wang, C.; Li, C.; Feng, T.; Li, X.; Wang, C.; Yao, Y.; Zhang, Z.; Chen, X. 2021. Analysis of the postharvest storage characteristics of the new red-fleshed apple cultivar 'meihong'. *Food Chemistry*, 354: 129470.

[23] Ren, Q.-Q.; Huang, X.-R.; Liu, G.-C.; Jun, O.-Y.; Li, M.-T.; Chen, H.; Zhao, Y.-D.; Chen, W. 2015. A field-compatible technique using an electrochemical sensing microbundle for real-time and simultaneous in vivo measurement of hydrogen peroxide, nitric oxide, and pH under drought stress. *Sensors Actuators, B*. 220: 743–748.

[24] Doménech-Carbó, A.; Domínguez, I; Hernández, P.; Gavara, R. 2015. Electrochemical tomato (*Solanum lycopersicum* L.) characterization using contact probe *in situ* voltammetry. *Food Chemistry*. 127: 318–325.

[25] Doménech-Carbó, A.; Dias, D. 2022. *In situ* electrochemical monitoring of ROS influence in the dynamics of ascorbic acid and polyphenolic compounds in apple fruits. *Food Chemistry*. 374: artic. 131818.

[26] Coseteng, M.Y.; Lee, C.Y. 1987. Changes in apple polyphenoloxidase and polyphenol concentrations in relation to degree of browning. *Journal of Food Science*.52: 985–989.

[27] Cruz-Rus, E.; Amaya, I.; Sánchez-Sevilla, J.F.; Botella, M.A.; Valpuesta, V. 2011. Regulation of L-ascorbic acid content in strawberry fruits. *Journal of Experimental Botany*. 62: 4191–4201.

[28] Li, M.; Ma, F.; Liang, D.; Li, J.; Wang, Y. 2010. Ascorbate biosynthesis during early fruit development is the main reason for its accumulation in kiwi. *PLoS One*. 5: artic. e14281.

[29] Karban, R.; Baldwin, I.T.; Baxter, K.J.; Laue, G.; Felton, G.W. 2000. Communication between plants: induced resistance in wild tobacco plants following clipping of neighboring sagebrush. *Oecologia*. 125: 66–71.

[30] De Moraes, M.; Mescher, M.C.; Tumlinson, J.H. 2001. Caterpillar-induced nocturnal plant volatiles repel conspecific females. *Nature*. 410: 577–580.

[31] Pare, P.W.; Tumlinson, J.H. 1999. Plant Volatiles as a Defense against Insect Herbivores. *Plant Physiology*. 121: 325–332.

[32] Wu, J.Q.; Baldwin, I.T. 2009. Herbivory-induced signalling in plants: perception and action. *Plant Cell & Environment*. 32: 1161–1174.

[33] Arimura, G.-I.; Ozawa, R.; Shimoda, T.; Nishioka, T.; Boland, W.; Takabayashi, J. 2000. Herbivory-induced volatiles elicit defence genes in lima bean leaves. *Nature*. 406: 512–515.

[34] Adler, F.R. 2011. Plant signalling: the opportunities and dangers of chemical communication. Adler, *Biology Letters*. 7: 161–162.

[35] Matignon, L.; Marietou Lo, M.; Monpierre, M.; Vicentini, M.; Valencia, D.P.; Palmeira-Mello, M.; Sylvestre, M.-N.; Runeau, L.; Sylvestre, M.; Doménech-Carbó, A.; Benfodda, Z.; Meffre, P.; Cebrián-Torrejón, G. 2023. Phytochemical and Biological Study of Trophic Interaction Between Pseudosphinx Tetrio L. Larvae and Allamanda Cathartica L. *Plants*. 12: 520.

[36] Doménech-Carbó, A.; Cebrián-Torrejón, G.; Lopes-Souto, A.; Martins de Moraes, M.; Jorge-Kato, M.; Fechine-Tavares, J.; Barbosa-Filho, J. M.. Electrochemical ecology: VIMP monitoring of plant defense against external stressors. *RSC Advances* 2015, **5**, 61006–61011.

[37] Moramdin-Gianetti, A.A.; Pin, A.R.; Santo Pietro, N.A.; de Oliveira, H.C.; Soares Mendes-Giannini, M.J.; Alecio, A.C.; Kato, M.J.; de Oliveira, J.E.; Furlan, M., 2010. Composition and antifungal activity against Candida albicans, C. parapsilopisis, Candida krusei and Cryptococcus neoformans of essential oils from leaves of Piper and Peperomia species. *Journal of Medicinal Plants Research*. 4: 1810–1814.

[38] Cebrián-Torrejón, G.; Kablan, L.; Ferreira, M.E.; Rodríguez de la Cruz, D.; Doménech-Carbó, A.; Vera de Bilbao, N.; Rojas de Arias, A.; Figadère, B.; Poupon, E.; Fournet, A. 2015. Harvesting canthinones: identification of the optimal seasonal point of harvest of Zanthoxylum chiloperone leaves as a source of 5-methoxycanthin–6-one. *Natural Products Research*. 29: 2054–2058.

[39] Ortiz-Miranda, A.S.; König, P.; Kahlert, H.; Scholz, F.; Osete-Cortina, L.; Doménech-Carbó, M.T.; Doménech-Carbó, A. 2016. Voltammetric analysis of Pinus needles with physiological, phylogenetic, and forensic applications. *Analytical and Bioanalytical Chemistry*. 408: 4943–4952.

[40] Pérez, A.; Cebrián-Torrejón, G.; Montoya, N.; Piquero-Cilla, J.; Sanders, C.J.; Doménech-Carbó, A.; Machado, W. 2021. Organic matter redox state driven by specific sources in mangrove sediments: A case study from Peruvian ecosystems. *Journal of Marine Science and Engineering*. 9: artic. 1438.

[41] Singh, B.; Kaur, P.; Singh, G.R.D.; Ahuja, P.S. 2008. Biology and chemistry of *Ginkgo biloba*. *Fitoterapia*. 79: 401–418.

[42] Del Tredici, P. 1991. Ginkgos and people: a thousand years of interaction. *Arnoldia*. 51: 2–15.

[43] Doménech-Carbó, A.; Prieto-Mossi, J.; Montoya, N. 2024. Age- and sex-dependent electrochemistry of *Ginkgo biloba* leaves. *Electrochemistry Communications*. 165: 107758.

ADDITIONAL LITERATURE

Kinetic Aspects

Dynamic Pathway of Plant Compounds

Bi, X.; Zhang, J.; Chen, C.; Zhang, D.; Li, P.; Ma, F. 2014. Anthocyanin contributes more to hydrogen peroxide scavenging than other phenolics in apple peel. *Food Chemistry*. 152: 205–209.

Chai, Y., Chaiab, Y., Li, A., Wai, S. C., Song, C., Zhao, Y., Duan, Y., Zhang, B., Lin, Q. (2020). Cuticular wax composition changes of 10 apple cultivars during postharvest storage. *Food Chemistry*, 324; 126903.

Dowdle, J.; Ishikawa, T.; Gatzek, S.; Rolinski, S.; Smirnoff, N. 2007. Two genes in Arabidopsis thaliana encoding GDP-L-galactose phosphorylase are required for ascorbate biosynthesis and seedling viability. *Plant Journal*. 52: 673–689.

Li, C., Wei, M., Ge, Y., Zhao, J., Chen, Y., Hou, J., Cheng, Y., Chen, J., Li, J. (2020). The role of glucose-6-phosphate dehydrogenase in reactive oxygen species metabolism in apple exocarp induced by acibenzolar-S-methyl. *Food Chemistry*, 308; 125663.

Loewus F. A. (1999). Biosynthesis and metabolism of ascorbic acid in plants and of analogs of ascorbic acid in fungi. *Phytochemistry*, 52: 193–210.

McGhie, T.K.; Hunt, M.; Barnett, L.E. 2005. Cultivar and growing region determine the antioxidant polyphenolic concentration and composition of apples grown in New Zealand. *Journal of Agricultural and Food Chemistry*. 53: 3065–3070.

Plant Communication and Plant Electrochemistry

Gomi, K.; Yamasaki, Y.; Yamamoto, H.; Akimitsu, K. 2003. Characterization of a hydroperoxide lyase gene and effect of C_6-volatiles on expression of genes of the oxylipin metabolism in Citrus. *Journal of Plant. Physiology*. 2003, 160: 1219–1231.

Kost, C.; Heil, M. 2006. Herbivore-induced plant volatiles induce an indirect defence in neighbouring plants. *Journal of Ecology*. 94: 619–628.

Paschold, A.; Halitschke, R.; Baldwin, I.T. 2006. Using 'mute' plants to translate volatile signals. *Plant Journal*. 45: 275–291.

Sivager, G.; Calvez, L.; Bruyere, S.; Boisne-Noc, R.; Hufnagel, B.; Cebrián-Torrejón, G.; Doménech-Carbó, A.; Gros, O.; Ollitrault, P.; Morillon, R. 2022. Better tolerance to Huanglongbing is conferred by tetraploid Swingle citrumelo roostock and is influenced by the ploidy of the scion. *Frontiers in Plant Science*. 13: artic.1030862.

Soriano, P.; Estrelles, E.; Martínez, M.I.; Doménech-Carbó, A.; Galié, M.; Biondi, E. 2022. Environmental predictors of seed germination in two Halocnemum species from Mediterranean (Balearic, Tyrrenic and Adriatic) and Red Sea coastal salt marshes. *Seed Science Research*. 32, 246–263.

11 Electrochemotaxonomy and Phylogenetics

11.1 INTRODUCTION

The distribution of animals and vegetables into different groups forming a hierarchically organized structure of species, genera, families, etc., i.e., into different taxonomic levels, is one of the angular stones of biology. Apart from its obvious practical interest in agriculture, for example, characterizing crop varieties, the identification of taxonomic units is an essential scientific target in all botanic fields. Since contemporary species result from a complex evolutionary process taking place along geological times, taxonomic studies are inherently linked to the phylogenetic evolution of the different taxa.

In this chapter, we will study how electrochemical techniques can contribute to taxonomic and phylogenetic studies based on their sensitivity to subtle changes in the chemical composition of vegetal matter as well as their capacity to analyze the response of plant components to external inputs.

11.2 ELECTROCHEMICAL TAXONOMY

The grouping of living organisms into different taxonomic units was classically based on morphological and/or functional criteria. Currently, this task is focused on genetic analysis using the identification of nucleotide sequences of nuclear and chloroplast regions [1]. There are cases, however, in which chimaeric sequences are obtained, resulting in discrepancies between molecular and morphological/functional classifications. This situation appears, for instance, in the Rosids [2, 3], Guamatelaceae [4], and Hydatellaceae [5] families.

A complementary way to classify living organisms is to study their chemical composition. The so-called chemotaxonomy has received considerable attention in the botanic domain. Here, the determination of the composition of plants and plant extracts, extensively used to determine their nutritional, pharmacological, and chemoecological properties, also has taxonomic value [6, 7].

For taxonomic purposes, some bioactive phenolic components such as anthocyanins, flavonols, procyanidins, and phenolic acids have been intensively studied [8, 9]. This has been prompted by the wide accessibility of gas and liquid chromatographies coupled to mass spectroscopy (GC-MS and HPLC-MS, respectively) and other analytical techniques. Of course, the chemical composition of the plants varies from one part to another (fruits, leaves, roots) and depends on age, sex, illumination, water uptake, stress conditions, etc. This means that the content of polyphenolic, compounds, although influenced mainly by genotype, is also dependent somewhat on the cultivation method and environmental conditions [10].

Despite these limitations, the composition of different plant structures has been used for chemotaxonomic purposes. A case of particular interest is that of cuticular waxes because their composition varies sharply from one species to another [11]. Extracts of cuticular waxes consist of homologous series of long and very-long-chain aliphatics (fatty acids, aldehydes, primary and secondary alcohols, ketones, and alkanes of chain lengths C20–C36, as well as C38–C70 alkyl esters, triterpenoids, tocopherols) [12]. In general, primary metabolites, those involved in the fundamental metabolic pathways, such as water-soluble polysaccharides, are of "universal" occurrence, but they are of low taxonomic value compared to secondary metabolites [13]. Among others, alkaloids [14] and fatty acids [15] have been extensively used for chemotaxonomic purposes. In favorable cases,

DOI: 10.1201/9781032697826-11

the presence of a compound or a family of compounds permits taxa characterization. For instance, a part of Caryophyllales synthesize betalains pigments while other species of the same order synthesize anthocyanins, and these two types of pigments are mutually exclusive [16]. At a more restricted taxonomic level, Sinigrin, an S-glycoside, characterizes the Cruciferae, Moringaceae, and Capparaceae families [17]. In most cases, taxonomic information can be extracted from quantitative data, i.e., from the proportions of different vegetal components. This deals with the so-called metabolomic analysis prompted by the development of multivariate chemometric methods. The scope of such methods is under expansion based on new molecular and macromolecular analysis techniques [18].

Several of these components are electroactive under "ordinary" electrochemical conditions so that they can provide well-defined voltammetric responses either in the solution phase or in the solid state (see Chapter 4). This voltammetry, as discussed in Chapter 9, offers distinctive features in the presence of ROS.

In this context, the first approach to chemotaxonomy and its coupled phylogentic implications explicitly based on electrochemical profiling was reported in 2015 [19]. Here, fresh ethanolic and acetone extracts (25–50 μL) of leaves of mature plants of different species of Rosids (including species of the Ragales, Rosales, Fabales, Celastrales, Malpighiales, Malvales, Brassicales, Sapindales, Myrtales, and Geraniales orders) were evaporated on the surface of glassy carbon electrodes (surface area 0.071 cm^2). The resulting microparticulate deposit contained a mixture of solid plant components whose voltammetric response in contact with aqueous acetate buffer solution at pH 4.75 was recorded using linear potential scan and pulsed techniques. Figure 11.1 compares the square wave voltammograms recorded for ethanolic extracts of *Eonymus japonicus* Thunb (Celastrales, Celatraceae family), *Lobularia maritima* Desv. (Brassicales, Brassicaceae family), *Koelrenteria bipinnata* Franch. (Sapindales, sapindaceae family), and *Begonia acerifolia* Kunth (Cucurbitales, Begoniaceae family). The positive-going potential scans show a series of anodic peaks that can be attributed to the oxidation of polyphenolic plant components (see Chapter 4). The voltammetric profiles differ significantly for different species, thus suggesting that there is a significantly different composition of electroactive species.

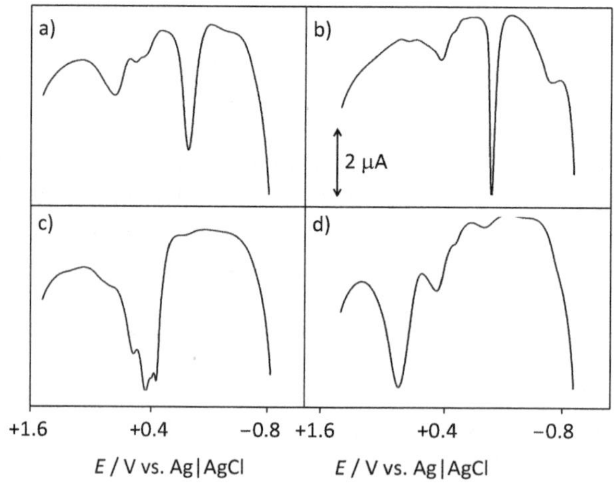

FIGURE 11.1 Square wave voltammograms of microparticulate films of ethanolic extracts of leaves of, (a) *Eonymus japonicus,* (b) *Lobularia maritima*, (c) *Koelrenteria bipinnata*, and (d) *Begonia acerifolia* deposited on GCE. Electrolyte solution: 0.25 M HAc/NaAc aqueous buffer at pH 4.75. Potential scan initiated at −1.05 V vs. Ag/AgCl in the positive direction, potential step increment 4 mV, square wave amplitude 25 mV, frequency 5 Hz. (Reproduced from ref. [19], with permission. Copyright the Royal Society of Chemistry 2015. Note that the axes are opposite to the IUPAC convention.).

The solid-state electrochemistry approach used here permits the record of intense voltammetric signals. Although several components of the extract can pass more or less extensively to the solution phase during the voltammetric measurements, the voltammetric response is dominated by the proton-assisted, solid-state oxidation of sparingly soluble components. This was studied using scanning electrochemical microscopy (SECM) experiments, as illustrated in Figure 11.2, where the redox competition strategy [20] was applied. Here it was studied the microparticulate deposit resulting from the evaporation of an ethanolic extract of *Ficus obscura* Blume (Rosales, Moraceae family) on the surface of a Pt electrode (surface area 0.018 cm^2). As the electrolyte, a 2.0 mM $K_4Fe(CN)_6$ solution in 0.25 M HAc/NaAc aqueous buffer at pH 4.75 was used. As described in Chapter 4, the images in Figure 11.2 show an "electrochemical topography" representative of the changes in conductivity of the substrate and the distance between it and the tip electrode that moves through the electrolyte solution close to the substrate electrode. The tip potential (E_T) was maintained at a constant value of 0.30 V vs. Ag/AgCl. At this potential, the solution-phase oxidation of $Fe(CN)_6^{4-}$ to $Fe(CN)_6^{3-}$ occurs under essentially reversible, diffusion-controlled conditions, and the fluctuations in the tip current are representative of the aforementioned changes in physical topography and conductivity.

When no potential inputs are applied to the substrate (E_S = 0.00 V), the insulating particulate deposit appears as negative feedback features (yellow-orange-red colors in the original image, clear areas here) contrasting with the positive feedback (blue color in the original map color, dark areas here) corresponding to the region of the Pt substrate electrode directly exposed to the electrolyte

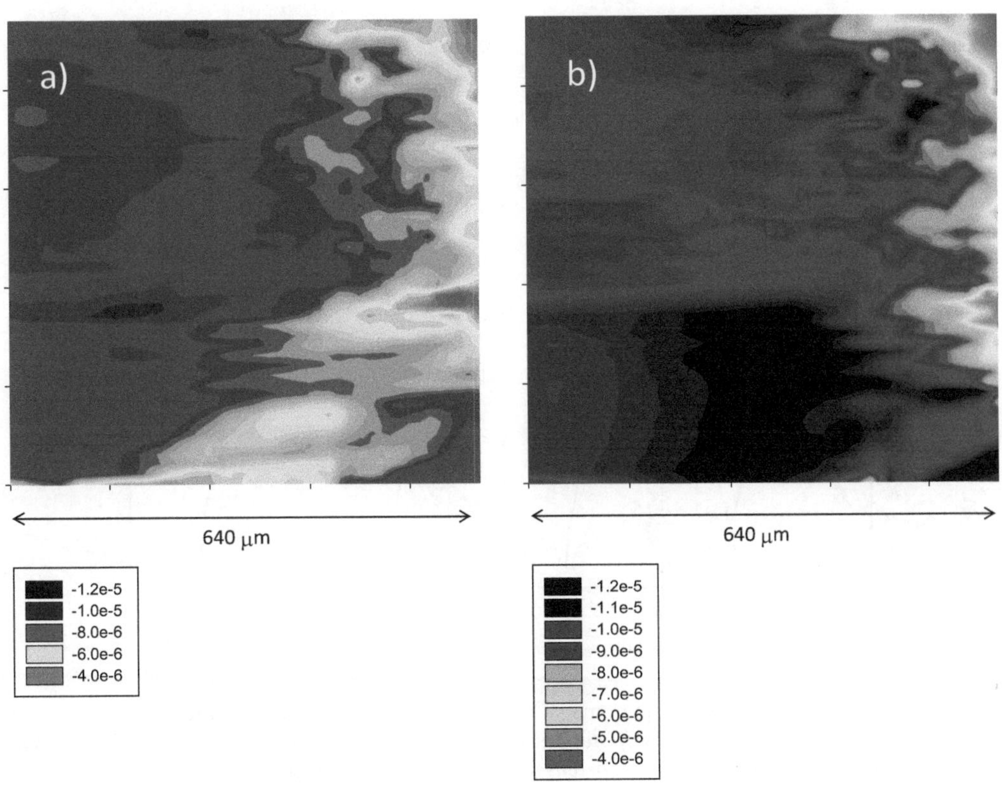

FIGURE 11.2 SECM color maps recorded on a microparticulate deposit resulting from the evaporation of an ethanolic extract of *Ficus obscura* leaf on the surface of a Pt electrode. Electrolyte solution: 2.0 mM $K_4Fe(CN)_6$ plus 0.25 M HAc/NaAc aqueous buffer at pH 4.75. (a) E_T = 0.30 V; E_S = 0.00 V; (b) E_T = 0.30 V; E_S = 0.60 V. (Adapted from ref. [19], with permission. Copyright the Royal Society of Chemistry 2015.).

solution (Figure 11.2a). When a potential sufficiently positive to promote the oxidation of the polyphenolic components is applied to the substrate ($E_S = 0.60$ V; Figure 11.2b), a significant portion of the negative feedback features vanish. This feature denotes that the solid substrate is being oxidized. In several regions, the change is minimal, thus denoting that a presumably topotactic solid-to-solid redox transformation occurs. In other regions, the change in electrochemical topography is drastic, thus suggesting that oxidative dissolution processes take place. In both cases, consistently with the interpretation of ion-assisted solid-state electrochemical processes [21, 22], the redox process starts at the solid particle/base electrode/electrolyte solution three-phase boundary. From here, it expands through the solid particles, which progressively dissolve as oxidized forms.

Although replicate experiments revealed differences in the voltammetric pattern depending on the harvesting time, maturation level, and other factors, a satisfactory repeatability was obtained when equivalent conditions of sampling were applied. Similar results were obtained when extracts from plant seeds were studied [23]. Figure 11.3 compares the voltmmograms of three species of the Rosaceae family: *Potentilla argentea* L., *Sarcopoterium spinosum* (L.) Spach, and *Agrimonia eupatoria* L, and three species of the Lamiaceae family: *Salvia valentina* Vahl, *Lavandula multifida* L., and *Lavandula canariensis* (L.) Mill. A common pattern can be recognized for each family.

Interestingly, when the taxonomic differences decreased, the voltammetric profile of the different species became increasingly similar. This can be seen in Figure 11.4, in which the voltammograms of three species of the order Fabales – namely, *Genista tinctoria* L., *Quillaja saponaria* Molina, and *Dorycnium pentaphyllum* Scop. are depicted. This figure compares the voltammograms recorded for ethanolic and acetone extracts. Here, small but significant differences in the current/potential profile can be seen between each species when comparing the extracts with these solvents.

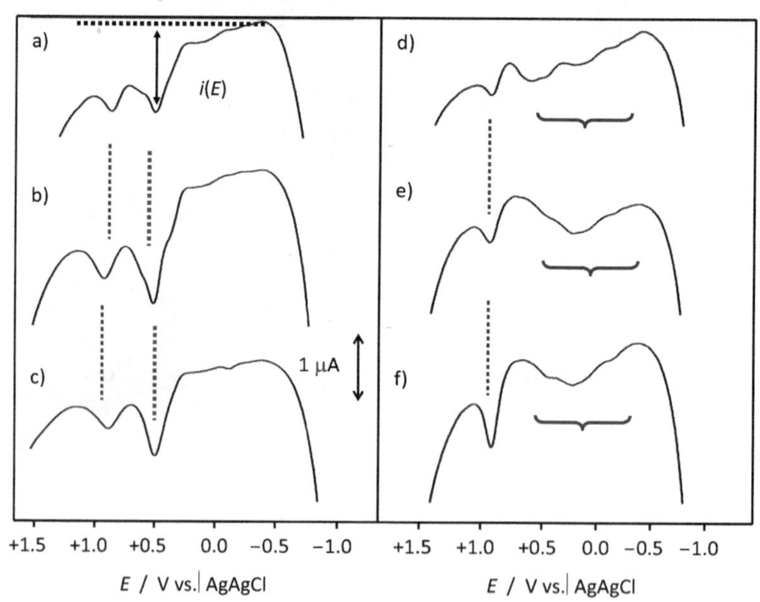

FIGURE 11.3 Square wave voltammograms of microparticulate deposits from the ethanolic extracts of seeds of, (a) *Potentilla argentea*, (b) *Sarcopoterium spinosum*, (c) *Agrimonia eupatoria*, (d) *Salvia valentina*, (e) *Lavandula multifida*, (f) *Lavandula canariensis* on GCE. Electrolyte: 0.25 M aqueous acetate buffer at pH 4.75. Potential scan initiated at −0.85 V in the positive direction; potential step increment 4 mV; square wave amplitude 25 mV; frequency 5 Hz. The horizontal dotted line in (a) marks the baseline used for current measurements whereas the vertical dotted lines mark the most prominent voltammetric signals. (Adapted from ref. [23], with permission. Copyright Wiley 2017. Note that the axes are opposite to the IUPAC convention.).

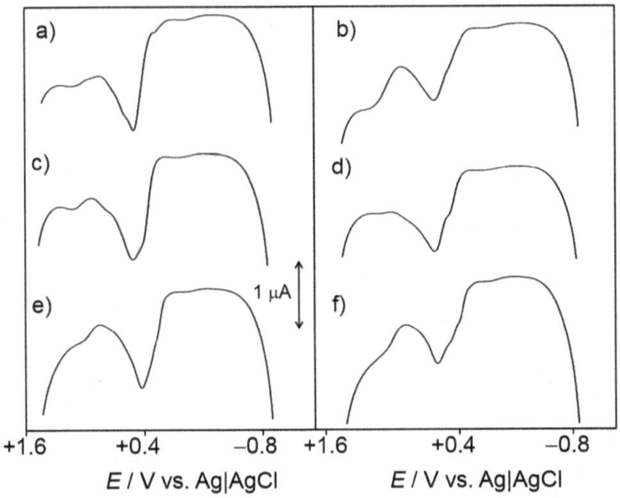

FIGURE 11.4 Square wave voltammograms of microparticulate deposits from extracts of leaves of (a, b) *Genista tinctoria*, (c, d) *Quillaja saponaria*, and (e, f) *Dorycnium pentaphyllum* leaves in ethanol (a, c, e) and acetone (b, d, f) on GCE. Electrolyte: 0.25 M aqueous acetate buffer at pH 4.75. Potential scan initiated at −1.05 V in the positive direction, potential step increment 4 mV, square wave amplitude 25 mV, frequency 5 Hz. (Reproduced from ref. [19], with permission. Copyright the Royal Society of Chemistry 2015. Note that the axes are opposite to the IUPAC convention.).

The aforementioned features are consistent with the idea that, depending on the solvent polarity and other factors, a different set of electroactive compounds is extracted and, hence, the voltammograms of the deposits from ethanol and acetone extracts are slightly different. This is an aspect of interest regarding the establishment of quantitative criteria for characterizing different taxonomic units. The most direct possibility consists of the measurement of the currents at selected potentials, $i(E)$, adopting a given criterion for the measurements. Since there is no easy way to control the net amount of vegetal components adhered to the electrode surface, normalized voltammograms were used taking the ratio between the current $i(E)$ measured at the potential E and the current at a given potential, usually that at which the maximum peak current was recorded, $i(E_{max})$. Then, the normalized currents, $I(E)$ (= $i(E)/i(E_{max})$), define a normalized voltammogram.

Accordingly, different normalized voltammograms can be defined for each species varying the solvent used for extraction purposes and/or the composition of the electrolyte solution. This allows us to define each individual species using a matrix of normalized currents, $I(E)$. If each voltammogram is constituted by currents measured at N potentials and separate data for K solvents (or K electrolyte solutions) are recorded, a matrix of $N \times K$ components is finally obtained. This permits the application of bivariant and multivariant chemometric techniques to this matrix of data.

Figure 11.5 illustrates a graphical method to visualize the differences between plant species. Here the values of the normalized currents measured in the voltammograms recorded for acetone extracts, $I(E)_{Acet}$, are plotted vs. the corresponding currents measured for ethanolic extracts, $I(E)_{EtOH}$, for four *Boehmeria cylindrica* (L.) Sw.(Rosales, Urticaceae family), *Quillaja saponaria* Molina (Fabales, Quillajaceae family), *Ricinus communis* L. (Malpighiales, Euphorbiceae family), and *Oxalis articulata* Savigny (Geraniales, Oxalidaceae family) [19]. The obtained plots define species-characteristic curves which can complement principal component analysis (PCA) and hierarchical cluster analysis (HCA) multivariant techniques [19, 23].

Another graphical form can be made by adding a third series of normalized currents and combining it through a color code. Figure 11.6 depicts the color maps obtained for the species in Figure 11.5 combining the normalized currents recorded in contact with aqueous acetate buffer using leaf

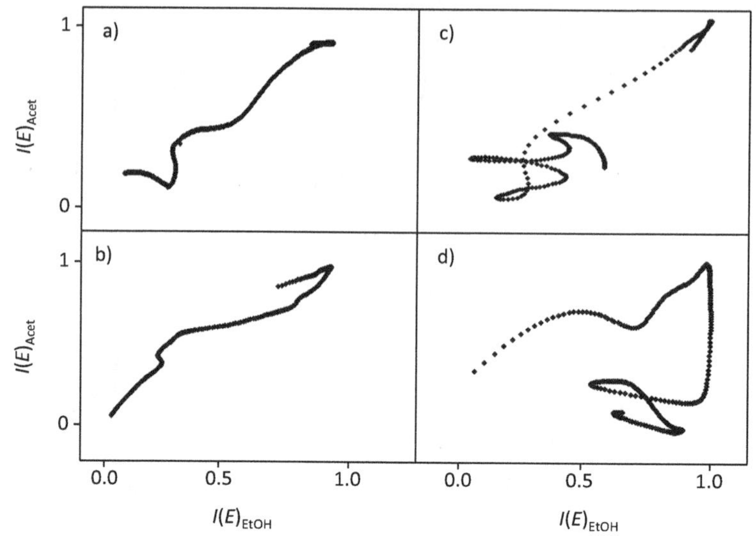

FIGURE 11.5 Plots of $I(E)_{Acet}$ vs. $I(E)_{EtOH}$ for films of leaf extracts of, (a) *Boehmeria cylindrica*, (b) *Quillaja saponaria*, (c) *Ricinus communis* and d) *Oxalis articulata* immersed into 0.25 M aqueous acetate buffer at pH 4.75. Data from square wave voltammograms at sample-modified GCEs. Potential scan initiated at −0.85 V vs. Ag/AgCl in the positive direction, potential step increment 4 mV, square wave amplitude 25 mV, frequency 5 Hz. Potentials were taken at intervals of 4 mV. (Adapted from ref. [19], with permission. Copyright the Royal Society of Chemistry 2015.).

extracts in ethanol, $I(E)_{EtOH}$, acetone, $I(E)_{Acet}$, and chloroform, $I(E)_{Chlor}$. The values of this last quantity are converted to 'color topography' according to the code indicated in the figure. Again, clear differences are visualized for the different species. Analyses of these types have been applied to fungal species [24] and different plant groups [25, 26].

The aforementioned schemes can also be applied to taxonomic studies upon recording the voltammetric response of selected vegetal organs or systems. This is the case of fungal spores, which can be adhered to GCEs from ethanolic suspensions of macerated fungi specimens illustrated in Figure 11.7. Here, the voltammograms of spores of *Aspergillus carbonarius* (Bainier) Thom, and *Aspergillus ochraceus* Wilhelm are shown illustrating the repeatability obtained in replicate voltammograms and the similarity of the voltammetric profiles [24].

The observed electrochemistry can be attributed to compounds released from the vegetal matter during the attachment of the fungal colony to the base electrode, but also to some transmembrane electrochemistry. This second possibility is supported by field-emission scanning electron microscopy (FESEM) images obtained for deposits of spores onto carbon plates submitted to oxidative potential inputs in contact with aqueous acetate buffer. As shown in Figure 11.8, the original spores of *Aspergillus flavus* Link and *Aspergillus carbonarius* appear as globular conidia with smooth spiny ornamentation [24]. After applying a potential sufficiently high to promote electrochemical oxidation processes such as in Figure 11.7, the shape of the conidia became cuboid or increasingly irregular, often showing irregular pores. The formation of pores suggests that there is ion-channeled transport through the spore crust, a situation to some extent parallel to that observed in metal-organic frameworks [22].

There are a variety of compounds potentially responsible for the observed voltammetric response eventually resulting in the permeation/degradation of the pore membrane. These include, among others, anthraquinonic and naphtapyrone metabolites containing polyphenol units (fonsecinone A, nigerone, aspergillin). One of the possible electroactive fungal components is melanin, a polymeric substance existing in cell walls constituted by dihydroxyindole units. Its electrochemical oxidation is a multi-electron transfer that involves both the *o*-phenol and indolic units as schematized in Figure 11.9 [24].

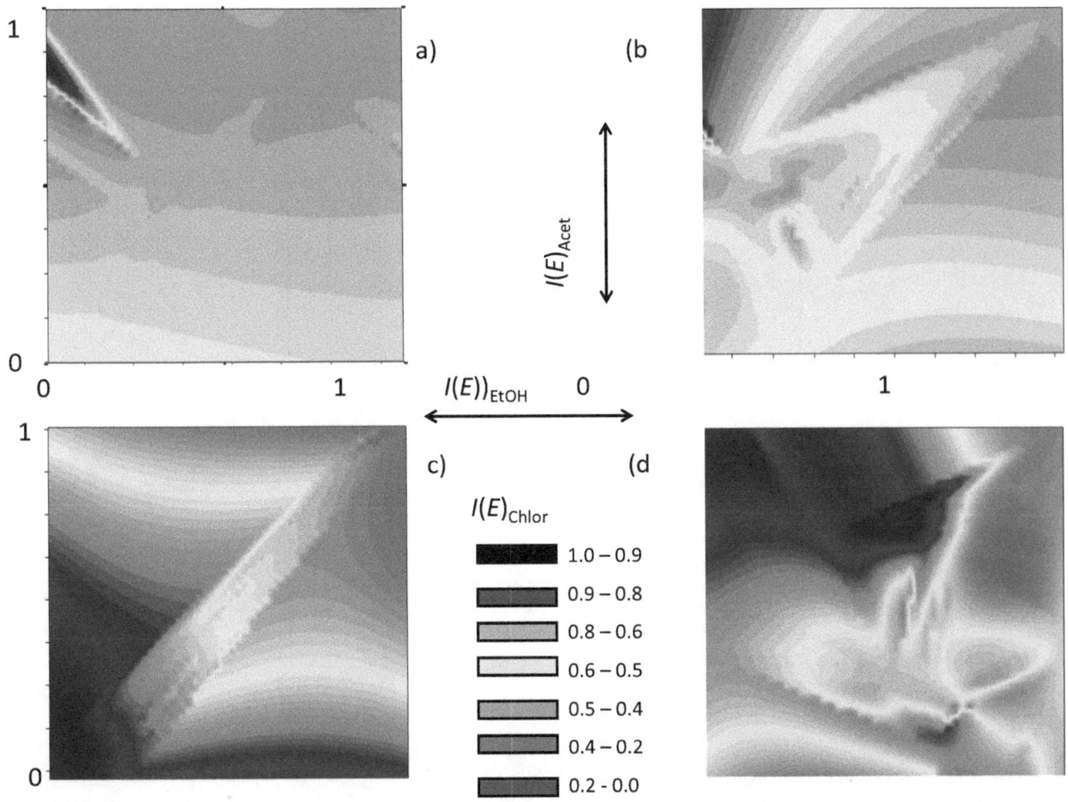

FIGURE 11.6 Color maps taking the values of $I(E)_{Acet}$, $I(E)_{EtOH}$, and $I(E)_{Chlor}$ for leaf extracts of, (a) *Boehmeria cylindrica*, (b) *Quillaja saponaria*, (c) *Ricinus communis*, and (d) *Oxalis articulata* in contact with 0.25 M aqueous acetate buffer at pH 4.75. Data from square wave voltammograms at sample-modified GCEs in conditions such as in Figure 11.5.

11.3 ADAPTIVE MECHANISMS

Electrochemical data can also be used to obtain information on adaptive mechanisms. These reflect the capacity of plants to change in the external environment; for instance, the accommodations of the germination timing to diverse seasonal cues [27] and to adopt different germination strategies [28, 29]. The case of gynodioecy, the co-occurrence of hermaphrodite and female individuals in a given (gynomonoecious) species, has been studied by electrochemical methods in the case of *Cortaderia selloana* (Schult. & Schult. f.) Asch. & Graebn. (Pampas grass) [30]. This is a neotropical species native to South America belonging to the Poaceae family, which is considered invasive in Western Europe.

Sexuality in plants is regulated by genetic, epigenetic, and physiological mechanisms. In principle, unisexuality in flowering plants would be the result of the evolution from a hermaphrodite ancestor, and this feature should have occurred multiple times across the different lineages of the angiosperms [31, 32]. The maintenance of gynodioecy would be the result of the differential reproductive success between sexes. Then, the male fitness of hermaphroditic flowers in mixed populations would be higher than that in purely hermaphroditic ones [33]. According to this hypothesis, the existence of females among hermaphrodites would serve to maintain a high out-crossing rate in the population. Different influencing factors, including floral architectural effects, resource competition, and pollination success, have been studied [34, 35].

The voltammetric response of ethanolic extracts of leaves of *C. selloana* permitted to discriminate between female and hermaphrodite specimens as illustrated in Figure 11.10 [30]. The cyclic

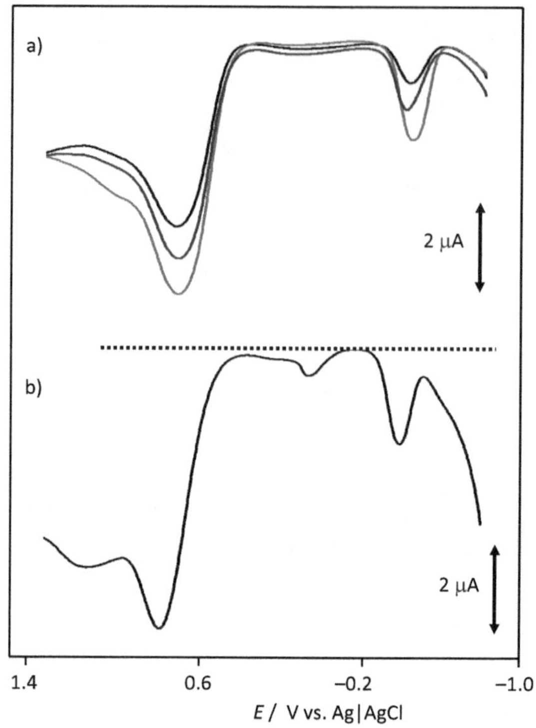

FIGURE 11.7 Square wave voltammograms of deposits of spores on GCE corresponding to (a) *Aspergillus carbonarius* and (b) *Aspergillus ochraceus* in contact with 0.25 M HAc/NaAc aqueous solution at pH 4.75. Potential scan initiated at −0.85 V in the positive direction, potential step increment 4 mV, square wave amplitude 25 mV, frequency 5 Hz. Three replicate experiments are superimposed in (a) while the baseline used for current measurements is illustrated in (b). (Adapted from ref. [24], with permission. Copyright Elsevier 2018. Note that the axes are opposite to the IUPAC convention.).

voltammograms of microparticulate deposits of ethanolic extracts of leaves of *C. selloana* in with contact air-saturated aqueous acetate buffer are dominated by the oxidation of polyphenolic compounds between 0.0 and 1.2 V vs. Ag/AgCl (A_{PL}) and the gross cathodic peak corresponding to the reduction of dissolved oxygen (C_{ox}).

As described in detail in Chapter 10, when the potential is scanned beyond the C_{ox} signal and returned to positive potentials, new voltammetric peaks appear resulting from the interaction of the parent polyphenolic components with ROS generated electrochemically. Varying the potential scan rate and the ROS-generating potential, HO·, HO_2·, and $O_2·^-$ reactive species can be generated in different proportions. Analysis of voltammetric data indicated that female and hermaphrodite specimens have different reactivity with ROS under the studied conditions [30].

One possible interpretation of voltammetric data derives from the abundance of sitosterol and related electroactive compounds [36–38] in *Cortaderia* species. Such species can be oxidized by free radicals at C-7 (see Schematics in Figure 11.11) abstracting the bound hydrogen atom. Lipid peroxy radicals or alkoxyl radicals from polyunsaturated fatty acids can produce this process so that the C-7-centered radical intermediate of sitosterol has a half-life long enough to react with a ground state molecular oxygen molecule, forming a hydroperoxy radical and ultimately epoxy derivatives [39]. Similar processes have been described associated with the radical scavenging activity of quercetin and other flavonoids [40, 41]. Sterols appear to act as constituents of a primitive cellular defense against O_2 and ROS. Accordingly, the formation of these components would be an ancestral

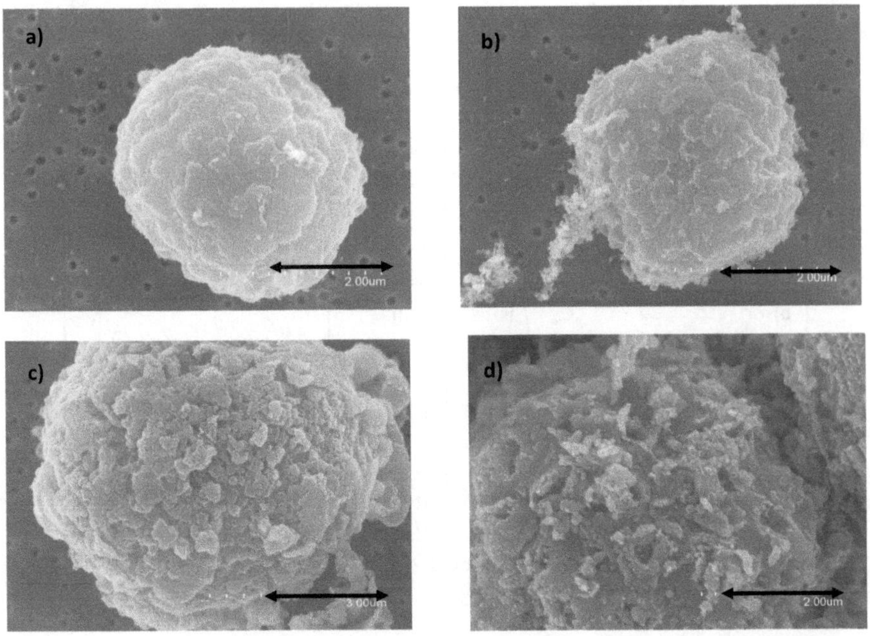

FIGURE 11.8 FESEM images of spores of (a, b) ***Aspergillus flavus*** and (c, d) ***Aspergillus carbonarius*** on a graphite plate before (a, c) and after (b, d) application of an oxidative potential input of 0.80 V vs. Ag/AgCl during 5 min to fungi samples attached to graphite electrodes in contact with 0.25 M HAc/NaAc aqueous buffer at pH 4.75. The scale arrows correspond to 2 μm. (Adapted from ref. [24], with permission. Copyright Elsevier 2018.).

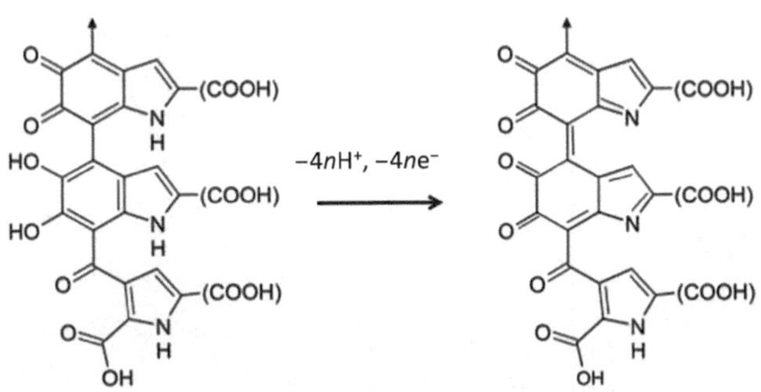

FIGURE 11.9 Scheme for the electrochemical oxidation of melanin involving both phenolic and indolic motifs. (Reproduced from ref. [24], with permission. Copyright Elsevier 2018.).

adaptive response to the rise of O_2 in the terrestrial atmosphere [42]. The voltammetric results permitted the proposal of the hypothesis that, given the different reactivity with ROS of female and hermaphrodite plants, the coexistence of both types of plant specimens would enhance the flexibility of the population in its defense against oxidative stress [30].

11.4 PHYLOGENETIC ASPECTS

Voltammetric data can be correlated with phylogenetic trees. The simplest way consists of the consideration of the presence/absence of voltammetric peaks at selected potentials or intervals of

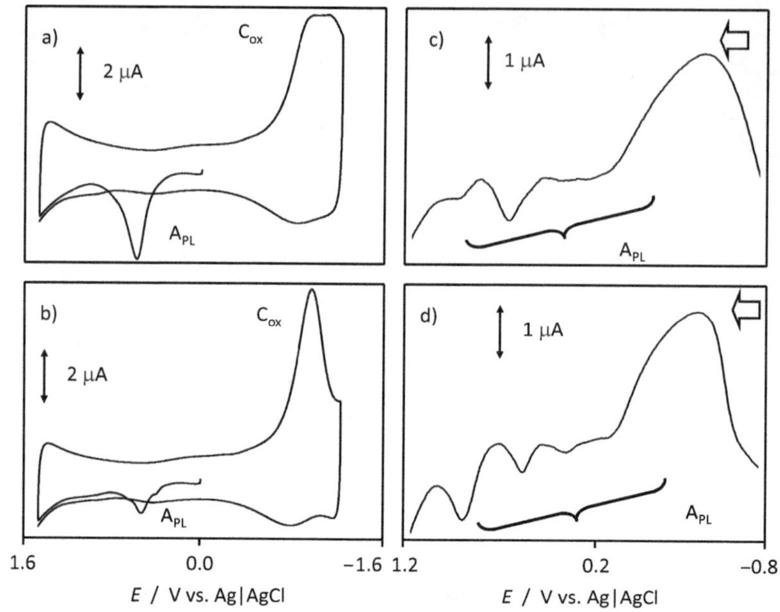

FIGURE 11.10 Cyclic (a, b) and square wave (c, d) voltammograms of microparticulate deposits from ethanolic extracts of leaves of *Cortaderia selloana* on a glassy carbon electrode. (a, c) Female specimen; (b, d) hermaphrodite specimen. Electrolyte: air-saturated 0.25 M aqueous acetate buffer at pH 4.75. (a, b) Potential scan rate 50 mV s^{-1}, (c, d) potential scan initiated at −1.25 V in the positive direction, potential step increment 4 mV; square wave amplitude 25 mV, frequency 10 Hz. Reproduced from ref. [30], with permission. Copyright Elsevier 2018. Note that the axes are opposite to the IUPAC convention.

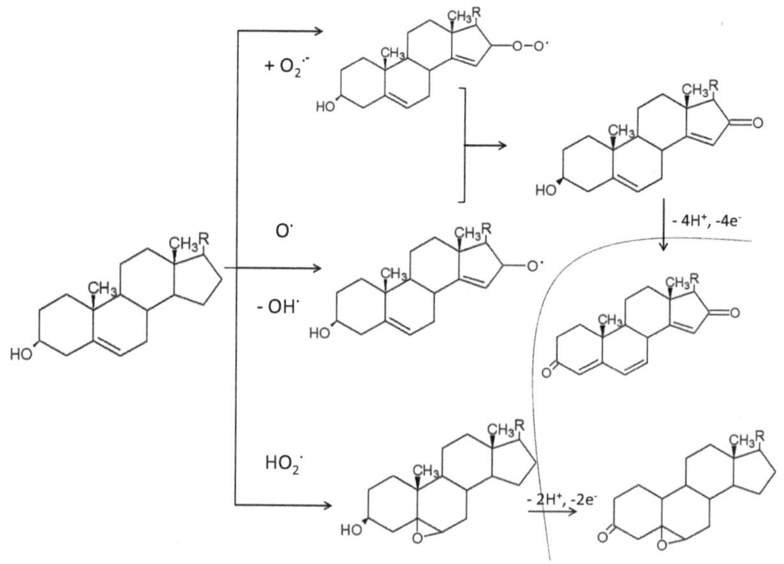

FIGURE 11.11 Possible reactions with ROS and coupled electrochemical processes associated with sitosterol, a component of ***Cortaderia*** leaves, oxidation. The dotted line separates ROS reactions from electrochemical processes. (Reproduced from ref. [30], with permission. Copyright Elsevier 2018.).

potential. Figure 11.12 shows a representative example corresponding to the appearance of voltammetric peaks in the potential range between 0.25 and 0.55 V vs. Ag/AgCl for species of different families of the Rosales order in conditions such as described in Figure 11.1.

These data can be reorganized as a cascade scheme able to be correlated with phylogenetic trees derived from genetic data. Figure 11.13 shows the grouping of data in Figure 11.12 according to a possible evolutionary scheme [19]. The starting point is the absence of voltammetric peak at 0.40 V in all families with the unique exception of the Rosaceae one. This feature suggests that this family was separated early from the others so that the phylogenetic tree of the Rosales order shows an initial partition into two branches. The other than Rosaceae families subsequently diverge into different branches as schematized in Figure 11.13. Remarkably, the resulting "electrochemical tree" is entirely consistent with the phylogenetic tree reported for the Rosales order from conventional genetic analysis [43].

The analysis of voltammetric data can be refined by applying HCA techniques to the matrix of data used to draw the diagrams in Figures 11.5 and 11.6, based on currents measured at intervals of 4 mV. These techniques have been applied to discriminate the genetic lines of the Asparagus genus represented by the *Asparagus*, *Protasparagus*, and *Myrsiphyllum* subgenera [23] and the species of *Aspergillus* genus [24], among others [25, 26, 44]. The electrochemical phylogenetic tree resulting in this last case is depicted in Figure 11.14. This scheme is consistent with the reported phylogenetic tree [45].

11.5 SIGNALING PATHWAYS AND PHYLOGENETICS

A second line of study of phylogenetic aspects results from the consideration of plant signaling mechanisms described in Chapter 10. As described here, the defense response of flowering plants to herbivory and pathogeny involves the mediation of signaling compounds. Among them, jasmonic

	E_p (mV vs. Ag/AgCl) →	550	500	450	400	350	300	250
Moraceae	*Ficus obscura*		●			●	●	●
	Maclura pomifera		●			●	●	
Urticaceae	*Boehmeria cylindrica*	●	●			●	●	
	Parietaria judaica	●	●			●	●	●
	Urtica dioica	●	●				●	
Cannabaceae	*Celtis australis*		●				●	
	Humulus lupulus		●				●	
	Trema orientalis		●				●	
Ulmaceae	*Zelokva carpinifolia*	●		●			●	
Eleagnaceae	*Elaeagnus×submacrophylla*	●				●		●
Rhamnaceae	*Paliurus spina-christi*					●		
	Rhamnus alaternus					●		●
	Zizipus jujuba					●		
Rosaceae	*Potentilla reptans*				●			
	Rosa canina				●			
	Rubus ulmifolium				●			

FIGURE 11.12 Voltammetric peaks appearing at intervals of potentials of 50 mV between 0.25 and 0.55 V vs. Ag/AgCl in square wave voltammograms of microparticulate deposits prepared from ethanolic extracts of mature leaves of species of the Rosales order on glassy carbon electrode. Data from voltammograms in conditions such as in Figure 11.1. (Reproduced from ref. [19], with permission. Copyright the Royal Society of Chemistry 2015.)

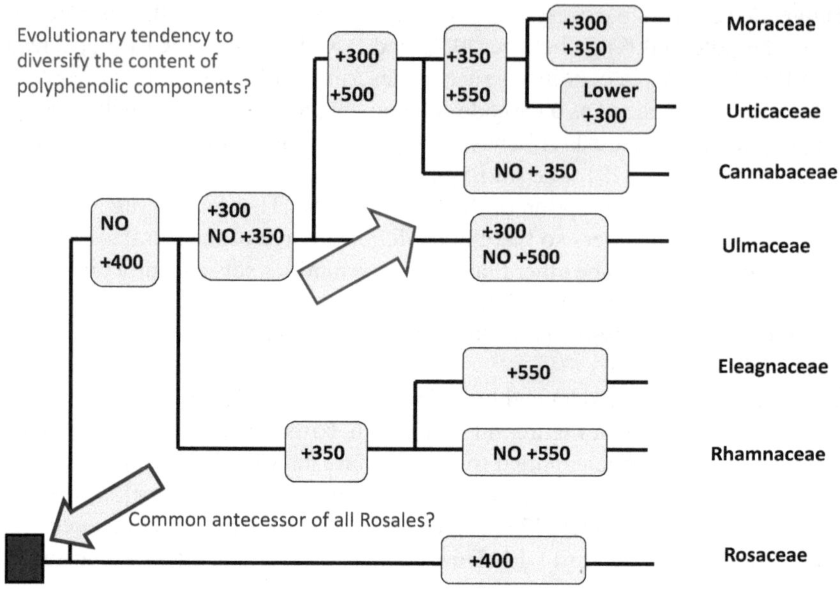

FIGURE 11.13 "Electrochemical tree" of the Rosales order based on the appearance/disappearance of relevant voltammetric signals as indicated in Figure 11.11. The resulting tree is consistent with that proposed in ref. [43]. (Adapted from ref. [19], with permission. Copyright the Royal Society of Chemistry 2015.)

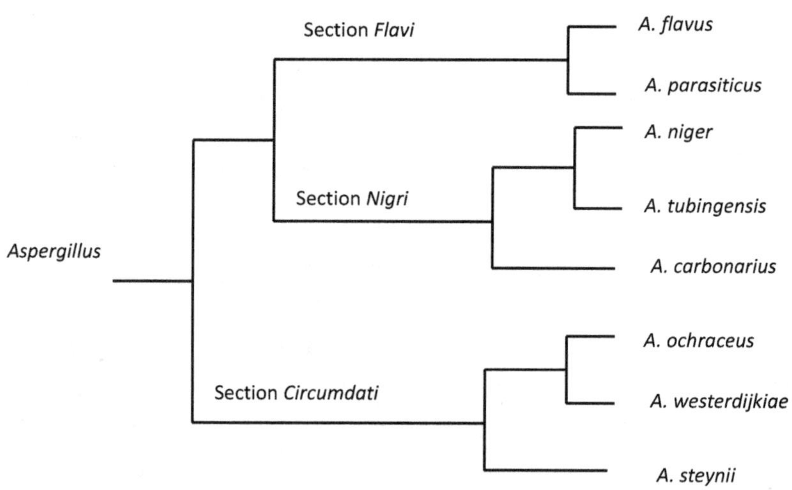

FIGURE 11.14 Dendrogram from HCA of normalized currents measured in the voltammograms of fungi samples in conditions such as described in Figure 11.7. The electrochemical tree is coincident with the phylogenetic tree of ***Aspergillus*** species from DNA sequencing reported in ref. [44]. (Reproduced from ref. [24], with permission. Copyright Elsevier 2018.).

acid (JA) and salicylic acid (SA) are of particular interest [46]. These compounds are electrochemically active and can operate independently [47] and can even produce antagonic signaling pathways [48–50].

These features have an evolutionary dimension [51] that can be summarized in terms of the elucidation of the role of the defense response in plant evolution and the role of the different defense responses in the evolution of the different taxonomic groups. These matters have to be revised in the context of a complex evolutionary scenario: herbivory would have promoted a change of plant metabolism from growth-oriented to defense-oriented [52] so that the phylogeny and plants and insects would be inextricably linked [53].

The change in the defense pathways can have different evolutionary meanings. Among other possibilities, this change may occur in a given species (with or without replacing the pre-existing defense pathway) with and without speciation. A simplified scheme of three of the possible situations is depicted in Figure 11.15. Here, an idealized phylogenetic tree in which one parent species gives rise to two new ones at different geological times is depicted. It is assumed that the parent species possesses a given ancestral signaling/defense pathway symbolized by black lines. One possible situation is, as schematized in Figure 11.15a, that the parent species experiences a change to a second signaling/defense path without the formation of a new species. This second path, symbolized by dotted lines, replaces (or coexists with) the primitive signaling/defense path and is transmitted to a new species formed from the primitive one. A second case is represented in Figure 11.15b. Here, the primitive species retain the ancestral defense pathway while a second is formed incorporating the new pathway as a result of a mutation (or series of mutations). In this case, speciation, the change in the signaling/defense pathway is associated with sufficiently drastic genetic changes to result in the formation of a new species. Figure 11.15c schematizes the case in which convergence occurs. Here, two species experience the change in the defense pathway without speciation at different times during their evolution.

The aforementioned scheme can be applied not only at the taxonomic level of species, but also at, in principle, other taxonomic levels. For our purposes, the relevant point to underline is that electrochemical data, as yielding information on defense pathways can in turn possess phylogenetic information. This was accessed via *in planta* electrochemistry of succulent plants of three monocots (*Aloe vera* (L.) Burm.f., *Asphodelus fistulosus* L., and *Bulbine frutescens* (L.) Willd., order Asparagales) and six core eudicots species, these last divided into two orders: Saxifragales (*Crassula ovata* Druce, *Sedum sediforme* (Jacq.) Pau, and *Graptopetalus paraguayense* (N.E.Br.) E. Walther) and Caryophyllales (*Aptenia cordifolia* (L.f.) Schwantes, *Delosperma cooperi* (Hook.f.) L.Bolus, and *Faucaria tigrina* (Haw.) Schwantes) [54, 55].

The essential idea is that the insertion of electrodes into the subcuticular region of plant leaves (see inset in Figure 11.16) permits monitoring plant defense against mechanical aggression. The production of H_2O_2 was monitored using Pt microelectrodes and the evolution of JA and SA using

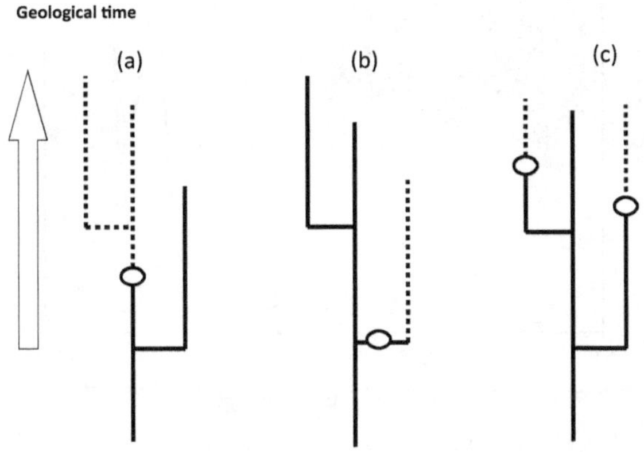

FIGURE 11.15 Influence of a change in the defense pathway during the evolution of two taxonomic units. The 'old' defense pathway is represented by continuous lines and the 'new' defense pathway by dotted lines. Simplified scheme illustrating three possible cases: (a) change in the defense pathway without speciation; (b) the change in the defense pathway is associated with the formation of a new species; (c) the change in the defense pathway occurs independently in two separate species (convergence).

graphite ones. Figure 11.16 shows the voltammograms recorded at Pt electrode on a leaf of *Faucaria tigrina* at different times. The main signal (C_{ox}) corresponds to the reduction of H_2O_2 (R_{ox}) at ca. −0.4 V and has been largely used for studying plant physiology [56, 57].

Figure 11.17 shows the voltammograms recorded at graphite microelectrodes inserted on leaves of *Crassula ovata* and *Delosperma cooperi* at intervals of 5 min. Here, two potential ranges were used to test the plant defense response limited to the mechanical stress and accompanying it by the electrochemical generation of ROS. In the first case, the potential scan was initiated at 0.0 V vs. Ag/AgCl in the positive direction. In the second, the potential scan was initiated at a value negative enough (−0.80 V) to produce the reduction of dissolved oxygen, thus generating ROS.

The voltammograms exhibit a multiple peak profile in which the signals at ca. 0.80 V (A_1) and 1.45 V (A_2) can be mainly attributed to the respective oxidations of SA and JA. Then, the time variation in the local concentration of these species in the vicinity of the working electrode can be represented by the variation of intensity of the corresponding voltammetric peak. Figure 11.18 depicts typical current vs. time curves for the processes C_{ox} and A_1. To compensate for the effect of the variable depth reached by the electrodes in replicate voltammetric measurements, the peak currents ($i(C_{ox})$, $i(A_1)$) were normalized relative to the maximum peak current ($i(C_{ox})_{max}$) and the initial peak current ($i(A_1)_{init}$) measured in each experiment, respectively.

The peculiar pattern obtained for the $i(C_{ox})/i(C_{ox})_{max}$ vs. time and $i(A_1)/i(A_1)_{inic}$ vs. time representations can be interpreted, assuming a homogeneous-like behavior, adopting a kinetic model equivalent to that described in section 10.5. Accordingly, the signaling compound (A) is formed in a competing way from two different precursors (P, Q) further experiencing degradation to, in principle, common product(s) (B): Assuming that all reactions proceed under pseudo-first-order conditions the time variation of the concentrations of P, Q, A and B (c_P, c_Q, c_A, c_B, respectively) can be represented by Eq. (10.30) which can be also written as

$$\frac{c_A}{c^o{}_A} = 1 + \frac{k_P\left(c_P^o / c^o{}_A\right)}{k_D - k_P}\left[e^{-k_Pt} - e^{-k_Dt}\right] + \frac{k_Q\left(c_Q^o / c^o{}_A\right)}{k_D - k_Q}\left[e^{-k_Qt} - e^{-k_Dt}\right] \quad (11.1)$$

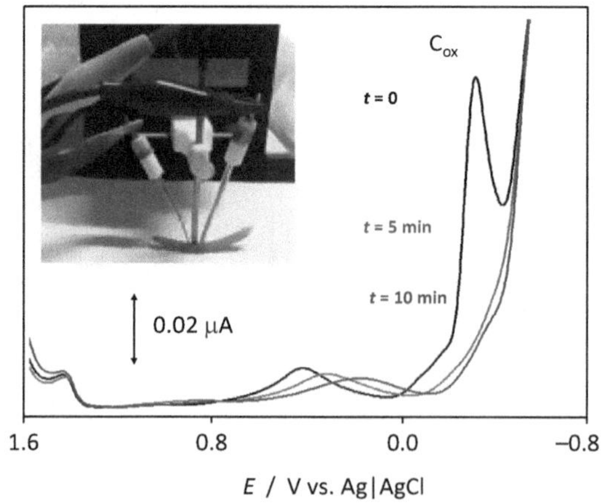

FIGURE 11.16 Square wave voltammograms at platinum microelectrode inserted into a leaf of ***Faucaria tigrina*** at different times after electrode insertion. Potential scan initiated at 1.6 V vs. Ag/AgCl in the negative direction, potential step increment 4 mV, square wave amplitude 25 mV, frequency 10 Hz. Inset: Photographic image of the experimental arrangement used for *Delosperma cooperi*. (Adapted from ref. [55], with permission. Copyright Wiley 2023. Note that the axes are opposite to the IUPAC convention.).

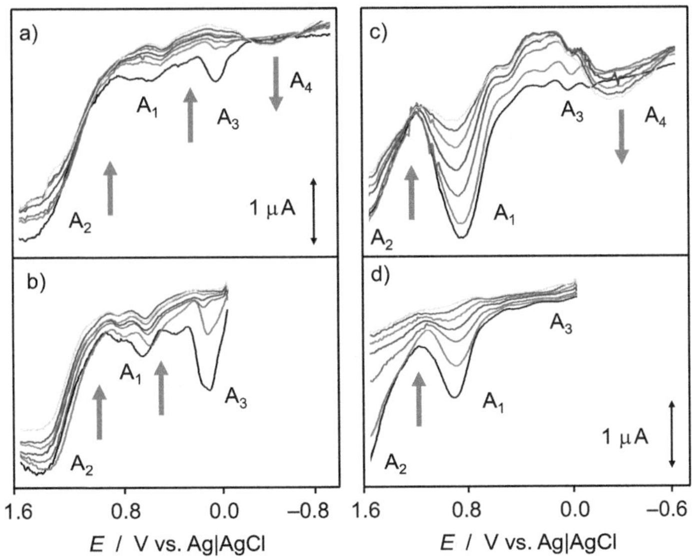

FIGURE 11.17 Square wave voltammograms at graphite microelectrode inserted into leaves of, (a, b) ***Crassula ovata*** and (c, d) *Delosperma cooperi*. The voltammograms were recorded at intervals of 5 min after the insertion of the microelectrode setup. Potential scan initiated at, (a, c) –0.80 and (b, d) 0.00 V vs. Ag/AgCl in the positive direction; potential step increment 4 mV, square wave amplitude 25 mV; frequency 10 Hz. The arrows mark the variation in the intensity of selected peaks in successive voltammograms. (Adapted from ref. [55], with permission. Copyright Wiley 2023. Note that the axes are opposite to the IUPAC convention.).

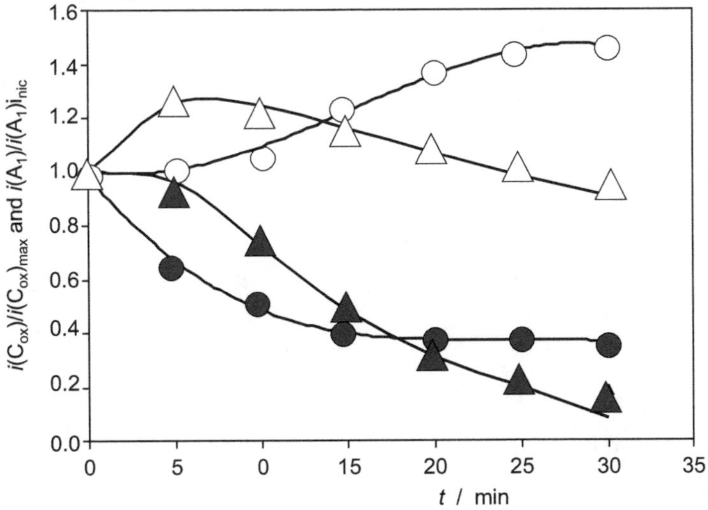

FIGURE 11.18 Time variation of the $\boldsymbol{i}(C_{ox})/\boldsymbol{i}(C_{ox})_{max}$ (circles) and $\boldsymbol{i}(A_1)/i(A_1)_{inic}$ (triangles) ratios for ***Aloe vera*** (circles and triangles) and *Aptenia cordifolia* (solid circles and solid triangles). Experimental data correspond to averaged values from three independent replicate voltammograms recorded in conditions such as in Figures 11.16 and 11.17 (i.e., without electrochemical generation of ROS). The continuous lines correspond to the theoretical predictions from Eq. (11.8) using appropriate values of the rate constants (see text). Error bars are omitted for clarity. (Adapted from ref. [55], with permission. Copyright Wiley 2023.).

with k_P, k_Q, k_D being the rate constants of the respective reactions. Several particular cases are of interest. When only one precursor (P) exists, Eq. (11.1) reduces to

$$\frac{c_A}{c^o{}_A} = 1 + \frac{k_P\left(c_P^o / c^o{}_A\right)}{k_D - k_P}\left[e^{-k_P t} - e^{-k_D t}\right], \tag{11.2}$$

and the concentrations of P and B can be approached by the expressions

$$\frac{c_P}{c^o{}_P} = e^{-k_P t} \tag{11.3}$$

and

$$\frac{c_B}{c^o{}_B} = 1 + \frac{k_P\left(c_P^o / c^o{}_B\right)}{k_D - k_P}\left[\frac{1}{k_P}\left(1 - e^{-k_D t}\right) - \frac{1}{k_D}\left(1 - e^{-k_D t}\right)\right], \tag{11.4}$$

respectively. When $k_D >> k_Q$. Then Eq. (10.30) can be approximated by

$$\frac{c_A}{c^o{}_A} \approx 1 + \frac{k_P\left(c_P^o / c^o{}_A\right)}{k_D - k_P}\left[e^{-k_P t} - e^{-k_D t}\right] + \frac{k_Q\left(c_Q^o / c^o{}_Q\right)}{k_D} e^{-k_Q t}. \tag{11.5}$$

The kinetic study can be made from voltammetric data assuming that the peak currents are proportional to the concentration of the respective electroactive species. This implies (see Chapter 10) the introduction of electrochemical constants whose effect can be avoided taking currents normalized to the current at time zero [54, 55]. In the case in which all involved compounds are electroactive and display separate voltammetric peaks, the respective peak ratios will be (for the simplest case of a unique precursor compound and a unique degradation compound)

$$\frac{i(A)}{i^\circ(A)} = 1 + \frac{k_P\left(c_P^o / c^o{}_A\right)}{k_D - k_P}\left[e^{-k_P t} - e^{-k_D t}\right], \tag{11.6}$$

$$\frac{i(P)}{i^\circ(P)} = e^{-k_P t}, \tag{11.7}$$

$$\frac{i(B)}{i^\circ(B)} = 1 + \frac{k_P\left(c_P^o / c^o{}_B\right)}{k_D - k_P}\left[\frac{1}{k_P}\left(1 - e^{-k_D t}\right) - \frac{1}{k_D}\left(1 - e^{-k_D t}\right)\right]. \tag{11.8}$$

This treatment allows estimating the values of the rate constants k_P, k_Q, k_D for the generation of H_2O_2, as well as for the production of SA and JA, in this case under with and without electrochemical ROS generation [55]. The competition between two signaling (P- and Q- pathways in the following) introduces a certain complexity in the analysis of data. Table 11.1 summarizes the k_Q, k_P, k_D values estimated for JA evolution.

Despite this complexity, definite correlations can be established from the values of the rate constants, as illustrated in Figure 11.19, where the values of k_Q determined for SA (k_Q(SA)) under conditions of no ROS generation are plotted vs. the values of k_Q determined for H_2O_2 ($k_Q(H_2O_2)$) for the studied plants. One can see in this figure that the species of the Caryophyllales family define a straight line separated from the ill-defined tendency lines corresponding to the species of the Saxyfragales and Asparagales orders. This suggests that the H_2O_2 production and the SA-mediated pathway are coupled. In contrast, there is no definite correlation between the values of k_Q(JA) vs. $k_Q(H_2O_2)$. On comparing the values of k_Q calculated from voltammograms with and without ROS generation, the most relevant result was the absence of correlations for JA while the unique reasonable correlation was obtained for SA in the case of Asparagales.

This approach can be considered regarding the phylogenetic evolution of plants and insects [58–60] schematized in Figure 11.20. Hypothetically, the signaling pathways could have appeared successively accompanying the parallel plant-insect evolution following any of the schemes described in Figure 11.15. Considering the antagonism between the JA and the SA pathways [48–50], and others such as the abscisic acid pathway [61], several hypotheses are possible. For instance, the JA signaling pathway could have been an ancestral defense mechanism that appeared in an early stage of plant evolution. But this pathway could be a remnant (a "living fossil") of early plants or could have replaced any previous, ancestral defense mechanism.

Under this view, voltammetric data recorded for species of the Caryophyllales, Saxyfragales, and Asparagales orders can be interpreted in terms of two disruptions in the plant defense response accompanying their phylogenetic evolution. The separation between monocots (Asparagales) and eudicots (Caryophyllales and Saxyfragales) would involve the abrupt variation of the SA signaling Q-type path sensitivity to ROS. The subsequent separation of the Caryophyllales from other eudicots would be associated with the strong coupling between the H_2O_2 generation and the Q-type SA signaling.

One of the possible evolutionary schemes relating to these taxa is depicted in Figure 11.21. Notice that this scheme adopts a speciation model equivalent to that seen in Figure 11.14b, assuming that the changes in the signaling pathway modes are associated with separation between taxa. This does not necessarily occur, and other evolutionary schemes such as those in Figures 11.15b and 11.15c can also be hypothesized. Although much more research is needed to establish definite conclusions regarding all these matters, currently available data illustrate the capabilities of electrochemical methods to provide information potentially useful in the phylogenetic domain using simple experimental arrangements and conventional electrochemical techniques.

TABLE 11.1
Kinetic Parameters Determined from JA Oxidation Peak (A_2) Recorded Under Conditions Electrochemical ROS Generation. From Square Wave Voltammograms at Graphite Microelectrodes in Conditions Such as Described in Figure 11.16

Plant	k_Q (min^{-1})	k_P (min^{-1})	k_D (min^{-1})
Aloe vera	0.002 ± 0.001	0.0015 ± 0.0005	0.10 ± 0.03
Asphodelus fistulosus	0.005 ± 0.002	0.0020 ± 0.0005	0.10 ± 0.03
Bulbine frutescens	0.004 ± 0.002	0.0025 ± 0.0005	0.10 ± 0.03
Crassula ovata	0.003 ± 0.001	0.060 ± 0.0005	0.30 ± 0.03
Graptopetalum paraguayense	0.004 ± 0.001	0.050 ± 0.0005	0.25 ± 0.03
Sedum sediforme	0.003 ± 0.001	0.055 ± 0.0005	0.30 ± 0.03
Aptenia cordifolia	0.026 ± 0.004	0.020 ± 0.004	0.020 ± 0.006
Delosperma cooperi	0.025 ± 0.002	0.020 ± 0.003	0.020 ± 0.004
Faucaria tigrina	0.024 ± 0.002	0.020 ± 0.003	0.020 ± 0.004

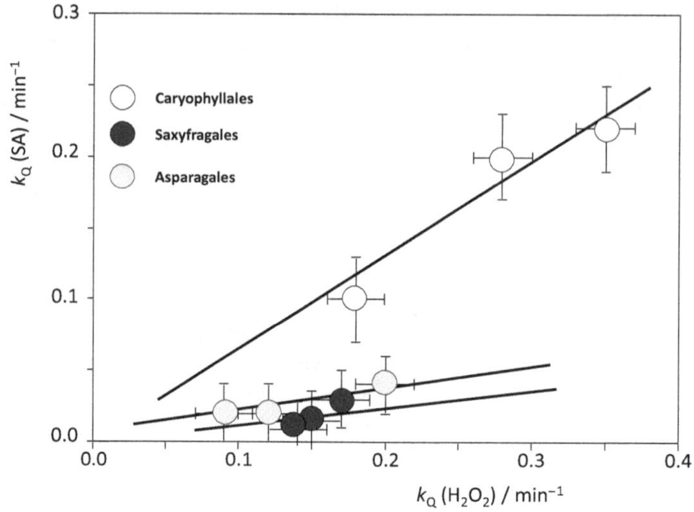

FIGURE 11.19 Plots of: a) k_Q(SA) vs. k_Q(H_2O_2) and b) k_Q(JA) vs. k_Q(H_2O_2) for species of the Caryophyllales, Saxyfragales and Asparagales orders. Rate constant values determined from voltammetric data such as in Figure 11.17. (Adapted from ref. [55], with permission. Copyright Wiley 2023.).

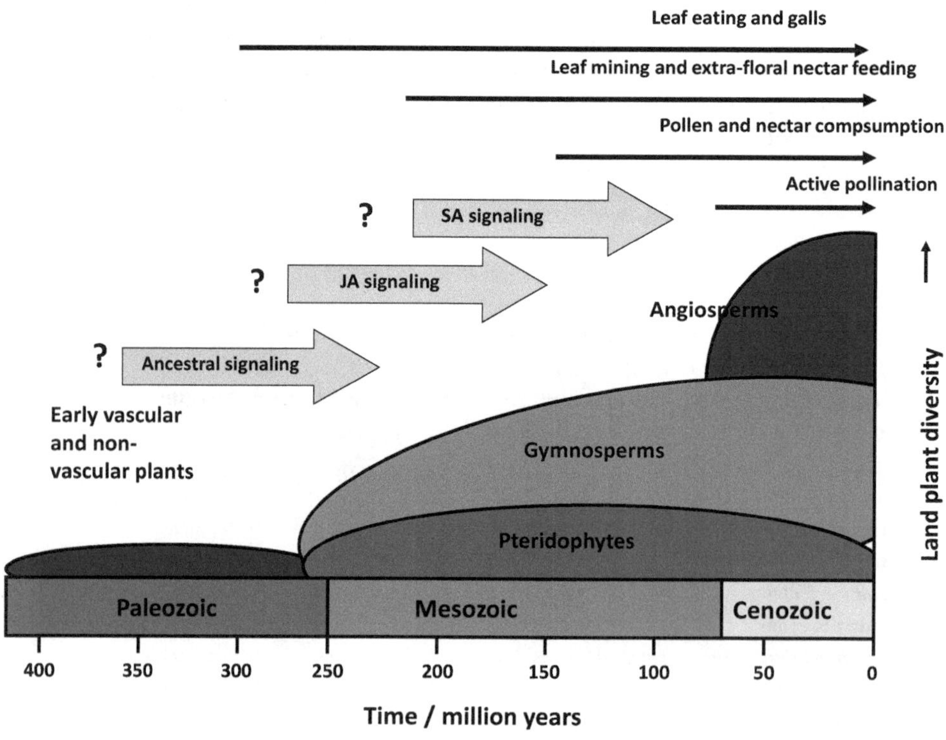

FIGURE 11.20 Evolutionary development of plant classes and their relationships with insects adapted from refs. [58–60]. (Adapted from ref. [54], with permission. Copyright Wiley 2021.).

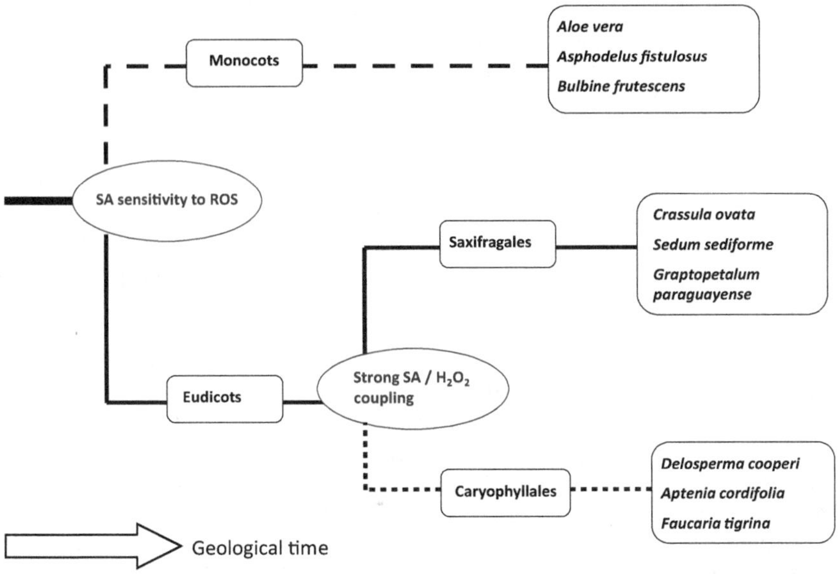

FIGURE 11.21 One of the possible hypothetical schemes for the evolutionary development of the plant taxa listed in Table 11.1. Events determining taxa separation correspond to the Q-type signaling pathways. (Adapted from ref. [55], with permission. Copyright Wiley 2023.).

REFERENCES

[1] APG III (The angiosperm phylogeny group), 2009. An update of the Angiosperm Phylogeny Group classification for the orders and families of flowering plants: APG III. *Botanic Journal of the Linnean Society*. 161: 105–121.

[2] Ravi, V.; Khurana, J.P.; Tyagi, A.K.; Khurana, P. 2007. Rosales sister to Fabales: Towards resolving the rosid puzzle. *Molecular Phylogenetic and Evolution*. 44: 488–493.

[3] Potter, D.; Eriksson, T.; Evans, R.C.; Oh, S.; Smedmark, J.E.E.; Morgan, D.R.; Kerr, M.; Robertson, K.R.; Arsenault, M.; Dickinson, T.A.; Campbell, C.S. 2007. Phylogeny and classification of Rosaceae. *Plant Systematics and Evolution*. 266: 5–43.

[4] Oh, S.-H.; Potter, D. 2006. Description and phylogenetic position of a new angiosperm family, Guamatelaceae, inferred from chloroplast *rbcL*, *atpB*, and *matK* sequences. *Systematic Botany*. 31: 730–738.

[5] Saarela, J.M.; Rai, H.S.; Doyle, J.A.; Endress, P.K.; Mathews, S.; Marchant, A.D.; Briggs, B.; Graham, S.W. 2007. Hydatellaceae identified as a new branch near the base of the angiosperm phylogenetic tree. *Nature*. 446: 312–315.

[6] Todd, A. 1961. Natural product chemistry -retrospect and prospect. *Pure and Applied Chemistry*. 2: 359–366.

[7] Tundis, R.; Peruzzi, L.; Menichini, F. 2014. Phytochemical and biological studies of Stachys species in relation to chemotaxonomy: A review. *Phytochemistry*. 102: 7–39.

[8] Vanhoenacker, G.; Van Rompaey, P.; De Keukeleire, D.; Sandra, P. 2002. Chemotaxonomic features associated with flavonoids of cannabinoid-free cannabis (Cannabis sativa subsp. sativa L.) in relation to hops (Humulus lupulus L.). *Natural Product Letters*. 16: 57–63.

[9] Wollenweber, E.; Stevens, J.F.; Dörr, M.; Rozefelds, A.C. 2003. Taxonomic significance of flavonoid variation in temperate species of Nothofagus. *Phytochemistry*. 65: 1125–1131.

[10] Vogl, C.R.; Hartl, A. 2003. Production and processing of organically grown fiber nettle (*Urtica dioica*) and its potential use in the natural textile industry: a review. *American Journal of Alternative Agriculture*. 18: 119–128.

[11] Buchhaus, C.; Herz, H.; Jetter, R. 2007. Chemical composition of the epicuticular and intracuticular wax layers on adaxial sides of *Rosa canina* leaves. *Annals of Botany*. 100: 1557–1564.

[12] Jetter, R.; Kunst, L.; Samuels, A.L. 2006. Composition of plant cuticular waxes, in: Riederer, M.; Müller, C. Eds., *Biology of the Plant Cuticle*. Blackwell Pub., Oxford, pp 182–215.

[13] Singh, R. 2016. Chemotaxonomy: A Tool for Plant Classification. *Journal of Medicinal Plant Studies*. 4: 90–93.

[14] Szabó, L.F. 2008. Rigorous biogenetic network for a group of indole alkaloids derived from strictosidine. *Molecules*. 13: 1875–1896.

[15] Lahlou, A.; Lyashenko, S.; Chileh-Chelh, T.; Belarbi, E.-H.; Torres-García, I.; Álvarez-Corral, M.; Rodríguez-García, I.; Rincón-Cervera, M. A.; Guil-Guerrero, J.L. 2023. Fatty acid profiling in the genus *Pinus* in relation to its chemotaxonomy and nutritional or pharmaceutical properties. *Phytochemistry*. 206: artic. 113517.

[16] Brockington, S.F.; Walker, R.H.; Glower, B.J.; Soltis, P.S.; Soltis, D.E. 2011. Complex pigment evolution in the Caryophyllales. *New Phytologist*. 190: 854–864.

[17] Kim, S.J.; Kawaguchi, S.; Watanabe, Y. 2003. Glucosinolates in vegetative tissues and seeds of twelve cultivars of vegetable turnip rape (*Brassica rapa* L.). *Soil Science and Plant Nutrition*. 49: 337–346.

[18] Zidorn, C. 2019. Plant chemophenetics - A new term for plant chemosystematics/plant chemotaxonomy in the macro-molecular era. *Phytochemistry*. 163: 147–148.

[19] Doménech-Carbó, A.; Ibars, A.M.; Prieto-Mossi, J.; Estrelles, E.; Scholz, F.; Cebrián-Torrejón, G.; Martini, M. 2015. Electrochemistry-based chemotaxonomy in plants using the voltammetry of microparticles methodology. *New Journal of Chemistry*. 39: 7421–7428.

[20] Karnicka, K.; Eckhard, K.; Guschin, D.A.; Stoica, L.; Kulesza, P.J.; Schumann, W. 2007. Visualisation of the local bio-electrocatalytic activity in biofuel cell cathodes by means of redox competition scanning electrochemical microscopy (RC-SECM), *Electrochemistry Communications*. 9: 1998–2002.

[21] Scholz, F.; Schröder, U.; Gulabowski, R.; Doménech-Carbó, A. 2014. *Electrochemistry of Immobilized Particles and Droplets*, 2nd edit. Springer, Berlin-Heidelberg.

[22] Doménech-Carbó, A. 2010. *Electrochemistry of Porous Materials*. Taylor & Francis, Boca Raton.

[23] Doménech-Carbó A.; Ibars, A.M.; Prieto-Mossi, J.; Estrelles, E.; Doménech-Carbó, M.T.; Ortiz-Miranda, A.S.; Lee, Y. 2017. Access to phylogeny from voltammetric fingerprints of seeds: the *Asparagus* case. *Electroanalysis*. 29: 643–650.

[24] Mateo, E.M.; Gómez, J.V.; Montoya, N.; Mateo-Castro, R.; Gimeno-Adelantado, J.V.; Jiménez, M.; Doménech-Carbó, A. 2018. Electrochemical identification of toxigenic fungal species using solid-state voltammetry strategies. *Food Chemistry*. 267: 91–100.

[25] Fu, L.; Zheng, Y.; Zhang, P.; Zhang, H.; Wu, M.; Zhang, H.; Wang, A.; Su, W.; Chen, F.; Wu, J.; Cai, W.; Lin, C.-T. 2019. An electrochemical method for plant species determination and classification based on fingerprinting petal tissue. *Bioelectrochemistry*. 129: 199–205.

[26] Xu, Y.; Lu, Y.; Zhang, P.; Wang, Y.; Zheng, Y.; Fu, L.; Zhang, H.; Lin, C.-T.; Yu, A. 2020. Infrageneric phylogenetics investigation of Chimonanthus based on electroactive compound profiles. *Bioelectrochemistry*. 133: artic. 107455.

[27] Donohue, K.; Dorn, L.; Griffith, C.; Kim, E.; Aguilera, A.; Polisetty, C.R.; Schmitt, J. 2005. Environmental and genetic influences on the germination of *Arabidopsis thaliana* in the field. *Evolution*. 59: 740–757.

[28] Estrelles, E.; Biondi, E.; Galiè, M.; Mainardi, F.; Hurtado, A.; Soriano, P. 2015. Aridity level, rainfall pattern and soil features as key factors in germination strategies in salt-affected plant communities. *Journal of Arid Environments*. 117: 1–9.

[29] Estrelles, E.; Prieto-Mossi, J.; Escribá, M.C.; Ferrando, I.; Ferrer-Gallego, P.; Laguna, E.; Ibars, A.M.; Soriano, P. 2018. Hydroperiod length as key parameter controlling seed strategies in Mediterranean salt marshes: The case of *Halopeplis amplexicaulis*. *Flora*. 249: 124–132.

[30] Doménech-Carbó, A.; Montoya, N.; Soriano, P.; Estrelles, A. 2018. An electrochemical analysis suggests role of gynodioecy in adaptation to stress in *Cortaderia selloana*. *Current Plant Biology*. 16: 9–14.

[31] Torices, R.; Anderberg, A.A. 2009. Phylogenetic analysis of sexual systems in Inuleae (Asteraceae). *American Journal of Botany*. 96: 1011–1019.

[32] Aryal, R.; Ming, R. 2014. Sex determination in flowering plants: Papaya as a model system. *Plant Science*. 217–218: 56–62.

[33] Delph, L.F.; Mutikainen, P. 2013. Testing why the sex of the material parent affects seedling survival in a gynodioecious species. *Evolution*. 57: 231–239.

[34] Bai, Y.P.; Zhang, Y.W.; Gituru, R.W.; Zhao, J.M.; Li, J.D. 2011. Sexual differences in reproductive characters and pollinator attractiveness in gynodioecious *Glechoma longituba* (Lamiaceae). *Plant Species Biology*. 26: 33–42.

[35] Miyake, T.; Satake, I.; Miyake, K. 2017. Sex-biased seed predation in gynodioecious *Dianthus superbus* var. *longicalycinus* (Capryophyllaceae) and differential influence of two seed predator species on the floral traits. *Plant Species Biology*. 33: 42–50.

[36] Cesarino, I.; Hümmelgen, I.A. 2015. An additional tool towards overcoming absence of specificity of carbon nanostructure-based electrochemical sensors-application to estriol and estradiol detection and distinction. *Journal of Solid State Electrochemistry*. 19: 3035–3050.

[37] Perez, C.; Fábio, R.S.; Codognoto, L. 2015. Voltammetric determination of 17α ethinylestradiol hormone in supply dam using BDD electrode. *Journal of Solid State Electrochemistry*. 20: 2471–2478.

[38] Brocenschi, R.F.; Rocha-Filho, R.C.; Li, L.; Swain, G.N. 2014. Comparative electrochemical response of estrone at glassy-carbon, nitrogen-containing tetrahedral amorphous carbon and boron-doped diamond thin-film electrodes. *Journal of Electroanalytical Chemistry*. 712: 207–214.

[39] Murphy, R.C.; Johnson, K.M. 2008. Cholesterol, Reactive Oxygen Species, and the Formation of Biologically Active Mediators. *Journal of Biological Chemistry*. 283: 15521–15525.

[40] Krishnamachari, V.; Levine, L.H.; Paré, P.W. 2002. Flavonoid oxidation by the radical generador AIBN: a unified mechanism for quercetin radica scavenging. *Journal of Agricultural and Food Chemistry*. 50: 4357–4363.

[41] Doménech-Carbó, A.; Gavara, R.; Hernández, P.; Domínguez, I. 2015. Contact probe voltammetry for *in situ* monitoring of the reactivity of phenolic tomato (*Solanum lycopersicum* L.) compounds with ROS. *Talanta*. 144: 1207–1215.

[42] Galea, A.M.; Brown, A.J. 2009. Special relationship between sterols and oxygen: Were steraols an adaptation to aerobic life?. *Free Radical Biology and Medicine*. 47: 880–889.

[43] Zhang, S.D.; Soltis, D.E.; Yang, Y.; Li, D.Z.; Yi, T.S. 2011. Multi-gene analysis provides a well-supported phylogeny of Rosales. *Molecular Phylogenetics and Evolution*. 60: 21–28.

[44] Ortiz-Miranda, A.S.; König, P.; Kahlert, H.; Scholz, F.; Osete-Cortina, L.; Doménech-Carbó, M.T.; Doménech-Carbó, A. 2016. Voltammetric analysis of Pinus needles with physiological, phylogenetic, and forensic applications. *Analytical and Bioanalytical Chemistry*. 408: 4943–4952.

[45] Peterson, S.W. 2008. Phylogenetic analysis of *Aspergillus* species using DNA sequences from four *loci*. *Mycologia*. 100: 205–226.

[46] Agrawal, A.A. 2011. Current trends in the evolutionary ecology of plant defence, *Functional Ecology*. 25: 420–432.

[47] Ballaré, C.L. 2011. Jasmonate-induced defenses: a tale of intelligence, collaborators and rascals. *Trends in Plant Science*. 16: 249–257.

[48] Koornneef, A.; Leon-Reyes, A.; Ritsema, T.; Verhage, A.; Den Otter, F.C.; Van Loon, L.C.; Pietersen, C.M.J. 2008. Kinetics of salicylate-mediated suppression of jasmonate signaling reveal a role for redox modulation. *Plant Physiology*. 147: 1358–1368.

[49] Kawazu, K.; Mochizuki, A.; Sato, Y.; Sugeno, W.; Murata, M.; Seo, S.; Mitsuhara, I. 2012. Different expression profiles of jasmonic acid and salicylic acid inducible genes in the tomato plant against herbivores with various feeding modes. *Arthropod-Plant Interactions*. 6: 221–230.

[50] Anderson, J.P.; Badruzsaufari, E.; Schenk, P.M.; Manners, J.M.; Desmond, O.J.; Ehlert, C.; Maclean, D.J.; Ebert, P.R.; Kazan, K. 2004. Antagonistic interaction between abscisic acid jasmonate-ethylene signalling pathways modulates defense gene expression disease resistance in *Arabidopsis*. *Plant Cell*. 16: 3460–3479.

[51] Chen, D.; Wilkinson, C.R.M.; Watt, S.; Penkett, C.J.; Toone, W.M.; Jones, N.; Bahler, J. 2008. Multiple pathways differentially regulate global oxidative stress responses in fission yeast. *Molecular Biology of the Cell*. 19: 308–317.

[52] Bolton, M.D. 2009. Primary metabolism and plant defense—Fuel for the FIRE. *Molecular Plant Microbe Interactions*. 22: 487–497.

[53] Labandeira, C.C. 1997. Insect Mouthparts: Ascertaining the Paleobiology of Insect Feedinbg Strategies. *Annual Review of Ecology, Evolution and Systematics*. 28: 153–193.

[54] Doménech-Carbó, A.; Dias, D.; Donnici, M. 2021. In vivo electrochemical monitoring of signaling transduction of plant defense against stress in leaves of *Aloe vera* L. *Electroanalysis*. 33: 1024–1032.

[55] Doménech-Carbó, A.; Dias, D.; Prieto-Mossi, J.; Montoya, N. 2023. *In situ* electrochemical monitoring of signaling transduction in plants. Implications on phylogenetic aspects of defense response. *Electroanalysis*. 35: artic. 202300087.

[56] Ren, Q.-Q.; Huang, X.-R.; Liu, G.-C.; Jun, O.-Y.; Li, M.-T.; Chen, H.; Zhao, Y.-D.; Chen, W. 2015. A field-compatible technique using an electrochemical sensing microbundle for real-time and simultaneous in vivo measurement of hydrogen peroxide, nitric oxide, and pH under drought stress. *Sensors and Actuators B*. 220: 743–748.

[57] Ai, F.; Chen, H.; Zhang, S.-H.; Liu, S.-Y.; Wei, F.; Dong, X.Y.; Cheng, J.K.; Huang, W. H. 2009. Real-time monitoring of oxidative burst from single plant protoplasts using microelectrochemical sensors modified by platinum nanoparticles, *Analytical Chemistry*. 81: 8453–8458.

[58] Niklas, K. J., Ti Ffney, B. H., Knoll, A. H., in *Phanerozoic Diversity Patterns: Profiles in Macroevolution*; J. W. Valentine, Ed.; Princeton University Press: Princeton, 1985; pp. 97–128.

[59] Schönenberger, J.; Friis, E.M.; Matthews, M.L.; Endress, P.K. 2001. Cunoniaceae in the Cretaceous of Europe: evidence from fossil flowers. *Annals of Botany*. 88: 423–437.

[60] Fürstenberg-Hägg, J.; Zagrobelny, M.; Bak, S. 2013. Plant defense Against Insect Herbivores. *International Journal of Molecular Sciences*. 14: 10242–10297.

[61] Anderson, J.P.; Badruzsaufari, E.; Schenk, P.M.; Manners, J.M.; Desmond, O.J.; Ehlert, C.; Maclean, D.J.; Ebert, P.R.; Kazan, K. 2004. Antagonistic interaction between abscisic acid and jasmonate-ethylene signaling pathways modulates defense gene expression and disease resistance in Arabidopsis. *Plant Cell*. 16: 3460–3479.

ADDITIONAL LITERATURE

Chemotaxonomy

Cuénoud, P.; Savolainen, V.; Chatrou, L.W.; Powell, M.; Grayer, R.J.;Chase, M.weM.W. 2002. olecular phylogenetics of Caryophyllales based on nuclear 18SrDNA and plastid *rbcL*, *atpB*, and *matK* DNA sequences. *American Journal of Botany*. 89: 132–144.

Seberg, O.; Petersen, G.; Davis, J.I.; Chris Pires, J.; Stevenson, D.W.; Chase, M.W.; Fay, M.F.; Devey, D.S.; Jørgensen, T.; Sytsma, K.J.; Pillon, Y. 2012. Phylogeny of the Asparagales based on three plastid and two mitochondrial genes. *American Journal of Botany*. 99: 875–889.

Soltis, D.E.; Mort, M.E.; Latvis, M.; Mavrodiev, E.V.; O'Meara, B.C.; Soltis, P.S.; Gordon Burleigh, J.; Rubio de Casas, R. 2013. Phylogenetic relationships and character evolution analysis of Saxyfragales using a supermatrix approach. *American Journal of Botany*. 100: 916–929.

General Aspects of Evolution and Phylogenetics

Carroll, S.B. 1995. Homeotic genes and the evolution of arthropods and chordates. *Nature*. 376: 479–485.

Darwin, C. 1901. *On the Origin of Species by Means of Natural Selection or the Preservation of Favoured Races in the Struggle for Life*. John Murray, London.

Eldredge, N.; Gould, S.J. 1972. Puntuated equilibria: an alternative to phyletic gradualism, in Schopf, T.J.N. Ed. *Models in Paleobiology*. Freeman & Cooper, San Francisco, pp. 82–115.

Gould, S.J.; Eldredge, N. 1977. Punctuated equilibria: the tempo and mode of evolution reconsidered. *Paleobiology*. 3: 115–151.

Simpson, G.G. 1984. *Tempo and Mode in Evolution*. Columbia Univ. Press. New York.

Wasternack, C. 2007. Jasmonates: an update on biosynthesis, signal transduction and action in plant stress response, growth and development. *Annals of Botany (Lond.)*. 100: 681–697.

Gymnodecy

Bertin, R.I.; Connors, D.B.; Kleinman, H.M. 2010. Differential herbivory on disk and ray flowers of gynomonoecious asters and goldenrods (Asteraceae). *Biological Journal of the Linnean Society*. 101: 544–552.

Brunet, J. 1996. Male reproductive success and variation in fruit and seed set in *Aquilegia caerulea* (Ranunculaceae). *Ecology*. 77: 2458–2471.

Huang, S.Q.; Tang, L.L.; Yu, Q.; Guo, Y.H. 2004. Temporal floral sex allocation in protogynous *Aquilegia yabeana* contrasts with protandrous species: support for the mating environment Hypothesis. *Evolution*. 58: 1131–1134.

Wise, M.J.; Coffey, L.E.; Abrahamson, W.G. 2008. Nutrient stress and gall flies interact to affect floral-sex ratio in gynomonoecious *Solidago altissima* (Asteraceae). *American Journal of Botany*. 95: 1233–1239.

Plant Defense Pathways

Merati, M.J.; Hassanpour, H.; Niknam, V.; Mirmasoumi, M. 2014. Exogenous application of penconazole regulates plant growth and antioxidative responses in salt-stressed *Mentha pulegium* L. *Journal of Plant Interactions*. 9: 791–801.

Mittler, R. 2016. ROS are god. *Trends in Plant Science*. 16: 300–309.

Noctor, G.; Foyer, C.H. 2017. Uptdate redox compartmentation intracellular redox compartmentation and ROS-related communication in regulation and signalling. *Plant Physiology*. 171: 1591–1592.

Palma, F.; Lluch, C.; Iribarne, C.; García-Garrido, J.N.; García, N.A.T. 2009. Combined effect of salicylic acid and salinity on some antioxidant activities, oxidative stress and metabolite accumulation in Phaseolus vulgaris. *Plant Growth Regulation*. 58: 307–316.

Sorahinobar, M.; Niknam, V.; Ebrahimzadeh, H.; Soltanloo, H.; Behmanesh, M.; Enferadi, S.T. 2016. Central role of salicylic acid in resistance of wheat against *Fusarium graminearum*. *Journal of Plant Growth Regulation*. 35: 477–491.

12 Nanoscale Plant Electrochemistry

12.1 INTRODUCTION

Electrochemistry at the nanoscale can be seen as the ensemble of electrochemical processes constrained to regions of nanometric size. This is a rapidly expanding area of research prompted by the development of micro and nanoelectrodes and the accompanying technology. In a more restricted sense, nanoscale electrochemistry is focused on living cells. This is of interest for the study of processes such as cellular cytosolic or secretory release rate, and the physiological effects of different stimuli [1]. Single-cell electrochemistry includes separate single cells and single cells in cell populations, and it can be extended to cell cultures, biofilms, and organelles, briefly anticipated in Chapter 4.

Electrochemical techniques of single-cell observation add to standard analytical techniques. These include fluorescence microscopy techniques, flow cytometry, capillary electrophoresis, microfluidic chip techniques, and mass spectrometry [2]. Apart from voltammetry and amperometry at nanoelectrodes, electrochemical techniques in this field are electrochemiluminescence (ECL), scanning electrochemical microscopy (SECM), and scanning ion conductivity microscopy (SICM), as well as resistive pulsed techniques and others [1, 3]. In this chapter, the essential guidelines and applications in plant science will be treated.

12.2 ELECTROCHEMISTRY OF CELL SUSPENSIONS

The electrochemistry of cellular suspensions can be directed to the detection of compounds passing to the solution phase after cell lysis. In general, this results in extremely low concentrations so that the solution phase electrochemical strategies for trace analysis apply. Given the structural, compositional, and metabolic complexity of cells, there is a great variety of components to be detected by means of *in vitro* electroanalysis. The research has been mostly concentrated in animal cells where the electrochemistry of components released in cell lysis can be attributed to oxidation of purines in the cytoplasm [4] and the activity of enzymes [5]

The redox-mediated detection of different bacteria in cell suspensions using microelectrodes has been recently described [6–8]. The performance of electrochemical measurements can be enhanced by the use of redox mediators such as N,N,N′,N′-tetramethyl-para-phenylenediamine (TMPD). TMPD is oxidized by the cytochrome c oxidase in target bacteria [6]. Upon application of an appropriate constant potential input, the redox mediator is regenerated at the electrode, producing a reductive current spike.

Three basic situations involving redox mediation have been described. These situations and the corresponding current transients are schematized in Figure 12.1 [7]. The former involves the adhesion of the bacteria to the electrode surface blocking the oxidation/reduction of the redox probe. The second approach consists of the use of a reversible redox probe in its oxidized form (e.g. $Fe(CN)_6^{3-}$) accompanied by the application of a potential at which this species cannot be reduced. Then, the redox probe is reduced by the bacteria and the generated reduced form ($Fe(CN)_6^{4-}$) is oxidized at the electrode yielding a cathodic current spike. In the third case, the oxidized form of the redox couple is again put into the solution (e.g. $Ru(NH_3)_6^{3+}$) but now the applied potential is sufficiently cathodic to reduce it (to $Ru(NH_3)_6^{2+}$). Here the bacteria regenerate the oxidized form of the redox probe and amplifies the cathodic current [7].

DOI: 10.1201/9781032697826-12

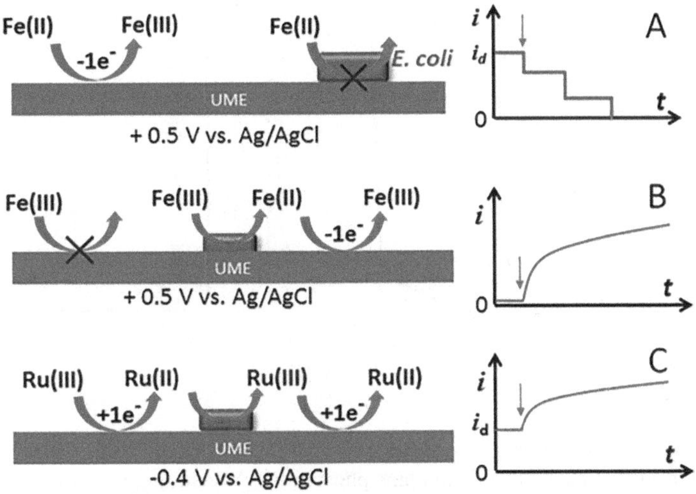

FIGURE 12.1 Schematic representation of different cases of electrochemical monitoring of single bacterial events (left side) and current/time records under the application of the indicated constant potential inputs (right side). (a) Blocking of $Fe(CN)_6^{4-}$ oxidation at the microelectrode surface. (b) Bacteria reduce $Fe(CN)_6^{3-}$ to $Fe(CN)_6^{4-}$, which is oxidized at the microelectrode. (c) $Ru(NH_3)_6^{3+}$ is reduced to $Ru(NH_3)_6^{2+}$ at the microelectrode and it is regenerated by the bacteria. The arrows indicate the moment when the bacteria are added to the solution of the redox probe. (Reproduced from ref. [7], with permission. Copyright the American Chemical Society 2018.).

Other strategies involve cell metabolism that can catalyze exogenous reactions to produce specific substances. Indeed, bacteria can bind Cu^{2+} ions and reduce them to Cu(I) species through their redox enzyme cascade. The generated Cu^{+} can catalyze several reactions allowing for target recognition, signal triggering, and signal amplification. The copper-catalyzed azide-alkyne cycloaddition reaction [9] has been exploited for voltammetric *Escherichia coli* detection using a nanocomposite based on azide-functionalized Fe_3O_4 nanoparticles-alkyne and black phosphorus-Au nanoparticles sheets deposited onto GCE [10].

Extracellular Cu^{2+} passes into the space between the outer and inner cytoplasmic membranes through the channel protein on the first membrane. Then, the cupric reductase NDH-2 reduces Cu^{2+} to Cu^{+}, some of which diffuse back between the outer and inner cytoplasmic membranes transported by the Cu^{+}-translocating ATPase CopA. The released Cu^{+} triggers the azide-alkyne cycloaddition whose product interacts with Fe_3O_4 resulting in a voltammetric signal at ca. 0.5 V vs. Ag/AgCl (see Figure 12.2) whose intensity increases as the concentration of *E. coli* increases [10].

12.3 ELECTROCHEMISTRY OF IMMOBILIZED CELLS

An alternative to electrochemistry in cell suspensions but based on a similar approach consists of the immobilization of bacteria onto the electrode surface. This is achieved via the formation of self-assembled monolayers (SAMs) using thiols. Thiols form stable SAMs on gold electrodes and can be functionalized to include a binding unit capable of binding with any bacterial antibody [11]. The binding of cells to gold electrodes has been described to enable the detection of specific target bacteria [6, 12]. This method exploits the affinity of thiol units for gold. As schematized in Figure 12.3a, a biotin-thiol and spacer-thiol SAM was formed on the gold surface, accompanied by blockage of unspecific binding sites with bovine serum albumin (BSA). The subsequent incubation with neutravidin replaces BSA at the biotinylated thiols and the addition of commercially available polyclonal anti-*Escherichia coli* antibody produces, via linking to neutravidin, the immobilization of the bacteria from a cell suspension. Again, electrochemical detection of *E. coli* was carried out with the aid

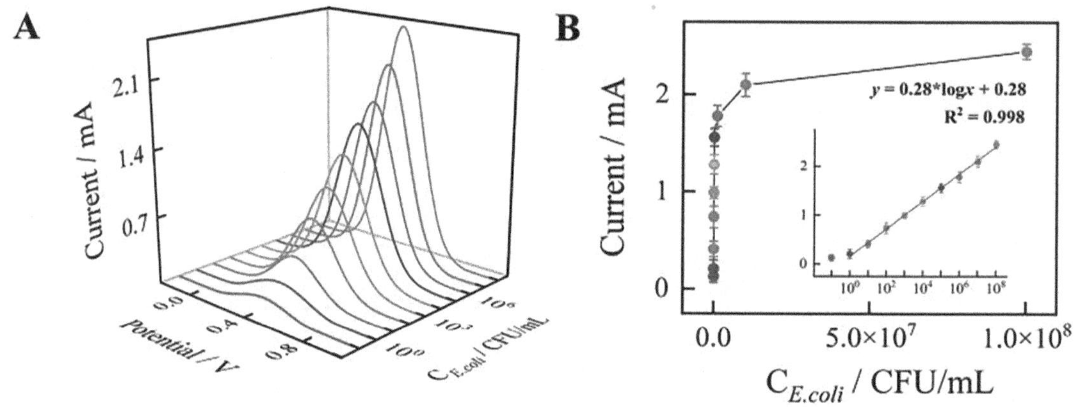

FIGURE 12.2 (a) Fast scan voltammetric records signals of an electrochemical sensor based on azide-functionalized Fe_3O_4 nanoparticles-alkyne and black phosphorus-Au nanoparticles in the presence of different ***E. coli*** concentrations (from 0.1 to 10^8 CFU/mL). (b) Variation of peak current on the logarithm of ***E. coli*** concentration. (Reproduced from ref. [10], with permission. Copyright Elsevier 2023.).

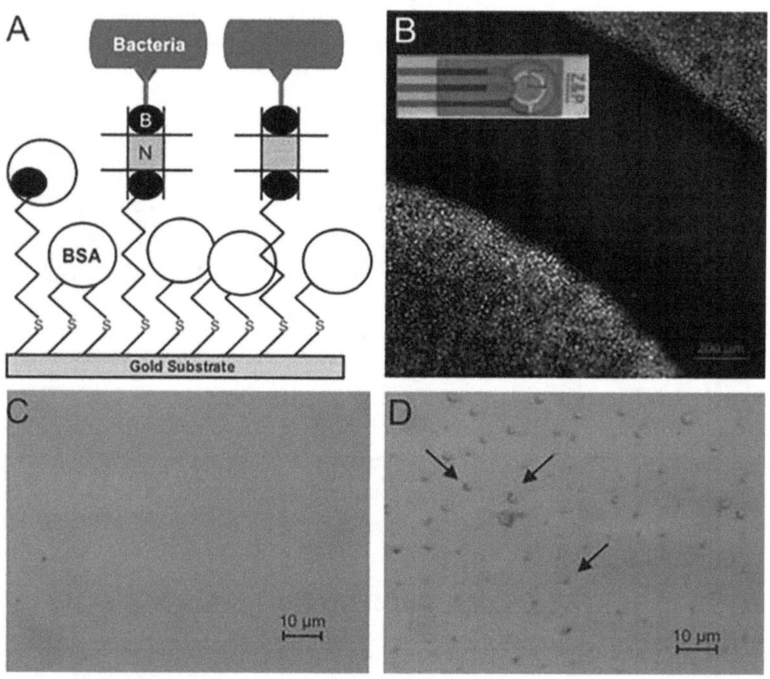

FIGURE 12.3 (**a**) Scheme for the immobilization of *Scherichia coli* bacteria on gold electrodes. The biotinylated antibody is represented as red segments. (b) Fluorescence image of a screen-printed electrode substrate where immobilized, biotinylated fluorescent beads have replaced the antibody. (c, d) Optical micrographs of bare (c) and a functionalized (d) gold electrode with immobilized bacteria (marked by arrows). (Reproduced from ref. [12], with permission. Copyright the American Chemical Society 2019.).

of a redox probe, TMPD, which reacts with bacterial cytochrome c oxidase [12]. The method is effective for attaching bacteria onto gold surfaces as evidenced by fluorescence imaging of screen-printed electrodes where gold bands alternate with another substrate. An illustrative example is shown in Figure 12.3b obtained upon replacement of the antibody by biotinylated fluorescent beads. The effective immobilization of *E. coli* bacteria is evidenced by comparing the optical microscopy images of a bare gold electrode in Figure 12.3c and the fully functionalized electrode in Figure 12.3d.

In this case, chronoamperometry was used as the electrochemical detection mode because of its inherently high sensitivity. The experiment was conducted by applying a reductive potential step responsive to the oxidized form of TMPD generated by the bacteria. Figure 12.4 shows the recorded current/time variations for the model organism *Bacillus subtilis* submitted to the oxidase test [6]. The positive test (a) gives rise to a significant current increase (relative to the unmodified gold electrode) resulting in an increase in current that is dependent on the number of monoloayers of the bacteria cultured on agar deposited onto the gold electrode.

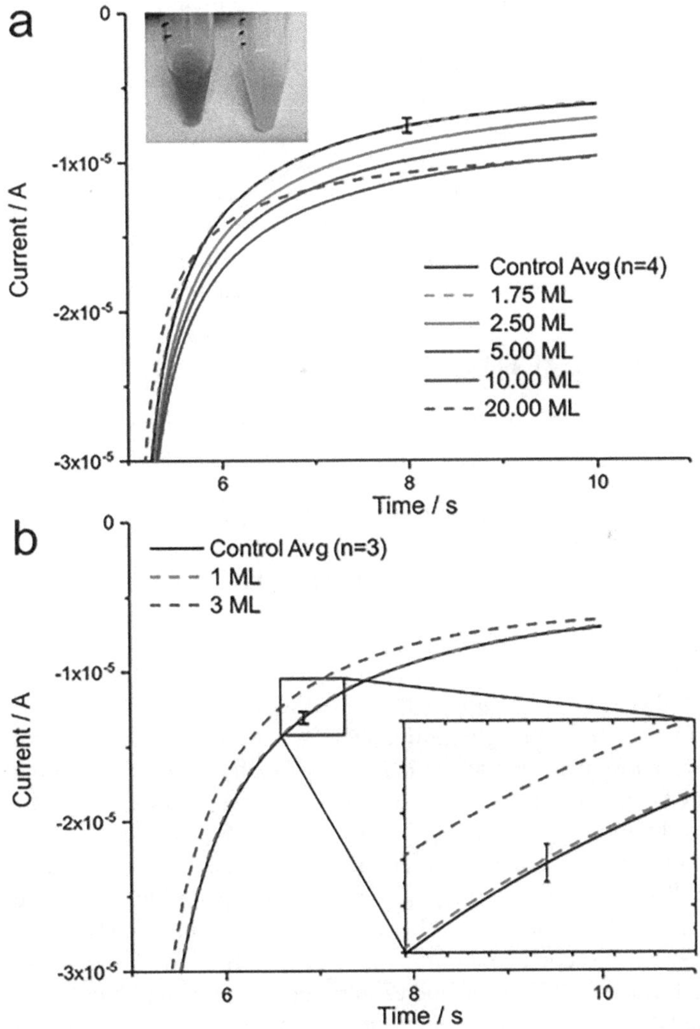

FIGURE 12.4 Chronoamperometry current/time curves recorded in 1.7 mM TMPD-BF_4 solution accompanying the model organism *Bacillus subtilis* oxidase test. (a) The positive test (inset test tube left) results in an increase in current. (b) The negative test (test tube right) is not accompanied by a current increase. A potential of 250 mV was applied for 5 s followed by 5 s of potential step of −150 mV. Error bars represent three times the standard deviation. (Reproduced from ref. [6], with permission. Copyright the Royal Society of Chemistry 2017.).

Impedimetric analysis has been also reported as a detection mode. Figure 12.5a shows the impedance spectra (Nyquist plots) recorded at T4-bacteriophage-modified gold electrode in the presence of different concentrations of *E. coli* [13]. Here, bacteriophage T4, a natural bioreceptor for living bacteria cells, was attached to the base gold electrode through cysteamine assembly. Impedance spectra were obtained using $Fe(CN)_6^{3-}/Fe(CN)_6^{4-}$ as a redox probe in PBS at biological pH. The Nyquist plots consist of a depressed capacitive loop at high and middle frequencies ending with a linear segment at low frequencies. These impedance spectra were modeled with a modified Randles circuit (see Figure 1.9) replacing the double-layer capacitor with a constant phase element. Fitting experimental data to this equivalent circuit results in charge transfer resistance values (R_{ct}) that increase on increasing the concentration of *E. coli*. Such values increase proportionally to the logarithm of bacteria concentration, as can be seen in Figure 2.5b.

The aforementioned method was complemented by returning partially to the approaches in section 12.1 using the so-called loop-mediated isothermal amplification [13]. The method consists of the amplification of the *E. coli Tuf* gene after lysis of the bound bacterial cells. Linear potential scan voltammetry was used as a detection mode by exploiting the electroactive character of the intercalating reagent Hoechst 33258 ({[20(4-hydroxyphenyl)-5-(4-methyl-1-piperazinyl)-2,50-bis(1H-benzimidazole)]}) used in the process. Figure 12.6 shows how the peak current for Hoechst oxidation decreases upon increasing the concentration of *E. coli*. Now, the peak current decrease becomes proportional to the logarithm of the concentration of bacteria [13].

Antibodies are also used as bio-receptors for this type of analysis, but it has been reported that they offer some drawbacks associated with low capturing efficiency of the cross-reactivity with nontarget bacterial cells and instability to environmental fluctuations [14, 15].

12.4 ELECTROCHEMICAL IMAGING OF CELLS

Several electrochemical methods are applied in molecular detection and bioimaging at the single-cell and subcellular levels [1]. ECL, SECM, and SICM are the three main techniques applied for imaging purposes.

ECL imaging is based on the promotion of a substrate species to an excited state by applying a voltage or current to an electrode. Then the excited species returns to its ground state emitting an optical signal [16]. The imaging system is then formed by combining optical and electrochemical devices. Two types of ECL processes can be distinguished: annihilation type and co-reaction-type, both schematically depicted in Figure 12.7.

In both ECL pathways, the exciting voltage is applied to two substances, reactant, and co-reactant, to generate an excited state whose return to the ground state will produce the luminous signal. In the annihilation-type ECL there is a single luminophore species (Figure 12.7a). In co-reactive ECL, a bimolecular reaction between a co-reactant and the luminophore species takes place. Luminophores are usually selected among Ru and Ir complexes (typically tris(2,2′-bipyridyl) ruthenium(II), $Ru(bpy)_3^{2+}$) [17], and luminol and other hydrazide compounds [18].

Due to the high sensitivity, temporal resolution, and nanoscale spatial resolution, ECL techniques are extensively applied to single-cell imaging [19]. Two main configurations are applied, positive, and shadow label-free ECL. In the former, the cells are labeled with the luminophore, thus obtaining luminescent images. In the second configuration, cell organelles hinder the diffusion of the reagents so dark images are formed [20]. Recent studies introduce bimodal and bicolor ECL strategies [21].

$[Ru(bpy)_3]^{2+}$ binds to proteins and nucleic acids by affinity interactions or functional groups and is used to observe single-cell surface proteins, cell motility states, and intercellular transport. Under ordinary conditions, the luminescent luminophore-substrate adducts are located a few microns from the electrodes. Then, the ECL signals are produced in the cell membrane region, while the electrically generated free radicals are concentrated near the cell. Accordingly, ECL events are confined to the cell boundary, the so-called ECL surface-limited imaging [22]. To overcome this limitation, the use of inkjet-printed disposable electrodes has been reported [23].

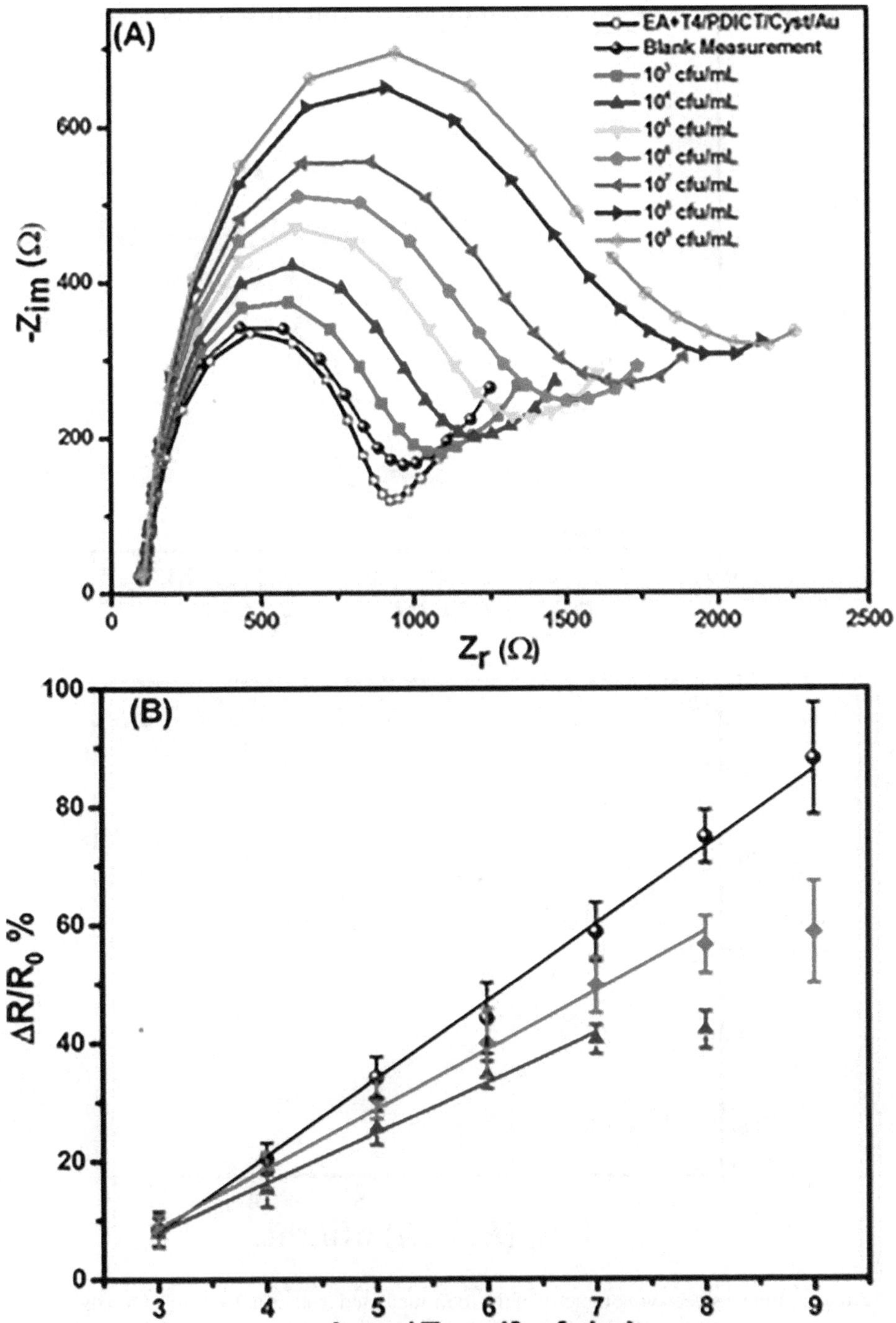

FIGURE 12.5 (a) Nyquist plots recorded at T4-bacteriophage-modified gold electrode at different concentrations of ***E. coli*** in 10 mM $Fe(CN)_6^{3-}/Fe(CN)_6^{4-}$ solution in 10 mM PBS (pH 7.4): 0 (●), 103 (■), 104 (▲), 105 (▼), 106 (◆), 107 (◀), 108 (▶), and 109 (⊕) cfu/mL. (b) Variation of the charge transfer resistance (ΔR_{ct}) with the concentration of ***E. coli*** in PBS (●), skim milk (▲), and lake water (◆). ΔR_{ct} (%) values obtained from the fitting of three independent impedance data sets. (Reproduced from ref. [13], with permission. Copyright the American Chemical Society 2013.).

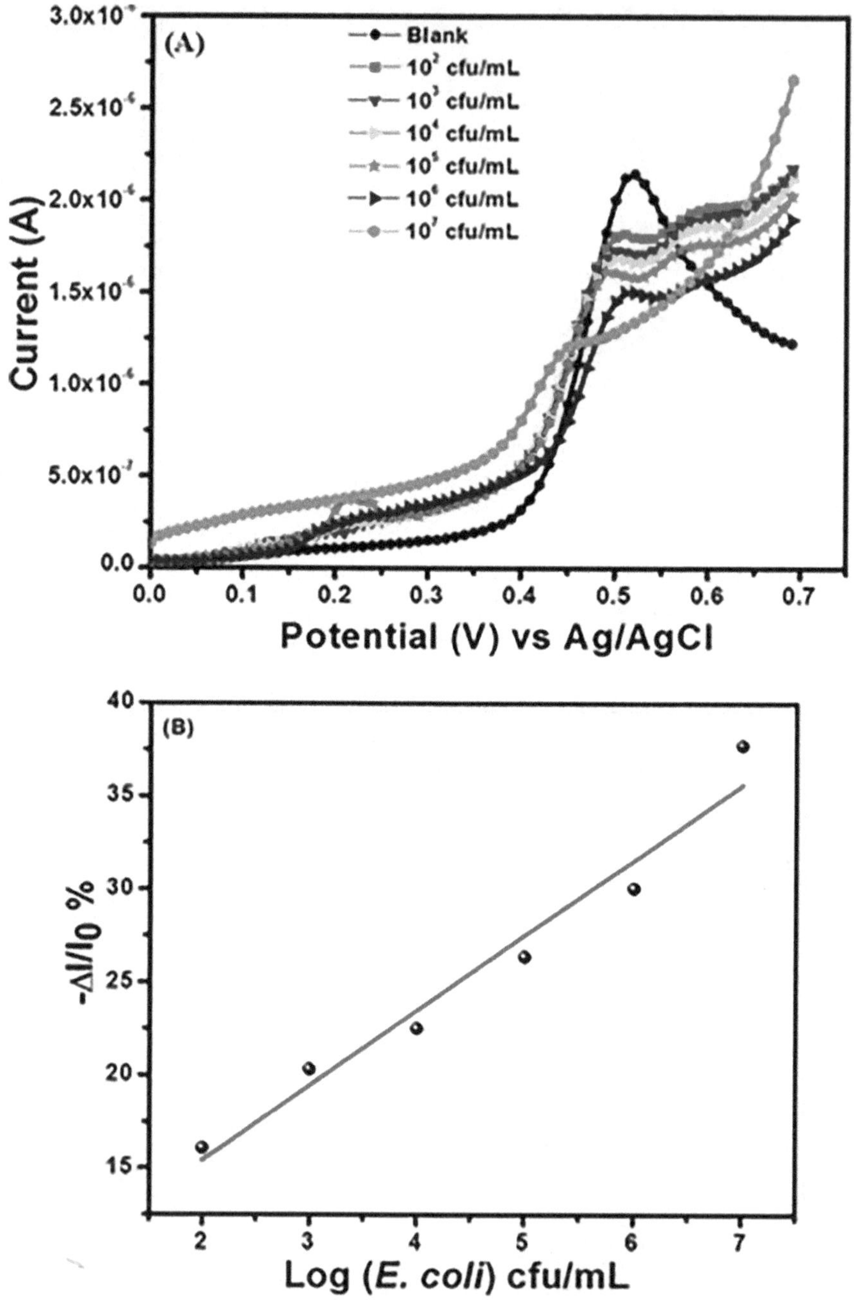

FIGURE 12.6 (a) Linear sweep voltammetry of the loop-mediated isothermal amplification method described in the text corresponding to the oxidation of the intercalating agent in the presence of different concentrations of *E. coli*. (b) Variation of the peak current with the concentration of bacterial cells in the solution. (Reproduced from ref. [13], with permission. Copyright the American Chemical Society 2013.).

The luminol-H_2O_2 system is also used for ECL imaging. Specific cell biomolecules react through the corresponding enzymatic mediation to produce H_2O_2, which in turn induces luminol luminescence, thus prompting the detection of the biomolecules [24].

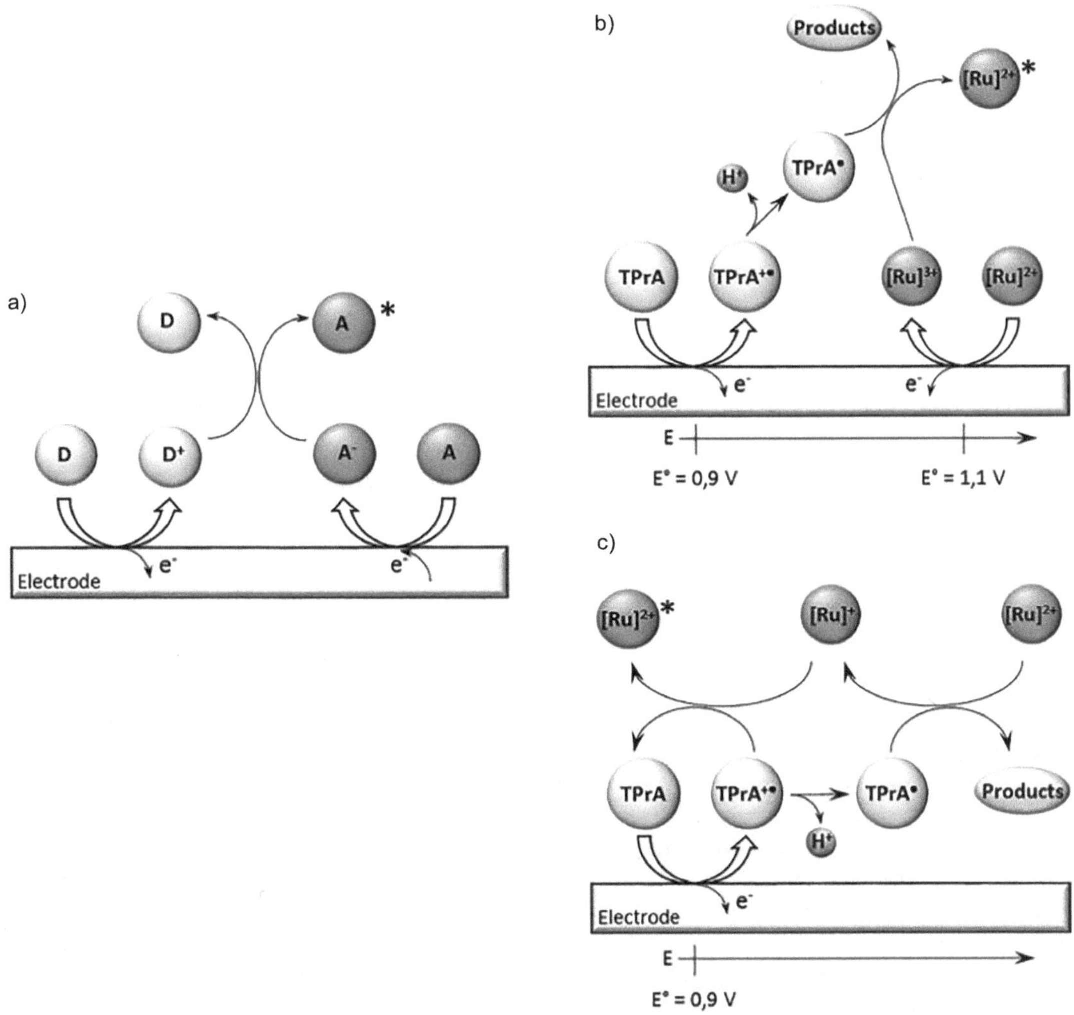

FIGURE 12.7 Schemes corresponding to ECL processes of (a) annihilation type, b, c: co-reaction type involving tri-n-propylamine (TPrA) and $Ru(bpy)_3^{2+}$ when both TPraA and the Ru complex are oxidized (b) and when only TPrA is oxidized (c). (Adapted from ref. [16], with permission. Copyright Elsevier 2018.).

Recent developments include the application of bipolar electrochemistry [25]. Here, bipolar electrodes are available for undergoing anodic and cathodic reactions simultaneously [26]. The experimental arrangement consists of a pair of driving electrodes connected to a potentiostatic power supply and a bipolar electrode embedded in between the driving electrodes. Figure 12.8 shows a scheme comparing conventional and bipolar electrochemistry [27]. In a conventional two-electrode cell that contains a high concentration of supporting electrolyte (>0.1 M), the potential difference applied between the electrodes produces electrical double layers upon rearranging ions at the electrode/solution interfaces. At appropriate potentials, redox reactions take place at the electrodes in a couple of nanometer thicknesses of the layers. In a low concentration of supporting electrolytes, electrical double layers are not fully formed, and redox reactions do not readily proceed at the electrode surfaces. Then a conductor placed in between the electrodes can simultaneously yield oxidation and reduction at its terminals.

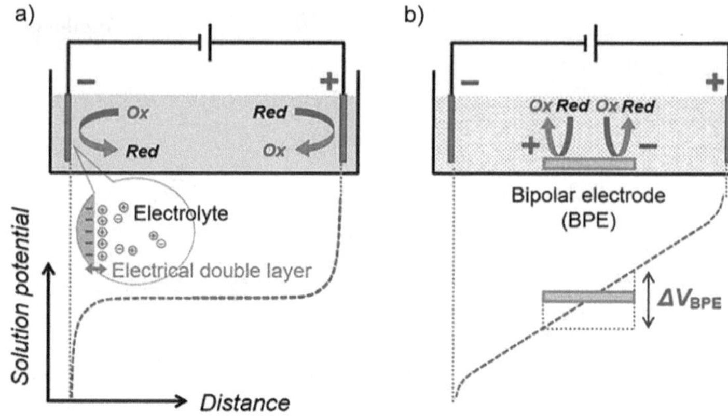

FIGURE 12.8 Scheme for (a) a conventional two-electrode electrolysis cell and (b) a bipolar electrochemistry electrolysis cell. (Adapted from ref. [27], with permission. Copyright the American Chemical Society 2019.).

For cell analysis, a bipolar electrode system can be achieved by depositing a platinum layer at the tip of a capillary placed into the cell. The target molecules in the cytoplasm were transported into the capillary to induce a bipolar ECL signal from the luminescent reagent [28].

The SECM technique, introduced in Chapter 1, can be seen as a label-free imaging method that obtains spatially resolved current signals. This method has been applied to studies of cell sample morphology, membrane permeability, and cellular activity, as well as to detect metabolic molecules and their cellular surface distribution. As a paradigmatic issue, the study of vivo topography and photosynthetic electron transport of individual guard cells in *Tradescantia fluminensis* has been reported [29]. The simplest operation mode involves a constant separation between the tip electrode and the surface of the substrate under examination, but different operation modes are available. Among them, the voltage-switching mode has been used for imaging individual cells using nanoelectrodes (5–100 nm sized) with diffusion-impeded feedback [30].

In the SICM technique, a glass or quartz nanopipette filled with electrolyte solution is used as the scanning probe accompanied by a quasi-reference counter electrode. A bias potential is established between the probe electrode and another electrode in the bulk solution to generate an ionic current through the end of the nanopipette [31]. In these conditions, the ionic current can be used as feedback to position the nanopipette near the surface of the substrate and obtain topographical information. A bias potential is applied between two quasi-reference counter electrodes, placed in the nanopipette and in the bulk electrolyte solution, respectively, to drive a current through the nanopipette. This is quite sensitive to the separation between the nanopipette and the substrate, as well as to changes in local ion conductivity. Figure 12.9 presents a scheme of SICM operation [32]. Here, the nanopipette tip acts as a quasi-reference counter electrode (QRCE) connected to a second QRCE in the bulk electrolyte solution and a potential difference, V_{Tip}, is established between them (Figure 12.9a). Figure 12.9b shows the typical variation with time of nanopipette height and V_{Tip} as applied at each position of the nanopipette, which is displaced near the bacteria surface [32].

Figure 12.10 shows SICM topography, normalized current, and local charge density maps of *E. coli* adhered to poly-L-lysine in contact with KCl solution buffered at pH 7 [32]. The bacterium is approximately 2 μm in length and 1 μm in height. Current data were processed via the finite element simulation method converting it into charge density treating the bacterium as a planar charged impermeable insulator. Then, reflecting the Gram-negative character of the bacteria, the local charge density is negative over the bacterium and positive over the polymer background. Noticeably, the surface charge density differs significantly between the central apex of the bacterium and its edges.

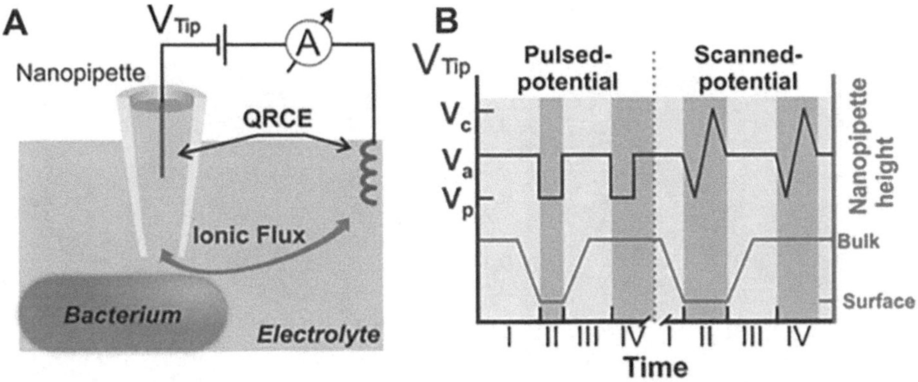

FIGURE 12.9 (a) Scheme of SICM operation. V_{Tip} is the potential difference established between the nanopipette QRCE and the QRCE in bulk electrolyte solution. (b) Time variation of the nanopipette height (red line, right y-axis) and V_{Tip} (blue line, left y-axis). Two different V_{Tip}-time profiles were applied: pulsed (left half) and scanned (right half). V_a is the approach potential; V_p the pulsed potential, which is one of the limits of the triangular sweep; and V_c the other limit of the linear cyclic potential sweep. Typical values are $V_a = 50$ mV, $V_p = -500$ mV, and $V_c = 500$ mV. (Reproduced from ref. [32], with permissions. Copyright the American Chemical Society 2020.).

Different SICM modes have been proposed. These include equal and different electrolyte solutions in the nanopipette and the external electrolyte solution, as well as modulated feedback types [33] and hopping scan regimes [34]. This technique can be combined with others, including SECM [35] and AFM [36].

12.5 INTRACELLULAR ELECTROCHEMISTRY

Single-cell measurements can either be performed by positioning micro- or nanoelectrodes near the cell surface with the aid of a micropositioning device and a patch-clamp amplifier of a microinjection system or by inserting the electrodes into the cell [1]. The fabrication of the ultramicroelectrodes of dimensions between µm and nm with different geometries and materials has facilitated their extensive within-the-cell use.

Interestingly, micro- and ultramicroelectrodes offer advantages to macro electrodes [1]. They improve signal-to-noise ratio relative to macro electrodes because they enhance faradaic currents. This effect results from the fact that diffusion toward microelectrodes is hemispherical or spherical (rather than planar at disk conventional macro electrodes), thus producing higher mass transfer rates. Microelectrodes also suffer low ohmic drop (*iR*) distortions because of the low magnitude of the measured currents. This allows for measuring in solutions without adding supporting electrolytes. Another advantage is that microelectrode and ultramicroelectrodes provide fast response times. This is due to the small values of double-layer capacitances (*C*) and, hence, of the time constants (*RC*).

Another interesting aspect of micro- and nanoelectrodes is the possibility of chemical/electrochemical modification aimed to enhance their sensitivity and selectivity. This aspect, which deals with general analytical strategies, will be treated in Chapter 13.

Nanoelectrodes can be inserted into a single-cell, subcellular locations, and/or a single organelle without disrupting the cellular homeostasis and microenvironment [37]. Several methods, namely ionic current rectification, resistive pulsing, and chrono-analysis have been developed for increasing the performance of nanoelectrodes in studies on single cells [2].

Chronoelectrochemistry analysis involves well-known electrochemical techniques based on the record of the variation with time of a given quantity such as current (chronoamperometry), charge (chronocoulometry), and potential (chronopotentiometry). These methods have been abundantly employed in studies around intracellular oxidative stress molecules [38].

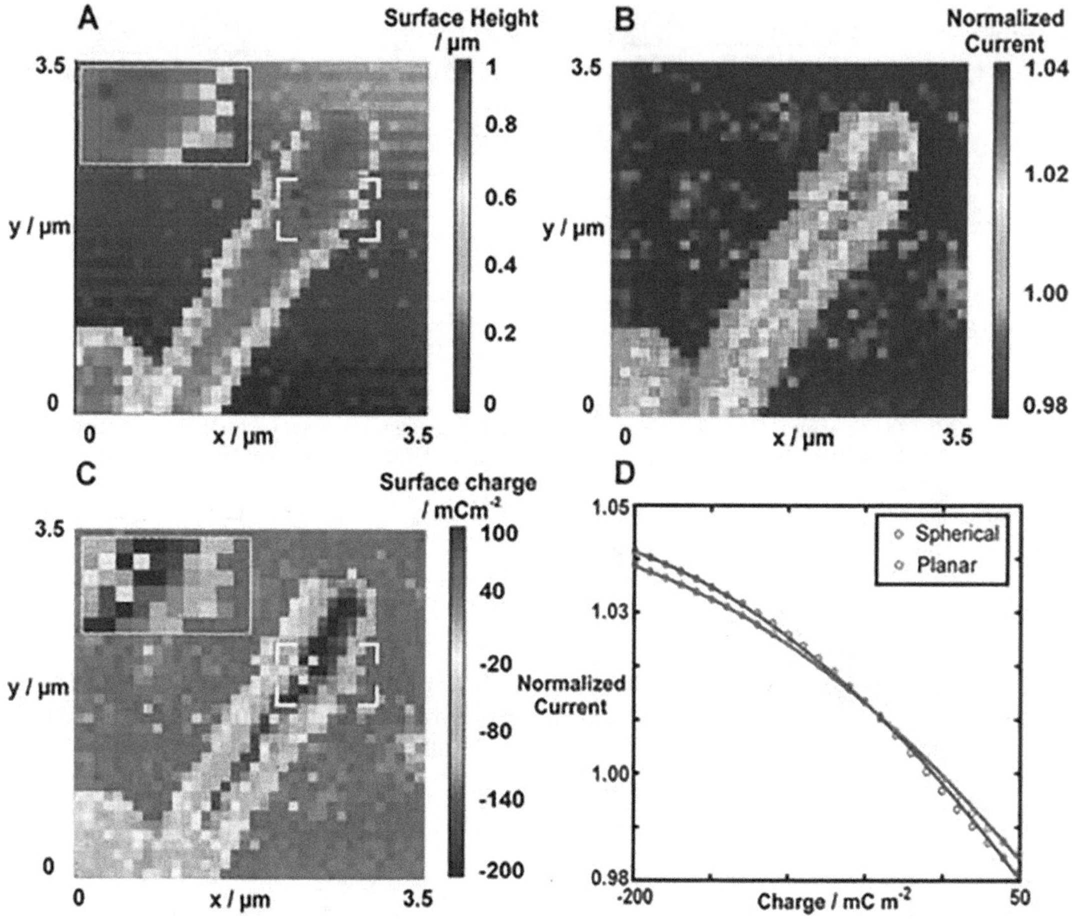

FIGURE 12.10 SICM topography (a), normalized current (b), and local charge density (mC m^{-2}) map of *Escherichia coli* bacteria adhered to ploy-L-lysine in 50 mM tris buffered KCl solution at pH 7. $V_p = -500$ mV and $V_a = 50$ mV. Nanopipette inner radius 93 nm, hopping step separation 100 nm, scan time 80 min. (c) Finite element method simulation map. (d) Comparison of the model simulation results using a planar (pale gray, red in the original figure) and spherical (dark gray, blue in the original figure) cell form. (Reproduced from ref. [32], with permission. Copyright the American Chemical Society 2020.).

Ionic current rectification (or electrochemical rectification) consists of a nonlinear current-potential response caused by the asymmetric migration of anions and cations through a nano-channel [39, 40]. This can be constituted by a conical capillary with a pore size of 20 nm–20 µm filled with a suitable electrolyte such as KCl. Then a diffuse double layer is generated, resulting in asymmetric current-voltage responses through the tip. Apart from the usual factors (pH, electrolyte concentration), this response depends on the capillary shape and size, surface charge, applied voltage, and other factors.

The resistive pulse technique is based on the application of a potential difference between two electrodes inside and outside a capillary immersed in an electrolyte solution, causing a resistive pulse signal [41]. This effect, discovered by Coulter in 1953, is the basis for classical cytometry measurements [42]. The pulse signal yields information on the shape and size of particles, their concentration, and surface charge of application for the study of intracellular components.

Intracellular electrochemistry can also exploit electroporation methods. This involves the increase of cell uptake of foreign molecules upon permeabilization of the plasma membrane by the application of pulsed electric fields [43]. The permeation effect depends on the intensity of the electric field, pulse duration (nanoseconds to milliseconds), and frequency. Depending on the values of such parameters, the electroporation can be reversible or irreversible. In the first case, the cells survive, whereas in the second, cellular death occurs. Reversible electroporation is an effective method for delivering pharmaceuticals inside the live cells.

The application of a high-voltage electric pulse can produce a variety of processes accompanying the permeabilization of the cell membrane. When the potential difference is large enough and the pulse duration is longer than the double-layer charging time, faradaic processes can occur with the concomitant change in the chemical composition [44].

12.6 SINGLE-ENTITY ELECTROCHEMISTRY

The term "single-entity electrochemistry" covers a series of phenomena having in common the detection of individual events occurring in the interaction with electrodes of entities of nanoscopic size [45]. The Coulter counter, the patch-clamp experiments (see Chapter 2), single-cell electrochemistry [46], and ultimately, single-molecule electrochemistry [47], are relevant insights marking the development of single-entity electrochemistry. Recently, collision electrochemistry, which originated in adsorption studies [48], is receiving considerable attention. Typically, this field is centered on the detection of the collisions of nanoparticles, clusters, etc., on metallic micro- or nanoelectrodes electrodes, but its implementation in single-cell studies is increasingly studied.

Impact events can be described assuming that electron transfer takes place by through-space tunneling of an electron between the particle and the electrode. Ignoring possible adsorption effects, the quadratic value of the free path normal to the electrode surface, $\langle \lambda_x^2 \rangle$ can be related to the mean three-dimensional free path, λ, as $\langle \lambda_x^2 \rangle = \lambda^2/3$. If the particle is initially at the origin and undergoes a given number of trajectories of lengthλ, one can take [8]

$$\left\langle \lambda_x^2 \right\rangle = 2Dt = \frac{\lambda^2}{3}\frac{t}{t_{\mathrm{MFP}}}, \tag{12.1}$$

where D is the diffusion coefficient and t_{MFP} is the collision time. Then, introducing the mean particle velocity v,

$$D = \frac{\lambda^2}{6t_{\mathrm{MFP}}} = \frac{\lambda v}{6}. \tag{12.2}$$

Assuming that the size of the particle/bacteria is larger than the solvent molecules, the Brownian motion must satisfy the energy equipartition principle,

$$\frac{1}{2}mv^2 = \frac{3}{2}k_B T, \tag{12.3}$$

and the Stokes-Einstein equation for diffusion,

$$D = \frac{k_B T}{6\pi\eta r}, \tag{12.4}$$

where k_B is Boltzmann's constant, r the particle (assumed to be spherical) radius, m is the particle mass, and η is the viscosity coefficient. Combining the previous equations, and assuming that the particles are homogeneous spheres of densityρ, the three-dimensional free path and the time t_{MFP} can be obtained as [8]

$$\lambda = \frac{2}{3\eta}\sqrt{\frac{\rho r k_B T}{\pi}} \tag{12.5}$$

$$t_{\text{MFP}} = \frac{\lambda}{v} = \frac{4}{9}\frac{\rho r^2}{\eta}, \tag{12.6}$$

respectively.

The impact events appear as sharp current transients in chronoamperometric curves recorded in highly diluted solutions of the aforementioned entities [49, 50]. In the case of bacterial suspensions, the responses described in section 12.2 can result in different current/time transients. An example of a blocking pattern can be seen in Figure 12.11, where current/time curves recorded at Pt microelectrode for *Escherichia coli* in $Na_4Fe(CN)_6$ solution of various concentrations are shown [7].

The current transients appear as current decreases over the background current. The magnitude of the decrease diminishes by lowering the concentration of the redox probe in the solution. This can be interpreted as balancing the surface blocking effect with the regeneration of $Fe(CN)_6^{4-}$ effect also exerted by the bacteria. At a high redox probe concentration, the bacterial reduction rate is slow relative to the blocking effect, but as the redox mediator concentration decreases, the contribution of the regeneration effect becomes more significant [7].

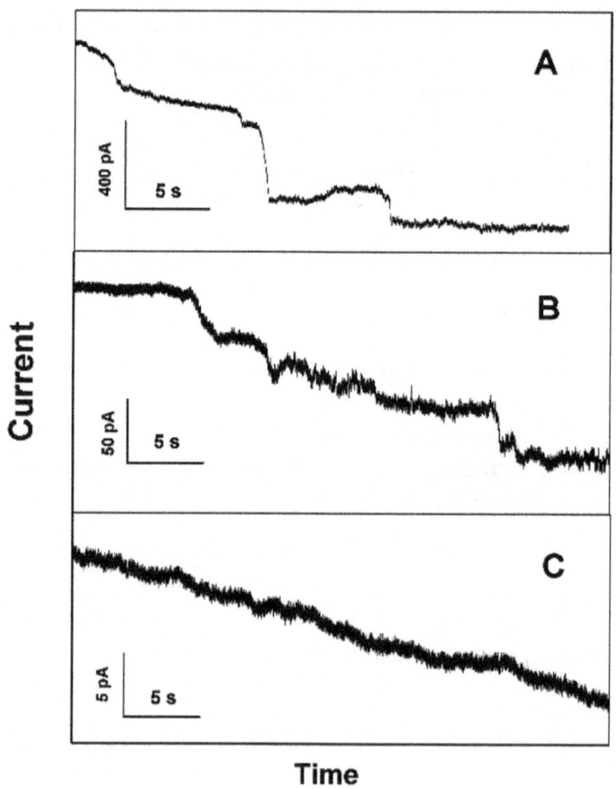

FIGURE 12.11 Chronoamperometric responses recorded at Pt microelectrode (25 μm diameter) upon application of a constant potential of 0.5 V vs. SCE at a 16.6 pM (1 × 10^{10} CFU/mL) suspension of *E. coli* in 10 mM PBS containing $Na_4Fe(CN)_6$ in concentrations of 50 (A), 10 (B), and 1 (C) mM. Applied potential 0.50 V vs. Ag/AgCl. (Reproduced from ref. [7], with permission.).

Chronoamperometric experiments in suspensions of *Stenotrophomonas maltophilia*, a Gram-negative bacterium, are illustrated in Figure 12.12 [7]. Unlike *E. coli*, the cells of *S. maltophilia* are positively charged, thus modifying the electrostatic interaction with the electrode surface. Figure 12.12a shows the current/time recorded at Pt microelectrode in suspensions of *S. maltophilia* in a $Fe(CN)_6^{3-}$ solution by applying a potential (0.5 V vs. Ag/AgCl) at which no direct reduction of the redox probe occurs. In contrast with the *E. coli* blockage transients, the current of *S. maltophilia* decreases continuously with minor transient decreases separated by long times. This can be interpreted by assuming that *S. maltophilia* cells reacts with $Fe(CN)_6^{3-}$, thus decreasing the local concentration of the redox probe in the vicinity of the electrode surface.

Figure 12.12b shows the chronoamperometric record using $Ru(NH_3)_6^{3+}$ as the redox probe and applying a potential where it is reduced to $Ru(NH_3)_6^{2+}$. The complex current/time pattern can be interpreted in terms of the competition between two opposite processes. On the one hand, the blocking effect exerted by bacteria cells, by the second hand, the bacterial regeneration of $Ru(NH_3)_6^{3+}$ from the $Ru(NH_3)_6^{2+}$ electrochemically produced, but other processes (dissolved oxygen, cellular respiration) can influence the observed chronoamperometric response [7].

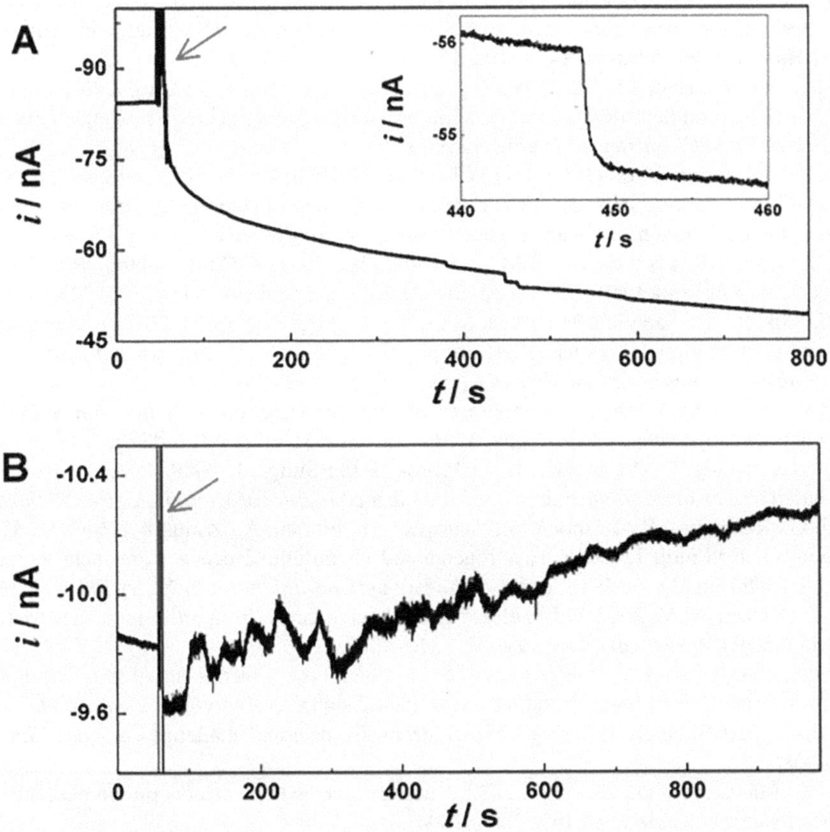

FIGURE 12.12 Chronoamperometric responses recorded at Pt microelectrode (25 μm diameter) in suspensions of *S. maltophilia* (approx. 16.6 pM) with the same in solutions containing (a) 100 mM $Fe(CN)_6^{3-}$ and (b) 10 mM $Ru(NH_3)_6^{3+}$. Greyarrows correspond to the moment of the bacteria addition to the solution. Applied potential –0.4 V vs Ag/AgCl. The arrows indicate the moment when the bacteria are added to the solution of the redox probe. The inset in (a) corresponds to the magnification of one transient step. (Reproduced from ref. [7], with permission. Copyright the American Chemical Society 2018.).

REFERENCES

[1] Schulte, A.; Schuhmann, W. 2007. Single-cell electrochemistry. *Angewandte Chemie International Edition*. 46: 8760–8777.

[2] Huang, K.; Wang, Y.H.; Zhang, H.; Wang, T.Y.; Liu, X.H.; Liu, L.; Jiang, H. 2023. Application and outlook of electrochemical technology in single-cell analysis. *Biosensors & Bioelectronics*. 242: artic. 115741.

[3] Zhang, J.; Zhou, J.; Pan, R.; Jiang, D.; Burgess, J.D.; Chen, H.-Y. 2018. New frontiers and challenges for single-cell electrochemical analysis. *ACS Sensors*. 3: 242–250.

[4] Li, J.; Lin, R.; Wang, Q.; Gao, G.; Cui, J.; Liu, J.; Wu, D. 2014. In situ electrochemical assessment of cytotoxicity of chlorophenols in MCF-7 and HeLa cells. *Analytical Biochemistry*. 456, 1–5.

[5] Feng, J.; Luo, G.A.; Jian, H.Y.; Wang, R.G.; An, C.C. 2000. Voltammetric Behavior of Tumor Cells U937 and Its Usefulness in Evaluating the Effect of Caffeic Acid. *Electroanalysis*. 12: 513–516.

[6] Kuss, S.; Tanner, E.E.L.; Ordovas-Montanes, M.; Compton, R.G. 2017. Electrochemical recognition and quantification of cytochrome c expression in *Bacillus subtilis* and aerobe/anaerobe *Escherichia coli* using N,N,N′,N′-tetramethyl-para-phenylene-diamine (TMPD) *Chemical Science*. 8: 7682–7688.

[7] Gao, G.; Wang, D.; Brocenschi, R.; Zhi, J.; Mirkin, M.V. 2018. Toward the Detection and Identification of Single Bacteria by Electrochemical Collision Technique. *Analytical Chemistry*. 90: 12123–12130.

[8] Dickinson, E.J.F.; Rees, N.V.; Compton, R.G. 2012. Nanoparticle–electrode collision studies: Brownian motion and the timescale of nanoparticle oxidation. *Chemical Physics Letters*. 528: 44–48.

[9] Gao, Z.; Zhang, E.; Zhao, H.; Xia, S.; Bai, H.; Huang, Y.; Wang, S. 2022. Bacteria- Mediated Intracellular Click Reaction for Drug Enrichment and Selective Apoptosis of Drug-Resistant Tumor Cells. *ACS Applied Materials & Interfaces*. 14: 12106–12115.

[10] Zhou, H.; Guo, W.; Hao, T.; Xie, J.; Wu, Y.; Jiang, X.; Hu, Y.; Wang, S.; Guo, Z. 2023. Electrochemical sensor for single-cell determination of bacteria based on target-triggered click chemistry and fast scan voltammetry. *Food Chemistry*. 417: artic. 135906.

[11] Maalouf, R.; Fournier-Wirth, C.; Coste, J.; Chebib, H.; Saïkali, Y.; Vittori, O.; Errachid, A.; Cloarec, J.-P.; Martelet, C.; Jaffrezic-Renault, N. 2007. Label-free detection of bacteria by electrochemical impedance spectroscopy: comparison to surface plasmon resonance. *Analytical Chemistry*. 79: 4879–4886.

[12] Kuss, S.; Couto, R.A.S.; Evans, R.M.; Lavender, H.; Tang, C.C.; Compton, R.G. 2019. Versatile Electrochemical Sensing Platform for Bacteria. *Analytical Chemistry*. 91: 4317–4322.

[13] Tlili, C.; Sokullu, E.; Safavieh, M.; Tolba, M.; Ahmed, M.U.; Zourob, M. 2013. Bacteria screening, viability, and confirmation assays using bacteriophage-impedimetric/loop-mediated isothermal amplification dual-response biosensors. *Analytical Chemistry*. 85: 4893–4901.

[14] Ruan, C.; Yang, L.; Li, Y. 2002. Immunobiosensor chips for detection of *Escherichia coil* O157:H7 using electrochemical impedance spectroscopy. *Analytical Chemistry*. 74: 4814–4820.

[15] Dover, J.E.; Hwang, G.M.; Mullen, E.H.; Prorok, B.C.; Suh, S.J. 2009. Recent advances in peptide probe-based biosensors for detection of infectious agents. *Journal of Microbiological Methods*. 78: 10–19.

[16] Valenti, G.; Rampazzo, E.; Kesarkar, S.; Genovese, D.; Fiorani, A.; Zanut, A.; Palomba, F.; Marcaccio, M.; Paolucci, F.; Prodi, L. 2018. Electrogenerated chemiluminescence from metal complexes-based nanoparticles for highly sensitive sensors applications. *Coordination Chemistry Reviews*. 367: 65–81.

[17] Bruce, D.; Richter, M.M. 2002. Green electrochemiluminescence from ortho-metalated tris(2-phenylpyridine)iridium(III). *Analytical Chemistry*. 74: 1340–1342.

[18] Marquette, C.A.; Blum, L.J. 1999. Luminol electrochemiluminescence-based fibre optic biosensors for flow injection analysis of glucose and lactate in natural samples. *Analytica Chimica Acta*. 381: 1–10.

[19] Chen, H.-Y. 2020. Bioanalysis in single cells: current advances and challenges. *Science China Chemistry*. 63: 564–588.

[20] Zhang, Z.; Ma, C.; Xu, Q.; Zhu, J.-J. 2022. Recent progress in electrochemiluminescence microscopy analysis of single cells. *Analyst*. 147: 2884–2894.

[21] Kněžević, S.; Kerr, E.; Goudeau, B.; Valenti, G.; Paolucci, F.; Francis, P.S.; Kanoufi, F.; Sojic, N. 2023. Bimodal electrochemiluminescence microscopy of single cells. *Analytical Chemistry*. 95: 7372–7378.

[22] Wang, L.; Liu, D.; Sun, Y.; Su, J.; Jin, B.; Geng, L.; Song, Y.-Y.; Huang, X.; Yang, M. 2018. Signal-on electrochemiluminescence of self-ordered molybdenum oxynitride nanotube arrays for label-free cytosensing. *Analytical Chemistry*. 90: 10858–10864.

[23] Voci, S.; Goudeau, B.; Valenti, G.; Lesch, A.; Jović, M.; Rapino, S.; Paolucci, F.; Arbault, S.; Sojic, N. 2018. Surface-confined electrochemiluminescence microscopy of cell membranes. *Journal of the American Chemial Society*. 140: 14753–14760.

[24] Xu, J.; Huang, P.; Qin, Y.; Jiang, D.; Chen, H. 2016. Analysis of intracellular glucose at single cells using electrochemiluminescence imaging. *Analytical Chemistry*. 88: 4609–4612.
[25] Wang, Y.-Z.; Xu, C.-H.; Zhao, W.; Guan, Q.-Y.; Chen, H.-Y.; Xu, J.-J. 2017. Bipolar electrode based multicolor electrochemiluminescence biosensor. *Analytical Chemistry*. 89: 8050–8056.
[26] Fosdick, S.E.; Knust, K.N.; Scida, K.; Crooks, R.M. 2013. Bipolar electrochemistry. *Angewandte Chemie International Edition*. 52: 10438–10456.
[27] Shida, N.; Zhou, Y.; Inagi, S. 2019. Bipolar Electrochemistry: A Powerful Tool for Electrifying Functional Material Synthesis. *Accounts of Chemical Research*. 52: 2598–2608.
[28] Wang, Yuling; Jin, R.; Sojic, N.; Jiang, D.; Chen, H. 2020. Intracellular wireless analysis of single cells by bipolar electrochemiluminescence confined in a nanopipette. *Angewandte Chemie International Edition*. 59: 10416–10420.
[29] Tsionsky, M.; Cardon, Z.G.; Bard, A.J.; Jackson, R.B. 1997. Photosynthetic Electron Transport in Single Guard Cells as Measured by Scanning Electrochemical Microscopy. *Plant Physiology*. 113: 895–901.
[30] Takahashi, Y.; Shevchuk, A.I.; Novak, P.; Babakinejad, B.; Macpherson, J.; Unwin, P.R.; Shiku, H.; Gorelik, J.; Klenerman, D.; Korchev, Y.E.; Matsue, T. 2012. Topographical and electrochemical nanoscale imaging of living cells using voltage-switching mode scanning electrochemical microscopy. *Proceedings of the National Academy of Sciences USA*. 109: 11540–11545.
[31] Perry, D., Page, A.; Chen, B.; Frenguelli, B.G.; Unwin, P.R. 2017. Differential-Concentration Scanning Ion Conductance Microscopy. *Analytical Chemistry*. 89: 12458–12465.
[32] Cremin, K.; Jones, B.A.; Teahan, J.; Meloni, G.N.; Perry, D.; Zerfass, C.; Asally, M.; Soyer, O.S.; Unwin, P.R. 2020. Scanning Ion Conductance Microscopy Reveals Differences in the Ionic Environments of Gram-Positive and Negative Bacteria. *Analytical Chemistry*. 92: 16024–16032.
[33] McKelvey, K.; Perry, D.; Byers, J.C.; Colburn, A.W.; Unwin, P.R. 2014. Bias Modulated Scanning Ion Conductance Microscopy. *Analytical Chemistry*. 86: 3639–3646.
[34] Takahashi, Y.; Murakami, Y.; Nagamine, K.; Shiku, H.; Aoyagi, S.; Yasukawa, T.; Kanzaki, M.; Matsue, T. 2010. Topographic imaging of convoluted surface of live cells by scanning ion conductance microscopy in a standing approach mode. *PhysChemChemPhys*. 12: 10012–10017.
[35] Takahashi, Y.; Shevchuk, A.I.; Novak, P.; Murakami, Y.; Shiku, H.; Korchev, Y.E.; Matsue, T. 2010. Simultaneous noncontact topography and electrochemical imaging by SECM/SICM featuring ion current feedback regulation. *Journal of the American Chemical Society*. 132: 10118–10126.
[36] Shi, X.; Qing, W.; Marhaba, T.; Zhang, W. 2020. Atomic force microscopy - Scanning electrochemical microscopy (AFM-SECM) for nanoscale topographical and electrochemical characterization: Principles, applications and perspectives. *Electrochimica Acta*. 332: artic. 135472.
[37] Ying, Y.-L.; Hu, Y.-X.; Gao, R.; Yu, R.-J.M.; Gu, Z.; Lee, L.P.; Long, Y.-T. 2018. Asymmetric nanopore electrode-based amplification for electron transfer imaging in live cells. *Journal of the American Chemical Society*. 140: 5385–5392.
[38] Hu, K.; Relton, E.; Locker, N.; Phan, N.T.N.; Ewing, A.G. 2021. Electrochemical measurements reveal reactive oxygen species in stress granules. *Angewandte Chemie International Edition*. 60: 15302–15306.
[39] Siwy, Z.S. 2006. Ion-current rectification in nanopores and nanotubes with broken symmetry. *Advanced Functional Materials*. 16: 735–746.
[40] Ai, Y.; Zhang, M.; Joo, S.W.; Cheney, M.A.; Qian, S. 2010. Effects of electroosmotic flow on ionic current rectification in conical nanopores. *Journal of Physical Chemistry C*. 114: 3883–3890.
[41] Heaton, I.; Platt, M. 2019. Peptide nanocarriers for detection of heavy metal ions using resistive pulse sensing. *Analytical Chemistry*. 91: 11291–11296.
[42] Robinson, J.P. 2013. Wallace H. Coulter: decades of invention and discovery. *Cytometry*. 83A: 424–438.
[43] Kašėta, V.; Kaušylė, A.; Kavaliauskaitė, J.; Peterykitė, M.; Stirkėė, A.; Biziulevičienė, G. 2020. Detection of intracellular biomarkers in viable cells using millisecond pulsed electric fields. *Experimental Cell Research*. 389: artic. 111877.
[44] Chafai, D.E.; Mehle, A.; Tilmatine, A.; Maouche, B.; Miklavcic, D. 2015. Assessment of the electrochemical effects of pulsed electric fields in a biological cell suspension. *Bioelectrochemistry*. 106: 249–257.
[45] Baker, L.A. 2018. Perspective and Prospectus on Single-Entity Electrochemistry. *Journal of the American Chemical Society*. 140: 15549–15559.
[46] Leszczyszyn, D.J.; Jankowski, J.A.; Viveros, O.H.; Diliberto, E.J.; Near, J.A.; Wightman, R.M. 1990. Nicotinic receptor-mediated catecholamine secretion from individual chromaffin cells – chemical evidence for exocytosis. *Journal of Biological Chemistry*. 265: 14736–14737.

[47] Fan, F.R.F.; Bard, A.J. 1995. Electrochemical detection of single molecules. *Science*. 267: 871–874.
[48] Quinn, B.M.; van't Hof, P.; Lemay, S.G. 2004. Time-resolved electrochemical detection of discrete adsorption events. *Journal of the American Chemical Society*. 126: 8360–8361.
[49] Kwon, S.J.; Fan, F.R.F.; Bard, A.J. 2010. Observing Iridium Oxide (IrO_x) Single Nanoparticle Collisions at Ultramicroelectrodes. *Journal of the American Chemical Society*. 132: 13165–13167.
[50] Zhou, Y.-G.; Rees, N.V.; Compton, R.G. 2011. Electrode–nanoparticle collisions: The measurement of the sticking coefficient of silver nanoparticles on a glassy carbon electrode. *Chemical Physics Letters*. 514: 291–293.

ADDITIONAL LITERATURE

Electrochemistry of Cells

Castle, L.M.; Schuh, D.A.; Reynolds, E.E.; Furst, A.L. 2021. Electrochemical Sensors to Detect Bacterial Foodborne Pathogens. *ACS Sensors*. 6: 1717–1730.
Chen, J.; Andler, S.M.; Goddard, J.M.; Nugen, S.R.; Rotello, V.M. 2017. Integrating recognition elements with nanomaterials for bacteria sensing. *Chemical Society Reviews*. 46: 1272–1283.
Wang, J.-T. Ge, L.-Y.; Yuan, X.; Wang, Z.-F.; Li, J.-L.; Guo, X.-L.; Wu, D.-M.; Liu, J.-G. 2012. Detection of the cell viability and proliferation using two-signal electrochemical method. *Analyst*. 137: 3230–3233.
Wu, J.; Wang, R.; Lu, Y.; Jia, M.; Yan, J.; Bian, X. 2019. Facile Preparation of a Bacteria Imprinted Artificial Receptor for Highly Selective Bacterial Recognition and Label-Free Impedimetric Detection. *Analytical Chemistry*. 91: 1027–1033.

Electroluminiscence

Ding, H.; Guo, W.; Su, B. 2020. Imaging cell-matrix adhesions and collective migration of living cells by electrochemiluminescence microscopy. *Angewandte Chemie International Edition*. 59: 449–456.
Dolci, L.S.; Zanarini, S.; Ciana, L.D.; Paolucci, F.; Roda, A. 2009. Development of a new device for ultrasensitive electrochemiluminescence microscopy imaging. *Analytical Chemistry*. 81: 6234–6241.
Engstrom, R.C.; Johnson, K.W.; DesJarlais, S. 1987. Characterization of electrode heterogeneity with electrogenerated chemiluminescence. *Analytical Chemistry*. 59: 670–673.
Ma, C.; Cao, Y.; Gou, X.; Zhu, J.-J. 2020. Recent progress in electrochemiluminescence sensing and imaging. *Analytical Chemistry*. 92: 431–454.
Valenti, G.; Scarabino, S.; Goudeau, B.; Lesch, A.; Jović, M.; Villani, E.; Sentic, M.; Rapino, S.; Arbault, S.; Paolucci, F.; Sojic, N. 2017. Single cell electrochemiluminescence imaging: from the proof-of-concept to disposable device-based analysis. *Journal of the American Chemical Society*. 139: 16830–16837.
Villani, E.; Inagi, S. 2021. Mapping the distribution of potential gradient in bipolar electrochemical systems through luminol electrochemiluminescence imaging. *Analytical Chemistry*. 93: 8152–8160.
Xu, J.; Huang, P.; Qin, Y.; Jiang, D.; Chen, H.-Y. 2016. Analysis of intracellular glucose at single cells using electrochemiluminescence imaging. *Analytical Chemistry*. 88: 4609–4612.

Bipolar Electrochemistry

Fan, Y.; Han, C.; Zhang, B. 2016. Recent advances in the development and application of nanoelectrodes. *Analyst*. 141: 5474–5487.
Goodridge, F.; King, C.J.H.; Wright, A.R. 1977. Performance studies on a bipolar fluidised bed electrode. *Electrochimica Acta*. 22: 1087–1091.
Ulrich, C.; Andersson, O.; Nyholm, L.; Björefors, F. 2008. Formation of molecular gradients on bipolar electrodes. *Angewandte Chemie International Edition*. 47: 3034–3036.

Imaging Techniques

Bard, A.J., Fan, F.R.F., Kwak, J.; Lev, O. 1989. Scanning electrochemical microscopy. Introduction and principles. *Analytical Chemistry*. 61: 132–138.
Comstock, D.J.; Elam, J.W.; Pellin, M.J.; Hersam, M.C. 2010. Integrated ultramicroelectrode–nanopipet probe for concurrent scanning electrochemical microscopy and scanning ion conductance microscopy. *Analytical Chemistry*. 82: 1270–1276.
Hansma, P.K.; Drake, B.; Marti, O.; Gould, S.A.; Prater, C.B. 1989. The scanning ion-conductance microscope. *Science*. 243: 641–643.

Nebel, M.; Grutzke, S.; Diab, N.; Schulte, A.; Schuhmann, W. 2013. Visualization of oxygen consumption of single living cells by scanning electrochemical microscopy: the influence of the faradaic tip reaction. *Angewandte Chemie International Edition*. 52: 6335–6338.

O'Connell, M.A.; Wain, A.J. 2015. Combined electrochemical-topographical imaging: a critical review. *Analytical Methods*. 7: 6983–6999.

Paulose Nadappuram, B.; McKelvey, K.; Al Botros, R.; Colburn, A.W.; Unwin, P.R. 2013. Fabrication and characterization of dual function nanoscale pH-Scanning Ion ConductanceMicroscopy (SICM) probes for high resolution pH mapping. *Analytical Chemistry*. 85: 8070–8074.

Other Techniques

Huang, L.; Fang, M.; Cupp-Sutton, K.A.; Wang, Z.; Smith, K.; Wu, S. 2021. Spray-capillary-based capillary electrophoresis mass spectrometry for metabolite analysis in single cells. *Analytical Chemistry*. 93: 4479–4487.

Olefirowicz, T.M.; Ewing, A.G. 1990. Capillary electrophoresis in 2 and 5 microns diameter capillaries: Application to cytoplasmic analysis. *Analytical Chemistry*. 62: 1872–1876.

Inside Cell Electrochemistry

Fan, Y.; Han, C.; Zhang, B. 2016. Recent advances in the development and application of nanoelectrodes. *Analyst*.141: 5474–5487.

Jiang, H.; Qi, Y.-T.; Wu, W.-T.; Wen, M.-Y.; Liu, Y.-L.; Huang, W.-H. 2020. Intracellular monitoring of NADH release from mitochondria using a single functionalized nanowire electrode. *Chemical Science*. 11: 8771–8778.

Qi, Y.-T.; Jiang, H.; Wu, W.-T.; Zhang, F.-L.; Tian, S.-Y.; Fan, W.-T.; Liu, Y.-L.; Amatore, C.; Huang, W.-H. 2022. Homeostasis inside single activated phagolysosomes: quantitative and selective measurements of submillisecond dynamics of reactive oxygen and nitrogen species production with a nanoelectrochemical sensor. *Journal of the American Chemical Society*. 144: 9723–9733.

Wei, C.; Bard, A.J.; Feldberg, S.W. 1997. Current rectification at quartz nanopipet electrodes. *Analytical Chemistry*. 69: 4627–4633.

Zheng, X.T.; Li, C.M. 2012. Single cell analysis at the nanoscale. *Chemical Society Reviews*. 41: 2061–2071.

Microelectrodes and Nanoelectrodes

Cox, J.T.; Zhang, B. 2012. Nanoelectrodes: recent advances and new directions. *Annual Review of Analytical Chemistry*. 5: 253–272.

Liu, Y.; He, R.; Zhang, Q.; Chen, S. 2010. Theory of electrochemistry for nanometer-sized disk electrodes. *Journal of Physical Chemistry C*. 114: 10812–10822.

Amatore, C., Grun, F., Maisonhaute, E., 2003. Electrochemistry within a Limited Number of Molecules: Delineating the Fringe Between Stochastic and Statistical Behavior. *Angewnadte Chemie International Edition*. 20: 4944–4947.

Matsue, T. 2013. Bioimaging with Micro/Nanoelectrode systems. *Analytical Sciences*. 29: 171–179.

Ying, Y.L.; Ding, Z.; Zhan, D.; Long, Y.T. 2017. Advanced electroanalytical chemistry at nanoelectrodes. *Chemical Science*. 8: 3338–3348.

Single-entity Electrochemistry

Amatore, C.; Arbault, S.; Guille, M.; Lemaitre, F. 2008. Electrochemical monitoring of single cell secretion: vesicular exocytosis and oxidative stress. *Chemical Reviews*. 108: 2585–2621.

Chen, Y.; Wang, D.; Liu, Y.; Gao, G.; Zhi, J. 2021. Redox activity of single bacteria revealed by electrochemical collision technique. *Biosensors and Bioelectronics*. 176: artic. 112914.

Couto, R.A.S.; Chen, L.; Kuss, S.; Compton, R.G. 2018. Detection of *Escherichia coli* bacteria by impact electrochemistry. *Analyst*. 143: 4840–4843.

Ellison, J.; Batchelor-McAuley, C.; Tschulik, K.; Compton, R.G. 2014. The use of cylindrical micro-wire electrodes for nano-impact experiments; Facilitating the sub-picomolar detection of single nanoparticles. *Sensors and Actuators B*. 200: 47–52.

Sepunaru, L.; Tschulik, K.; Batchelor-McAuley, C.; Gavish, R.; Compton, R.G. 2015. Electrochemical detection of single *E. coli* bacteria labeled with silver nanoparticles. *Biomaterials Science*. 3: 816–820.

Tavaddod, S.; Charsooghi, M.A.; Abdi, F.; Khalesifard, H.R.; Golestanian, R. 2011. Probing passive diffusion of flagellated and deflagellated *Escherichia coli*. *European Physics Journal E*. 34: 1–7.

13 Electroanalysis of Plants/Plants in Electroanalysis

13.1 OVERVIEW

Electrochemistry offers a vast amplitude of analytical techniques applicable to plant science, agriculture, food chemistry, and related fields. In its classical meaning, electrochemical sensing is based on the conversion of the concentration signal of a chemical species into electrical signals such as potential, charge, current, impedance, etc., on the surface of an electrode for quantification [1]. For the sake of simplicity, the most extensively used can be grouped into potentiometric, amperometric, voltammetric, and impedimetric, although this does not exhaust the range of available methodologies [2]. In the aforementioned research and application fields, such techniques can be applied for *in vitro* and *in vivo* assays and can cover a variety of analytical demands. In the following, electroanalytical techniques will be presented according to the analytical function/objective rather than from a formal electrochemical perspective. Several previous aspects should be considered.

As described in Chapter 1, potentiometric techniques involve the measurement of the equilibrium potential under conditions of zero current (electromotive force) of a species-selective electrode in contact with the problem solution. This technique, widely used in plant science (see Chapters 2 and 3), permits it to cover a wide range of concentrations but requires high selectivity. Its scope has been increased by the development of solid contact ion-selective electrodes [3]. Voltammetric, amperometric, and impedimetric methods are concerned with similar demands of sensitivity and selectivity. Of course, general analytical demands concerning accuracy and precision, repeatability and reproducibility, traceability, etc. must be accomplished by all analytical methodologies.

13.2 ELECTROCATALYSIS

Voltammetric and amperometric methods have in common the record of the current flowing through an electrode under the application of selected potential inputs. The most frequent approach is the use of electrodes with sensitive and selective modifiers (eventually immobilizing enzymes or other biocomponents) and involves the occurrence of electrocatalytic effects. Generically, electrocatalysis consists of the increase of the standard rate constant of the electrode reaction, resulting in the increase of the faradaic current and the shift of the electrode potential to lower overpotentials relative to the noncatalyzed process [4]. As far as the current enhancement can be masked by other effects, the potential shift criterion is preferred [5].

The classical electrocatalytic process involves a regeneration/reaction scheme. The reduction or oxidation of the substrate species is kinetically hindered. Taking for simplicity the monoelectronic reduction of the oxidized form of the substrate, S^{ox}, to its reduced counterpart, S^{rd}, this means that the process

$$S^{ox} + e^{-} \rightarrow S^{rd} \qquad (13.1)$$

does not occur at an appreciable rate under the accessible conditions of experimentation. The catalyst is a species that, in these conditions, experiences a facile one-electron reduction

$$Cat^{ox} + e^{-} \rightarrow Cat^{rd} \qquad (13.2)$$

under the conditions of operation and can react rapidly with the substrate

DOI: 10.1201/9781032697826-13

$$Cat^{rd} + S^{ox} \rightarrow Cat^{ox} + S^{rd}. \tag{13.3}$$

This reaction regenerates the oxidized form of the catalyst. As a result, the voltammetric response of the Cat^{ox}/Cat^{rd} pair is enhanced, yielding higher voltammetric peak currents. The previous reaction scheme is typical of homogeneous electrocatalysis, where all species are in the solution phase. A similar scheme can be applied when the catalyst or the substrate is immobilized onto the electrode. Figure 13.1 shows a scheme of the electrocatalytic process in which a substrate is reduced with the electrochemical regeneration of the catalyst [6]. Notice that the catalyst can be an adsorbed species, a continuous film, a microparticulate deposit of the solid catalyst, or a catalyst attached to an inert microporous support. In the last cases, the catalytic process also involves charge transport through the solid particle or film. This is constrained by charge conservation requirements and can require the coupling of processes of electron hopping between immobile redox centers and permeation by electrolyte cations or anions [6].

In antioxidant sensing, it is frequent to use electrode modifiers whose effect is to enhance the intensity of the voltammetric peaks of electroactive analytes or a redox probe. This can be due to a reaction/regeneration scheme in which the electroactive catalyst is electrochemically regenerated but also to other pathways generically termed electrocatalytic. In several cases, the electrode modifier binds to the analyte, lowering the energy barrier for electron transfer, thus resulting in current enhancement in voltammetric experiments. In several cases, the electrode modifier acts as a preconcentrating agent increasing the effective concentration of the analyte in the electrode surface. This confinement effect, characteristic of porous electrodes, enhances the voltammetric currents but, strictly, cannot be conceived as a "true" electrocatalytic effect [7].

Also frequently, the electron transfer properties of the electrode modifier are characterized using cyclic voltammetric and electrochemical impedance spectroscopy applied to the modified electrode in contact with a redox probe typically $Fe(CN)_6^{3-}/Fe(CN)_6^{4-}$ in aqueous solution. This is a valid practice that must be taken with caution because the electron transfer properties of the analyte may differ substantially from those of the redox probe [8, 9]. Similarly, it is frequent in literature to attribute the variations in the peak-to-peak separation of the reversible redox probe to the increase/decrease of the rate constant, ignoring the influence of uncompensated ohmic drops (which can be particularly significant in modified electrodes) producing the same effects [6].

Accordingly, electrocatalytic sensing of individual components in samples from plants, foods, beverages, etc. possesses large capabilities but also involves important difficulties. In this context, a detailed knowledge of the electrochemical response of analytes and electrode modifiers plays an

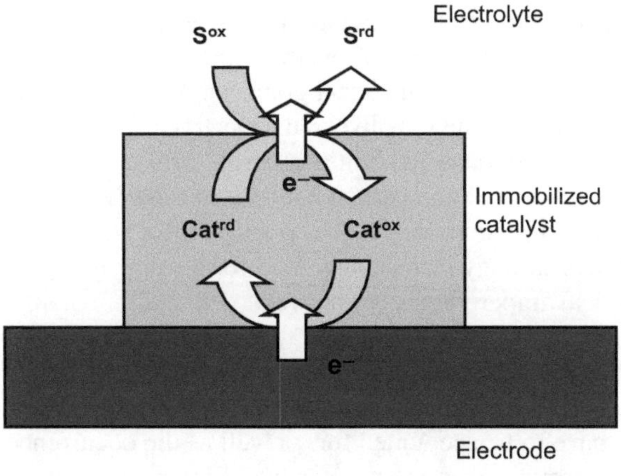

FIGURE 13.1 Schematic representation of an electrocatalytic process involving redox mediation between a substrate in the solution phase and a catalyst immobilized onto the electrode surface.

essential role. Notice that, although frequently the electrochemistry of antioxidants is described in terms of diffusion-controlled processes, most of the oxidative pathways of antioxidants (see Chapter 4) are more complex. Then their description as electron transfer controlled by diffusion and uncomplicated by coupled reactions is merely a rough approximation. The consideration of the electrochemical mechanism is also important as far as it can influence the electrocatalytic pathway.

The aforementioned ideas can be extended to impedimetric techniques, concentrated in electrochemical impedance spectroscopy measurements. Usually, in this type of sensing, the changes in the electrical properties of biorecognition units immobilized onto the working electrode surface are monitored. Variations in impedance are quite sensitive to the interaction between the analyte and the immobilized receptor (including antibody-antigen reactions) [10].

13.3 DETECTION OF INDIVIDUAL PLANT COMPONENTS

Most electroanalytical methods are directed to identify and quantify individual chemical components of plants. These complement the multicomponent methodologies classically used to study vegetal matter, high-performance liquid chromatography-mass spectrometry (HPLC-MS), fluorescence spectroscopy (FLD), gas chromatography-mass spectrometry (GC-MS), micellar electrokinetic capillary chromatography, capillary electrophoresis (CE), and enzyme-linked immunosorbent assay (ELISA).

Ex situ analysis performed in extracts from plants and plant-derived products such as foods, beverages, and infusions cover a plethora of compounds and electrochemical sensors. The fraction of such methods aimed at evaluating the antioxidant capacity or determining groups of compounds (for instance, polyphenols [11]) was described in Chapter 7, while ROS sensing was treated in Chapter 9.

Ascorbic acid (vitamin C) is a paradigmatic example of the enormous variety of electroanalytical methods suitable for the study of individual components of vegetal matter [12]. The analytical performance is conditioned by the need for high sensitivity and high selectivity because ascorbic acid is accompanied in real samples with other potentially interfering components. Potentiometric methods involve the ascorbate oxidase catalyzed reaction in the presence of oxygen [13], or horseradish peroxidase immobilized on the surface of an ion-sensitive field effect transistor (ISFET) in the presence of H_2O_2 [14]. Here, peroxidase reduces H_2O_2 to water in the presence of ascorbic acid. The consumption of ascorbic acid during the enzyme-catalyzed reduction of hydrogen peroxide causes a local pH to increase in the electrode, which is detected by the ISFET. Equivalent nonenzymatic potentiometric sensors have been reported using RuO_2- and MnO_2-modified electrodes [15]. The scope of potentiometric sensors has been expanded by the use of novel contact electrodes and molecularly imprinted polymers [16].

Voltammetric and amperometric techniques have been extensively applied for ascorbic acid detection. These include a variety of strategies for electrode modification and/or electrode pretreatment, including polymers, metal nanoparticles, different graphene forms, quantum dots, enzymes, and others. The voltammetric response is quite sensitive to the electrode modification, as illustrated in Figure 13.2, where the cyclic voltammograms for 6 mM ascorbic acid at unmodified GCE (b1), and GCE modified with carbon nanotubes (b2) and carbon nanoplatelets derived from ground cherry (*Physalis peruviana* L.) husks (b3) are superimposed [17]. The third sensor was obtained by carbonization of a plant precursor, a preparative strategy increasingly used for sensing and energy storage devices [18].

Figure 13.3 shows typical amperometric graphs recorded on fresh orange juice under the application of a constant oxidative potential high enough to promote the oxidation of ascorbic acid under diffusion-controlled conditions. The successive additions of 10 μM ascorbic acid result in sharp current steps, which are typical of amperometric procedures. As for voltammetric sensing, interference, and matrix effects have to be accounted for, as well as the occurrence of secondary reactions, electrode fouling, and adsorption phenomena. Repeatability, reproducibility, and interference tests must be applied. Apart from accuracy and precision, robustness plays an essential role for practical sensing purposes.

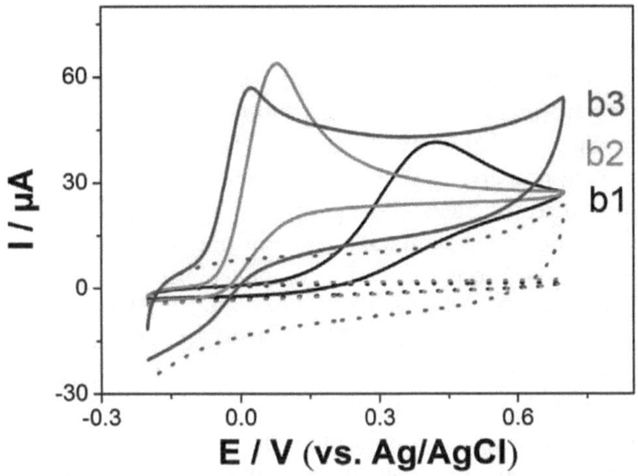

FIGURE 13.2 Cyclic voltammograms recorded for a 6 mM ascorbic acid solution in 0.1 M PBS at pH 7.0 at unmodified GCE (b1), GCE modified with carbon nanotubes (b2), and carbon nanoplatelets derived from ground cherry (***Physalis peruviana***) husks. Dotted lines correspond to voltammetric responses in the absence of ascorbic acid. Potential scan rate 10 mV s^{-1}. (Adapted from ref. [17], with permission. Copyright Elsevier 2017.).

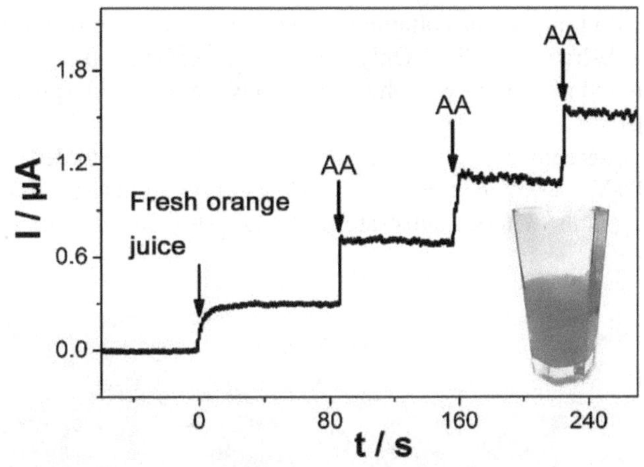

FIGURE 13.3 Amperometric curves recorded on fresh orange juice upon successive additions of 10 µM ascorbic acid. Applied potential 0.026 V vs. SCE. (Adapted from ref. [17], with permission. Copyright Elsevier 2017.).

The same philosophy and techniques can be applied for the detection of plant nutrients, signaling compounds (treated in previous chapters), contaminants, elicitors of defense responses [19], metabolic inhibitors [20], and plant growth stimulators [21], among others. Of course, strategies for electrode modification can also be applied to *in situ* analysis of plants and cells, eventually cultured on the electrode surface.

13.4 IN SITU DETECTION OF BIOMARKERS OF PLANT HEALTH AND STRESS

There are a variety of biomarkers associated with plant growth and health, as well as plant stress, phytohormones such as salicylic acid and indole-3-acetic acid in particular. Electrochemical sensing is being increasingly studied, including direct electrochemical, immunoelectrochemical, and photoelectrochemical methods [22] in the context of sustainable agriculture and the so-called precision agriculture. Here, sensors cover soil analysis (sensing for temperature, moisture, conductivity, and/or salinity level), air analysis (pressure, temperature, humidity, wind speed), and plant analysis (temperature, wetness, moisture, CO_2, photosynthesis) [23]. In general, plant-based functional sensors detect a change in any plant property or a signal generated via genetic expression of biomarkers in turn detected directly or indirectly via detectable products generated in trigger reactions.

Figure 13.4 shows a device for voltammetric monitoring stress in *Aloe vera* L. The sensor is inserted into the plant leaves and consists of three graphene electrodes, which were laser-engraved on a polyimide sheet [24]. The sensor is responsive to salicylic acid, producing "clean" square wave voltammetric responses in leaf homogenates (Figure 13.5) and inserted into plant leaves.

Voltammetric sensing on plants in parts of plants deposited onto graphitic electrodes has been reported for sensing and mapping phytohormones [25].

An example of indirect sensing is drought monitoring in genetically modified tobacco plants based on the expression of the β-D-glucuronidase enzyme. The enzyme is expressed under drought conditions and monitored using the enzyme-substrate (4-nitrophenyl β-D-glucuronide) reaction. This generates an electrochemically active product, 4-nitrophenol, which displays a well-defined oxidative voltammetry.

Figure 13.6a compares the cyclic voltammetric responses recorded at a three-electrode chip fabricated on a 4-inch silicon wafer covered with a thermal oxide layer on watered (W+) and unwatered (W−) tobacco plants after injection of PBS (S−) or PBS plus 4-nitrophenyl β-D-glucuronide substrate (S+). One can see that voltammograms of watered plants, both with and without substrate (W+,S+, and W+,S−), do not present voltammetric signals. The same result is obtained for unwatered plants without substrate (W−,S−). Only with the nitrophenyl β-D-glucuronide substrate is injected into the unwatered plant (W−,S+), the oxidation wave for 4-nitro phenol at ca. 0.4 V vs. Ag/AgCl is recorded [26].

These results are in agreement with chronoamperometric data recorded upon application of a constant potential of 0.4 V. As shown in Figure 13.6b, the chronoamperograms of the control watered plants and unwatered plants without substrate, although showing different baselines, present a

E.

FIGURE 13.4 Photographic image of a voltammetric sensor for salicylic acid attached to leaves of ***Aloe vera***. (Adapted from ref. [24], with permission. Copyright the Royal Society of Chemistry 2024.).

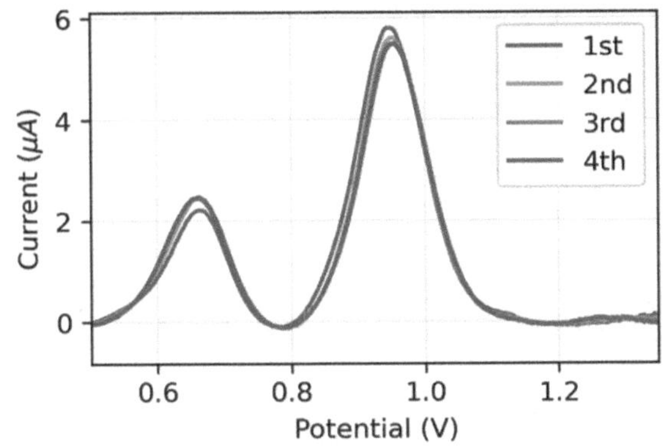

FIGURE 13.5 Square wave voltammograms (four successive scans are superimposed) recorded for a graphene-based sensor in contact with a homogenate of Philodendron Brasil leaves. (Adapted from ref. [24], with permission. Copyright the Royal Society of Chemistry 2024.).

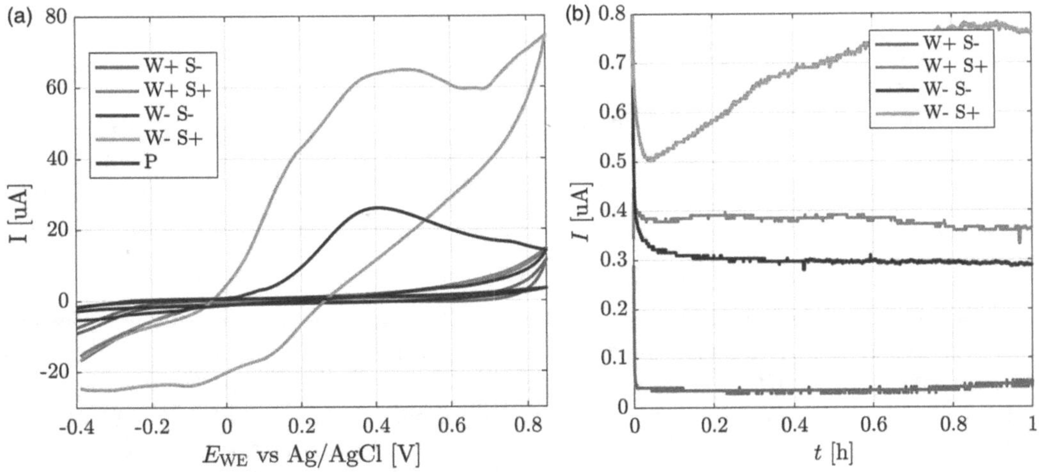

FIGURE 13.6 (a) Cyclic voltammograms recorded at a three-electrode chip fabricated on a 4-inch silicon wafer covered with a thermal oxide layer on watered (W+) and unwatered (W−) genetically modified tobacco plants injected with PBS plus (S+) and without (S−) 4-nitrophenyl β-D-glucuronide substrate. P denotes the cyclic voltammogram of 4-nitro phenol in PBS; potential scan rate 100 mV s^{-1}. (b) Corresponding chronoamperograms recorded at potentials ca. 0.4 V. (Adapted from ref. [26], with permission. Copyright Elsevier 2022.).

common sharp current decrease, reducing to an essentially horizontal current/time pattern. In contrast, the unwatered plants injected with the substrate produce a progressive current increase a few minutes after the initial abrupt current decay.

Chronoamperometric data can be interpreted based on the Michaelis-Menten model for enzymatic kinetics. Here, the substrate S reacts with the enzyme E to give an adduct ES, which decomposes into a final product P regenerating the enzyme.

$$S + E \xleftrightarrow{k_1\,(k_{-1})} ES \xrightarrow{k_2} E + P \tag{13.4}$$

Here, three rate constants (k_1, k_{-1}, and k_2) are involved. It is assumed that, under stress, the plant generates locally the enzyme and is transported by the plant fluids near the electrode surface. If the first step proceeds rapidly (quasi-equilibrium conditions), the concentration of substrate-enzyme complex, c_{SE}, satisfies the relationship:

$$c_{SE} = \frac{c_E * c_S}{c_S + K_m}, \tag{13.5}$$

where c_E* is the total enzyme concentration and c_S is the substrate concentration. K_m is the Michaelis-Menten constant, equal to $(k_1+k_2)/k_{-1}$. It will be assumed that only the final product P is electrochemically responsive, being oxidized at the electrode surface. The rate of product formation can be approximated by the expression

$$\frac{dc_P}{dt} \approx \frac{k_2 c_E * c_S}{c_S + K_m}. \tag{13.6}$$

This equation leads to two limiting situations. When there is a large concentration of substrate ($c_S >> K_m$), it tends to

$$\frac{dc_P}{dt} \approx k_2 c_E *, \tag{13.7}$$

corresponding to a zero-order kinetics. In the opposite case ($c_S << K_m$), Eq. (13.6) tends to

$$\frac{dc_P}{dt} \approx \frac{k_2 c_E * c_S}{K_m}, \tag{13.8}$$

which corresponds to a first-order kinetics.

The previous equations apply to a process taking place in a homogenous solution. However, under in situ measurement in plants, the product can be generated in the volume of the plant organelle and must be transported by plant fluids toward the electrode surface. Accordingly, the rate of the electrochemical oxidation process depends on the applied potential, the rate of enzymatic catalysis, and the transport properties. A simplified scheme can be applied assuming that the current measured in a chronoamperometric experiment at an electrical potential large enough to promote P oxidation under diffusion-controlled conditions, the current i, can be approximated by

$$i \approx nFD_{eff}S_{eff}\frac{c_P}{\sqrt{D_{eff}t}}. \tag{13.9}$$

Here, n represents the electron stoichiometry of the P oxidation process, S_{eff} is the effective electrode area, and D_{eff} is the effective coefficient of diffusion of P. This equation can be seen as equivalent to that corresponding to a redox process under semi-infinite diffusion conditions (the Cottrell equation).

Figure 13.7 shows an interpretative scheme for measurements in Figure 13.6. The product P is generated in the leaf and is transported to the electrode surface. The typical pore diameter (typically, ca. 30 μm) and the inter-pore distance ranges between 10 and 100 μm. The aforementioned semi-infinite diffusion conditions apply when the distance between the pores and the electrode (estimated as 10^{-2}–10^{-1} cm [26]) is longer than the diffusion length, L_{diff} (= $(D_{eff}t)^{1/2}$). Accordingly, the characteristic time for typical values of diffusivity, $D_{eff} \approx 10^{-6}$ cm^2 s^{-1}, will be ten thousands of seconds.

Chronoamperometric curves in Figure 13.6b can be approached by assuming that Eq. (13.9) must incorporate a delay time (t_d) representative of the transport through the leaf and a term (i_b) associated with the oxidation of other products rather than P on the electrode surface, and replacing the time exponent −1/2 by an exponent x [26]:

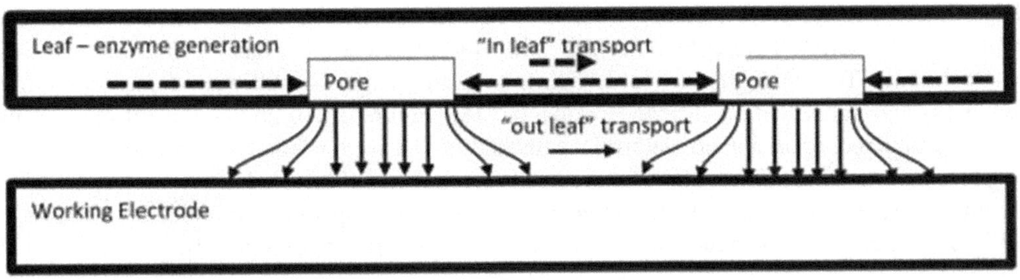

FIGURE 13.7 Scheme for the interpretation of chronoamperometric curves in Figure 13.6. The product P is generated in the leaf bulk and is transported to the electrode surface through the leaf pores. (Reproduced from ref. [26], with permission. Copyright Elsevier 2022.).

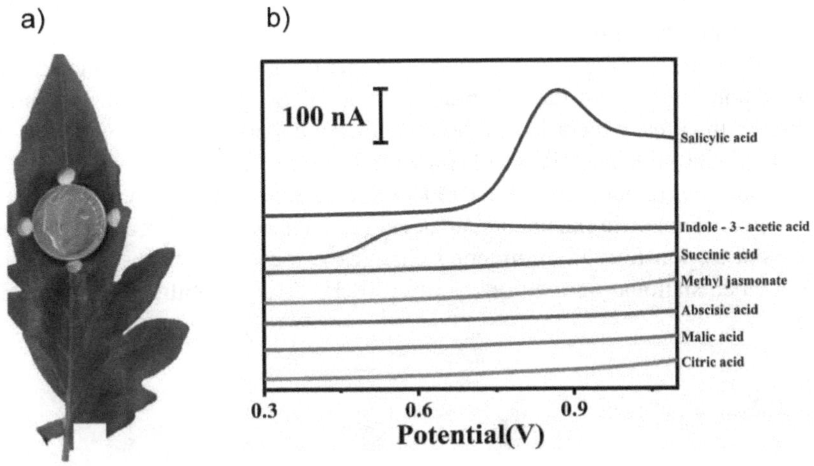

FIGURE 13.8 (a) Image of a tomato leaf with a nickel coin. (b) Square wave voltammograms recorded for salicylic acid, indole-3-acetic acid, succinic acid, methyl jasmonate, abscisic acid, malic acid, and citric acid in 0.2 M phosphate buffer solutions at pH 7.5. Potential scan initiated at 0.2 V vs. Ag/AgCl in the positive direction, potential step increment 5 mV, square wave amplitude 25 mV, frequency 20.0 Hz, quiet time 50 s. (Adapted from ref. [27], with permission. Copyright Elsevier 2019.).

$$i \approx a\left(t - t_{\mathrm{d}}\right)^{x} + i_{\mathrm{b}}, \tag{13.10}$$

where a is a coefficient depending on the rate of generation of the enzyme. In the case of the cell-on-chip mounted on the leaves of tobacco plants, experimental data can be reproduced satisfactorily for $x = ½$ [27].

Electrochemical sensing can be made, as previously noted, by sampling on leaf surfaces and transferring onto carbon tape-based electrodes. Figure 13.8a shows a photographic image of a tomato leaf incorporating a nickel coin (US 5-ct) for positioning the sampling points [27]. The sensor was fabricated through pencil trace by hand drawing onto a base ITO glass electrode and produced voltammetric records, as illustrated in Figure 13.8b. This figure shows the square wave voltammograms for salicylic acid, indole-3-acetic acid, succinic acid, methyl jasmonate, abscisic acid, malic acid, and citric acid. One can see that only the two first compounds produce voltammetric peaks and that these appear at clearly separated potentials. Accordingly, the sensor could detect salicylic acid and then monitor the infection of tomato plants by *Pseudomonas syringae* pv. Tomato DC3000 (pst DC3000), a fungus [27].

Paper-based devices are increasingly studied. Figure 13.9 shows an example also aimed at the voltammetric detection of salicylic acid [28]. The working electrode consists of a piece of carbon tape attached to ITO glass, the ensemble being covered by nafion/multiwalled carbon nanotubes (MWCNTs) and submitted to further treatments [28]. Figure 13.10 shows the differential pulse voltammograms recorded at such electrodes attached to diseased, normal, and phytoene desaturase gene silent tomato leaves. The voltammetric peak corresponds to the oxidation of salicylic acid and its height increases significantly from the normal to the tomato specimen infected by *Botrytis cinerea* Whetzel. In contrast, the gene silent specimen displays a lower voltammetric signal for salicylic acid oxidation.

Figure 13.11 shows another example of chronoamperometric sensing [29] devoted to the detection of the heat-induced expression of the β-glucuronidase enzyme in genetically transformed *Nicotiana tabacum* L. plants. The sensor is based on silicon wafers where gold working and auxiliary electrodes and an electroplating Ag/AgCl reference electrode were imprinted [29]. Nitrophenol is the product of the reaction between the enzyme and the substrate, and, as previously noted, it can be electrochemically detected at potentials ca. 0.7 V vs. Ag/AgCl. The assays were performed with the substrate PhG using the wild-type cells lacking the β-glucuronidase enzyme (GUS−) and transgenic (GUS+) cells producing the enzyme injecting either phosphate buffer or phenolphthalein beta-D glucuronide. Upon injection of the substrate into the GUS+ plant, there is an oxidative current significantly higher than those recorded in the two control experiments, GUS+ plants without phenolphthalein β-D-glucuronide and GUS− plants with that substrate.

Impedimetric sensors are based on the use of EIS as a detection mode, where different parameters can be used to detect changes in the plant state at a wide interval of time scales. An essential condition to detect plant response to environmental changes is the repeatability and reproducibility of impedance data when stationary conditions apply [30, 31]. The capabilities of impedance analysis

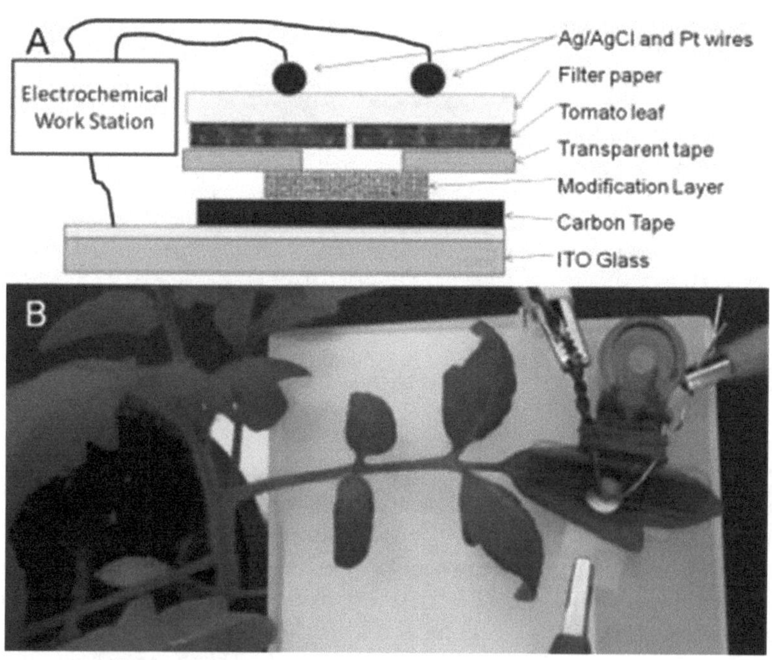

FIGURE 13.9 (a) Schematic cross section and (b) photographic image of paper-based device for in situ detection of salicylic acid in a tomato leaf. A clasp with the reference (Ag/AgCl) and auxiliary (Pt wire) electrodes were used to attach the filter paper, and a carbon tape working electrode to the tomato leaf. (Adapted from ref. [28], with permission. Copyright Elsevier 2014.).

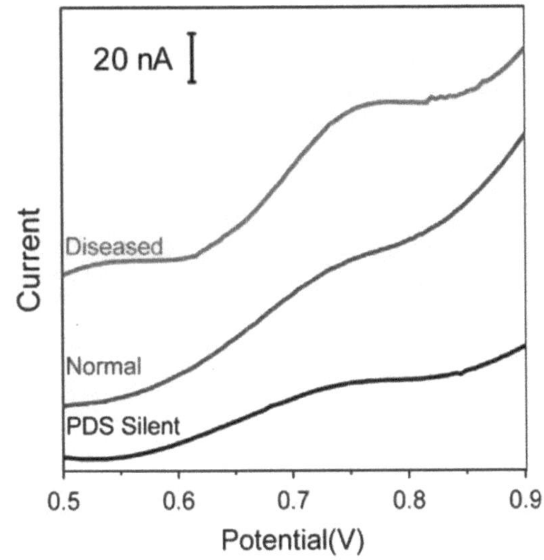

FIGURE 13.10 Differential pulse voltammetry at MWCNTs/nafion modified carbon tape in paper-based devices anchored to diseased (infected by ***Botrytis cinerea***), normal, and phytoene desaturase gene silent tomato leaves. Potential scan initiated at 0.4 V vs. Ag/AgCl in the positive direction; potential increment 4 mV, pulse amplitude 50 mV, pulse width 0.2 s, sampling width 0.067 s, pulse period 0.5 s, quiet time 50 s. (Adapted from ref. [28], with permission. Copyright Elsevier 2014.).

are illustrated in Figure 13.12. In this figure, the variation of the real part of the total impedance (a) and the (minus) phase angle (b) on the logarithm of the frequency in impedance spectra is recorded *in vivo* at *Nicotiana tabacum* plants over ten days [30]. The spectra were measured using needles and Kelvin clips inserted into the plant stem. The data offered consistent results, with small day-to-day variations of the total impedance and almost negligible differences in the phase angle.

In situ detection of biomarkers of plant health and stress is currently under intensive research. Recent developments include a variety of plant-wearable electrochemical, photoelectrochemical, and immunoelectrochemical sensors [22].

13.5 DETECTION OF CONTAMINANTS, POLLUTANTS AND ADULTERATIONS

A significant fraction of electroanalytical research is devoted to the detection of contaminants and pollutants in plants, foods, beverages, etc. In the context of environmental studies, contaminants are hazardous chemical species passing to the environment in measurable quantities because of natural phenomena and/or human activity. Pollutants are contaminants that produce perceptible effects in living organisms and/or objects of social interest [32].

Apart from the global aspects such as climatic change directly associated with anthropogenic activity, the knowledge of the mechanisms determining the pollutants uptake, accumulation, resistance, and release by plants is important for agriculture, food science, and biomedical applications. During their evolution, plants have developed different mechanisms of accumulation (and hyperaccumulation), tolerance, and exclusion in response to pollutant stress [33]. The use of plants for environmental remediation is another aspect of considerable interest [34]. Sensing has increasing importance in the context of precision agriculture and the so-called process analytical technology [35].

Accumulation of toxic metals and metalloids in the cultivated lands where they pass to crops and alimentary chains is an important environmental problem [36]. The risks can be condensed into a series of indexes. The index of geoaccumulation, F_{geo}, describes the relationship between the total concentration of problem metal in the sample and the average concentration of the same in the study area [37]:

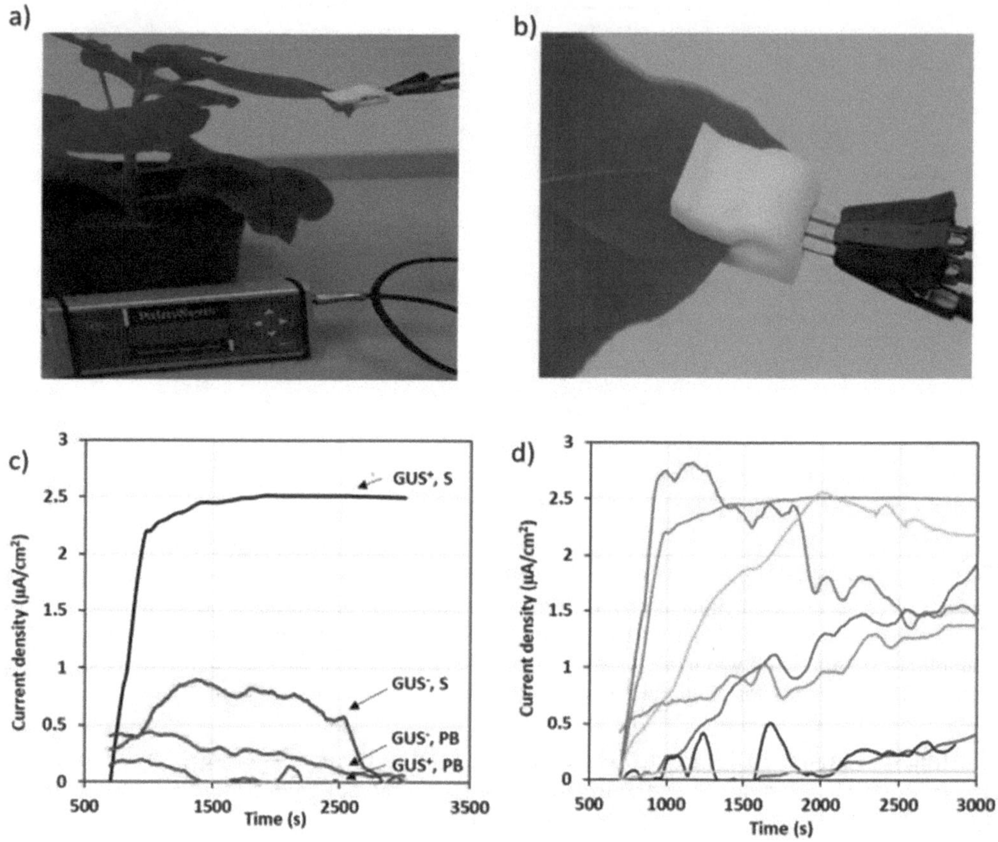

FIGURE 13.11 (a, b) Scheme of the connections of a chip to the leaves of the tobacco plant. (c, d) Chronoamperometric monitoring of the generation of nitrophenol in the reaction of β-glucuronidase enzyme. The applied potential was 0.7 V vs. Ag/AgCl. (c) Current/time curves for different GUS– and GUS+ ***N. tabacum*** plants with substrate phenolphthalein beta-D glucuronide (S) in phosphate buffer and phosphate buffer (PB), (d) Chronoamperometric measurements on GUS+ plant with phenolphthalein beta-D glucuronide at different leaf and different positions of the leaf. (Adapted from ref. [29], with permission. Copyright Elsevier 2018.).

$$F_{\text{geo}} = \left(\frac{c_{\text{i}}}{Kb_{\text{i}}} \right), \tag{13.11}$$

where c_i represents the concentration of the pollutant i in the sample, b_i is the average concentration of the pollutant in the region under study, and K is an empirical coefficient whose value is generally near 1.5. The pollution index, F_{poll}, balances the maximum and average concentration of the i-pollutant [38],

$$F_{\text{poll}} = \sqrt{\frac{1}{2}\left[\left(\frac{c_{\text{i}}}{s_{\text{i}}} \right)_{\text{max}}^2 - \left(\frac{c_{\text{i}}}{s_{\text{i}}} \right)_{\text{aver}}^2 \right]}, \tag{13.12}$$

with s_i being the standard or reference value of the concentration of the pollutant.

The ecological risk index, R_{ecol}, combines the effect of pollutant concentration, toxicity level, and environmental sensitivity. This index is expressed as [39]

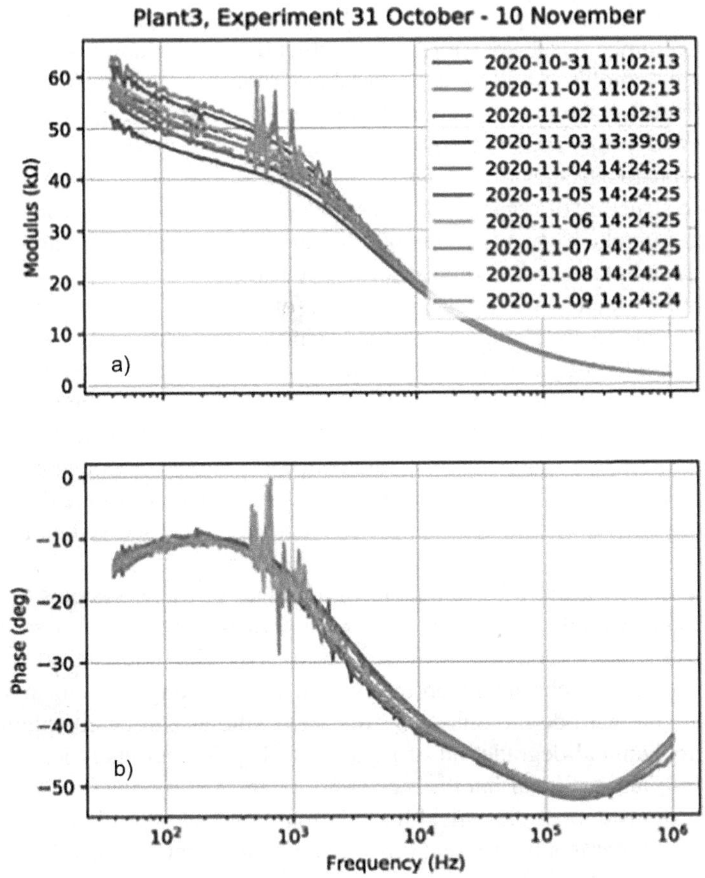

FIGURE 13.12 Variation of the real part of the total impedance (a) and the (minus) phase angle (b) on the logarithm of frequency recorded ***in vivo*** at ***Nicotiana tabacum*** plants over ten days. (Adapted from ref. [30], with permission. Copyright Elsevier 2022.).

$$R_{\text{ecol}} = \sum_{i=1}^{N} \frac{T_{\text{i}} c_{\text{i}}}{S_{\text{i}}}. \tag{13.13}$$

Here, T_{i} is the toxicity response coefficient of the pollutant i.

Finally, the bioconcentration factor (BCF) expresses the ability of crops to absorb and enrich heavy metals from soil. This factor is defined as the ratio between the concentration of the pollutant (mg kg^{-1}) in the ripe crop (including corn, oats, soybeans, broad beans, potato, and beans), c_{crop}, and its concentration in the corresponding soil, c_{soil} [40].

$$BCF = \frac{c_{\text{crop}}}{c_{\text{soil}}} \tag{13.14}$$

A variety of analytical methods have been devised for determining the concentration of heavy metals in crops and soils [41]. These are dominated currently by inductively coupled plasma mass spectrometry, but traditional electroanalytical methods for trace metal analysis can also be used.

Organic pollutants include polychlorinated dibenzo-*p*-dioxins and polychlorinated dibenzofurans, polychlorinated biphenyls, bisphenol A, and herbicides but also antibiotics from anthropogenic sources. These are commonly found in agricultural environments and constitute a serious environmental concern. Plants uptake of organic pollutants either from soil or air as schematized in Figure 13.13. The uptake and translocation of organic pollutants depend on their physicochemical properties, the biological characteristics of the plants, and the environmental media [42]. The molecular mass and molecular structure of organic pollutants, related to their volatility and hydrophobicity, are determinants of their uptake by plants. Compounds that have low volatility can only be absorbed by roots, while volatile pollutants can also be absorbed by leaves.

The aforementioned factors influence the kinetics of the absorption of pollutants through the root/soil and leaf/air interfaces. For quantification purposes, the partition coefficients of the pollutant between octanol and water and between octanol and air are used, respectively [43]. The redox extractable lipid content, the carbohydrate content of plant cell walls, the amount and composition of epicuticular waxes, the permeability and pH of biomembranes, the electrical gradients across such membranes, and others (transpiration stream concentration factor) are plant-characteristic factors determining the uptake of pollutants. The main environmental factors are the type and pH of the background electrolyte, the organic content, the concentration of dissolved carbon in the soil, and the temperature [42].

Electroanalytical methods are of application for a variety of living systems (generically termed biota [44]) and are increasingly studied for the detection of pollutants in foods and beverages [45, 46]. As in the case of plant components previously treated, electroanalytical techniques include potentiometry, voltammetry, chronoamperometry, and impedimetry accompanying an enormous variety of strategies of electrode modification. Analytical studies can be complemented with those aimed at the electrochemical degradation of pollutants [47]. Recent analytical tendencies involve electronic noises [48] and artificial intelligence [49]. An interesting electroanalytical aspect is the detection of adulterations in pharmaceutical formulations made of natural products. Here, synthetic products are used as adulterants in different phytotherapeutic formulations [50].

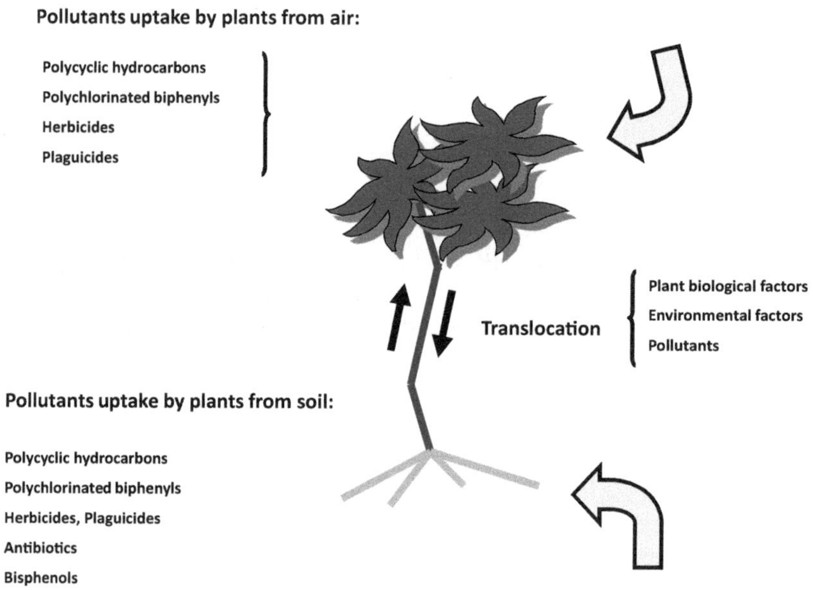

FIGURE 13.13 Scheme for the uptake and translocation of organic pollutants in plants.

13.6 SOLID-STATE TECHNIQUES

As noted in previous chapters, solid-state electrochemical techniques focused on the voltammetry of immobilized particles methodology are of application in the context of plant science for different purposes. These include the characterization of tea varieties [51] and herbal formulations [52]. Figure 13.14 compares the square wave voltammetric responses of the microparticulate deposits resulting from the evaporation of ethanolic extracts of several commercial herbal formulations on glassy carbon electrodes in contact with aqueous phosphate buffer at neutral pH. The voltammograms at laboratory and portable types of equipment were equivalent and consisted of oxidation signals associated with polyphenolic components and characterized the different nominal species, *Caralluma adscendens* (Roxb.) Haw (a, b), *Phaseolus vulgaris* L. (c, d), and *Cordia ecalyculata* Vell. (e, f).

The solid-state nature of the involved electrochemical processes can be tested by using SECM. Figure 13.15 shows the result of the application of different potential inputs to a microparticulate deposit of the ethanolic extract of a commercial herbal formulation of *Caralluma adscendens* on Pt substrate. The redox probe was a $K_4Fe(CN)_6$ solution in acetate buffer at pH 4.75. Fixing the tip potential at 0.30 V vs. Ag/AgCl, the oxidation of $Fe(CN)_6^{4-}$ to $Fe(CN)_6^{3-}$ occurs under diffusion-limited conditions so the fluctuations in the tip current map the variations in the topography and conductivity of the substrate. The particles from the herbal formulation appear as negative feedback features (in dark gray) marked over the background (pale gray) corresponding to the uncovered Pt basal electrode. When no potential was applied to the substrate, the color map reflected the topography of the particles. When an oxidative potential capable of oxidizing the herbal components (see Figure 13.14) is applied, the profile of the particles changes significantly in the borderline.

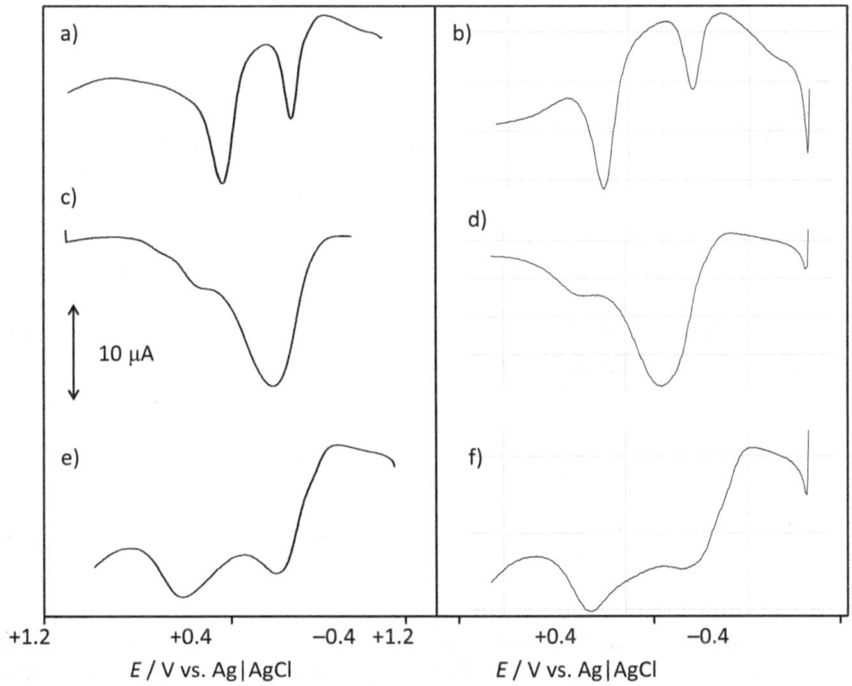

FIGURE 13.14 Square wave voltammograms of microparticulate deposits of ethanolic extracts of commercial herbal formulations of *Caralluma adscendens* (a, b); *Phaseolus vulgaris* (c, d), and *Cordia ecalyculata* (e, f) on glassy carbon electrode at the laboratory (a, c, e) and portable (b, d, f) equipment. Electrolyte: air-saturated 0.10 M aqueous potassium phosphate buffer at pH 7.00. Potential scan initiated at −0.25 V in the positive direction, potential step increment 4 mV, square wave amplitude 25 mV, frequency 5 Hz. (Reproduced from ref. [52], with permission. Copyright the Royal Society of Chemistry 2015.).

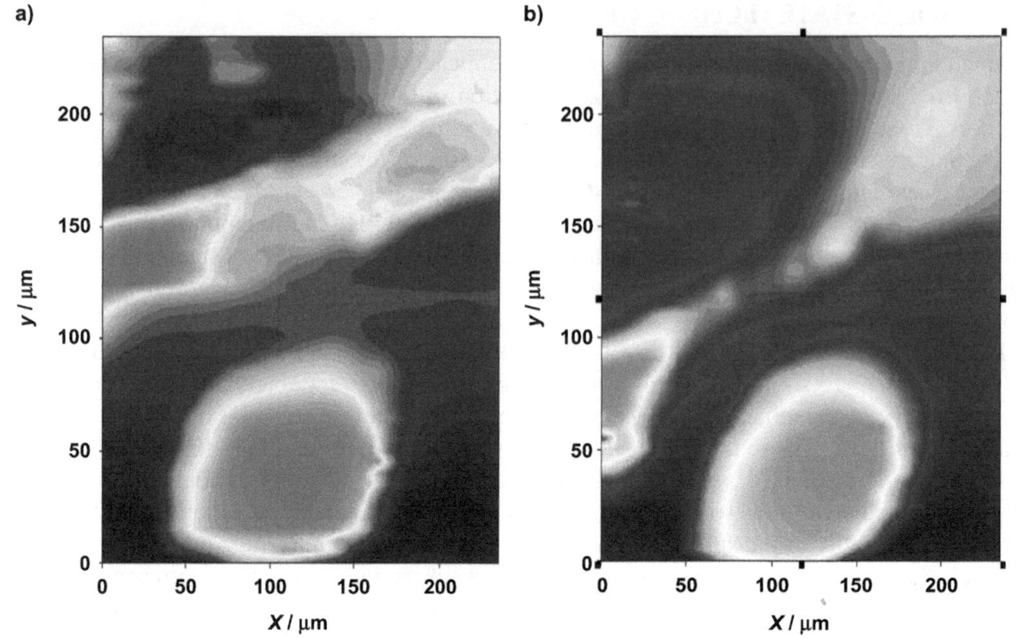

FIGURE 13.15 SECM map colors of a microparticulate deposit of the ethanolic extract of a commercial herbal formulation of *Caralluma adscendens* on Pt substrate in contact with 2.0 mM $K_4Fe(CN)_6$ plus 0.25 M HAc/NaAc, pH 4.75. The tip potential was fixed at 0.30 V vs. Ag/AgCl, and the substrate potential was at 0.0 V (a) and 0.30 V (b).

This feature is consistent with the occurrence of solid-state oxidations, which are initiated at the particle/base electrode/electrolyte three-phase boundary [53] (see also Figure 4.11 in Chapter 4).

Solid-state techniques have also been reported for screening [54] and quantification [55] of pharmacologic adulterant classes in herbal formulations. The simplest approach involves the determination of the relative amount of the components of the sample. In the most favorable situation, two components of interest, A and B, display separated, well-defined voltammetric peaks. Then their molar or mass ratio (m_A/m_B) in the sample can be obtained from the ratio between the respective peak currents (i_{pA}/i_{pB}) or peak areas (A_{pA}/A_{pB}), assuming linearity in the current response on the net number of particles of each component in the electrode, i.e. [52],

$$\frac{m_A}{m_B} = h\left(\frac{i_{pA}}{i_{pB}}\right) = H\left(\frac{A_{pA}}{A_{pB}}\right). \tag{13.15}$$

This procedure requires, however, previous calibration for determining the values of the electrochemical constants h, H, which depend on the electrochemical conditions (potential scan rate, etc.) and the pair of compounds under study.

In the case of overlapping peaks, we can measure the currents (i_I, i_{II}) at two potentials I and II, at which both components (A, adulterant, and B, herbal component) are responsive. Assuming, as before, linear responses, the measured currents will be

$$i_J = g_{JA}m_A + g_{JB}m_B \quad J = I, II, \tag{13.16}$$

where g_{JA}, g_{JB}, are the respective electrochemical coefficients of response. Notice that m_A, m_B represent the masses or mole numbers of the respective compounds attached to the electrode. Although the masses m_A, m_B vary from one experiment to another, their ratio will be representative of their proportion in the herbal sample. Then, the m_A/m_B ratio can be calculated from the i_I/i_{II} ratio providing that the ratio of the response coefficients is known from previous calibration experiments, as [52]

$$\frac{m_A}{m_B} = \frac{\left(\frac{g_{IIB}}{g_{IA}}\right)\left(\frac{i_I}{i_{II}}\right) - \left(\frac{g_{IB}}{g_{IA}}\right)}{1 - \left(\frac{g_{IIA}}{g_{IA}}\right)\left(\frac{i_I}{i_{II}}\right)}. \tag{13.17}$$

Absolute quantification can be obtained by enriching the solid sample with a known amount of an external standard displaying a voltammetric signal separated from the signals of A and B [53, 55].

Other reported analytical applications of solid-state electrochemistry include studies for screening antibacterials [56, 57], and detecting and monitoring the production of melatonin and other related compounds in different *Saccharomyces* strains [58]. The identification of the optimal seasonal point of harvest of plants [59], and the estimate of organic matter content in sediments aimed at their characterization [60], among other applications, has been reported.

13.7 PLANTS IN ELECTROANALYSIS

Given the interest in the development of nontoxic, high-yielding, and environmentally friendly nanomaterials for diverse applications, the use of extracts obtained from different plant tissues is being studied as sources of such nanomaterials. Although seeds, leaves, tubers, barks, roots, and fruits can be used, leaves are the preferred sources due to their ready availability [61]. Many applications involve the exploitation of the reducing properties of plant extracts to reduce the precursor compounds to metal nanoparticles [62]. A second, increasingly extended, application is the preparation of porous carbonaceous materials via carbonization of vegetal sources [63].

Apart from the aforementioned, plant-derived materials can be applied directly as electrode modifiers. This is the case of plant materials of the genus *Sphagnum*, which have been embedded into carbon paste matrices. Plant materials absorb cations preferentially from dilute solutions, thus acting as effective preconcentration agents [64]. A similar strategy has been reported for the bioaccumulation of gold at antheral cells from the flower *Datura inoxia* Mill. [65].

13.8 CHIRALITY

Chirality appears at different structural and functional levels of plants. Many plant materials such as climbing tendrils, snail shells, twisted leaves, and even flower petals, can exhibit chiral growth [66]. The mechanisms underlying macroscopic chirality remain unclear, but they can be attributed to the asymmetry of growth, deswelling, surface stresses, and other phenomena ultimately associated with anisotropic atomic or molecular structures [67]. Accordingly, a variety of electroanalytical strategies have been devised for discriminating enantiomers. These include chiral solvents, chiral supporting electrolytes, and/or chiral electrodes. The most obvious creation of chirality in electrodes is via molecularly imprinted polymers, but there are different forms of generation of chiral electrode surfaces – namely, growth of metal oxide films on metal surfaces, adsorption of chiral complexes onto clays and electrode modification with solids incorporating chiral centers [68].

Generally, the enantiomer species produce identical or quite similar voltammetric responses. Then, the enantioselectivity derives from the different intensities of the voltammetric response. This current-based approach is widely extended in recent electroanalytical literature, but its application to real enantiomer recognition is difficult. Future research should expand conventional electrochemical enantiomer recognition to connect with structural and functional chirality in plants.

REFERENCES

[1] Bakker, E.; Telting-Diaz, M. 2002. Electrochemical sensors. *Analytical Chemistry*. 74: 2781–2800.

[2] Scholz, F. Ed. 2022. *Electroanalytical Methods*. Springer, Berlin.

[3] Cheong, Y.H.; Ge, L.; Lisak, G. 2021. Highly Reproducible Solid Contact Ion Selective Electrodes: Emerging Opportunities for Potentiometry - A Review. *Analytica Chimica Acta*. 1162: artic. 338304.

[4] Bard, A.J.; Faulkner, L.R. 2001. *Electrochemical methods*, 2nd edn. John Wiley & Sons, New York

[5] Bard, A.J.; Inzelt, G.; Scholz, F. Eds. 2008. *Electrochemical Dictionary*. Springer, Berlin-Heidelberg, p. 185.

[6] Doménech-Carbó, A. 2021. *Electrochemistry of Porous Materials*, 2nd edn. Taylor & Francis, Boca Raton.

[7] Streeter, I.; Wildgoose, G.G.; Shao, L.D.; Compton, R.G. (2008) Cyclic voltammetry on electrode surfaces covered with porous layers: An analysis of electron transfer kinetics at single-walled carbon nanotube modified electrodes. *Sensors and Actuators B*. 133: 462–466.

[8] Chen, P.; McCreery, R.L. 1996 Control of electron transfer kinetics at glassy carbon electrodes by specific surface modification. *Analytical Chemistry*. 68: 3958–3965.

[9] Davies, T.J.; Hyde, M.E.; Compton, R.G. 2005 Nanotrench arrays reveal insight into graphite electrochemistry. *Angewandte Chemie International Edition*. 44: 5121–5126.

[10] Umapathi, R.; Ghoreshian, S.M.; Sonwal, S.; Rani, G.M.; Huh, Y.S. 2022 Portable electrochemical sensing methodologies for on-site detection of pesticide residues in fruits and vegetables. *Coordination Chemistry Reviews*. 453: artic. 214305.

[11] Piovesan, J.V.; Jost, C.L.; Spinelli, A. 2015. Electroanalytical determination of total phenolic compounds bysquare-wave voltammetry using a poly(vinylpyrrolidone)-modifiedcarbon-paste electrode. *Sensors & Actuators B*. 216: 192–197.

[12] Pisoschi, A.M.; Pop, A.; Serban, A.I.; Fafaneata, C. 2014. Electrochemical methods for ascorbic acid determination. *Electrochimica Acta*. 121: 443–460.

[13] Fernandes, J.C.B.; Kubota, L.T.; de Oliveira Neto, G. 1999. Potentiometric biosensor for l-ascorbic acid based on ascorbate oxidase of natural source immobilized on ethylene–vinylacetate membrane. *Analytica Chimica Acta*. 385: 3–12.

[14] Volotovsky, V.; Kim, N. 1998. Ascorbic acid determination with an ion-sensitive field effect transistor-based peroxidase biosensor. *Analytica Chimica Acta*. 359: 143–148.

[15] Luo, X.-L.; Xu, J.-J.; Zhao, W.; Chen, H.-Y. 2004. Ascorbic acid sensor based on ion-sensitive field-effect transistor modified with MnO_2 nanoparticles. *Analytica Chimica Acta*. 512: 57–61.

[16] Tonelli, D.; Ballarin, B.; Guadagnini, L.; Mignani, A.; Scavetta, E. 2011. A novel potentiometric sensor for l-ascorbic acid based on molecularly imprinted polypyrrole. *Electrochimica Acta*. 56: 7149–7154.

[17] Li, X.; Wang, Y.; Liu, J.; Sun, M.; Bo, X.; Wang, H.-L.; Zhou, M. 2017. Amperometric ascorbic acid biosensor based on carbon nanoplatelets derived from ground cherry husks. *Electrochemistry Communications*. 82: 139–144.

[18] Wang, L.; Zhang, Q.; Chen, S.; Xu, F.; Chen, S.; Jia, J.; Tan, H.; Hou, H.; Song, Y. 2014. Electrochemical sensing and biosensing platform based on biomass-derived macroporous carbon materials. *Analytical Chemistry*. 86: 1414–1421.

[19] Besson-Bard, A.; Griveau, S.; Bedioui, F.; Wendehenne, E. 2008. Real-time electrochemical detection of extracellular nitric oxide in tobacco cells exposed to cryptogein, an elicitor of defence responses. *Journal of Experimental Botany*. 59: 3407–3414.

[20] Kinpara, T.; Murakami, Y.; Yokoyama, K.; Tamiya, E. 2001. Electrochemical Monitoring of Plant Stress Responses. *Electroanalysis*. 13: 451–456.

[21] Janíková, L.; Chýlková, J.; Šelešovská, R.; Sedlák, M.; Váňa, J.; Dušek, L.; Bartáček, J. 2020. Electrochemical behavior of plant growth stimulator 1-naphthaleneacetic acid and its voltammetric determination using boron doped diamond electrode. *Journal of Electroanalytical Chemistry*. 859: artic. 113855.

[22] Li, Z.; Zhou, J.; Dong, T.; Xu, Y.; Shang, Y. 2021. Application of electrochemical methods for the detection of abiotic stress biomarkers in plants. *Biosensors and Bioelectronics*. 182: artic. 113105.

[23] Kim, M.-Y.; Lee, K.H. 2022. Electrochemical Sensors for Sustainable Precision Agriculture—A Review. *Frontiers in Plant Science*. 10: artic. 848320.

[24] Nejad, S.K.; Ma, H.; Al-Shami, A.; Soleimani, A.; Mohamed, M.A.; Dankwah, P.; H.J. Lee, Mousavi, M.P.S. 2024. Sustainable agriculture with LEAFS: a low-cost electrochemical analyzer of foliage stress. *Sensors & Diagnostics*. 3, 400–411.

[25] Sun, L.-J.; Zhou, J.-J.; Pan, J.-L.; Liang, Y.-Y.; Fang, Z.-J.; Xie, Y.; Yang, H.; Gu, H.-Y.; Bao, N. 2018. Electrochemical mapping of indole-3-acetic acid and salicylic acid in whole pea seedlings under normal conditions and salinity. *Sensors and Actuators B*. 276: 545–551.

[26] Desagani, D.; Jog, A.; Teig-Sussholz, O.; Avni, A.; Sacham-Diamand, Y. 2022. Drought monitoring in tobacco plants by in-vivo electrochemical biosensor. *Sensors and Actuators B*. 356: artic. 131357.

[27] Wang, H.-R.; Bi, X.-M.; Fang, Z.-J.; Yang, H.; Gu, H.-Y.; Sun, L.-J.; Bao, N. 2019. Real time sensing of salicylic acid in infected tomato leaves using carbon tape electrodes modified with handed pencil trace. *Sensors and Actuators B*. 286: 104–110.

[28] Sun, L.J.; Feng, Q.M.; Yan, Y.F.; Pan, Z.Q.; Li, X.H.; Song, F.M.; Yang, H.; Xu, J.J.; Bao, N.; Gu, H.Y. 2014. Paper-based electroanalytical devices for in situ determination of salicylic acid in living tomato leaves. *Biosensors & Bioelectronics*. 60: 154–160.

[29] Pandey, R.; Teig-Sussholz, O.; Schuster, S.; Avni, A.; Shacham-Diamand, Y. 2018. Integrated electrochemical Chip-on-Plant functional sensor for monitoring gene expression under stress. *Biosensors & Bioelectronics*. 117: 493–500.

[30] Garlando, U.; Calvo, S.; Barezzi, M.; Sanginario, A.; Ros, P.M.; Demarchi, D. 2022. Ask the plants directly: Understanding plant needs using electrical impedance measurements. *Computers and Electronics in Agriculture*. 193: artic. 106707.

[31] Kadan-Jamal, K.; Sophocleous, M.; Jog, A.; Deshagani, D.; Teig-Sussholz, O.; Georciou, J.; Ovni, A.; Sacham-Diamand, Y. 2020. Electrical Impedance Spectroscopy of plant cells in aqueous biological buffer solutions and their modelling using a unified electrical equivalent circuit over a wide frequency range: 4Hz to 20 GHz. *Biosensors & Bioelectronics*. 168: artic. 112485.

[32] Bowen, H.J.M. 1966. *Trace Elements in Biochemistry*. Academic Press, London.

[33] Angulo-Bejarano, P.I.; Puente-Rivera, J.; Cruz-Ortega, R. 2021. Metal and Metalloid Toxicity in Plants: An Overview on Molecular Aspects. *Plants*. 10: artic. 635.

[34] Pedrosa Gomes, M.; Gusmäo Pereira, G.; Qiu, B.-S.; Juneau, P. 2021. Editorial: Coping With Pollution – the Effects of Environmental Contaminants on Plant Growth and Physiology. *Frontiers in Plant Science*. 12: artic. 740802.

[35] Salt, D.E.; Smith, R.D.; Raskin, I. 1998. Phytoremediation. *Annual Review of Plant Physiology*. 49: 643–668.

[36] Xie, N.; Kang, C.; Ren, D.; Zhang, L. 2022. Assessment of the variation of heavy metal pollutants in soil and crop plants through field and laboratory tests. *Science of the Total Environment*. 811: artic. 152343.

[37] Muller, G. 1969. Index of geoaccumulation in sediments of the Rhine River. *GeoJournal*. 2: 109–118.

[38] Nemerow, N.L. 1974. *Scientific stream pollution analysis*. Script Book Co., New York, pp. 210–231.

[39] Hakanson, L. 1980. An ecological risk index for aquatic pollution control. A sedimentological approach. *Water Research*. 14: 975–1001.

[40] Huang, M.; Zhou, S.; Bo, S.; Zhao, Q. 2008. Heavy metals in wheat grain: assessment of potential health risk for inhabitants in Kunshan, China. *Science of the Total Environment*. 405: 54–61.

[41] Reeves, R.D.; Baker, A.J.M.; Jaffré, T.; Erskine, P.D.; Echevarria, G.; Van Der Ent, A. 2018. A Global Database for Plants That Hyperaccumulate Metal and Metalloid Trace Elements. *New Phytologist*. 218: 407–411.

[42] Zhang, C.; Feng, Y.; Liu, Y.-W.; Chang, H.-Q.; Li, Z.-J.; Xue, J.-M. 2017. Uptake and translocation of organic pollutants in plants: A review. *Journal of Integrative Agriculture*. 16: 1659–1668.

[43] Gao, Y.; Zhu, L.; Ling, W. 2005. Application of the partition-limited model for plant uptake of organic chemicals from soil and water. *Science of the Total Environment*. 336: 171–182.

[44] Álvarez-Ruiz, R.; Picó, Y. 2020. Analysis of emerging and related pollutants in aquatic biota. *Trends in Environmental Analytical Chemistry*. 25: artic. 111.

[45] Zhang, Y.; Lei, Y.; Lu, H.; Shi, L.; Wang, P.; Ali, Z.; Li, J. 2021. Electrochemical detection of bisphenols in food: A review. *Food Chemistry*. 346: artic. 128895.

[46] Jain, U.; Saena, K.; Hooda, V.; Balayan, S.; Singh, A.P.; Tikadar, M.; Chauhan, N. 2022. Emerging vistas on pesticides detection based on electrochemical biosensors – An update. *Food Chemistry*. 371: artic. 131126.

[47] Aragay, G.; Pino, F.; Merkoçi, A. 2012. Nanomaterials for sensing and destroying pesticides. *Chemical Reviews*. 112: 5317–5338.

[48] Gobbi, E.; Falasconi, M.; Zambotti, G.; Sberveglieri, V.; Pulvirenti, A.; Sberveglieri, G. 2015. Rapid diagnosis of Enterobacteriaceae in vegetable soups by a metal oxide sensor based electronic nose. *Sensors and Actuators B*. 207: 1104–1113.

[49] Yigit, E.; Sabanci, K.; Toktas, A.; Kayabasi, A. 2019. A study of visual features of leaves in plant identification using artificial intelligence techniques. *Computers and Electronics in Agriculture*. 156: 369–377.

[50] de Carvalho, L.M.; Martini, M.; Moreira, A.P.L.; de Lima, A.P.S.; Correia, D.; Falcao, T.; Garcia, S.C.; de Bairros, A.V.; do Nascimento, P.C.; Bohrer, D. 2011. Presence of synthetic pharmaceuticals as adulterants in slimming phytotherapeutic formulations and their analytical determination. *Forensic Science International*. 204: 6–11.

[51] Domínguez, I.; Doménech-Carbó, A. 2015. Screening and authentication of tea varieties based on microextraction-assisted voltammetry of microparticles. *Sensors and Actuators B*. 210: 491–499.

[52] Martini, M.; De Carvalho, L.M.; Blasco, A.; Doménech-Carbó, A. 2015. Screening and authentication of herbal formulations based on microextraction-assisted voltammetry of microparticles. *Analytical Methods*. 7: 5740–5747.

[53] Scholz, F.; Schröder, U.; Gulabowski, R.; Doménech-Carbó, A. 2014. *Electrochemistry of Immobilized Particles and Droplets*, 2nd ed. Springer, Berlin-Heidelberg.

[54] Doménech-Carbó, A.; Martini, M.; De Carvalho, L.M.; Doménech-Carbó, M.T.; Silva, M. 2013. Screening of pharmacologic adulterant classes in herbal formulations using voltammetry of microparticles. *Journal of Pharmaceutical and Biomedical Analysis*. 74: 194–204.

[55] Doménech-Carbó, A.; Martini, M.; De Carvalho, L.M.; Viana, C.; Doménech-Carbó, M.T.; Silva, M. 2013. Standard additions-dilution method for absolute quantification in voltammetry of microparticles. Application for determining psychoactive 1,4-benzodiazepine and antidepressants drugs as adulterants in phytotherapeutic formulations. *Journal of Pharmaceutical and Biomedical Analysis*.vol. 80: 159–163.

[56] Doménech-Carbó, A.; Maciuk, A.; Figadère, B.; Poupon, E.; Cebrián-Torrejón, G. 2013. Solid State Electrochemical Assay of Heme-Binding Molecules for Screening of Drugs with Antimalarial Potential. *Analytical Chemistry*. 85: 4014–4021.

[57] Brinvillier, D.; Barrast, M.; Couderc-Murillo, P.; Bono-Yagüe, J.; Rousteau, A.; Gómez Escribano, A.P.; Palmeira, M.V.; Doménech-Carbó, A.; Thesnor, V.; Passe-Coutrin, N.; Sylvestre, M.; Vázquez-Manrique, M.P.; Cebrián-Torrejón, G. 2022. Spectroscopic, electrochemical and biological assays of copper-binding molecules for screening of different drugs and plant extracts against neurodegenerative disorders. *ACS Omega*. 7: 16260–16269.

[58] Muñiz-Calvo, S.; Guillamón, J.M.; Domínguez, I.; Doménech-Carbó, A. 2017. Detecting and monitoring the production of melatonin and other related indole compounds in different *Saccharomyces* strains by solid state electrochemical techniques. *Food Analytical Methods*. 10: 1408–1418.

[59] Cebrián-Torrejón, G.; Kablan, L.; Ferreira, M.E.; Rodríguez de la Cruz, D.; Doménech-Carbó, A.; Vera de Bilbao, N.; Rojas de Arias, A.; Figadère, B.; Poupon, E.; Fournet, A. 2015. Harvesting canthinones: identification of the optimal seasonal point of harvest of Zanthoxylum chiloperone leaves as a source of 5-methoxycanthin–6-one. *Natural Products Research*. 29: 2054–2058.

[60] Cebrián-Torrejón, G.; Pérez, A.; Montoya, N.; Piquero-Cilla, J.; Saldarriaga, M.S.; Gutiérrez, D.; Sanders, C.J.; Machado, W.; Doménech-Carbó, A. 2019. Electrochemical characterization of mangrove sediments: a proposal of new proxies for organic matter oxidation. *Applied Geochemistry*. 101: 42–49.

[61] Mendes Hacke A.C.; Lima, D.; Kuss, S. 2022. Green synthesis of electroactive nanomaterials by using plant-derived natural products. *Journal of Electroanalytical Chemistry*. 922: artic. 116786.

[62] Jadoun, S.; Arif, R.; Jangid, N.K.; Meena, R.K. 2021. Green synthesis of nanoparticles using plant extracts: a review. *Environmental Chemistry Letters*. 19: 355–374.

[63] Wang, L.; Zhang, Q.; Chen, S.; Xu, F.; Chen, S.; Jia, J.; Tan, H.; Hou, H.; Song, Y. 2014. Electrochemical sensing and biosensing platform based on biomass-derived macroporous carbon materials. *Analytical Chemistry*. 86: 1414–1421.

[64] Wang, J.; Naser, N.; Darnall, D.; Gardea-Torresdey, J. 1992. Bioaccumulation and quantitation of metals at peat moss-modified electrodes. *Electroanalysis*. 4: 71–76.

[65] Wang, J.; Tian, B.; Rayson, G.D. 1992. Bioaccumulation and voltammetry of gold at flower-biomass modified electrodes. *Talanta*. 39: 1637–1642.

[66] Wang, J.-S.; Wang, G.; Feng, X.-Q.; Kitamura, T.; Kang, Y.-L.; Yu, S.-W.; Qin, Q.-H. 2013. Hierarchical chirality transfer in the growth of Towel Gourd tendrils. *Scientific Reports*. 3: artic. 3102.

[67] Chen, Z.; Majidi, C.; Srolovitz, D.J.; Haataja, J. 2011. Tunable helical ribbons. *Applied Physics Letters*. 98: artic. 011906.

[68] Doménech-Carbó, A.; Alarcón, J. 2007. Microheterogeneous Electrocatalytic Chiral Recognition at Monoclinic Vanadium-Doped Zirconias: Enantioselective Detection of Glucose. *Analytical Chemistry*. 79: 6742–6751.

ADDITIONAL LITERATURE

Sensing Plant Components

Gan, T.; Hu, C.; Chen, Z.; Hu, S. 2010. Direct Electrochemical Determination of Methyl Jasmonate in Wheat Spikelet at a Nano-Montmorillonite Film Modified Electrode by Derivative Square Wave Voltammetry. *Journal of Agricultural and Food Chemistry*. 58: 8942–8947.

Liu, Y.; Li, M.; Wang, G.; Long, Y.; Li, A.; Yang, B. (2019) In Situ Detection of Melatonin and Pyridoxine in Plants Using a CuO–Poly(L-lysine)/Graphene-Based Electrochemical Sensor. *ACS Sustainable Chemistry & Engineering*. 7: 19537–19545.

Sensing Plant Health and Stress

Artigas, J.; Jimenez, C.; Lemos, S.G.; Nogueira, A.R.A.; Torre-Neto, A.; Alonso, J. 2003. Development of a Screen-Printed Thick-Film Nitrate Sensor Based on a Graphite-Epoxy Composite for Agricultural Applications. *Sensors and Actuators B*. 88: 337–344.

Duhan, J.S.; Kumar, R.; Kumar, N.; Kaur, P.; Nehra, K.; Duhan, S. 2017. Nanotechnology: The New Perspective in Precision Agriculture. *Biotechnology Reports*. 15: 11–23.

Gieling, T.; Vanstraten, G.; Janssen, H.; Wouters, H. 2005. ISE and Chemfet Sensors in Greenhouse Cultivation. *Sensors Actuators B*. 105: 74–80.

Liu, D.; Li, M.; Li, H.; Li, C.; Wang, G.; Li, P.; Yang, B. 2021. Core-shell Au@Cu2O-graphene-polydopamine interdigitated microelectrode array sensor for in situ determination of salicylic acid in cucumber leaves. *Sensors and Actuators B*. 341: artic. 130027.

Okajima, M.; Nakagawa, H.; Sugiyama, M. 2023. Possibility of directly sensing plant stress under environment temperature changes using electrochemical impedance spectroscopy. *Japanese Journal of Applied Physics*.62: artic. 088002.

Pedersen, S.M.; Lind, K.M. 2017. *Precision Agriculture: Technology and Economic Perspectives*. Springer, Berlin.

Ren, Q.-Q.; Huang, X.-R.; Liu, G.-C.; Ou-Yang, J.; Li, M.-T.; Chen, H.; Zhao, Y.-D.; Chen, W. 2015. A field-compatible technique using an electrochemical sensing microbundle for real-time and simultaneous in vivo measurement of hydrogen peroxide, nitric oxide, and pH under drought stress. *Sensors and Actuators B*. 220: 743–748.

Shafi, U.; Mumtaz, R.; García-Nieto, J.; Hassan, S.A.; Zaidi, S.A.R.; Iqbal, N. 2019. Precision Agriculture Techniques and Practices: From Considerations to Applications. *Sensors*. 19: artic. 3796.

Sneha, M.; Ravindranath, N.A.; Jayaraman, V. 2024. Electrochemical Investigations of Plant Response to Salt Stress: Insights Through Impedance Studies of OECT-Based Biosensor. *Chemistry Europe*. 9: artic. e202302928.

Sun, L.-J.; Xie, Y.; Yan, Y.-F.; Yang, H.; Gu, H.-Y.; Bao, N. 2017. Paper-based Analytical Devices for Direct Electrochemical Detection of Free IAA and SA in Plant Samples with the Weight of Several Milligrams. *Sensors Actuators B*. 247: 336–342.

Contaminants, Pollutants

Alistair, B.; Paul, J.; Edward, J.S.; Chris, J.S.; Edward. S.; Len, S.L. 2006. Uptake of veterinary medicines from soils into plants. *Journal of Agricultural and Food Chemistry*. 54: 2288–2297.

Briggs, G.G.; Bromilow, R.H.; Evans, A.A. 1982. Relationships between lipophilicity and root uptake and translocation of non-ionised chemicals by barley. *Pesticide Science*.13: 495–504.

Clemens, S. 2006. Toxic metal accumulation, responses to exposure and mechanisms of tolerance in plants. *Biochimie*. 88: 1707–1719.

Collins, C.; Fryer, M.; Grosso, A. 2006. Plant uptake of non ionic organic chemicals. *Environmental Science and Technology*. 40: 45–52.

Collins, D.C. 2008. A semi-quantitative approach to deriving a model structure for the uptake of organic chemicals by vegetation. *International Journal of Phytoremediation*. 10: 371–377.

Gao, Y.; Zhu, L.; Ling, W. 2005. Application of the partition-limited model for plant uptake of organic chemicals from soil and water. *Science of the Total Environment*. 336: 171–182.

Gugoasa, L.A. 2019. Review-Electrochemical sensors for determination of the endocrine disruptor, bisphenol A. *Journal of The Electrochemical Society*. 167: Artic. 037506.

Harding Jr, L.W.; Phillip Jr, J.H. 1978. Polychlorinated biphenyl effects on marine phytoplankton photosynthesis and cell division. *Marine Biology*. 49: 93–101.

Korte, F.; Kvesitadze, G.; Ugrekhelidze, D.; Gordeziani, M.; Khatisashvili, G.; Buadze, O.; Zaalishvili, G.; Coulston, F. 2000. Organic toxicants and plants. *Ecotoxicology & Environmental Safety*. 47: 1–26.

Li, L.; Wang, Q.; Qiu, X.H.; Dong, Y.; Jia, S.L.; Hu, J.X. 2014. Field determination and QSPR prediction of equilibrium-status soil/vegetation partition coefficient of PCDD/Fs. *Journal of Hazardous Materials*. 76:278–286.

Muhammad, M.S. 2013. Human health risk from heavy metal via food crops consumption with wastewater irrigation practices in Pakistan. *Chemosphere*. 93: 2230–2238.

Paterson, S.; Mackay, D.; Tam, D.; Shiu, W.Y. 1990. Uptake of organic chemicals by plants: A review of processes, correlations and models. *Chemosphere*. 21: 297–331.

Ryan, J.A.; Bell, R.M.; Davidson. J.M.; O'Connor. G.A. 1988. Plant uptake of non-ionic organic chemicals from soils. *Chemosphere*. 17: 2299–2323.

Sterling, T.M. 1994. Mechanisms of herbicide absorption across plant membranes and accumulation in plant-cells. *Weed Science*. 42: 263–276.

Rezania, S.; Taib, S.M.; Md Din, M.F.; Dahalan, F.A.; Kamyab, H. 2016. Comprehensive review on phytotechnology: heavy metals removal by diverse aquatic plants species from wastewater. *Journal of Hazardous Materials*. 318: 587–599.

Verma, N.; Bhardwaj, A. 2015. Biosensor Technology for Pesticides—A review. *Applied Biochemistry and Biotechnology*. 175: 3093–3119.

Wang, W.; Xu, Y.; Cheng, N.; Xie, Y.; Huang, K.; Xu, W. 2020. Dual-recognition aptazyme-driven DNA nanomachine for two-in-one electrochemical detection of pesticides and heavy metal ions. *Sensors and Actuators B*. 321: artic. 128598

Zhang, H.; Chen, J.; Ni, Y.; Zhang, Q.; Zhao, L. 2009. Uptake by roots and translocation to shoots of polychlorinated dibenzo-p-dioxins and dibenzofurans in typical crop plants. *Chemosphere*. 76: 740–746.

Adulterations

Nunes-Vergara, L.J.; Bolo, S.; Olea-Azar, C.; Navarrete-Encina, P.A.; Squella, J.A. 1997. Cyclic voltammetric and EPR spectroscopic studies of benzodiazepines: alprazolam and flunitrazepam. *Journal of Electroanalytical Chemistry*. 436: 227–238.

Lozano-Chaves, M.E.; Palacios-Santander, J.M.; Cubillana-Aguilera, L.M.; Naranjo-Rodriguez, I.; Hidalgo-Hidalgo-de-Cisneros, J.L. 2006. Modified carbon-paste electrodes as sensors for the determination of 1,4-benzodiazepines: application to the determination of diazepam and oxazepam in biological fluids. *Sensors and Actuators B*. 115: 575–583.

Phytoremediation

Dalvi, A.A.; Bhalerao, S.A. 2013. Response of plants towards heavy metal toxicity: an overview of avoidance, tolerance and uptake mechanism. *Annals of Plant Sciences*. 2: 362–368.

Göhre, V.; Paszkowski, U. 2006. Contribution of the arbuscular mycorrhizal symbiosis to heavy metal phytoremediation. *Planta*. 223: 1115–1122.

Yan, A.; Wang, Y.; Tan, S.N.; Yusof, M.L.M.; Ghosh, S.; Chen, Z. 2020. Phytoremediation: A Promising Approach for Revegetation of Heavy-Metal Polluted Land. *Frontiers in Plant Science*.11: artic. 358.

Electrochemical Noses

Ampuero, S.; Bosset, J.O. 2003. The electronic nose applied to dairy products: a review. *Sensors and Actuators B*. 94: 1–12.

Peris, M.; Escuder-Gilabert, L. 2009. A 21st century technique for food control: electronic noses. *Analytica Chimica Acta*. 638: 1–15.

van den Berg, F.; Lyndgaard, C.F.; Sørensen, K.M.; Engelsen, S.B. 2013. Process analytical technology in the food industry. *Journal of Trends Food Science and Technology*. 31: 7–35.

Solid-state Electrochemistry

Cebrián-Torrejón, G.; Kablan, L.; Roffman A.B.; Ferreira, M.E.; Rodríguez de la Cruz, D.; Doménech-Carbó, A.; Vera de Bilabao, N.; Rojas de Arias, A.; Cech, N.B.; Figadère, B.; Poupon, E.; Fournet, A. 2017. Phytoelectrochemical analysis of *Zanthoxylum chiloperone*. *Phytochemical Analysis*. 28: 171–175.

Pérez, A.; Cebrián-Torrejón, G.; Montoya, N.; Piquero-Cilla, J.; Sanders, C.J.; Doménech-Carbó, A.; Machado, W. 2021. Organic matter redox state driven by specific sources in mangrove sediments: A case study from Peruvian ecosystems. *Journal of Marine Science and Engineering*. 9: artic. 1438.

Matignon, L.; Lo, M.M.; Monpierre, M.; Vicentini, M.; Valencia, D.P.; Palmeira-Mello, M.; Sylvestre, M.-N.; Runeau; Sylvestre, M.; Doménech-Carbó, A.; Befodda, Z.; Meffre, P.; Cebrián-Torrejón, G. 2023. Phytochemical and Biological Study of Trophic Interaction Between *Pseudosphinx Tetrio* L. *Larvae and Allamanda Cathartica L. Plants*. 12: artic. 520.

Martínez-Nieto, M.I.; González-Orenga, S.; Soriano, P.; Prieto-Mossi, J.; Larrea, E.; Doménech-Carbó, A.; Tofei, A.M.; Vicente, O.; Mayoral, O. 2022. Are Traditional Lima Bean (Phaseolus lunatus L.) Landraces Valuable to Cope with Climate Change? Effects of Drought on Growth and Biochemical Stress Markers. *Agronomy*. 12: artic. 1715.

Sivager, G.; Calvez, L.; Btuyere, S.; Boisne-Noc, R.; Hufnagel, B.; Cebrián-Torrejón, G.; Doménech-Carbó, A.; Gros, O.; Ollitrault, P.; Morillon, R. 2022. Better tolerance to Huanglongbing is conferred by tetraploid Swingle citrumelo roostock and is influenced by the ploidy of the scion. *Frontiers in Plant Science*. 13: artic.1030862.

Soriano, P.; Estrelles, E.; Martínez-Nieto, M.I.; Doménech-Carbó, A.; Galiè, M.; Biondi, E. 2022. Environmental predictors of seed germination in two Halocnemum species from Mediterranean (Balearic, Tyrrenic and Adriatic) and Red Sea coastal salt marshes. *Seed Science Research*. 32: 246–263.

Sussmann, R.A.; Gabriel, H.B.; Bofill; I.; Menchaca, D.S.; Yamaguchi, L.F.; Kato, M.J.; Doménech-Carbó, A.; Cebrián-Torrejón, G.; Kimura, E.A.; Katzin, A.M. 2019. Biosynthesis of heme O in intraerythrocytic stages of *Plasmodium falciparum* and potential inhibitors of this pathway. *Scientific Reports*. 9:: artic. 19261.

Sussmann, R.A.; Gabriel, H.B.; García, A.; Menchaca, D.S.; Yamaguchi, L.F.; Doménech-Carbó, A.; Cebrián-Torrejón, G.; Kimura, E.A.; Kato, M.J.; Bofill, I.; Crispín, M.; Katzin, A.M. 2022. Presence of phylloquinone in the intraerythrocytic stages of *Plasmodium falciparum*. *Frontiers in Plant Science*. 13: artic. 1030862.

Plants in Electroanalysis

Dulyaseree, P.; Fujishige, M.; Yoshida, I.; Toya, Y.; Banba, Y.; Tanaka, Y.-s.; Aoyama, T.; Phonyiem, M.; Wongwiriyapan, W.; Takeuchi, K.; Endo, M. 2017. Nitrogen-rich green leaves of papaya and *Coccinia grandis* as precursors of activated carbon and their electrochemical properties. *RSC Advances*. 7: 42064–42072.

Huston, M.; DeBella, M.; DiBella, M.; Gupta, A. 2021. Green synthesis of nanomaterials. *Nanomaterials*. 11: artic. 2130.

Zhou, M.; Dong, S. 2011. Bioelectrochemical interface engineering: toward the fabrication of electrochemical biosensors, biofuel cells, and self-powered logic biosensors. *Accounts of Chemical Research*. 44: 1232–1243.

Zhou, M.; Guo, S. 2015. Electrocatalytic interface based on novel carbon nanomaterials for advanced electrochemical sensors. *ChemCatChem*. 7: 2744–2764.

Chirality and Chiral Analysis

Darwin, C. 1865. On the movements and habits of climbing plants. *Botanical Journal of the Linnean Society*. 9: 1–118.

Isnard, S.; Silk, W.K. 2009. Moving with climbing plants from Charles Darwin's time into the 21st century. *American Journal of Botany*. 96: 1205–1221.

Kuhn, A.; Anson, F.C. 1996. Effects of chirality during electrochemical oxidation of 2,3 butanediol stereoisomers. *Journal of Electroanalytical Chemistry*. 410: 243–246.

Scholz, F.; Gulaboski, R.; Mirceski, V.; Langer, P. 2002. Quantification of the chiral recognition in electrochemically driven ion transfer across the interface water/chiral liquid. *Electrochemistry Communications*. 4: 659–662.

Vargas, A.; Santarossa, G.; Iannuzzi, M.; Baiker, A. 2008. Chiral Recognition on Catalytic Surfaces: Theoretical Insight in a Biomimetic Heterogeneous Catalytic System. *Journal of Physical Chemistry C*. 112: 10200–10208.

14 Vegetal Matter in Energy Production and Storage

14.1 INTRODUCTION

The origins of the electrical energy production from vegetal sources can be associated with the pioneering work of Potter who produced an electric current in a suspension of *Escherichia coli* and yeast [1]. Further studies produced bacterial fuel cells [2]. They can be seen as the precursors of contemporary microbial fuel cells (MFCs) [3] and, ultimately, a variety of bioelectrochemical systems aimed to produce, convert, and store energy [4].

Photosynthetic organisms can convert light energy into electrical energy using different bioelectrochemical systems (BES). In general, the aim is to oxidize components of high molecular mass to simplest compounds such as CO_2. BES can be grouped according to the type of organism used at the anode side in (a) photosynthetic, (b) photovoltaic, and (c) complex. The first type of devices uses anoxic photosynthetic organisms, whereas the second uses oxygenic photo organisms. In complex systems, two types of microorganisms, one anoxic photosynthetic, and the other a heterotrophic microorganism, are used [5].

The simplest approach for electro-vegetal devices consists of the insertion of different solid electrodes into fruits, pulps, or plant-derived pastes to exploit their rich electrolyte composition. This is illustrated in Figure 14.1, corresponding to the reproduction of the well-known copper/zinc battery in lemon [6]. There is an increasingly growing interest in the development of devices for sensing, energy production, and energy storage characterized by a large biodegradability, good mechanical flexibility, and stretchability [7].

There are various types of plant-based bioelectrochemical systems. These include microbial fuel cells and sediment microbial fuel cells. These can be associated with (or expanded to) bioelectrochemical systems aimed at chemical production, pollutant degradation, or desalination [4].

Batteries are devices that supply electrical energy based on the occurrence of electrochemical processes. They are constituted by two or more electrochemical cells providing a stable potential difference between their terminals. Primary batteries are intended to be used until exhausted, while secondary batteries (or accumulators) are rechargeable. A fuel cell is an electrochemical conversion device that operates under the continuous supply of fuel (hydrogen, natural gas, etc.) and an oxidant (oxygen, etc.) [8].

In general, batteries comprise a cathode (positive electrode, associated with reduction processes), an anode (negative electrode, associated with oxidation processes), an electrolyte, and a separator. The electrolyte provides ionic conductivity between the electrodes. The separator is an inert but ion-permeable physical barrier that prevents electrical shorting. The separator can be a porous material filled with an electrolyte or a layer of gelled electrolyte.

14.2 ENZYMATIC FUEL CELLS

The first modern approach to constructing batteries and fuel cells based on biomaterials is focused on enzymes. In these devices, chemical energy is transformed into electrical energy using biocatalyst enzymes [9]. Initiated in the 1960s [10], enzymes have high electrocatalytic performance but are unsustainable and less durable when compared to microbes. The use of microbes to generate electricity was conceived at the beginning of the 20th century [11] with subsequent developments [12, 13].

DOI: 10.1201/9781032697826-14

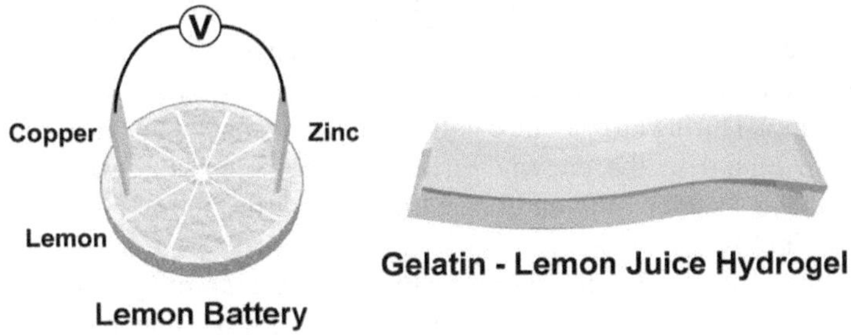

FIGURE 14.1 Image of the simplest plant-based energy device, the copper/zinc battery in lemon, accompanied by the image of a lemon juice hydrogel. (Adapted from ref. [6], with permission. Copyright Wiley 2021.).

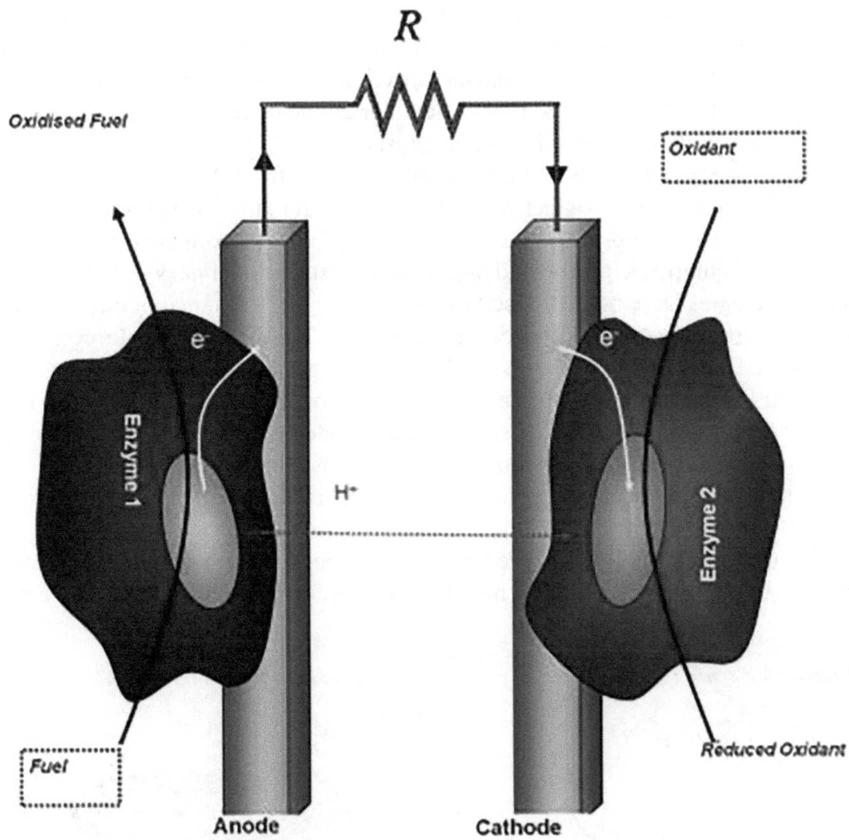

FIGURE 14.2 Scheme corresponding to a typical glucose-based enzymatic fuel cell. **(Adapted from ref. [14], with permission. Copyright Elsevier 2012.).**

Enzyme-based fuel cells are principally based on enzymes capable of catalytically oxidizing sugars or alcohols (e.g., glucose oxidase) or reducing oxygen (e.g., bilirubin oxidase). Figure 14.2 shows a scheme for the typical operation of enzymatic fuel cells [14]. In this membraneless cell, the fuel is oxidized by enzyme 1, providing electrons and protons. This process is coupled to the reduction of oxidant by enzyme 2, where electrons and protons are consumed.

In the simplest version, glucose is oxidized to gluconolactone, a two-electron process. The maximum ideal performance corresponds to the oxidation of glucose to CO_2 coupled to the reduction of O_2 to water [14]. Redox enzymes consist of an electroactive moiety (cofactor) bound to a protein component (apoenzyme). During enzymatic catalysis, the cofactor can be released from the enzyme's active site or remain bound to the enzyme structure. The electrons can be transferred directly between the enzyme and the electrode and/or be transported through mediator molecules, as schematized in Figure 14.3 [14].

The presence of the cofactor (in the case of glucose-oxidizing enzymes, usually nicotinamide adenine dinucleotide (NAD), flavin adenine dinucleotide (FAD), and pyrroloquinoline quinone (PQQ), ensures the electron transfer between the enzyme and the substrate. Figure 14.4 depicts the typical redox processes associated with these cofactors.

Direct electron transfer is only possible when the distance between the enzyme cofactor in the active site is close enough (ca. 2 nm) to the electrode surface because the electron transfer rate decays exponentially with such distance. Of course, not only the distance but also the orientation of the enzyme significantly influence the electron transfer rate. Mediated electron transfer involves the coupling of different electron transfer processes. Now, the thermodynamic redox potentials of the mediators determine the maximum potential difference produced by the cell. Notice that redox mediation (electron transfer between the enzyme active site and the mediator), to be thermodynamically favored, requires that the potential of the mediator couple is more positive (negative) than the redox potential of the oxidative (reductive) biocatalysis [14].

Figure 14.5 shows a typical cyclic voltammogram of GOx from *Aspergillus niger* immobilized onto a glassy carbon electrode in contact with phosphate buffer at pH 7.0 [15]. The voltammogram shows a well-defined, adsorption-mediated reversible couple. Here, both anodic and cathodic peaks show a characteristic symmetric profile and appear at almost identical peak potentials.

A variety of strategies have been devised to enhance the performance of enzymatic fuel cells. These include different anode and cathode materials and different enzyme combinations. An example is depicted in Figure 14.6, corresponding to a glucose biobattery with a solid-state, Prussian blue thin film cathode accompanying a glucose oxidase enzymatic anode immobilized on carbon paper substrates [16]. The Prussian blue cathode offers two interesting properties: (i) it bypasses the need for O_2 as the electron acceptor (and is concomitant with overpotential phenomena) and (ii) reacts with H_2O_2 produced by undesirable side reactions at oxidase-based bioanodes.

Prussian blue ($Fe^{III}_4[Fe^{II}_2(CN)_6]_3$ or $KFe^{III}[Fe^{II}(CN)_6]$ depending on the preparation procedures) is the parent term of a family of transition metal hexacyanoferrates characterized by a microporous

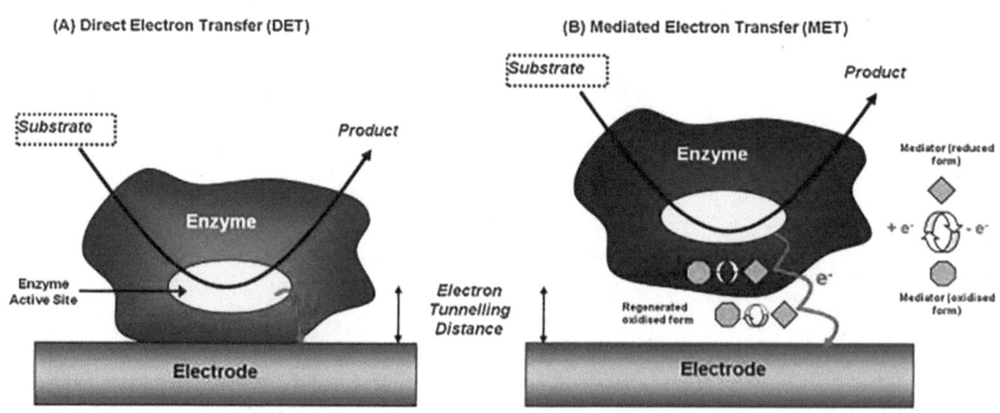

FIGURE 14.3 Schematic representation of direct electron transfer (a) and mediated electron transfer (b) between enzymes and electrodes. (Adapted from ref. [14], with permission. Copyright Elsevier 2012.).

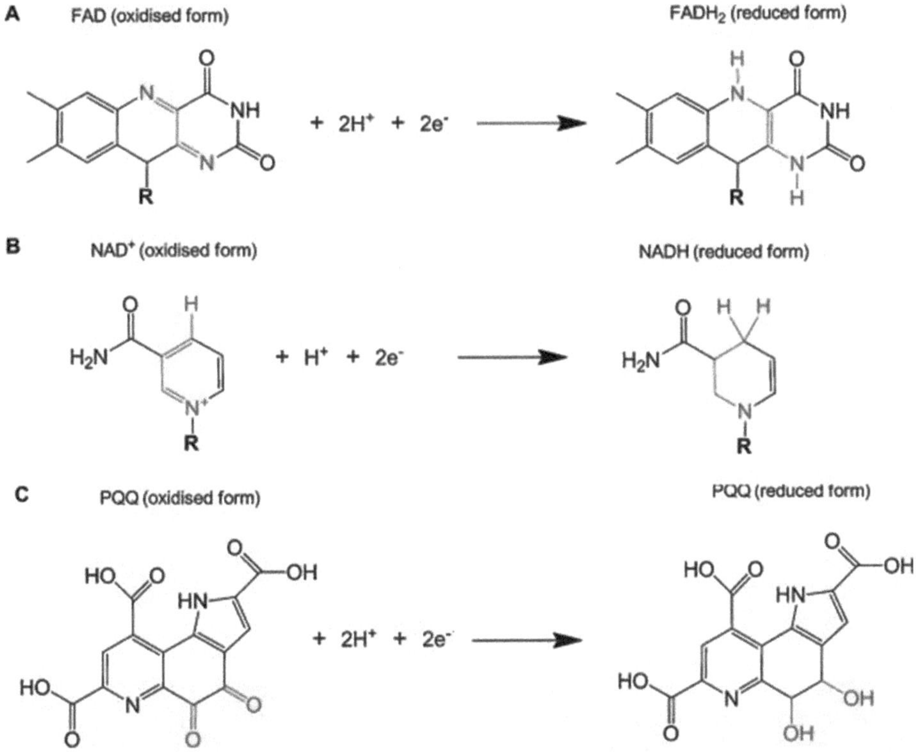

FIGURE 14.4 Redox processes associated with NAD, FAD, and PQQ factors of glucose oxidases. (Adapted from ref. [14], with permission. Copyright Elsevier 2012.).

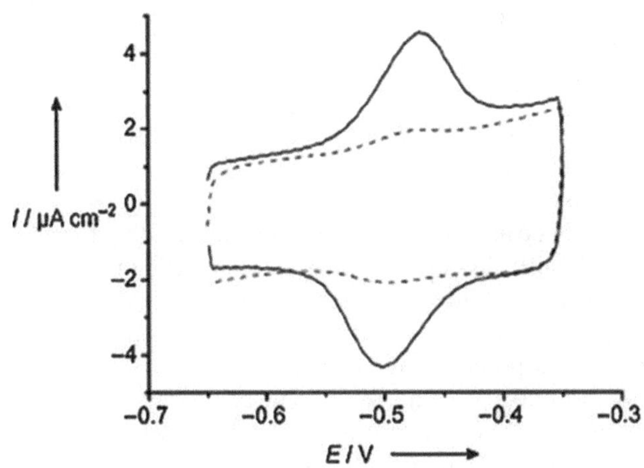

FIGURE 14.5 Cyclic voltammogram of GOx from ***Aspergillus niger*** adsorbed onto glassy carbon electrode in contact with aqueous phosphate buffer at pH 7.0. Potential scan rate 100 mV s^{-1}. (Adapted from ref. [15], with permission. Copyright Elsevier 2013.).

structure able to be permeated by cations. These ion intercalation properties have prompted rich solid-state electrochemistry [17, 18] and have been exploited in electrochemical sensing and metal-ion batteries [19].

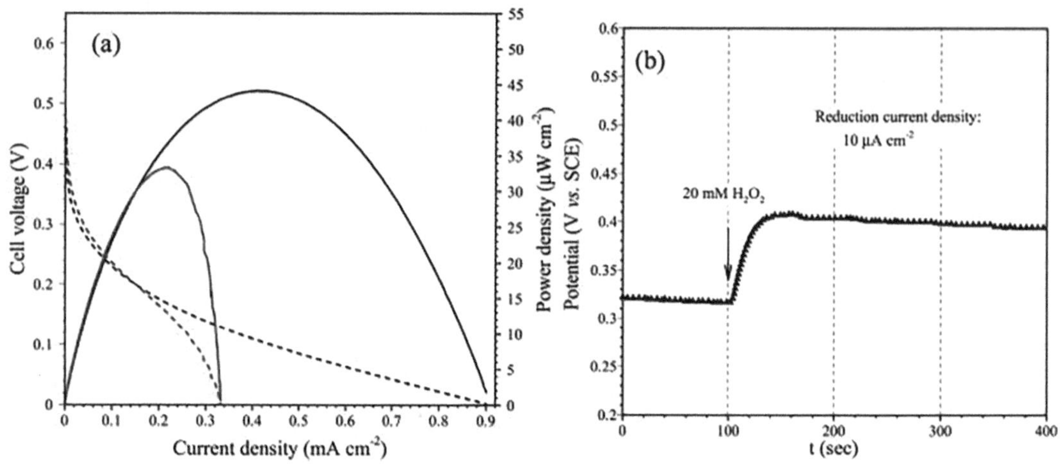

FIGURE 14.6 (a) Comparison of the performances of glucose/Prussian blue (black lines) and glucose/O_2 (gray lines) fuel cells. Power curves (continuous lines) and linear polarization graphs (dashed lines) measured in 0.1 M glucose and O_2-saturated 0.1 M glucose solutions, respectively. (b) Chronopotentiometric potential/time curve of the Prussian blue thin film cathode in the same solution being injected by H_2O_2 at a time of 100 s. (Adapted from ref. [16], with permission. Copyright Elsevier 2017.).

Figure 14.6a compares the performance of glucose/Prussian blue and glucose/O_2 fuel cells in terms of power curves and linear polarization graphs measured in 0.1 M glucose and O_2-saturated 0.1 M glucose solutions, respectively. The second was recorded by sweeping the voltage between the open-circuit potential of the cell (0.45 and 0.6 V vs. SCE, respectively) to 0 at a scan rate of 1 mV s^{-1}. One can see that the Prussian blue cathode significantly enhances the power output and the short-circuit current density relative to the conventional oxygen cell. The interaction between Prussian blue and H_2O_2 is illustrated in Figure 14.6b, where the chronopotentiometric curve was recorded for the cathode, showing a sharp potential increase when H_2O_2 was injected into the cell.

14.3 MICROBIAL FUEL CELLS

The use of microorganisms to generate electricity is an increasingly growing research field covering a wide spectrum of applications. As a result, an enormous variety of cell architectures and materials are under intensive research [20].

The biofilm substrates can be any type of organic matter, including from simple (glucose, carbohydrates) to complex (cellulose, lignins) components and agricultural, industrial, and domestic wastes [22]. Conjointly considered, this involves the processing and exploitation of biomass [23]. Figure 14.7 shows a scheme of a typical microbial, two-chambered fuel cell [23]. In the anodic compartment, microorganisms produce oxidative reactions on a substrate while chemical and/or microbial reductive processes take place in the cathodic compartment. Such compartments are usually separated by a proton or cation exchange membrane. The electrodes are interconnected through an external circuit incorporating an external resistor or load.

In general, microbial fuel cells involve the use of a biotic (generally bacteria) electrocatalyst at the anode coupled to oxygen reduction at the anode. The kinetic constraints to this second process constitute the main bottlenecks of this technology [21]. Relative to other low-temperature fuel cells such as the proton exchange membrane fuel cell, and the direct methanol fuel cell, microbial fuel cells offer the possibility of working at neutral pH and at ambient temperatures [20]. Of course, the operation causing a moderate environmental impact is another factor of interest concerning microbial fuel cells.

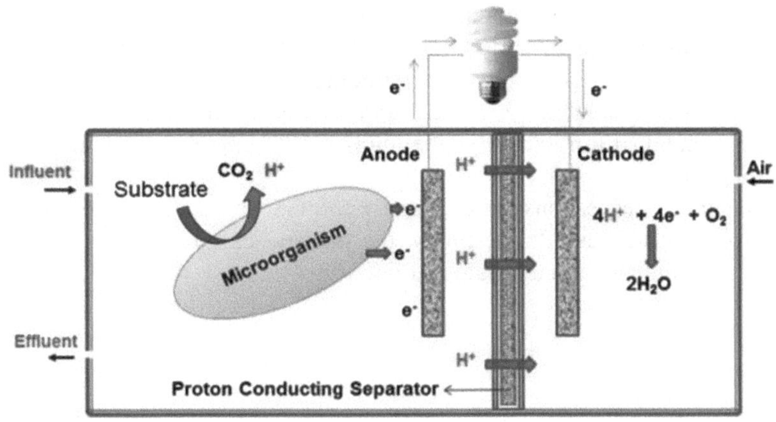

FIGURE 14.7 Scheme for a microbial fuel cell. (Adapted from ref. [23], with permission. Copyright Elsevier 2016.).

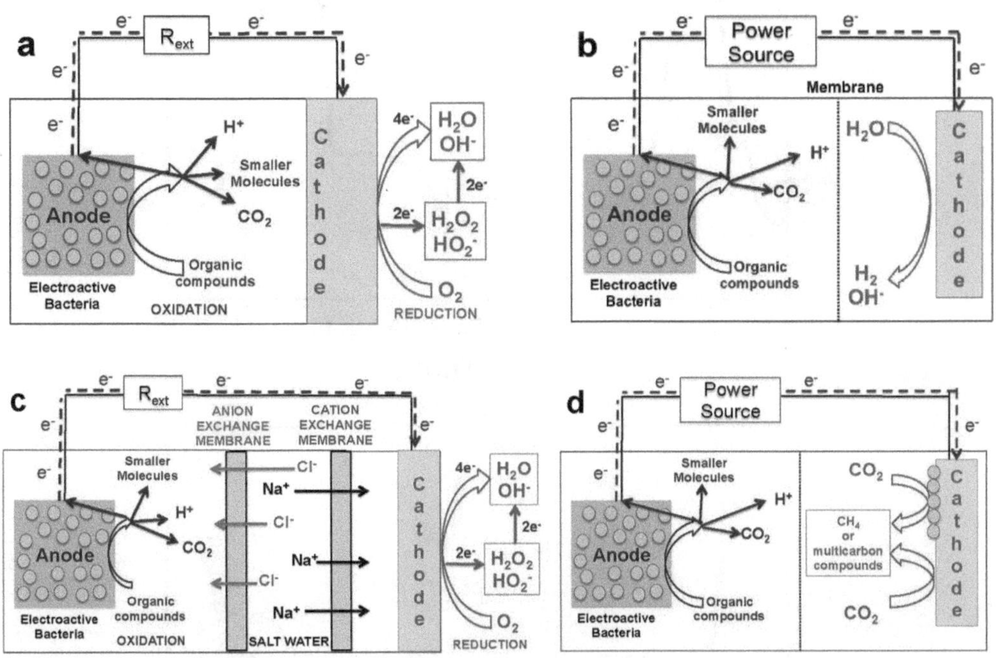

FIGURE 14.8 Schematic representation of microbial cells acting as (a) fuel cell, (b) electrolysis cell, (c) desalination cell, and (d) general electrosynthetic cell. (Adapted from ref. [20], with permission. Copyright Elsevier 2017.).

As previously noted, microbial cells are not limited to electric energy production, i.e., fuel cells. The same general principles and similar experimental arrangements can be applied to build electrolysis cells, electrosynthetic cells, and desalination cells. The schemes for the one-chamber type of the aforementioned cells are depicted in Figure 14.8 [20].

Microbial electrolysis cells are usually connected to an external source of electricity to produce hydrogen at the cathode. Here, a significant part of the energy comes from the chemical energy extracted from substrates oxidized at the anode, thus minimizing the contribution of the external source [24]. Similarly, electrosynthetic cells require the use of an external power source aiding the synthesis of any chemical at the cathodic compartment. Most efforts are concentrated on the so-called CO_2 activation, reducing the atmospheric component to methane, methanol, and/or other products [25]. The application of microbial cells to desalination involves the employment of ion-exchange membranes [26] expanding to the use of a variety of porous materials [27].

Figure 14.9 illustrates the response of *Geobacter soli* GSS01 in a dual chamber cell under successive renewals of the medium. Interestingly, the bacteria can act as an oxidant in an acetate medium and as a reducing agent in a nitrate medium [28].

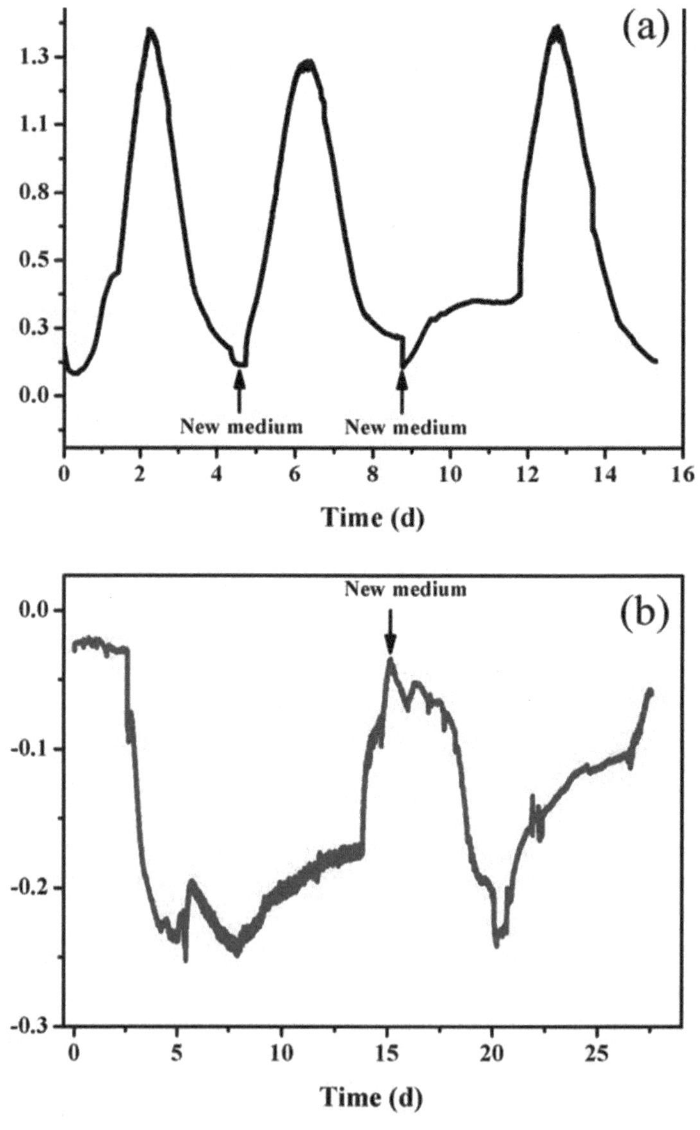

FIGURE 14.9 Time variation of the current density measured in a bacterial cell containing biofilms of ***Geobacter soli*** in contact with acetate and nitrate electrolytes that were periodically renewed. (Adapted from ref. [28], with permission. Copyright Elsevier 2017.).

14.4 PLANT FUEL CELLS

Plant fuel cells derive directly from microbial fuel cells using plant roots [4, 29]. Here, soil microorganisms around the plant roots use the plant's root exudates to generate electricity. The essential idea is that adding plants to microbial fuel cells can increase the effective biomass and avoid the need to harvest the plants in conventional microbial fuel cells [30].

Plant fuel cells consist of living plants accompanied by their supporting matrix where the anode is inserted and a cathode placed in air or an aqueous solution. The supporting matrices are flooded soils, paddy, wetland, sand, clay, silt, sediments, vermiculite, or other materials, such as graphite granules. Soils include natural soil, agricultural and forest soils, rice field/paddy soil, red soil, peat soil, salt marsh soil, compost soil mix, sediment/rhizodeposits, wetland sediment, garden soil, garden soil mixed with cow dung, and potting soil [31].

The performance of plant fuel cells depends on a variety of factors. It is primarily conditioned by rhizodeposite production and availability and the proliferation of rhizobia microbial species, as well as by the rate of substrate oxidation, electrons, and proton transfer to the electrodes. These factors are in turn influenced by growth medium, reactor configuration, oxygen reduction at the cathode, and other factors [32]. Since green plants are involved, the type of plant, type and intensity of the light, temperature, and type of accompanying microorganisms, are also important factors to be considered.

The substrate contains different microbial and archaeal species that form a microbial community in the biofilm attached to the electrodes [33]. Based on their ability to transfer electrons to the electrodes without exogenous mediators, the components of the microbiota can be classified as electricigens, anodophiles, and exoelectrogens [34].

The plants produce a series of organic substances that are stored or excreted at the root as rhizodeposits. These are formed through the loss of root cap or border cell, flow of organic carbon to root-associated symbionts, gas emission, solute leakage, and insoluble polymer release [35]. Rhizodeposits are constituted by exudates of sugars and organic acids, secretions of enzymes and other compounds, and dead plant cell materials (lysates), among others. Most of these substances can be oxidized by electroactive root-dwelling microorganisms to yield electrons. There are many other redox processes in the substrate such as the oxidation of reductants resulting from microbial activity (humic acids, iron (II), and sulfur compounds), oxidation of ammonia to nitrite/nitrate by ammonia-oxidizing bacteria, among others [36]. All these processes produce electrons, which can be transferred to the anode in the fuel cell.

Figure 14.10a shows a scheme of a plant fuel cell [29]. Formally, the electrons produced in the rhizosphere are captured by the anode and transferred to the cathode where a reduction process occurs. Generally, this consists of the reduction of oxygen to water. In certain cell configurations, the plants are also inserted in the cathodic chamber. Then, the O_2 released during photosynthesis can be directly used as a biocathode [37].

An interesting variant of plant fuel cells, macrophyte cells, is schematized in Figure 14.10b. Macrophytes are multicellular aquatic plants present in marine, estuarine or in riverine ecosystems. As photosynthetic organisms, the rhizodeposits and exudates of macrophytes can be utilized as fuels [38]. This system, schematized in Figure 14.10c, can be coupled to a water pollution control constituting wetland plant fuel cells [39].

The unique physiology of bryophytes, plants without vascular tissues, which are particularly tolerant to dehydration, has prompted their use in fuel cells. Although bryophytes have no roots, they possess hair-like rhizoids that bind them to soils and prevent nutrient loss. These plants can also be efficiently used in plant fuel cells (see scheme in Figure 14.10d) [40].

14.5 EXTRACELLULAR ELECTRON TRANSFER

The operation of microbial fuel cells and plant fuel cells involves the transfer of electrons between the microorganisms existing in biofilms and an electrode. This process can follow three basic pathways,

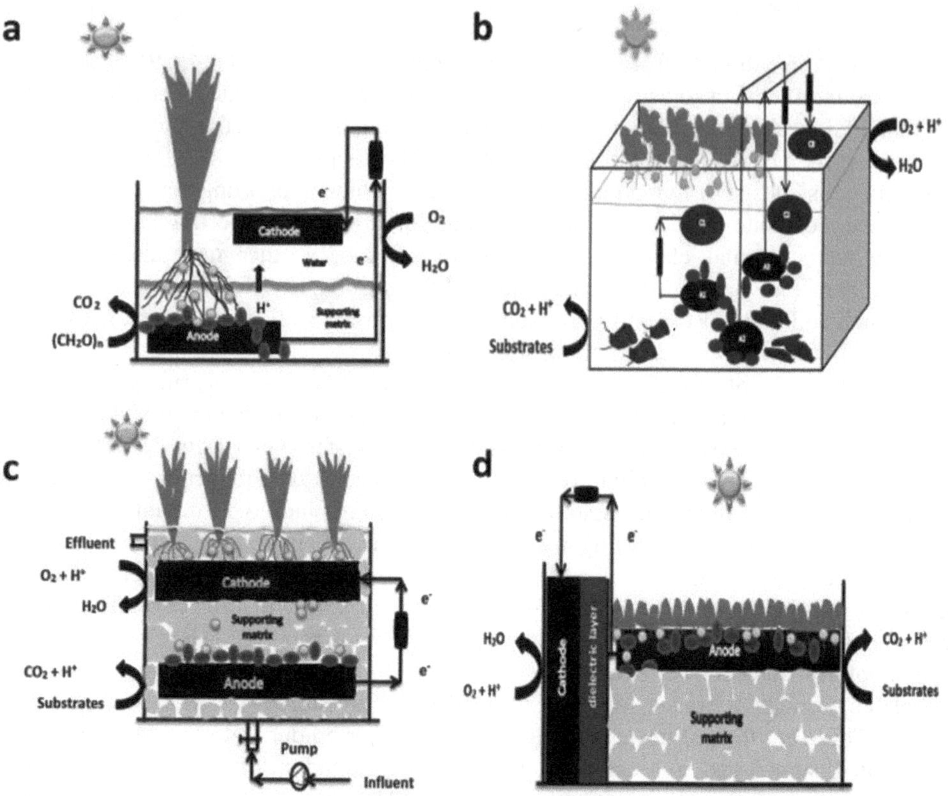

FIGURE 14.10 Schemes of (a) plant fuel cell, (b) macrophyte fuel cell, (c) constructed wetland macrophyte fuel cell, and (d) bryophyte fuel cell. (Adapted from ref. [29], with permission. Copyright Elsevier 2019.).

direct electron transfer, mediated electron transfer, and electron transfer mediated by extracellular conductive connections [20, 29], the so-called conductive pili or bacterial nanowires [41].

These processes are schematically depicted in Figure 14.11 [20]. The direct electron transfer between the electroactive microorganism and the electrode occurs by direct contact between its outer membrane and the electrode surface of the anode [42]. Here, the cytochrome c membrane proteins are responsible for the process of extracellular electron transfer.

The mediated or indirect electron transfer involves redox mediators in most cases produced by the microbiota, as has been evidenced for many bacteria [43]. This modality of electron transfer may also result from the oxidation of a by-product resulting from bacterial metabolism. This is the case of hydrogen produced by fermentative bacteria which is subsequently oxidized at the anode surface [44].

From the energetic point of view, two major metabolic pathways can be distinguished: respiration and fermentation. Respiration involves an external oxidant (O_2) and proceeds via a redox cascade where the energy is finally transferred to an external electron acceptor. In the simplest energetic approach, respiration can be summarized by the process of glucose oxidation.

$$C_6H_{12}O_6 + 6O_2 \rightarrow 6H_2O + 6CO_2 \tag{14.1}$$

Under anoxic conditions, anaerobic organisms utilize other terminal acceptors for respiration – namely, nitrate, sulfate, fumarate, or metal ions. Thermochemical considerations can be based on

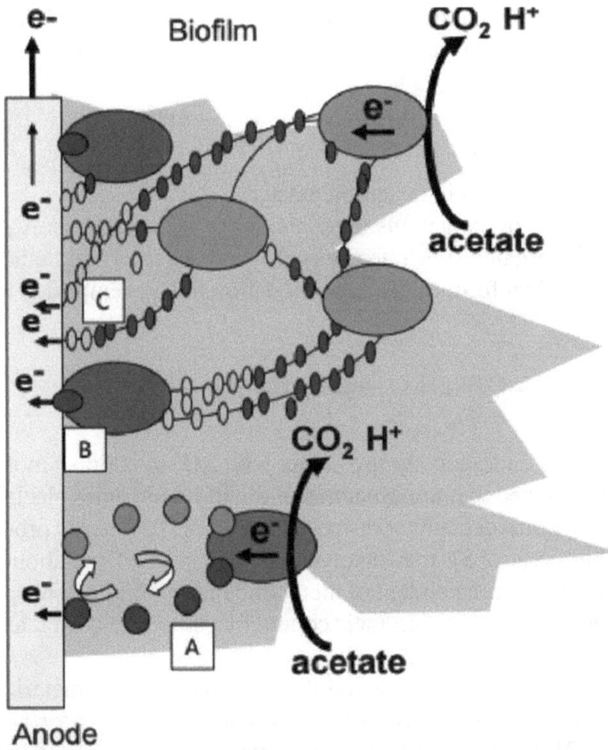

FIGURE 14.11 Schematic representation of different mechanisms of electron transfer in biofilms, including indirect transfer via mediators or fermentation products (a), direct transfer via cytochrome proteins (b), and direct transfer via conductive pili (c). (Adapted from ref. [20], with permission. Copyright Elsevier 2017.).

standard Gibbs energies, $\Delta G°$, and standard electrode potentials, $E°$, using the well-known Nernst equation,

$$\Delta G^{\circ} = -nF\left(E^{o}_{\text{red}} - E^{o}_{\text{oxid}}\right) \tag{14.2}$$

The aerobic respiration combining the O_2/H_2O (($E°(O_2/H_2O)$ = 0.82 V vs. SHE at pH 7.0) and CO_2/glucose ($E°CO_2$/glucose) = −0.43 V vs. SHE at pH 7.0) couples is particularly favored. This marks the maximum energy gain of the organism through the process described by Eq. (14.1), $\Delta G° = -2895$ kJ mol^{-1} [43]. This also corresponds to the maximum reversible voltage of a biofuel cell, 1.24 V [14], corresponding to the transfer of 24 electrons per glucose unit.

$$6CO_2 + 24H^+ + 24e^- \rightarrow C_6H_{12}O_6 + 6H_2O \tag{14.3}$$

Of course, in an enzymatic cell where glucose was oxidized to gluconolactone,

$$C_6H_{12}O_6 + 1/2O_2 \rightarrow C_6H_{10}O_6 + H_2O, \tag{14.4}$$

the number of transferred electrons per glucose is 2 and the maximum cell voltage becomes 1.18 V [14].

As can be inferred from standard electrode potentials, the energetic performance of anaerobic respiration processes is lower than that of aerobic oxidation. The most favorable is the sulfate/

hydrogen sulfide couple, whose standard electrode potential ($E°(SO_4^{2-}/SH^-) = -0.22$ V vs. SHE at pH 7.0) corresponds to the electron transfer process,

$$SO_4^{2-} + 9H^+ + 8e^- \rightarrow HS^- + 4H_2O, \tag{14.5}$$

which is a frequent respiratory path in anaerobic bacteria.

When there is no disposal of exogenous oxidants, many microorganisms perform fermentation. In this process, parts of the organic substrate act as electron acceptors and others as electron donors. This can be seen as a biological form of disproportionation [43], exemplified by the fermentation of glucose to acetate.

$$C_6H_{12}O_6 + 2H_2O \rightarrow 2CH_3COOH + 2CO_2 + 4H_2 \tag{14.6}$$

This process is much less energetically favorable, with $\Delta G° = -206$ kJ mol^{-1} [42].

Recently, it has been reported that the existence of direct interspecies electron transfer consisting of metabolic cooperation of interdependent microbes is suitable for energy production [45]. Contrary to conventional microbial syntrophy that involves the transport of metabolic products, the direct interspecies electron transfer is independent of metabolic intermediates. This phenomenon proceeds via electron flow between microbes using their conductive structures, such as pili and c-type cytochromes [46].

A photosensitive variant of this phenomenon has been recently characterized for the interaction between *Geobacter sulfurreducens* Caccavo Jr. et al. and *Prosthecochloris aestuarii* Gorlenko, a phototropic bacterium [47]. When co-cultured, the former oxidizes acetate ions and transfers electrons directly to the second, but this accepts the electrons for photosynthesis only under illumination.

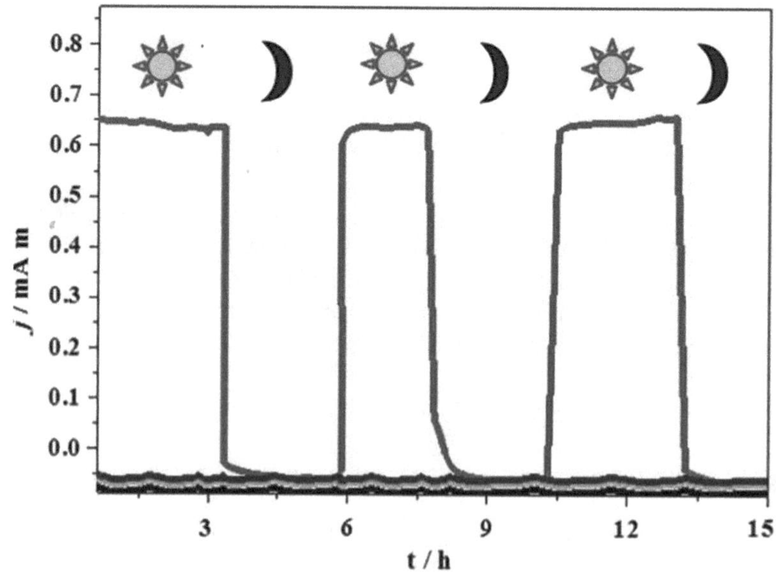

FIGURE 14.12 Current density response of a *P. aestuarii/G. sulfurreducens* cell compared to blank experiments on abiotic plus individual strains of *P. aestuarii* or *G. sulfurreducens* under alternate light/dark circles after 48-hour operation in a two-chamber microbial fuel cell under a CO_2 headspace. See details in the text. (Adapted from ref. [48], with permission. Copyright Elsevier 2019.).

Figure 14.12 shows the current density response of the abiotic, individual strains of *P. aestuarii* and *G. sulfurreducens* and their co-culture under alternate light/dark circles after 48-hour operation under a CO_2 headspace [48]. The experiments were performed on a microbial cell where the two chambers were separated by a proton exchange membrane. The anode served as the electron acceptor for *G. sulfurreducens*, which oxidized acetate, and the cathode donated electrons for *P. aestuarii*, which performed the photosynthetic reduction of CO_2 injected into the chamber headspace.

The experiment compares the aforementioned results with blank experiments with individual strains (abiotic anode + *P. aestuarii* biocathode, and *G. sulfurreducens bioanode* + abiotic cathode). As can be seen in this figure, only under illumination does the cell display a significant current, consistent with the idea that light-mediated direct interspecies electron transfer takes place.

14.6 ELECTRICAL ENERGY STORAGE (AND PRODUCTION)

Capacitors are classical systems for electrical energy storage constituted by a dielectric interposed between two metallic conductors of a large surface. In the last decades, there has emerged a novel class of capacitors, generically termed supercapacitors, with increased performance for electrical energy storage. These can be divided into electric double-layer capacitors (EDLs), faradaic capacitors (or pseudocapacitors), and hybrid capacitors [49]. The mechanism of charge storage in EDLs is based on the electrosorption of ions on porous materials. These form an electrical double layer on the of such materials, which includes an electrolyte diffusion layer, and a compact Helmholtz layer, and a space charge layer equivalent to that which characterizes semiconductor/electrolyte contacts [27].

Faradaic capacitors store charge from redox reactions on the electrode surface. These include capacitors associated with underpotential deposition [50], redox pseudocapacitance [51], and intercalation pseudocapacitance [52]. Finally, hybrid supercapacitors possess double-layer and faradaic charge storage processes.

Ideally, capacitors should display a purely capacitive voltammetric response with no faradaic signals. An example can be seen in Figure 14.13, corresponding to different porous carbon composites fabricated from tobacco leaf–based, carbon dot, filled rubber, wood chip–extracted lignin/polyacrylonitrile [53]. In Figure 14.13a-c, the voltammograms for specimens carbonized at temperatures of 900°C, 1,200°C, and 1,400°C are depicted, superimposing the voltammograms at different potential scan rates. In Figure 14.13d, the voltammograms at 5 mV s^{-1} recorded for the different materials are superimposed. One can see that the rectangular-like shape of the voltammogram becomes less distorted on increasing the carbonization rate, thus approaching the ideal capacitive response. Figure 14.14 shows a scheme of the preparative process, involving a carbonization step resulting in the formation of highly porous carbons.

In the case of faradaic capacitors, faradaic processes appear as peaked responses in the voltammograms. Figure 14.15a depicts an example corresponding to the cyclic voltammetry of a graphite/lignin (plus wood pulps) composite material in contact with 0.1 M $HClO_4$ [54]. The voltammogram shows cathodic and anodic signals attributed to quinone/catechol, and interconversions associated with hydrogen functionalities existing in the lignins.

Figure 14.15b shows the galvanostatic charge/discharge curves recorded for the same material and electrolyte at different gravimetric current densities (A g^{-1}). In contrast with the triangular-like potential/charge patterns expected for ideal capacitors, the recorded curves exhibit a plateau in the charging curve between 0.4 and 0.7 V and a second plateau in the discharge curve at the potential range between 0.5 and 0.15 V. The former can be assigned to the oxidation of catechol units to quinines, while the second can be attributed to the inverse process [54].

In a purely capacitive response, the maximum horizontal current, *i*, recorded in the voltammograms will increase linearly with potential scan rate, *v*, as,

$$i = Cv \tag{14.7}$$

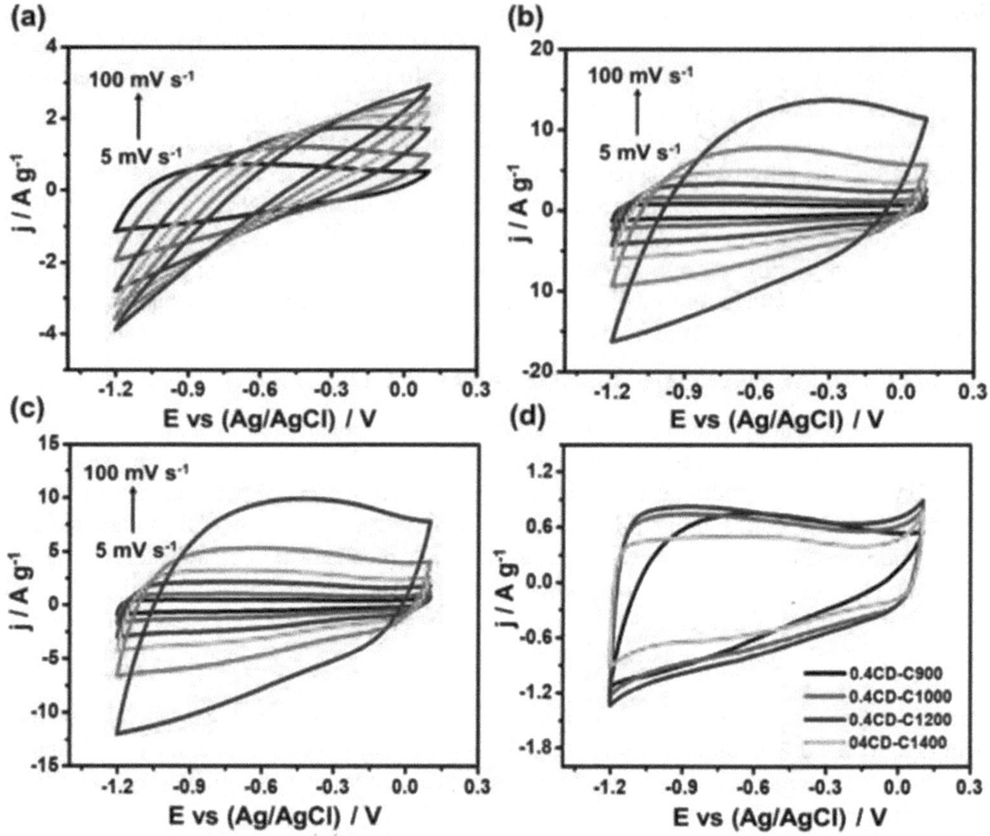

FIGURE 14.13 Cyclic voltammetry of porous carbons prepared by carbonization of vegetal matter at temperatures of (a) 900°C, (b) 1,200°C, and (c) 1,400°C in contact with 6 M KOH. Voltammograms at potential scan rates from 5 to 100 mV s^{-1} are superimposed. The voltammograms at 5 mV s^{-1} of materials carbonized at 900°C, 1,000°C, 1,200°C, and 1,400°C are superimposed in (d). (Adapted from ref. [53], with permission. Copyright Elsevier 2023.).

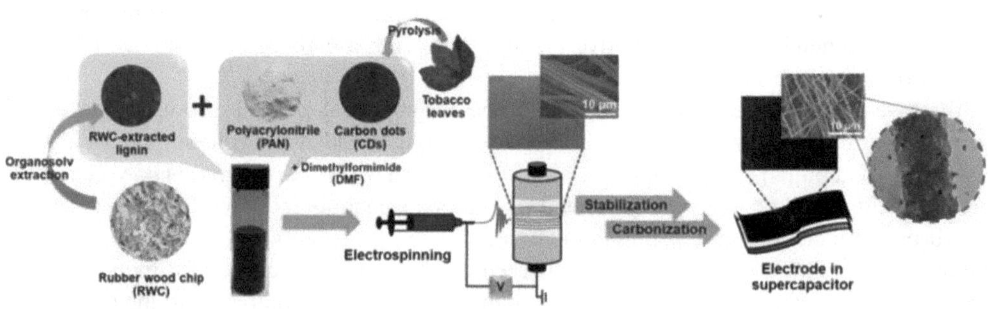

FIGURE 14.14 Scheme of the preparation of the porous carbon-based capacitor in Figure 14.13, prepared from tobacco leaf–based carbon dots filled rubber wood chip–extracted lignin/polyacrylonitrile. (Adapted from ref. [53], with permission. Copyright Elsevier 2023.).

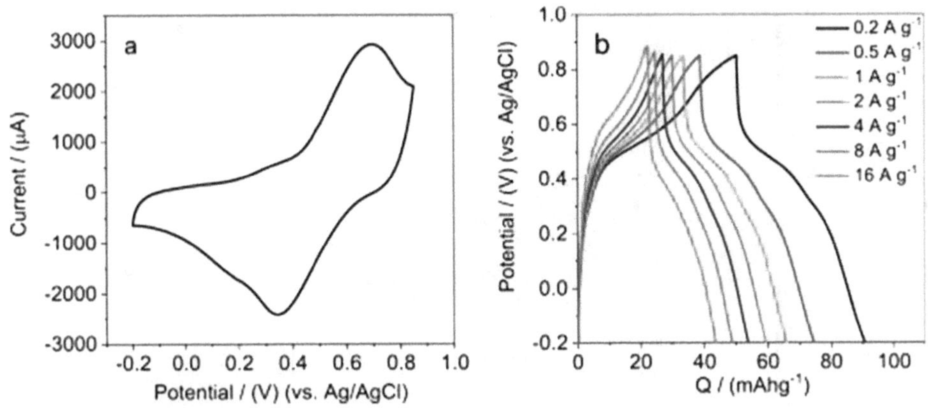

FIGURE 14.15 Cyclic voltammogram (a) and galvanostatic charge/discharge curves at different gravimetric current densities (b) of a graphite/lignin material in contact with 0.1 M $HClO_4$ aqueous solution. **(Adapted from ref. [54], with permission. Copyright Elsevier 2021.).**

where C is the electrode capacitance. Then, the specific (or gravimetric) capacitance, C_m, can be calculated as

$$C_m = \frac{i}{mv} \tag{14.8}$$

where m is the mass of active potential existing in the electrode. Since, in general, the voltammetric responses do not fit the ideal rectangular response, it is convenient to estimate the gravimetric capacitance by integrating the current-potential curves, i.e.,

$$C_m = \frac{\int idV}{mv\Delta V}, \tag{14.9}$$

where ΔV is the difference between the extreme potentials in the voltammogram. It is pertinent to note that the complete capacitor consists of two interfaces: metal/electrolyte. Since this is equivalent to two capacitors in series, the total capacitance will differ by a factor of 4 from the capacitance calculated from voltammetric data in a three-electrode cell.

The energy density, E_m, can be calculated as

$$E_m = \frac{\int idV}{m} \tag{14.10}$$

Taking into account the previous considerations, the specific energy density (E_m, W h kg^{-1}) of a capacitor working at a (difference of) potential U, will be,

$$E_m = \frac{CU^2}{2m} \tag{14.11}$$

The power density (P_m, W kg^{-1}) will be

$$P_m = \frac{E_m}{\Delta t} = \frac{iU}{4m} \tag{14.12}$$

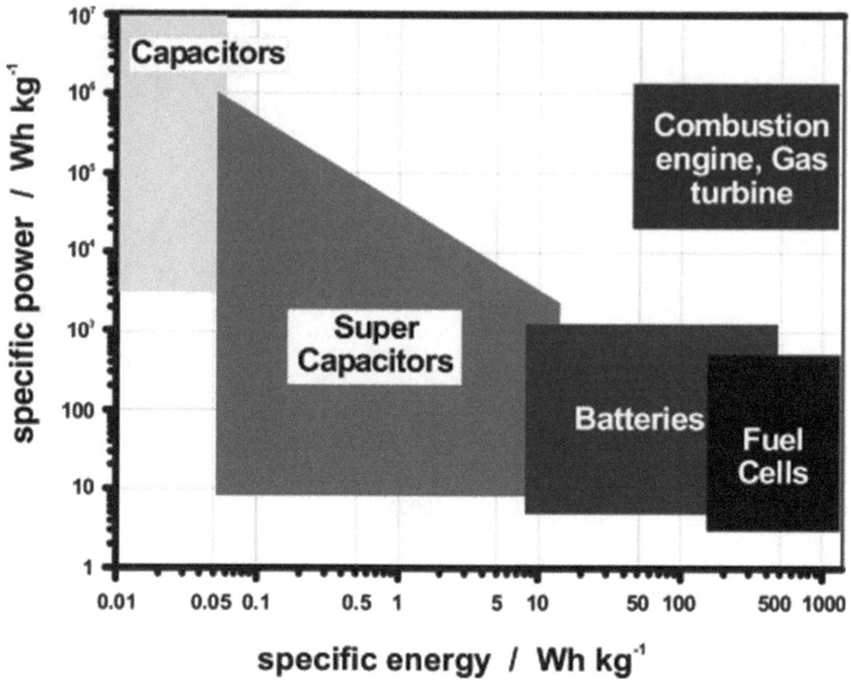

FIGURE 14.16 Ragone plot of the energy storage domains for different energy conversion systems. (Reproduced from ref. [8], with permission. Copyright the American Chemical Society 2004.).

where Δt is the discharge time. Volumetric rather than gravimetric densities can be equally defined. The energy density represents the energy content of the system, whereas the power density is representative of the rate capability of the same [8]. Batteries, capacitors, and fuel cells differ in their typical E_m and P_m values. Batteries and fuel cells can store high specific energy (typically around 150 W h kg^{-1}) with relatively low specific power (around 1.5 kW kg^{-1}). In contrast, capacitors store low specific energy (< 0.1 W h kg^{-1}) but possess high specific power (up to 1,000 kW kg^{-1}). These properties are visually condensed in the so-called Ragone plots [27, 55]. Figure 14.16 shows an example where the areas covered by classical capacitors, supercapacitors, batteries, and electrochemical fuel cells are compared with those of combustion engines [8].

Depending on the technological demands, research is focused on enhancing preferentially E_m and P_m values. Fast charging, cyclability, and reversibility are important properties to consider. Of course, economy, safety, and sustainability are crucial for the practical implementation of systems of electrical energy production and storage.

14.7 WEARABLE SYSTEMS AND OTHER DEVELOPMENTS

The fabrication of flexible and stretchable batteries is receiving considerable attention. These devices are characterized by high energy and power densities, long cycle life, and high rate capability [9]. Figure 14.17 shows an example of a flexible paper-based battery prepared from Mg-sputtered and Mo-sputtered paper substrate and gelatin-lemon juice gel electrolyte [6].

Among other possibilities, *in vivo* electrochemically assisted polymerization in the vascular system of a plant's leaves has been reported for testing the possibility of *in planta* biofuel cell electrode fabrication [57]. Figure 14.18a shows the image of a *Nicotiana tabacum* leaves after ten potential cycles following the injection of ethylenedioxythiophene-based trimers anchoring an Os(2,2′-bipyridine)2(1-(3-aminopropyl)-imidazole)Cl complex. An equivalent image of another leaf after the injection of the same product accompanied by glucose and glucose oxidase is shown in Figure 14.18b. Here, it is observed as a darkening in the color of the vascular system, which is consistent with the π-conjugation extension of the ethylenedioxythiophene segment.

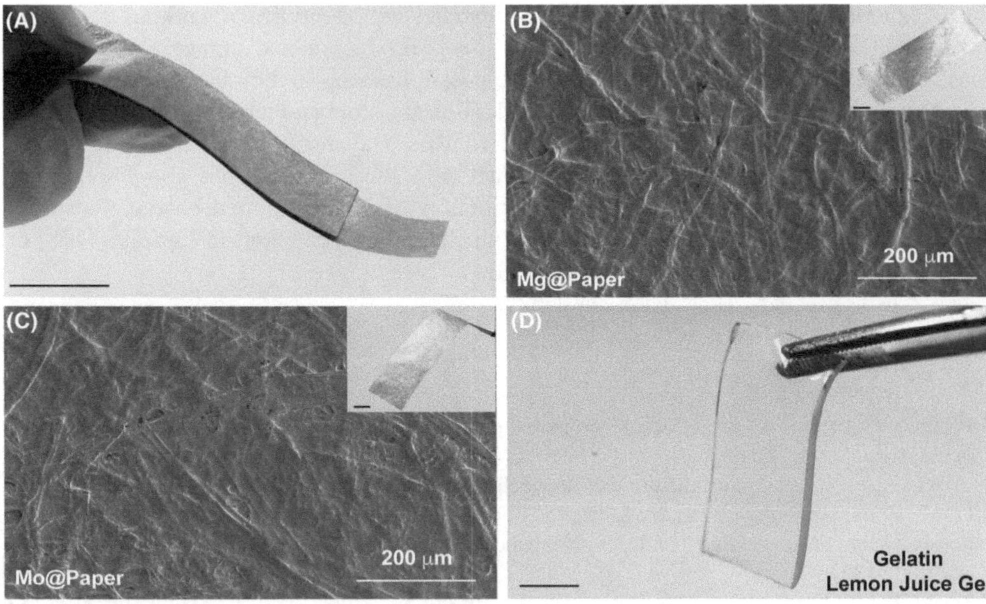

FIGURE 14.17 (a) Optical image of a flexible paper-based battery, (b) scanning electron microscopy (SEM)image of Mg-sputtered paper substrate (inset: optical image), (c) SEM image of Mo-sputtered paper substrate (inset: optical image), and (d) optical image of a gelatin-lemon juice gel electrolyte. (Adapted from ref. [6], with permission. Copyright Wiley 2020.).

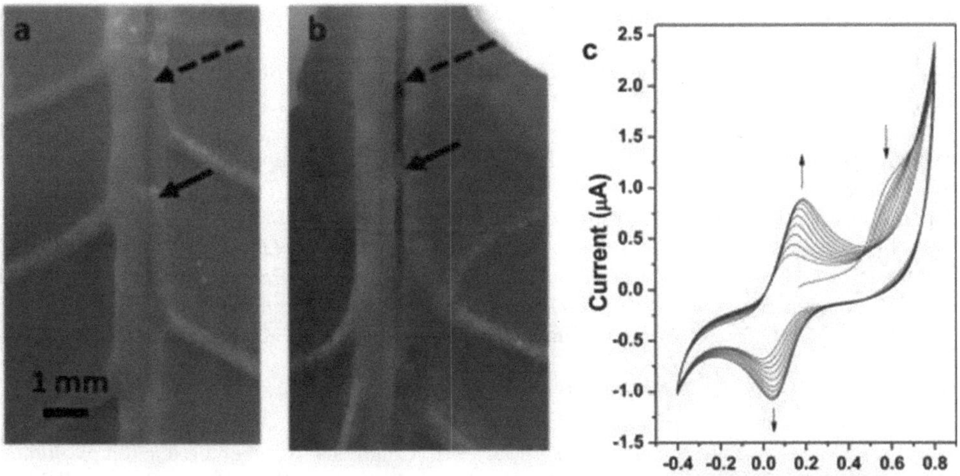

FIGURE 14.18 Photographic images of a ***Nicotiana tabacum*** leaf after 10 potential cycles following the injection of (a) ethylenedioxythiophene-based trimers anchoring an Os(2,2′ -bipyridine)2(1-(3-aminopropyl)-imidazole)Cl Os-complex, and (b) image of a second leaf after the injection plus 30 mM glucose plus 5 mg mL^{-1} GOx. The solid arrows represent the injection position, and the dashed arrows represent the working electrode position. (c) In vivo cyclic voltammograms recorded in a leaf as in (b). Potential scan rate 100 mV s^{-1}. (Adapted from ref. [57], with permission. Copyright Elsevier 2022.).

Figure 14.18c shows the repetitive cyclic voltammetry recorded in vivo after the injection of polymer precursors accompanied by glucose plus glucose oxidase. The advance of the polymerization process can be monitored by the decrease of the voltammetric wave at ca. 0.65 V vs. Ag/AgCl for the oxidation of the trimer precursor. This is accompanied by the progressive increase of the

Os-localized oxidation peak at ca. 0.20 V. This illustrates the possibility of working *in planta* with ultimate potential applications for sensing and/or energy production and storage.

These examples illustrate the amplitude of the possibilities opened by the intersection between plant science and electrochemistry for energy production and storage. Practical developments should be accompanied by a deeper understanding of fundamental electrochemical phenomena. This includes the translation to bacterial and plant fuel cells the considerations about self-discharge mechanisms (diffusion-controlled charge redistribution, resistance-limited charge redistribution, activation-controlled faradaic reaction coupled with charge redistribution) in capacitors [58] or ion intercalation and phase transitions in batteries [18].

REFERENCES

[1] Potter, M.C. 1911. Electrical effects accompanying the decomposition of organic compounds. *Proceedings of the Royal Society London*. 84: 260–276.

[2] Cogen, B. 1931. The bacterial culture as an electrical half-cell. *Journal of Bacteriology*. 21: 18–19.

[3] Li, M.; Zhou, M.; Tian, X.; Tan, C.; Mcdaniel, C.T.; Hassett, D.J.; Gu, T. 2018. Microbial fuel cell (MFC) power performance improvement through enhanced microbial electrogenicity. *Biotechnology Advances*. 36: 1316–1327.

[4] Sayed, E.T.; Abdelkareem, M.A.; Obaideen, K.; Elsaid, K.; Wilberforce, T.; Maghrabie, H.M.; Olabi, A.G. 2021. Progress in plant-based bioelectrochemical systems and their connection with sustainable development goals. *Carbon Resources Conversion*. 4: 169–183.

[5] McCormick, A.J.; Bombelli, P.; Bradley, R.W.; Thorne, R.; Wenzel, T.; Howe, C.J. 2015. Biophotovoltaics: oxygenic photosynthetic organisms in the world of bioelectrochemical systems. *Energy & Environmental Science*. 8: 1092–1109.

[6] Wang, Z.; Li, X.; Yang, Z.; Guo, H.; Tan, Y.J.; Susanto, G.J.; Cheng, W.; Yang, W.; Tee, B.C.K. 2020. Fully transient stretchable fruit-based battery as safe and environmentally friendly power source for wearable electronics. *EcoMat*. 3: artic. e12073.

[7] Esquivel, J.P.; Alday, P.; Ibrahim, O.A.; Fernández, B.; Kjeang, E.; Sabaté, N. 2017. A metal-free and biotically degradable battery for portable single-use applications. *Advanced Energy Materials*. 7: artic. 700275.

[8] Winter, M.; Brodd, R.J. 2004. What Are Batteries, Fuel Cells, and Supercapacitors? *Chemical Reviews*. 104, 4245–4270.

[9] Patra, S.; Verma, J.; Mishra, Y.K.; Kurinec, S.; Wang, Q.; Siväjärvi, M.; Tiwari, A. 2023. The positioning of biofuel cells-based biobatteries for net-zero energy future. *Journal of Energy Storage*. 72: artic. 107919.

[10] Yahiro, A.T.; Lee, S.M.; Kimble, D.O. 1964. Bioelectrochemistry: I. Enzyme utilizing bio-fuel cell studies. *Biochimica et Biophysica Acta (BBA) Specialized Section on Biophysical Subjects*. 88: 375–383.

[11] Vishwanathan, A.S. 2021. Microbial fuel cells: a comprehensive review for beginners. *3 Biotech*. 11:artic. 248.

[12] Schröder, U. 2011. Discover the possibilities: microbial bioelectrochemical systems and the revival of a 100-year–old discovery. *Journal of Solid State Electrochemistry*. 15:1481–1486.

[13] Karube, I.; Matasunga, T.; Suzuki, S.; Tsuru, S. 1976. Continous hydrogen production by immobilized whole cells of *Clostridium butyricum*. *Biochimica et Biophysica Acta (BBA) General Subjects*. 24: 338–343.

[14] Leech, D.; Kavanagh, P.; Suchmann, W. 2012. Enzymatic fuel cells: recent progress. *Electrochimica Acta*. 84: 223–234.

[15] Nasri, Z.; Shams, E. 2013. A glucose biosensor based on direct electron transfer of glucose oxidase immobilized onto glassy carbon electrode modified with nitrophenyl diazonium salt. *Electrochimica Acta*. 112: 640–647.

[16] Yazdi, A.A.; Preite, R.; Milton, R.D.; Hickey, D.P.; Minteer, S.D.; Xu, J. 2017. Rechargeable membraneless glucose biobattery: towards solid-state cathodes for implantable enzymatic devices. *Journal of Power Sources*. 343: 103–108.

[17] Dostal, A.; Meyer, B.; Scholz, F.; Schröder, U.; Bond, A. M.; Marken, F.; Shaw, S. J. 1995. Electrochemical study of microcrystalline solid prussian blue particles mechanically attached to graphite and gold electrodes: electrochemically induced lattice reconstruction. *Journal of Physical Chemistry*. 95: 2096–2103.
[18] Doménech-Carbó, A.; López, S.; Muñoz, E. 2024. Logistic Approximation to Ion-intercalation Electrochemistry: Application to Potassium-Copper(II) Hexacyanoferrate in DMSO electrolytes. *Journal of Physical Chemistry C*. 128: 1487–1495.
[19] Paolella, A.; Faure, C.; Timoshevskii, V.; Marras, S.; Bertoni, G.; Guerfi, A.; Vijh, A.; Armand, M.; Zaghib, K. 2017. A review on hexacyanoferrate-based materials for energy storage and smart windows: Challenges and perspectives. *Journal of Materials Chemistry A*. 5: 18919–18932.
[20] Santoro, C.; Arbizzani, C.; Erable, B.; Ieropoulos, I. 2017. Microbial cells: From fundamentals to applications. A review. *Journal of Power Sources*. 356: 225–244.
[21] Rismani-Yazdi, H.; Carver, S.M.; Christy, A.D.; Tuovinen, O.H. 2008. Cathodic limitations in microbial fuel cells: An overview. *Journal of Power Sources*. 180: 683–694.
[22] Pant, D.; Van Bogaert, G.; Diels, L.; Vanbroekhoven, K. 2010. A review of the substrates used in microbial fuel cells (MFCs) for sustainable energy production. *Bioresource Technology*. 101: artic. 1533–1543.
[23] Pandey, P.; Shinde, V.N.; Deopurkar, R.L.; Kale, S.P.; Patil, S.A.; Pant, D. 2016. Recent advances in the use of different substrates in microbial fuel cells toward wastewater treatment and simultaneous energy recovery. *Applied Energy*. 168: 706–723.
[24] Sleutels, T.H.J.A.; Lodder, R.; Hamelers, H.V.M.; Buisman, C.J.N. 2009. Improved performance of porous bio-anodes in microbial electrolysis cells by enhancing mass and charge transport. *International Journal of Hydrogen Energy*. 34: 9655–9661.
[25] Rabaey, K.; Girguis, P.; Nielsen, L.K. 2011. Metabolic and practical considerations on microbial electrosynthesis. *Current Opinion in Biotechnology*. 22: 371–377.
[26] Zhang, F.; He, Z. 2015. Scaling up microbial desalination cell system with a post-aerobic process for simultaneous wastewater treatment and seawater desalination. *Desalination*. 360: 28–34.
[27] Doménech-Carbó, A. 2021. *Electrochemistry of Porous Materials*, 2nd edit. Taylor & Francis, Boca Raton, chapters 7 and 14.
[28] Yang, G.; Huang, L.; You, L.; Zhuang, L.; Zhou, S. 2017. Electrochemical and spectroscopic insights into the mechanisms of bidirectional microbe-electrode electron transfer in *Geobacter soli* biofilms. *Electrochemistry Communications*. 77: 93–97.
[29] Kabutey, F.T.; Zhao, Q.; Wei, L.; Ding, J.; Antwi, P.; Quashie, F.K.; Wang, W. 2019. An overview of plant microbial fuel cells (PMFCs): Configurations and applications. *Renewable and Sustainable Energy Reviews*. 110: 402–414.
[30] Strik, D.P.; Hamelers, H.V.M. Snel, J.F.; Buisman, C.J. 2008. Green electricity production with living plants and bacteria in a fuel cell. *International Journal of Energy Research*. 32: 870–876.
[31] Jiang, D.; Li, B.; Jia, W.; Lei, Y. 2010. Effect of inoculum types on bacterial adhesion and power production in microbial fuel cells. *Applied Biochemistry and Biotechnology*. 160: 182–196.
[32] Goto, Y.; Yoshida, N.; Umeyama, Y.; Yamada, T.; Tero, R.; Hiraishi, A. 2015. Enhancement of electricity production by graphene oxide in soil microbial fuel cells and plant microbial fuel cells. *Frontiers in Bioengineering and Biotechnology*. 3: artic. 42.
[33] Chae, K.J.; Choi, M.J.; Lee, J.W.; Kim, K.Y.; Kim, I.S. 2009. Effect of different substrates on the performance, bacterial diversity, and bacterial viability in microbial fuel cells. *Bioresource Technology*. 100: 3518–3525.
[34] Logan, B.E.; Regan, J.M. 2006. Electricity-producing bacterial communities in microbial fuel cells. *Trends in Microbiology*. 14: 512–518.
[35] Jones, D.L.; Nguyen, C.; Finlay, R.D. 2009. Carbon flow in the rhizosphere: carbon trading at the soil-root interface. *Plant and Soil*. 321: 5–33.
[36] Tapia, N.F.; Rojas, C.; Bonilla, C.A.; Vargas, I.T. 2017. Evaluation of sedum as driver for plant microbial fuel cells in a semi-arid green roof ecosystem. *Ecological Engineering*. 108: 203–210.
[37] Chen, Z.; Huang, Y.C.; Liang, J.H.; Zhao, F.; Zhu, Y.G. 2012. A novel sediment microbial fuel cell with a biocathode in the rice rhizosphere. *Bioresource Technology*. 108: 55–59.
[38] Mohan, S.V.; Mohanakrishna, G.; Chiranjeevi, P.; Peri, D.; Sarma, P.N. 2010. Ecologically engineered system (EES) designed to integrate floating, emergent and submerged macrophytes for the treatment of domestic sewage and acid rich fermented-distillery wastewater: evaluation of long term performance. *Bioresource Technology*. 101: 3363–3370.

[39] Song, X.; Yan, D.; Liu, Z.; Chen, Y.; Lu, S.; Wang, D. 2011. Performance of laboratory-scale constructed wetlands coupled with micro-electric field for heavy metal-contaminating wastewater treatment. *Ecological Engineering*. 37: 2061–2065.

[40] Hubenova, Y.; Mitov, M. 2011. Bacterial mutalism in the mosses roots applicable in Bryophyta-microbial fuel cell. *Communications in Agricultural and Applied Biological Science*. 76: 63–65.

[41] Reguera, G.; McCarthy, K.D.; Mehta, T.; Nicoll, J.S.; Tuominen, M.T.; Lovley, D.R. 2005. Extracellular electron transfer via microbial nanowires. *Nature*. 435: 1098–1101.

[42] Bond, D.R.; Holmes, D.E.; Tender, L.M.; Lovley, D.R. 2002. Electrode-reducing microorganisms that harvest energy from marine sediments. *Science*. 295: 483–485.

[43] Schröder, U. 2007. Anodic electron transfer mechanisms in microbial fuel cells and their energy efficiency. *PhysChemChemPhys*. 9: 2619–2629.

[44] Niessen, J.; Schröder, U.; Scholz, F. Exploiting complex carbohydrates for microbial electricity generation – a bacterial fuel cell operating on starch. 2004. *Electrochemistry Communications*. 6: 955–958.

[45] Summers, Z.M.; Fogarty, H.E.; Leang, C.; Franks, A.E.; Malvankar, N.S.; Lovley, D.R. 2010. Direct exchange of electrons within aggregates of an evolved syntrophic coculture of anaerobic bacteria. *Science*. 330: 1413–1415.

[46] Liu, X.; Tremblay, P.L.; Malvankar, N.S.; Nevin, K.P.; Lovley, D.R.; Vargas, M. 2014. A *Geobacter sulfurreducens* strain expressing *Pseudomonas aeruginosa* type IV pili localizes OmcS on pili but is deficient in Fe(III) oxide reduction and current production. *Applied and Environmental Microbiology*. 80: 1219–1224.

[47] Ha, P.T.; Lindemann, S.R.; Shi, L.; Dohnalkova, A.C.; Fredrickson, J.K.; Madigan, M.T., Beyenal, H. 2017. Syntrophic anaerobic photosynthesis via direct interspecies electron transfer. *Nature Communications*. 8: artic. 13924.

[48] Huang, L.; Liu, X.; Tang, J.; Yu, L.; Zhou, S. 2019. Electrochemical evidence for direct interspecies electron transfer between *Geobacter sulfurreducens* and *Prosthecochloris aestuarii*. *Bioelectrochemistry*. 127: 21–25.

[49] Kotz, R.; Carlen, M. 2000. Principles and Applications of Electrochemical Capacitors. *Electrochimica Acta*. 45: 2483–2498.

[50] Herrero, E.; Buller, L.J.; Abruna, H.D. 2001. Underpotential Deposition at Single Crystal Surfaces of Au, Pt, Ag and Other Materials. *Chemical Reviews*. 101: 1897–1930.

[51] Sugimoto, W.; Iwata, H.; Yokoshima, K.; Murakami, Y.; Takasu, Y. 2015. Proton and Electron Conductivity in Hydrous Ruthenium Oxides Evaluated by Electrochemical Impedance Spectroscopy: The Origin of Large Capacitance. *Journal of Physical Chemistry B*. 109: 7330–7338.

[52] Ali, G.A.M.; Osama, A.F.; Makhlouf, S.A.; Yusoff, M.M.; Chong, K.F. 2014. Co_3O_4/SiO_2 nanocomposites for supercapacitor application. *Journal of Solid State Electrochemistry*. 18: 2505–2512.

[53] Supchocksoonthorn, P.; Hrimchum, N.; Budsirak, T.; Intaraprasit, S.; Thongsai, N.; Aussawasathien, D. 2023. Lignin based carbon fiber fabrics with hybrid doping approach as self-standing electrodes for supercapacitors. *Electrochimica Acta*. 437: artic. 141523.

[54] Li, L.; Solin, N.; Inganäs, O. 2021. Self-discharge study of lignin/graphite hybrid material electrodes. *Electrochimica Acta*. 371: artic. 137836.

[55] Christen, T.; Carlen, M.W. 2000. Theory of Ragone plots. *Journal of Power Sources*. 91: 210–216.

[56] Pang, S.; Gao, Y.; Choi, S. 2018. Flexible and stretchable biobatteries: monolithic integration of membrane-free microbial fuel cells in a single textile layer. *Advanced Energy Materials*. 8: artic. 1702261.

[57] Pham, J.; Forget, A.; Bridonneau, N.; Mattana, G.; Stavrinidou, E.; Zrig, S.; Piro, B.; Noel, V. 2022. In vivo electrochemically-assisted polymerization of conjugated functionalized terthiophenes inside the vascular system of a plant. *Electrochemistry Communications*. 137: artic. 107270.

[58] Andreas, H.A. 2015. Self-discharge in electrochemical capacitors: a perspective article. *Journal of the Electrochemical Society*. 162: A5047–A5053.

ADDITIONAL LITERATURE

Enzyme Electrochemistry and Enzymatic Fuel Cells

Courjean, O.; Gao, F.; Mano, N. 2009. Deglycosylation of Glucose Oxidase for Direct and Efficient Glucose Electrooxidation on a Glassy Carbon Electrode. *Angewandte Chemie-International Edition*. 48: 5897–5899.

Haque, S.U.; Duteanu, N.; Ciocan, S.; Nasar, A.; Inamuddin. 2021. A review: Evolution of enzymatic biofuel cells. *Journal of Environmental Management*. 298: artic. 113483.

Xiao, X.; Xia, H.-Q.;Wu, R.; Bai, L.; Yan, L.; Magner, E.; Cosnier, S.; Lojou, E.; Zhu, Z.; Liu, A. 2019. Tackling the Challenges of Enzymatic (Bio)Fuel Cells. *Chemical Reviews*. 119: 9509–9558.

Microbial Fuel Cells

Fernández-Marchante, C.M.; Asensio, Y.; León, L.F.; Villaseñor, J.; Cañizares, P.; Lobato, J.; Rodrigo, M.A. 2018. Thermally-treated algal suspensions as fuel for microbial fuel cells. *Journal of Electroanalytical Chemistry*. 814: 77–82.

Gajda, I.; Greenman, J.; Melhuish, C.; Ieropoulos, I. 2015. Self-sustainable electricity production from algae grown in a microbial fuel cell system. *Biomass & Bioenergy*. 82: 87–93.

Gao, Y.; Mohammadifar, M.; Choi, S. 2019. From microbial fuel cells to Biobatteries: Moving toward on-demand micro-power generation for Small-scale Single-Use Applications. *Advanved Materials Technologies*. 4: artic. 1970079.

Regmi, R.; Nitisoravut, R.; Ketchaimongkol, J. 2018. A decade of plant-assisted microbial fuel cells: looking back and moving forward. *Biofuels*. 9: 605–612.

Biomass Processing and Exploitation for Different Applications

Cabrera, J.; Irfan, M.; Dai, Y.; Zhang, P.; Zong, Y.; Liu, X. 2021. Bioelectrochemical system as an innovative technology for treatment of produced water from oil and gas industry: A review. *Chemosphere*. 285: artic. 131428.

Logan, B.E. 2005. Simultaneous wastewater treatment and biological electricity generation. *Water Science and Technology*. 52: 31–37.

Sayed, E.T.; Alawadhi, H.; Elsaid, K.; Olabi, A.G.; Adel Almakrani, M.; Bin Tamim, S.T.; Alafranji, G.H.M.; Abdelkareem, M.A. 2020. A Carbon-Cloth Anode Electroplated with Iron Nanostructure for Microbial Fuel Cell Operated with Real Wastewater. *Sustainability*. 12: artic. 6538.

Wang, M.; Wang, Z.; Hu, F.; Fan, L.; Zhang, X. 2020. Polyelectrolytes/α-Fe_2O_3 modification of carbon cloth anode for dealing with food wastewater in microbial fuel cell. *Carbon Resources Conversion*. 3: 76–81.

Wilberforce, T.; Sayed, E.T.; Abdelkareem, M.A.; Elsaid, K.; Olabi, A.G. 2021. Value added products from wastewater using bioelectrochemical systems: Current trends and perspectives. *Journal of Water Process Engineering*. 39: 101737.

Bioelectrosynthesis

Nath, I.; Chakraborty, J.; Lips, R.; Dekyvere, S.; Min, J.; Varma, R.S.; Verpoort, F. 2023. Hydrogen bond-mediated pH-universal electrocatalytic hydrogen production by conjugated porous poly-indigo. *Journal of Materials Chemistry A*. 11: 10699–10709.

Yang, E.; Omar Mohamed, H.; Park, S.-G.; Obaid, M.; Al-Qaradawi, S.Y.; Castaño, P.; Chon, K.; Chae, K.-J. 2021. A review on self-sustainable microbial electrolysis cells for electro-biohydrogen production via coupling with carbon-neutral renewable energy technologies. *Bioresource Technology*. 320: artic. 124363.

Phytoremediation

Ashraf, S.; Ali, Q.; Zahir, Z.A.; Ashraf, S.; Asghar, H.N. 2019. Phytoremediation: Environmentally sustainable way for reclamation of heavy metal polluted soils. *Ecotoxicology and Environmental Safety*. 174: 714–727.

Rezania, S.; Taib, S.M.; Din, M.F.M.; Dahalan, F.A.; Kamyab, H. 2016. Comprehensive review on phytotechnology: heavy metals removal by diverse aquatic plants species from wastewater. *Journal of Hazardous Materials*. 318: 587–599.

Yadav, K.K.; Gupta, N.; Kumar, A.; Reece, L.M.; Singh, N.; Rezania, S.; Khan, S.A. 2018. Mechanistic understanding and holistic approach of phytoremediation: a review on application and future prospects. *Ecological Engineering*. 120: 274–298.

Plant Fuel Cells

Helder, M.; Strik, D.P.B.T.B.; Hamelers, H.V.M.; Kuijken, R.C.P.; Buisman, C.J.N. 2012. New plantgrowth medium for increased power output of the plant-microbial fuel cell. *Bioresources Technology*. 104: 417–423.

Ju, X.; Wu, S.; Huang, X.; Zhang, Y.; Dong, R. 2014. How the novel integration of electrolysis in tidal flow constructed wetlands intensifies nutrient removal and odor control. *Bioresource Technology*. 169: 605–613.

Shen, X.; Zhang, J.; Liu, D.; Hu, Z.; Liu, H. 2018. Enhance performance of microbial fuel cell coupled surface flow constructed wetland by using submerged plants and enclosed anodes. *Chemical Engineering Journal*. 351: 312–318.

Xu, P.; Xiao, E.; Wu, J.; He, F.; Zhang, Y.; Wu, Z. 2019. Enhanced nitrate reduction in water by a combined bio-electrochemical system of microbial fuel cells and submerged aquatic plant *Ceratophyllum demersum*. *Journal of Environmental Sciences*. 78: 338–351.

Yadav, A.K. 2010. Design and development of novel constructed wetland cum microbial fuel cell for electricity production and wastewater treatment. In: Cerioni, A.M.; Motta, R. Eds. *Proceedings of 12th international conference on wetland systems for water pollution control (IWA)*. Palombi, Venice, p. 321.

Extracellular Electron Transfer

Catania, C.; Karbelkar, A.A.; Furst, A.L. 2021. Engineering the interface between electroactive bacteria and electrodes. *Joule*. 5: 743–747.

Niessen, J.; Harnisch, F.; Rosenbaum, M.; Schröder, U.; Scholz, F. Heat treated soil as convenient and versatile source of bacterial communities for microbial electricity generation. 2006. *Electrochemistry Communications*. 8: 869–873.

Capacitors

Conway, B.E. 1991. Transition from "supercapacitor" to "battery" behavior in electrochemical energy storage. *Journal of the Electrochemical Society*. 138: 1539–1548.

Sarangapani, S.; Lessner, P.; Forchione, J.; Griffith A.; Laconti, A.B. 1990. Advanced double layer capacitors. *Journal of Power Sources*. 29: 355–364.

Simon, P.; Gogotsi, Y. 2008. Materials for electrochemical capacitors. *Nature Materials*. 7: 845–854.

Zhai, S.; Karahan, H.E.; Wang, C.; Pei, Z.; Wei, L.; Chen, Y. 2019. 1D Supercapacitors for Emerging Electronics: Current Status and Future Directions. *Advanced Materials*. 32: artic. e1902387.

Wearable Cells and Biodegradable Batteries

Jia, X.; Wang, C.; Zhao, C.; Ge, Y.; Wallace, G.G. 2016. Toward biodegradable Mg–air bioelectric batteries composed of silk fibroin–polypyrrole film. *Advanced Functional Materials*. 26:1454–1462.

Liu, Y.; He, K.; Chen, G.; Leow, W.R.; Chen, X. 2017. Nature-Inspired Structural Materials for Flexible Electronic Devices. *Chemical Reviews*. 117: 12893–12941.

Pal, R.K.; Kundu, S.C.; Yadavalli, V.K. 2018. Fabrication of flexible, fully organic, degradable energy storage devices using silk proteins. *ACS Applied Materials and Interfaces*. 10: 9620–9628.

Wang, X.; Lu, X.; Liu, B.; Chen, D.; Tong, Y.; Shen, G. 2014. Flexible energy-storage devices: design consideration and recent progress. *Advanced Materials*. 26: 4763–4782.

Wu, H.; Zhang, Y.; Kjøniksen, A.; Zhou, X.; Zhou, X. 2021. Wearable Biofuel Cells: Advances from Fabrication to Application. *Advanced Functional Materials*. 31: artic. 2103976.

15 Plants and Environment

15.1 GENERAL ASPECTS

There is a great variety of interactions between vegetal species and their physical environment. These have been enormously influential on the geological time scale concerning the evolution of the atmosphere and soil formation [1]. The contemporary importance of these interactions is evident concerning pollution/remediation, greenhouse effect, and climatic change, as well as dealing with the specific interactions of vegetal species with technological materials. Here, microbial-assisted corrosion of metals and concrete deterioration by fungi are paradigmatic examples.

Microbiologically influenced metal corrosion was first observed in 1910 [2], and its importance was subsequently underlined [3, 4]. Many microorganisms are capable of producing corrosion effects, namely, bacteria, archaea, algae, molds, and fungi. These are not restricted to metals but may also involve plastic, wood, stone, concrete, etc.

In this chapter, we will concentrate our attention on bacterial-assisted metal corrosion and studies on corrosion inhibition where plant essential oils play an interesting role. Biodeterioration of nonmetallic materials will also be briefly treated. Electrochemical degradation of pollutants and electrochemical valorization of biomass complete the matters reflecting the interaction between vegetal biota and the environment treated here.

15.2 MICROBIOLOGICALLY INFLUENCED METAL CORROSION

As it is well-known, metal technology is continuously struggling against thermodynamics because the oxidation of metals to their oxides, sulfides, etc., is in general a thermodynamically favored process [5]. Under ordinary aerobic conditions, metal corrosion can be seen as an electrochemical process in which a humidity layer acts as the electrolyte between anodic and cathodic regions [6]. In anodic regions, the metal is oxidized to different species, while in cathodic regions, oxygen is reduced to water. Anodic and cathodic sites are spatially separated over atomic distances and randomly distributed on the metal surface. Figure 15.1 shows a simplified pictorial scheme exemplified for the case of iron. The anodic reaction consists of the Fe oxidation to Fe^{2+} ions passing to the humidity layer.

$$Fe \rightarrow Fe^{2+} + 2e^- \qquad (15.1)$$

The cathodic reaction can be represented as

$$O_2 + 4H^+ + 4e^- \rightarrow 2H_2O \qquad (15.2)$$

Notice that the nascent Fe^{2+} ions in the humidity layer experience a complex series of subsequent reactions to yield maghemite, magnetite, goethite, etc., minerals. Both the anodic and cathodic processes proceed under kinetic control, the access of oxygen to the metal surface usually being the rate-determining step [7, 8]. Under anaerobic conditions, other oxidants can replace oxygen.

The corrosion process is influenced by the physicochemical properties of the metal (including its chemical composition and microstructural properties) and its corrosion products, as well as the characteristics of the environment. As a result, there are different micro- and mesostructural corrosion types such as intergranular corrosion or pitting corrosion. The primary corrosion processes are

DOI: 10.1201/9781032697826-15

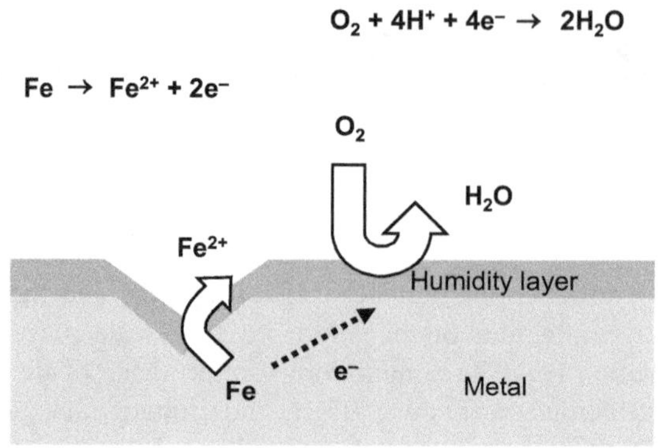

FIGURE 15.1 Simplified scheme for the "ordinary" aerobic electrochemical oxidation of iron.

TABLE 15.1
Standard Potentials of Different Couples Dealing with Iron Corrosion

Couple	E° (V vs. SHE)
$O_{2\,g} + 4H^+_{\ aq} + 4e^- \rightarrow H_2O$	+1.23
$NO_3^-{}_{aq} + 2H^+_{\ aq} + 2e^- \rightarrow NO_2^-{}_{aq} + H_2O$	+0.84
$SO_4^{2-} + 9H^+ + 8e^- \rightarrow HS^- + 4H_2O$	+0.25
$2H^+_{\ aq} + 2e^- \rightarrow H_{2\,g}$	0.00
$Fe^{2+}_{\ aq} + 2e^- \rightarrow Fe_{\ solid}$	−0.44

accompanied, depending on the corrosive environment, by pedological processes of re-crystallization, segregation, and cementation, etc., resulting in the formation of a frequently complex patina of corrosion products.

There are different modalities of microbiologically influenced (metal) corrosion (MIC) [9]:

a) Direct (or electrical) corrosion: Caused by extracellular electron transfer by microorganisms in direct contact with the metal surface. The mechanism of electron transfer involves cell surface enzymes or membrane redox proteins.
b) Indirect corrosion: Involving soluble electron transfer mediators released from the microorganisms.
c) Metabolite corrosion: Achieved directly or indirectly by metabolites produced by microorganisms.
d) Under deposit (or oxygen gradient) corrosion: Taking place around or under a deposit of corrosion products produced otherwise.
e) Crevice corrosion: A localized corrosion associated with narrow apertures or clearances formed between the metal surface and another surface.

Sulfate-reducing bacteria (SRB) are the most widely studied group of corrosion-influencing microorganisms, but nitrate-reducing bacteria (NRB), acid-producing bacteria (APB), sulfur-oxidizing bacteria (SOB), iron-oxidizing bacteria (IOB), and iron-reducing bacteria (IRB) are also involved in corrosion processes [10]. As can be seen in Table 15.1, where the standard potentials of different couples dealing with iron corrosion are listed, different oxidant species, from hydrogenions to nitrate and sulfate ions, are capable of oxidizing metallic iron. Similarly, the formation of different corrosion products is thermodynamically spontaneous under standard conditions. Table 15.2 summarizes several pertinent data.

Different mechanisms for MIC have been proposed. Apart from direct electron extraction, these include galvanic cells, H_2S, NH_3, or PH_3 bacterial secretion; differential aeration or chemical concentration cells; enzymatic oxygen reduction reactions; and acidification [11]. It must be taken into account that intermittent aerobic conditions may occur in real systems. This creates oxygen gradient corrosion or differential aeration cells [9]. In principle, intermittent aeration increases the corrosion rate of anaerobic corrosion [9].

15.3 BIOFILM FORMATION

MIC involves the interaction between the microorganisms, the medium, and the metal. The adhesion of microorganisms to solid surfaces is mediated by sensory organs (flagella, pili) and depends on several physiochemical factors, such as polar and ionic interactions, van der Waals forces, and hydrophobicity. Bacterial adhesion involves four steps: transport, reversible initial bonding, irreversible bonding, and colonization [12].

Biofilm adherence depends, among other factors, on surface charge, hydrophobicity, roughness, and topography at different size scales (macro, micro, and nano) of the metal surface. This process depends on the surface charge and hydrophobicity of the microorganism surface. The cell walls of most bacterial cells have many carboxyl and phosphate groups, thus resulting in an overall negative charge [13].

After biofilm adhesion to the substrate, bacteria begin to multiply locally forming microcolonies. The extracellular secretion of polymeric materials such as exopolysaccharides generates a multilayer grouping. Then, the maturation of the biofilm occurs, giving rise to a macro colony firmly attached to the substrate. This step ends with the detachment of the colony, reversion of bacteria to their planktonic state, and dispersion of the colony in the surrounding medium [14].

15.4 MECHANISMS OF BACTERIAL-INFLUENCED CORROSION

The effect of SRB at biological pHs is the production of HS^- formally via

$$SO_4^{2-} + 9H^+ + 8e^- \rightarrow HS^- + 4H_2O \tag{15.3}$$

Sulfide species react with Fe^{2+} ions generated in the initial iron oxidation (see Figure 15.1) forming FeS, a low solubility product:

$$Fe^{2+} + HS^- \rightarrow FeS + H^+ \tag{15.4}$$

This is schematized in Figure 15.2. Then, the overall cathodic reaction responsible for the electrochemical corrosion can be represented as

$$Fe + HS^- \rightarrow FeS + H^+ + 2e^- \tag{15.5}$$

TABLE 15.2
Standard Gibbs Energy Values (per Mole of Iron) for Selected Reactions Generating Iron Corrosion Products Under Anaerobic Conditions

Reaction	ΔG° (kJ mol⁻¹)
$Fe_{solid} + H_2O_{liq} \rightarrow H_{2\,g} + FeO_{solid}$	−14.2
$Fe_{solid} + 2H_2O_{liq} \rightarrow H_{2\,g} + Fe(OH)_{2\,solid}$	−15.6
$Fe_{solid} + H_2S_{g} \rightarrow H_{2\,g} + FeS_{solid}$	−67.4
$Fe_{solid} + 2H_2S_{g} \rightarrow 2H_{2\,g} + FeS_{2\,solid}$	−100.9
$3Fe_{solid} + 4H_2O_{liq} \rightarrow 4H_{2\,g} + Fe_3O_{4\,solid}$	−22.1

Under anaerobic conditions, the cathodic reaction is the hydrogen evolution reaction:

$$2H^+ + 2e^- \rightarrow H_2 \tag{15.6}$$

However, this process takes place via the initial formation of adsorbed hydrogen atoms (Volmer step).

$$H^+ + e^- \rightarrow H\bullet(ads) \tag{15.7}$$

That can be followed by the recombination of adsorbed hydrogen atoms (Tafel pathway), or by a second electron transfer (Heyrovsky pathway):

$$H\bullet(ads) + H^+ + e^- \rightarrow H_2 \tag{15.8}$$

The first approach to describe the action of SRBs was the proposal that they promote the abstraction (desorption) of adsorbed hydrogen atoms (the so-called cathodic depolarization mechanism) [10]:

$$SO_4^{2-} + 8H\bullet(ads) \rightarrow S^{2-} + 4H_2O \tag{15.9}$$

However, electrochemical data suggested that hydrogen atom desorption was not the rate-determining step of the corrosion process [15, 16].

A second model considered the generation of conductive iron sulfide where the HER process is favored relative to the base metal [17, 18]. This proposal has the drawback of the variable conducting properties of iron sulfide films, where troilite (FeS), greigite (Fe_3S_4), pyrite (FeS_2), and different nonstoichiometric minerals can be formed [19]. These display a relatively complex solid-state electrochemistry dominated by sulfur-based oxidation processes [20, 21].

This mechanistic scenario can be completed by considering the local acidification and the different extracellular electron transfer mechanisms seen in Chapter 14. The essential idea, schematized in Figure 15.3, is that the electrons released by the metal's oxidation are transported across the bacterium's cell wall, and the sulfate reduction would occur inside the cell [22]. In (a), the electrons are transported through the bacterial cell wall, whereas in (b), the electron transport chain occurs through the cell pili. The case of redox mediation is illustrated in (c).

Although there is controversy around the mechanisms of charge transport operating in MIC [10], similar mechanistic considerations can be applied to the NRB where biocatalytic cathodic nitrate

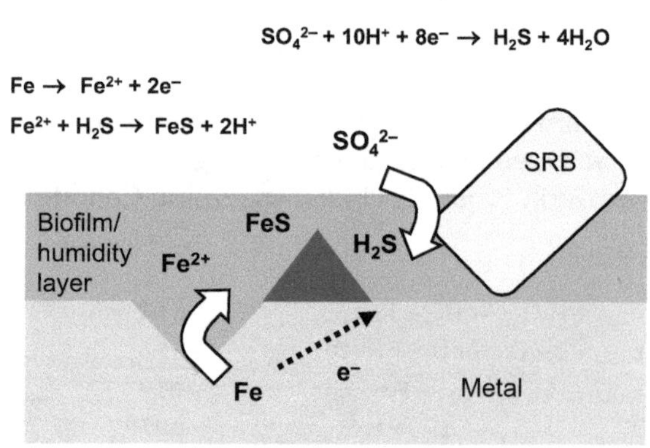

FIGURE 15.2 Simplified scheme for the anaerobic electrochemical oxidation of iron promoted by SRBs.

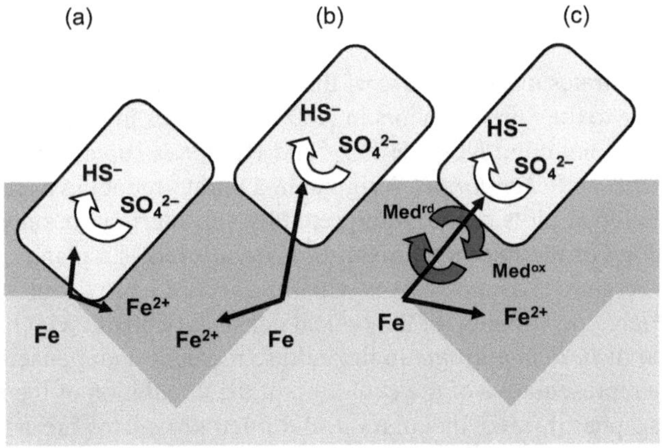

FIGURE 15.3 Schematic representation of three hypothetic extracellular electron transfer pathways for anaerobic iron oxidation promoted by SRBs. (a) Direct transport through the cell wall, (b) transport through the pili, and (c) transport through a mediator (either endogenous or exogenous).

reduction and acidification could be responsible for MIC [23]. Here, the principal process is nitrate-to-nitrite reduction:

$$NO_3^- + 2H^+ + 2e^- \rightarrow NO_2^- + H_2O \tag{15.10}$$

Under aerobic conditions, IOB produce a fast oxidation of Fe(II) to Fe(III). The overall process can be represented as [24]

$$4Fe^{2+} + O_2 + 4H^+ \rightarrow 4Fe^{3+} + 2H_2O \tag{15.11}$$

The corrosion pathway comprises a series of electron transporters (mainly enzymes, coenzymes, and cofactors involved in oxidative phosphorylation inside the bacterial cell membrane) to produce water. The proposed electron transfer chain can be represented as [24]

$$2e^- \rightarrow \text{Fe oxidase} \rightarrow \text{Rusticyanin} \rightarrow \text{Cytochrome C} \rightarrow \text{Cytochrome aa}_3 \rightarrow \text{Coenzyme Q} \rightarrow \text{NAND} \rightarrow O_2 \rightarrow O_2\bullet^- \rightarrow H_2O \tag{15.12}$$

accompanying the energy release for the growth of the bacterium.

It is interesting to note that in real systems, multibacterial populations exist, and synergistic effects can occur [25]. Another aspect to underline is that bacterial activity may produce a dual effect, favoring corrosion but also eventually inhibiting the process [25]. This second aspect will be treated in Section 15.6.

15.5 ELECTROCHEMICAL MONITORING OF MIC

Electrochemical techniques are routinely used in corrosion studies.

a) Open circuit potential (OCP) consists of the measurement of the difference of potential between the system under study, conformed as a working electrode, and a reference electrode, connected through a high-impedance voltmeter. The measured potential, E_{OCP}, is often called corrosion potential. (In the following text, a second definition of corrosion potential will be provided.) This technique has been applied to characterize the so-called metal ennoblement, a positive shift of E_{OCP} associated with the formation of a biofilm on the metal surface [26].

Figure 15.4 illustrates a classical use of the OCP technique. In this figure, E_{OCP} vs. time plot corresponding to the measurements in X80 steel probes immersed under sterile conditions in a medium containing NO_3^-, $H_2PO_4^-$ and HPO_4^{2-} is superimposed to that obtained in a culture of *Bacillus licheniformis* Weigmann, a nitrate-reducing bacterium [23].

The aforementioned plots can be interpreted by considering not only thermochemistry but also the biology of the microorganism. *B. licheniformis* is a planktonic species whose population in the culture medium evolves with time according to a four-step path illustrated in Figure 15.5. Here, the variation of the optical density at 600 nm with time, representative of the population of *B. licheniformis* in the culture medium under anaerobic conditions, is depicted. This is representative of the evolution of the population of the microorganism. In the first step (lag phase), after inoculation, the microorganisms are adapting to the new environment. This is followed by a step of fast population increment (log phase) until a maximum density is reached. After that, the population attains a stationary phase. Finally, the rapid consumption of nutrients produces the progressive decay of the population (decline phase).

These phases can be correlated with the E_{OCP} vs. time curves in Figure 15.4 in the light of the Pourbaix diagram in Figure 15.6, corresponding to the Fe/ NO_3^-/$H_2PO_4^-$/HPO_4^{2-} system and including the lines corresponding to the nitrate-to-nitrite and nitrite-to-NO reduction processes.

The E_{OCP} measured for X80 steel immersed under sterile conditions falls initially within the Fe^{2+} stability zone but increases rapidly, reaching an almost stationary value within the $Fe(OH)_2/FePO_4{\cdot}2H_2O$ region after seven days. This OCP behavior is consistent with the

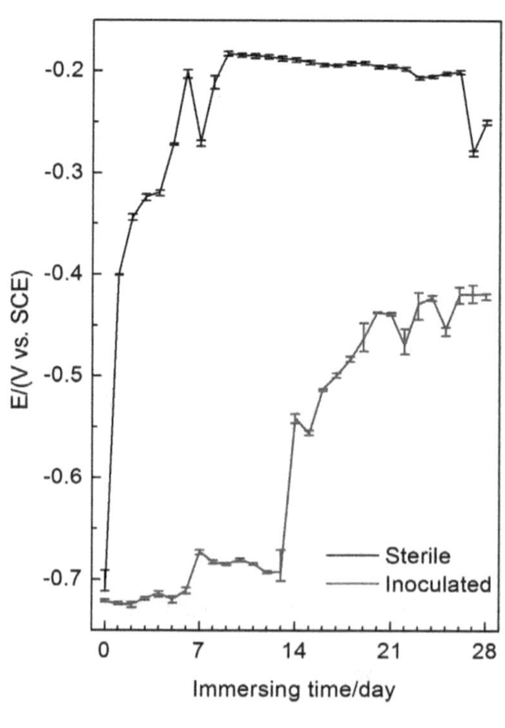

FIGURE 15.4 E_{OCP} vs. time plot for X80 steel immersed under sterile conditions or in *B. licheniformis* culture. Data recorded in 28 days are represented where error bars correspond to the range of variation in measurement for 60 min. (Adapted from ref. [23], with permission. Copyright Elsevier 2022.).

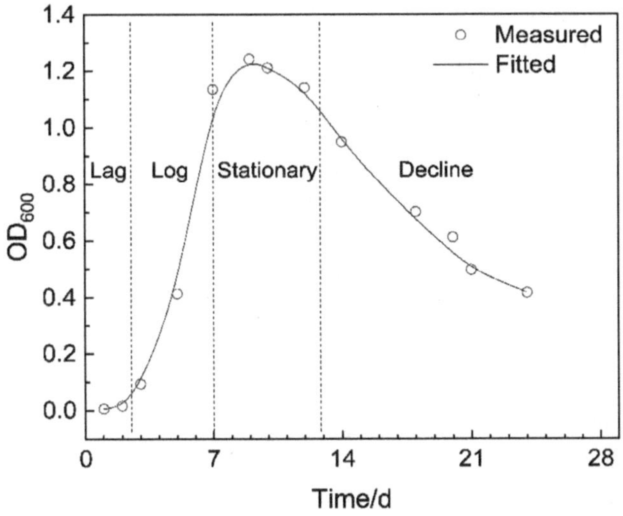

FIGURE 15.5 Variation of the optical density at 600 nm (OD_{600}) with the time of a culture of *B. licheniformis* under anaerobic conditions. This is representative of the evolution of the population of the microorganism. (Reproduced from ref. [23], with permission. Copyright Elsevier 2022.).

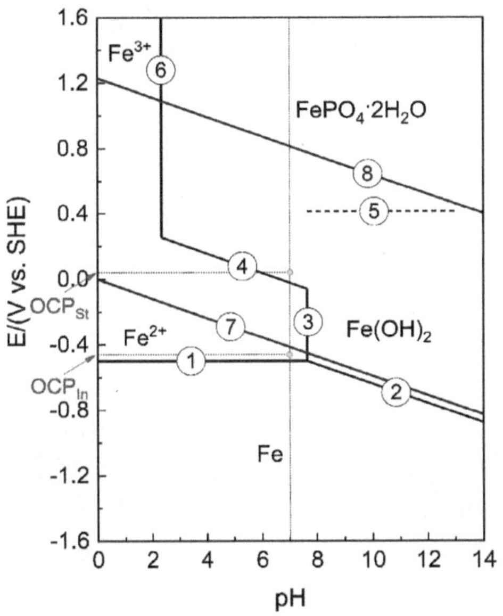

FIGURE 15.6 Pourbaix diagram of $Fe/H_2PO_4^-/HPO_4^{2-}$ system based on thermodynamic data. Lines 7 and 8 correspond to the nitrate-to-nitrite and nitrite-to-NO reduction processes in aqueous solution, respectively. The time evolution of E_{OCP} measured for X80 steel immersed under sterile conditions or in *B. licheniformis* culture in Figure 15.4, are marked as OCP_{st} and OCP_{in}, respectively. (Adapted from ref. [23], with permission. Copyright Elsevier 2022.).

expected formation of a $FePO_4{\cdot}2H_2O$ protective layer passivating the iron surface. In contrast, the E_{OCP} values measured for steel probes in the inoculated media, although they shift positively after 14 days, remain within the Fe^{2+} area; i.e., the corrosive conditions are maintained [23].

The previous example illustrates the capabilities of the OCP technique but also its limitations, as far as it does not permit discriminating between thermodynamic and kinetic effects influencing the overall corrosion process [9].

b) Polarization curves: Usually termed potentiodynamic polarization curves, they use an equivalent electrode configuration to OCP. Polarization curves are based on the record of the intensity passed (i) when a potential (E) is applied under steady-state conditions to force the oxidation of the metal of which the working electrode is made. In practice, the potential is varied linearly with time at a very low sweep rate (below 1 mV s^{-1}), and the results are presented as log$i$ vs. E plots or vice-versa.

In a typical polarization experiment, the potential is scanned in the positive direction from a sufficiently cathodic value, yielding a characteristic lambda-shaped record. Figure 15.7 shows a series of polarization graphs corresponding to carbon steel coupons in contact with sterile seawater and the same cultured with *Pseudomonas*, *Vibrio*, and a mixture of *Pseudomonas* and *Vibrio* bacteria [27]. At the initial highly cathodic potentials, hydrogen evolution takes place, and the current varies rapidly with the applied potential. This initial step is followed by a region under mass transport control with a slow current variation with the potential until an abrupt current decrease appears. Here, the metal is oxidized to metal ions, and the current reaches a minimum that defines an apparent corrosion potential. The zero current corresponds to the rest potential, which cannot be displayed in the logarithmic scale.

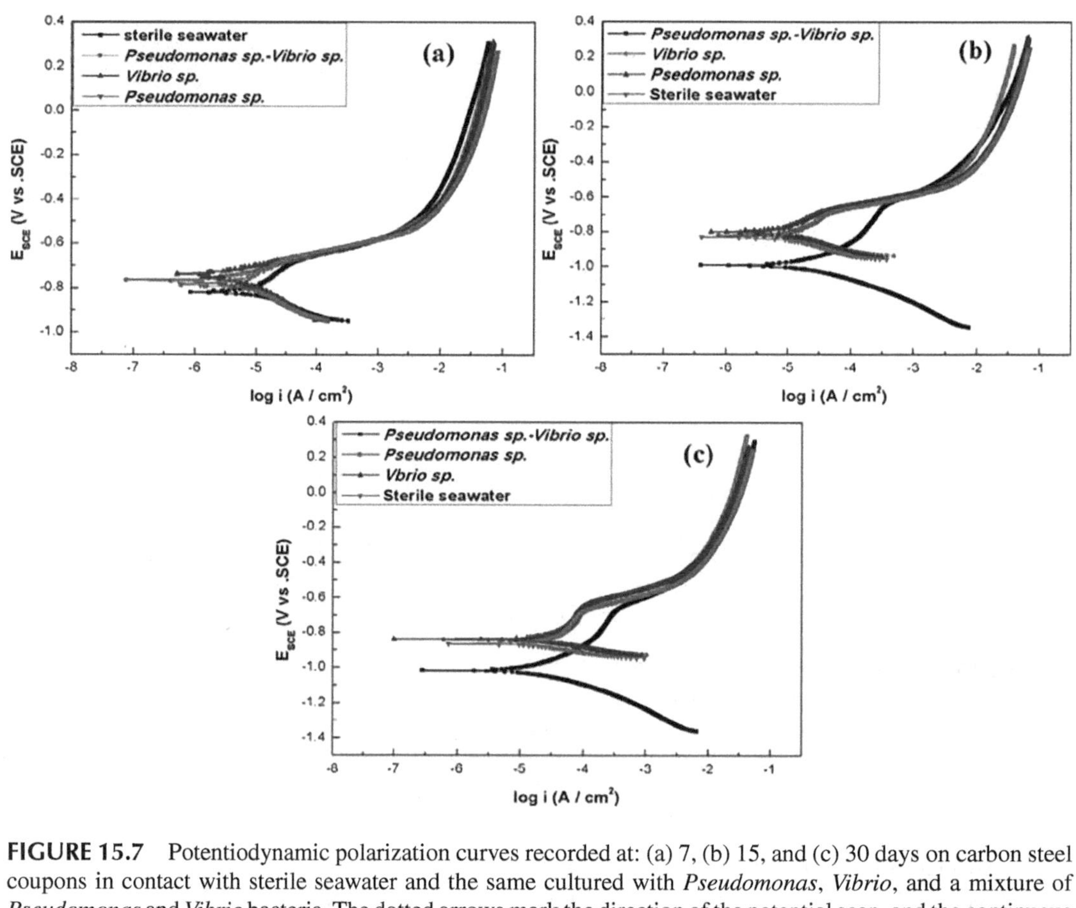

FIGURE 15.7 Potentiodynamic polarization curves recorded at: (a) 7, (b) 15, and (c) 30 days on carbon steel coupons in contact with sterile seawater and the same cultured with *Pseudomonas*, *Vibrio*, and a mixture of *Pseudomonas* and *Vibrio* bacteria. The dotted arrows mark the direction of the potential scan, and the continuous arrows mark the cathodic (below) and anodic (upper) Tafel regions. (Adapted from ref. [27], with permission. Copyrigth The American Chemical Society 2021.).

In the region of potentials previous (cathodic branch) and following (anodic branch) the sharp current minimum, there is a linear variation of log$i$ (or ln$i$) on E, defining the Tafel regions represented by equations of the following type:

$$i = i_{corr} \exp\left[-\frac{\alpha_{cat} nF\left(E - E_{corr}\right)}{RT}\right] \tag{15.13}$$

$$i = i_{corr} \exp\left[\frac{\alpha_{anod} nF\left(E - E_{corr}\right)}{RT}\right], \tag{15.14}$$

where E_{corr} represents the "true" corrosion potential, a parameter that characterizes the corrosivity of the metal in the medium, and i_{corr} the corrosion current. These are defined as the point of intersection between the anodic and cathodic Tafel branches. Corrosion resistance (or polarization resistance) is another widely used parameter defined as

$$R_p = \left(\frac{b_{cat} b_{an}}{b_{cat} + b_{an}}\right)\left(\frac{1}{i_{corr}}\right). \tag{15.15}$$

In this equation, the terms b_{anod} (= $RT/\alpha_{cat}nF$) and b_{cat} (= $RT/\alpha_{an}nF$) represent the Tafel slopes representative of the kinetics of the electron transfer process in the anodic and cathodic regions. Frequently, the current density (A cm^{-2}) rather than the current is used in the preceding representations and equations.

The same methodology is routinely used to study the kinetics of faradaic processes using current/potential data taken at the foot of reduction or oxidation peaks in linear potential scan voltammetry. Then, the Tafel slopes give mechanistic information of interest. As a typical example, the ideal Tafel slopes for the Volmer, Heyrovsky, and Tafel steps of HER are 120, 40, and 30 mV $decade^{-1}$, respectively.

After the abrupt current minimum, the current again varies slowly with the potential in a second region of mass control. In most metals, at sufficiently anodic potentials, a plateau appears corresponding to the formation of an oxide layer. This is the passive region where the metal becomes passivated. It is pertinent to note that all of the aforementioned processes are accompanied by others, in particular, by the charging/discharging of Helmholtz's double-layer. To minimize these undesired capacitive phenomena, slow potential scan rates are employed.

Data in Figure 15.7a provide E_{corr} values around −0.75 V vs. SCE for sterile and all cultured media, but the corrosion current density in the former group was initially slightly higher than that in the second [27]. This suggests that the corrosion of the coupons in the biotic environment is inhibited at the initial time (seven days). After 15 (Figure 15.7b) and 30 days (Figure 15.7c), E_{corr} of the coupons in the mixed *Pseudomonas* and *Vibrio* bacteria becomes ca. 200 mV more negative. This suggests that there is a synergistic effect of both types of bacteria enhancing the corrosivity of the carbon steel surface.

c) Electrochemical impedance spectroscopy (EIS): As described in Chapter 1, EIS provides impedance data as a function of the frequency of a low-amplitude, alternating potential signal. This is superimposed to an arbitrary constant bias potential so that the experiment may involve both faradaic and nonfaradaic processes but operating under stationary conditions.

Figure 15.8 shows the Nyquist plots recorded at different immersion times for carbon steel coupons in contact with sterile seawater and sterile seawater cultured with *Pseudomonas*, *Vibrio*, and a mixture of *Pseudomonas* and *Vibrio* bacteria [27]. In all cases, an apparently unique loop was recorded, whose diameter along the real impedance

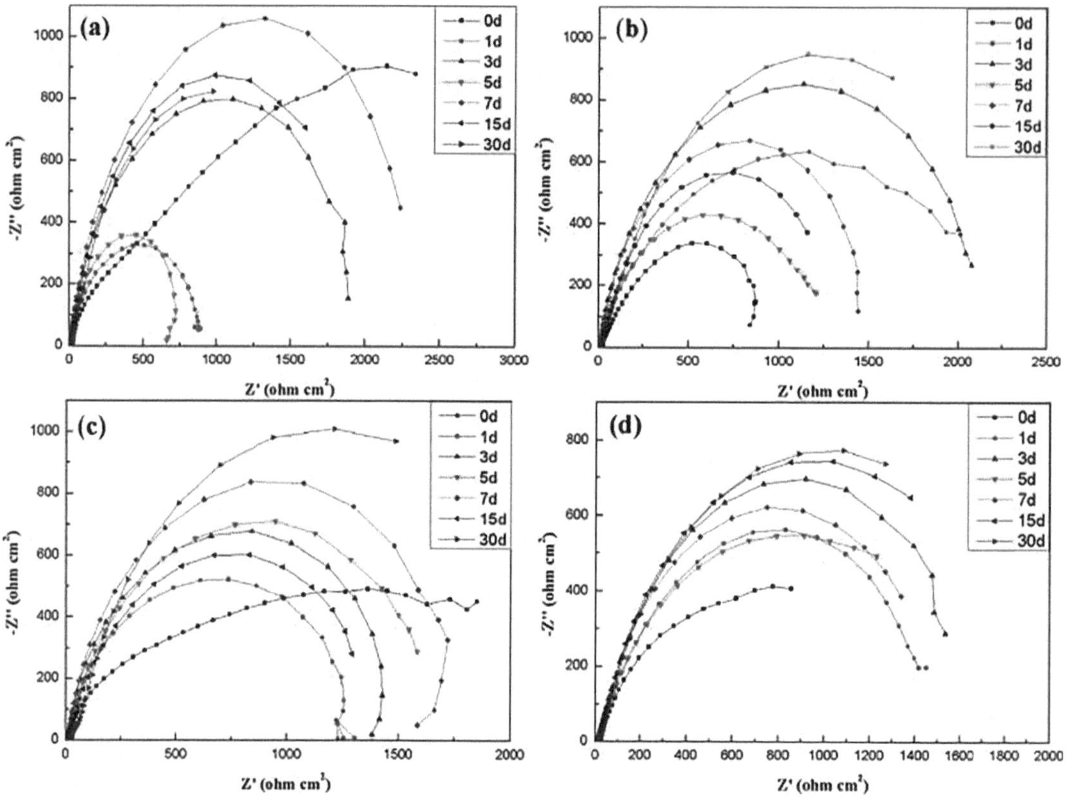

FIGURE 15.8 Nyquist plots from impedance spectra recorded at different immersion times for carbon steel coupons in contact with (a) a mixture of *Pseudomonas* and *Vibrio*, (b) *Vibrio*, (c) *Pseudomonas* in sterile seawater, and (d) sterile seawater. (Adapted from ref. [27], with permission. Copyright the American Chemical Society 2021.).

axis can be taken – roughly – as representative of the charge transfer resistance in turn related to the corrosion rate. In the presence of *Vibrio* alone and *Pseudomonas* alone, the diameter first increases, further decreases, and finally increases, reaching a maximum after 30 days of contact with the cultured seawater. These features suggest that a protective biofilm is initially formed and then subsequently dismantled, opening the door for faster corrosion at the end of the tested time period. In contrast, when the steel probe is exposed to *Pseudomonas* and *Vibrio* bacteria conjointly, the maximum resistance is acquired on the seventh day and further decreases monotonically. These responses differ from those in contact with sterile seawater, characterized by a continuous increase of the resistance.

However, data analysis is not straightforward, as indicated by the corresponding Bode plots of log|Z| vs. log$f$ and $-\varphi$ vs. log$f$ represented in Figure 15.9 [27]. Now, the variation of the phase angle with the frequency involves two maxima at middle and low frequencies, consistent with the equivalent circuit in Figure 15.10 [27]. This equivalent circuit reproduces satisfactorily experimental data using two parallel resistances plus capacitor associations in series between them and with the usual solution resistance. This equivalent circuit suggests that EIS data have to be analyzed, taking into account the factors influencing the impedance response: resistance of charge transfer at the metal/electrolyte interface, resistance of the biofilm, double-layer capacitance, and capacitance associated with the biofilm.

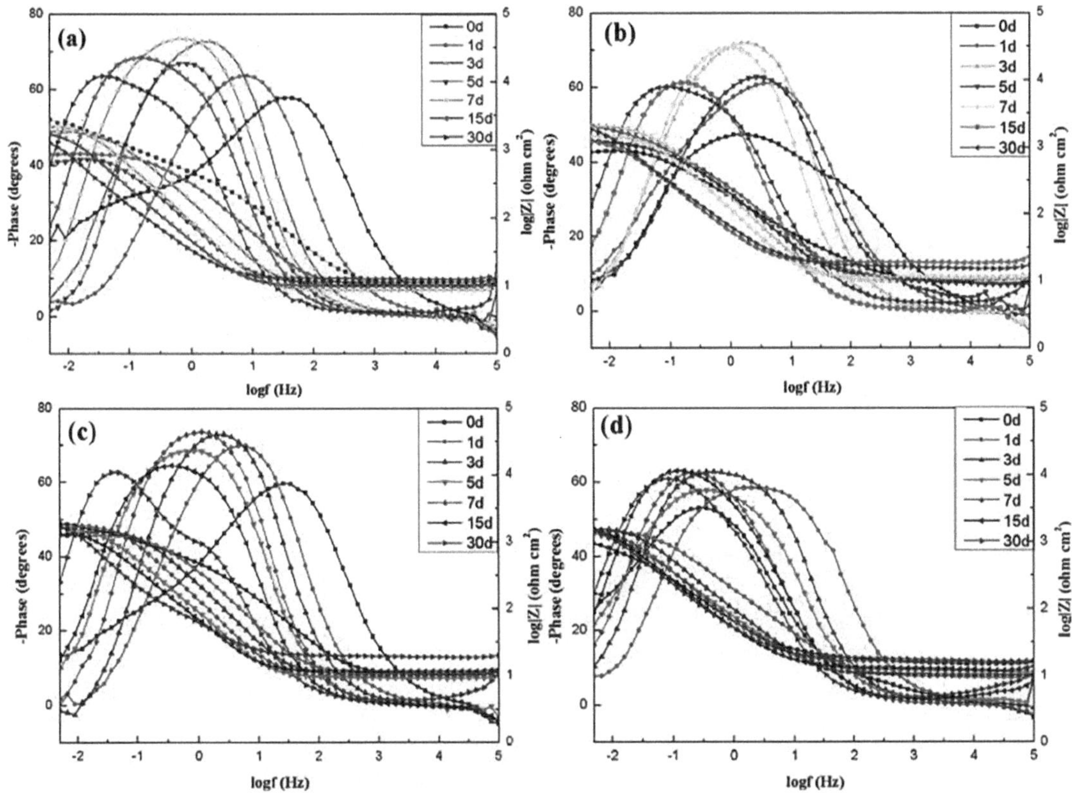

FIGURE 15.9 Bode plots of log|*Z*| vs. log*f* and –φ vs. log*f* from impedance spectra recorded at different immersion times for carbon steel coupons in contact with (a) a mixture of *Pseudomonas* and *Vibrio*, (b) *Vibrio*, (c) *Pseudomonas* in sterile seawater, and (d) sterile seawater. (Adapted from ref. [27], with permission. Copyright the American Chemical Society 2021.).

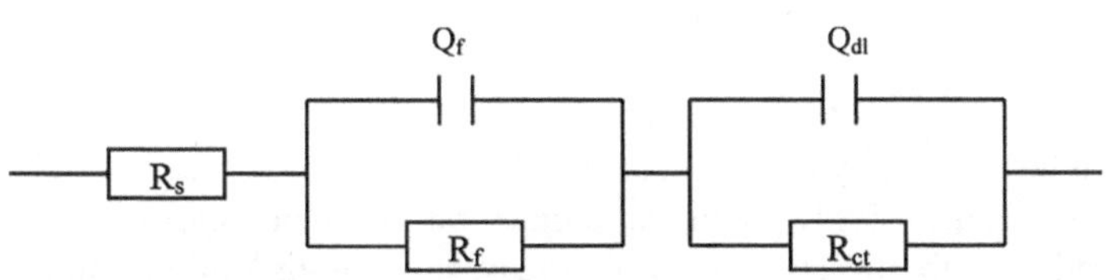

FIGURE 15.10 Equivalent circuit used to model EIS data in Figures 15.8 and 15.9. R_s: solution resistance, R_{ct}: charge transfer resistance, R_f: biofilm resistance, Q_{dl}: double-layer capacitance, Q_f: biofilm capacitance. (Reproduced from ref. [27], with permission. Copyright the American Chemical Society 2021.).

d) Scanning electrochemical microscopy (SECM): As introduced in Chapter 1, this technique offers the possibility of local analysis – at the micrometer scale – of the "electrochemical topography" of surfaces. In the study of biofilms and bacterial corrosion, SECM permits in situ and real-time monitoring of nutrients, metabolites, and quorum-sensing molecules [28]. An example of SECM study on bacterial corrosion is presented in Figure 15.11, corresponding to the corrosion of 304 stainless steel by thermoacidophilic archaeon *Metallosphaera cuprina* Liu et al. [29].

In this figure, the *z*-axis denotes the local concentration of Fe^{2+} ions in *M. cuprina* biofilms under different concentrations of nutrients. SECM experiments were performed in

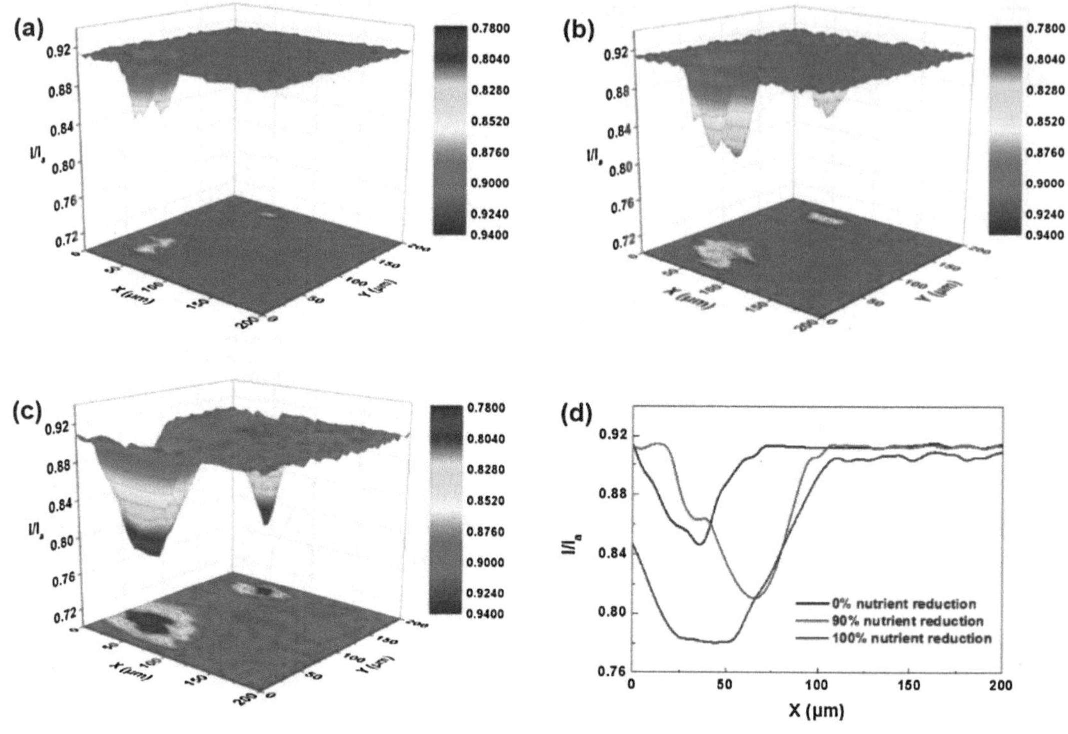

FIGURE 15.11 SECM images (a–c) and line scan results (d) of the Fe^{2+} distributions in the *M. cuprina* biofilm regions partially covering stainless steel for nutrient concentrations of (a) 1, (b) 0.2, and (c) 0 g L^{-1}. Experimental details in the main text. (Adapted from ref. [29], with permission. Copyright Elsevier 2021.).

sterilized PBS with exogenous $FeCl_2$ as a Fe^{2+} source applying a constant potential of 0.65 V where Fe^{2+} is oxidized to Fe^{3+}. The uniform gray surface corresponds to the regions of the metal surface not covered by the biofilm, whereas negative feedback features correspond to the fluctuations of Fe^{2+} concentration in the biofilm-covered areas [29].

The decrease in Fe^{2+} concentration in the biofilm region can be mainly attributed to the Fe^{2+} to Fe^{3+} oxidation metabolism of *M. cuprina* [30]. The oxidation of Fe^{2+} releases electrons that can be utilized by archaeon cells for carbon dioxide fixation and cell growth. As can be seen in Figure 15.11, when the concentration of organic nutrients in the culture medium is reduced, the tip current drop increases, reaching a minimum in the absence of nutrients. These features indicate that the consumption of Fe^{2+} by the *M. cuprina* biofilms increases when the nutrient concentration decreases, resulting in more Fe^{2+} being oxidized. The experiment in Figure 15.11 involves exogenous Fe^{2+}, but in its absence, *M. cuprina* takes Fe^{2+} via the anodic dissolution of pitting corrosion, thus accelerating the corrosion process.

The aforementioned system exemplifies the capability of SECM to provide information on biofilms and corrosion. Current studies are addressed to investigate, among other topics, cell-to-cell communication, microbial metabolism, and antimicrobial impact [28].

e) Electrochemical noise (EN): This technique is based on the record of the stochastic fluctuations of current and potential in electrochemical systems in time and frequency domains. Classically, measurements can be made under potentiostatic or galvanostatic control. EN measurements require a zero-resistance ammeter that can measure electrochemical potential noise and electrochemical current noise synchronously. Alternatively, EN can be performed through a potentiostat that can measure electrochemical current noise at an applied potential, or a galvanostat that can measure electrochemical potential noise at an applied current [31]. Noise sources include aliased noise, thermal noise, and noise generated at the metal/electrolyte interface, as well as noise generated by the instrumentation. Although data analysis is often difficult, EN has been extensively used in corrosion studies, including those centered on bacterial corrosion [32, 33].

15.6 OTHER BIODETERIORATION PROCESSES

A wide variety of materials undergo biodeterioration due to their interaction with vegetal species, namely, minerals, wood, concrete, and plastics [9]. From an electrochemical perspective, analysis of biodeterioration in paintings is potentially interesting [35, 36]. These are complex multicomponent systems where pigment grains are embedded into different natural or eventually synthetic binding media typically submitted to the action of bacteria and fungi.

The study of pictorial specimens can be done using solid-state voltammetric techniques (see Section 4.4), EIS, and SECM. Figure 15.12 shows the cyclic voltammograms, after semi-derivative convolution, of reconstructed paint films containing cadmium yellow (CdS) bound with egg-linseed oil emulsion inoculated with *Mucor rouxii* (fungus) and *Streptomyces cellulofans* (bacteria). The voltammograms were recorded in contact with an aqueous acetate buffer and showed cathodic signals corresponding to the reduction of CdS to Cd metal and its subsequent stripping oxidation. Reduction of CdS proceeds through several superimposed processes due to the interaction of the pigment grains with the binding media. The voltammetric profile is sensitive to the attack of bacteria and fungi, giving rise to distinctive voltammetric features [35].

15.7 BIOFILM CORROSION INHIBITION

Corrosion inhibition is an important social demand. The inhibition of MIC can follow several essential guidelines [37]:

a) Formation of a protective layer through physical barriers (coatings) or the production of an excess of extracellular polymeric substances. This requires the adherence of the material to the metal surface, blocking the access of corrosive substances (shielding effect).
b) Removal of chemical species that react with the metal surface, for example, oxygen consumption through aerobic respiration.
c) By interacting with the corrosive substances, converting them into non-corrosive ones.
d) Inhibition of the growth of microorganisms causing corrosion. The inhibitor can act as a biocide, destroying the microbial organisms.
e) Biofilm interactions, eventually including the generation of protective substances and the competition and negative synergies in multispecies biofilms [38]. Notice that biofilms can act as a diffusion barrier for corrosion agents, whereas microorganisms can synthesize metabolic products suppressing corrosion (e.g., siderophores) and/or consume oxygen, thus lowering their accessibility to the metal surface.

An enormous variety of corrosion inhibitors have been devised. These include surfactants, biocides, hydrophobic coatings, anti-protein coatings, chemical inhibitors and self-repairing systems, ozone, and nanomaterials. For our purposes, those involving plant extracts, most consisting of essential oils, are of particular interest [39].

The action of plant extracts is often linked to the presence of active constituents and in general, involves their adsorption on the metal surface. This can occur either by chemisorption or by physisorption. The former involves chemical interaction (electron exchange and chemical bond formation) between metal surface atoms and the inhibitor molecules. The second involves a physical interaction mechanism (generically, van der Waals forces).

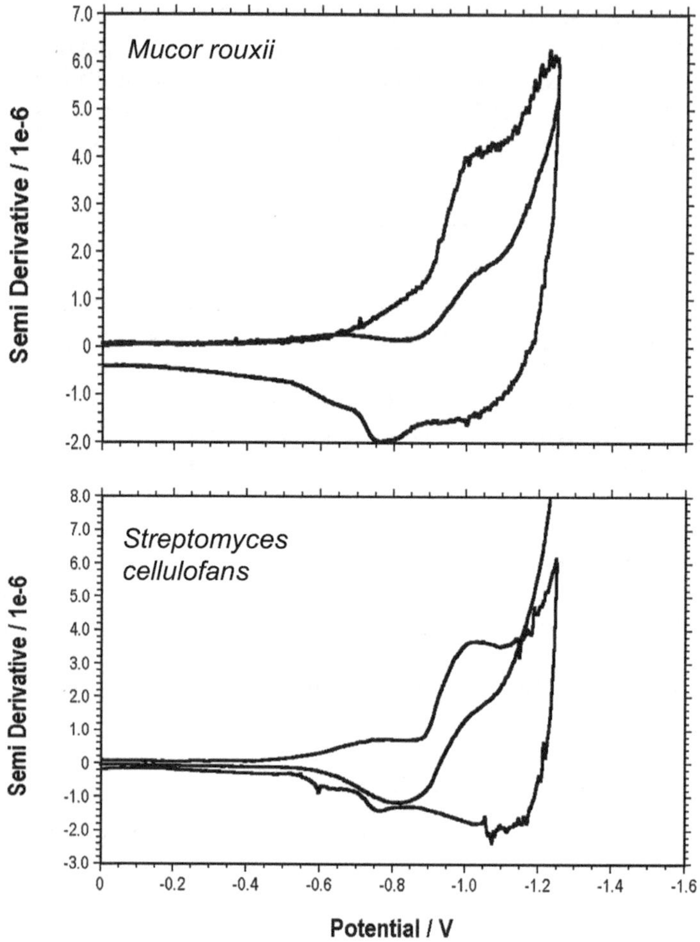

FIGURE 15.12 Cyclic voltammograms of reconstructed paint films containing cadmium yellow (CdS) bound with egg-linseed oil emulsion inoculated with (a) *Mucor rouxii* and (b) *Streptomyces cellulofans* in contact with air-saturated 0.25 M HAc/NaAc aqueous buffer at pH 4.75. Potential scan initiated at 0.05 V vs. Ag/AgCl in the negative direction, potential scan rate 50 mV s^{-1}. Semi-derivative convolution was used to enhance peak resolution. Note that the axes are opposite to the IUPAC convention.

The efficiency of corrosion inhibition is routinely monitored by the electrochemical techniques previously seen, in particular, by potentiodynamic polarization curves and EIS. Figure 15.13 illustrates the inhibition effect exerted on AA6063-T5 aluminum by parsley oil in different aqueous concentrations as tested in 3.5 %wt NaCl solution, typically used in electrochemical tests. The figure depicts the Nyquist representation of impedance spectra recorded in aluminum sheets in contact with the test solution enriched with different concentrations of the oil [40].

The detailed analysis of the impedance spectra requires consideration of diffusion layer resistance, accumulation resistance (associated with corrosion product accumulation), inhibitor film resistance, and charge transfer resistance. In addition, although dominated by a gross capacitive loop, the spectra show an inductive contribution at low frequencies. This contribution can be associated with the relaxation process of the adsorbates, the adsorption of dissolved aluminum ions, and the formation of the two-stage faradaic process of desorption-adsorption of the adsorbate in passive regions [40]. Despite these complications, the Nyquist plots reveal a well-defined tendency to increase the overall impedance of the system by increasing the concentration of the inhibitor oil. This is representative of the corrosion inhibition efficiency. This quantity is usually expressed in

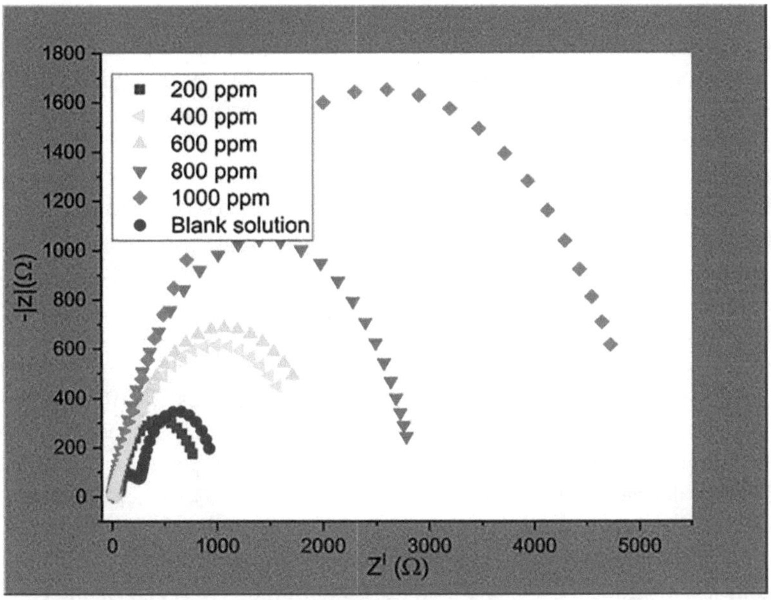

FIGURE 15.13 Nyquist plots corresponding to EIS on AA6063-T5 aluminum sheets immersed into 3.5 %wt NaCl solution containing different concentrations of parsley oil. (Adapted from ref. [40], with permission. Copyright the American Chemical Society 2022.).

terms of the charge transfer resistance, the total resistance, and/or the corrosion resistance determined in potentiodynamic polarization curves as

$$\eta(\%) = 100\left(\frac{R_{\text{in}} - R}{R_{\text{in}}}\right). \tag{15.16}$$

where R and R_{in} are the resistance measured in the absence and in the presence of inhibitor, respectively. Using this last technique, the inhibition efficiency can also be estimated from the corrosion current density in the absence, j, and in the presence, j_{in}, of the inhibitor as

$$\eta(\%) = 100\left(\frac{j - j_{in}}{j}\right). \tag{15.17}$$

15.8 ELECTROCHEMICAL DEGRADATION OF POLLUTANTS

Water scarcity and water pollution have become a problem affecting a significant proportion of the human population. The problem is aggravated by climatic changes and the increasing presence of the so-called emerging pollutants (resulting from pharmaceuticals, food additives, personal care products, etc.). Accordingly, the development of sustainable methods for degradation of pollutants concentrates much research effort.

Electrochemical degradation of pollutants appears as a promising methodology. A significant fraction of this research is focused on the oxidative degradation of organic pollutants in wastewater to CO_2, H_2O, N_2, etc. As far as residuals from vegetal and vegetal-derived products are involved, the electrochemical methodologies for pollutant degradation are of interest in the context of plant electrochemistry. This interest is reinforced by the possibility of developing degradation/energy cogeneration systems [41], as seen in Chapter 14.

The most widely approach for the electrochemical degradation of pollutants in waters is anodic oxidation [42]. This can be direct or indirect. In the first case, the pollutants are oxidized at the electrode surface, while in the second, the oxidation is mediated by an oxidizing species, either externally added or generated electrochemically [43]. Photocatalytically assisted systems exploiting semiconductor electrodes like TiO_2 complete the previous scenario [44].

Many factors influence the performance of the process, where the electrode material plays an important role. The most extensively studied include boron-doped diamond (BDD), Pt, RuO_2, IrO_2, PbO_2, and SnO_2 [45], but a great variety of materials, including doped transition metal oxides, different nanostructured materials, as well as multilayered electrodes, are under study [46].

Most oxidative pathways are based on the electrochemical generation of ROS, species with high oxidizing capacity, as evidenced by standard electrode potentials in Table 15.3.

Electrochemical water oxidation at the anode surface yields O_2 (oxygen evolution reaction, OER), in competence with the formation of ROS. All these processes are affected by kinetic restrictions strongly dependent on the electrode material and other experimental conditions (pH, temperature, etc.). Accordingly, the disposal of electrodes favoring ROS generation vs. OER is a general demand for pollution degradation.

Figure 15.14a schematizes another possible strategy usable in the electrochemical degradation of pollutants [47]. In this case, methylene blue, which is a compound classically taken as a model for this kind of studies, is degraded under the synergistic action of H_2O_2 and ozone (O_3), the so-called electro-peroxone methodology. Ozone is produced in an ozone generator and sparged into the solution accompanying, usually, the external addition of accompanying, usually, the external addition of H_2O_2. Alternatively, H_2O_2 can be generated in the cathode of the cell by reduction of dissolved oxygen.

Figure 15.14b compares the performance of different variants of the aforementioned procedure in terms of the total organic content (TOC) removed at different experimentation times in 0.05 M Na_2SO_4 aqueous solution containing 180 mg L^{-1} methylene blue. The TOC (determined by conventional analytical methods) increases with time being the combination of electrochemical production of H_2O_2 with ozonization (electro-peroxone method) the most efficient method.

Another strategy is based on the use of the Fenton reaction through the addition of Fe^{2+} ions and H_2O_2 to the problem solution. This results in the generation of hydroxyl radicals via

$$H_2O_2 + Fe^{2+} \rightarrow Fe^{3+} + HO\bullet + OH^- \tag{15.18}$$

The process can be catalytically promoted via electrochemical regeneration of ferrous ions in the cell cathode. The same regenerative strategy can be applied using the electrochemical reduction of O_2 to H_2O_2. Cathodic degradative processes can be produced by adding zerovalent iron nanoparticles to the cathode. These can promote the reduction of nitrate and nitrite species.

Given the variety of pollutant compounds and possible degradation approaches, the understanding of electrochemical mechanisms appears as one of the priority directions in immediate future research [45].

TABLE 15.3
Standard Electrode Potentials for Selected Oxygen Species in Aqueous Media

Process	E° (Vvs. SHE)	Process	E° (Vvs. SHE)
$HO\bullet_{aq} + H^+_{aq} + e^- \rightarrow H_2O$	2.722	$H_2O_2 + 2H^+_{aq} + 2e^- \rightarrow 2H_2O$	1.776
$O_g + 2H^+_{aq} + 2e^- \rightarrow H_2O$	2.421	$HO_2\bullet + 3H^+_{aq} + 3e^- \rightarrow 2H_2O$	1.684
$O_{3\,g} + 2H^+_{aq} + 2e^- \rightarrow O_2 + H_2O$	2.076	$O_{2\,g} + 4H^+_{aq} + 4e^- \rightarrow 2H_2O$	1.228

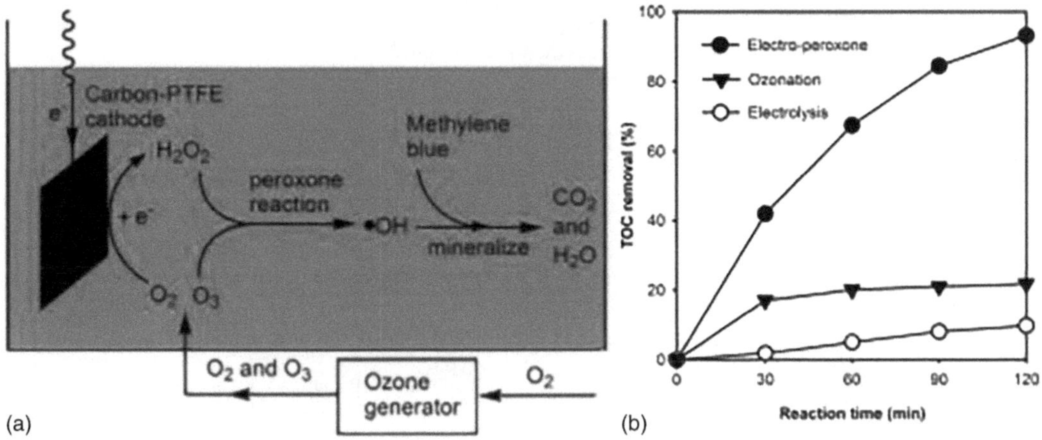

FIGURE 15.14 (a) Scheme for the electrochemical degradation of methylene blue using the electro-peroxone method in 0.05 M Na_2SO_4 aqueous solution containing 180 mg L^{-1} methylene blue. (b) Comparison of different variants of the above method showing the time variation of TOC. (Adapted from ref. [47], with permission. Copyright Elsevier 2013.).

15.9 BIOMASS VALORIZATION

Vegetal biomass is produced in significant amounts from agriculture, food industry, and other sources. The reuse of this biomass for the generation of chemical products and energy production and storage offers important issues for economic and environmental safety purposes [48].

A variety of chemical methodologies have been devised for biomass transformation. Among them, electrochemical valorization poses several interesting characteristics [49]:

(i) capability to operate directly on aqueous feedstocks,
(ii) no need for external reagents, and
(iii) operability, near ambient conditions, with highly flexible parameters.

The sources of vegetal biomass are mainly agricultural residues, crops dedicated to energy production, and food waste, with the chemical composition depending on the source. Reasonable averaged values on a dry basis are 47–51 wt% carbon, 5.8–6.0 wt% hydrogen, and 41–43 wt% oxygen. Sulfur, nitrogen, and other elements represent less than 0.15 wt%. Ash, less than 10 wt%, is mainly composed of inorganic compounds containing Ca, Mg, Si, K, Na, and trace metals [50].

A significant fraction of valorization processes are directed to the synthesis of chemical products. Figure 15.15 shows a scheme of the biorefining of lignocellulose-, lipid-, and protein-rich sources from energy crops and forestry residues, algae, and food waste [49]. The original vegetal sources can be deconstructed to yield a variety of chemicals that can be used as intermediates for obtaining other products. From the electrochemical point of view, electrochemical valorization processes can be seen within the general demands of organic electrosynthesis [51].

It must be taken into account that the translation from laboratory to plant scale is not straightforward. Here, high current densities and bulk conversion of reactants may differ substantially from the usual conditions in laboratory experiments. A variety of cell designs and operation modes have been devised, including different types of paired cells. In addition, electrochemical reactions may be coupled to unexpected side chemical and electrochemical reactions [49]. This means that the faradaic efficiency, i.e., the fraction of charge passed that effectively produces the desired faradaic process, may differ from 100%.

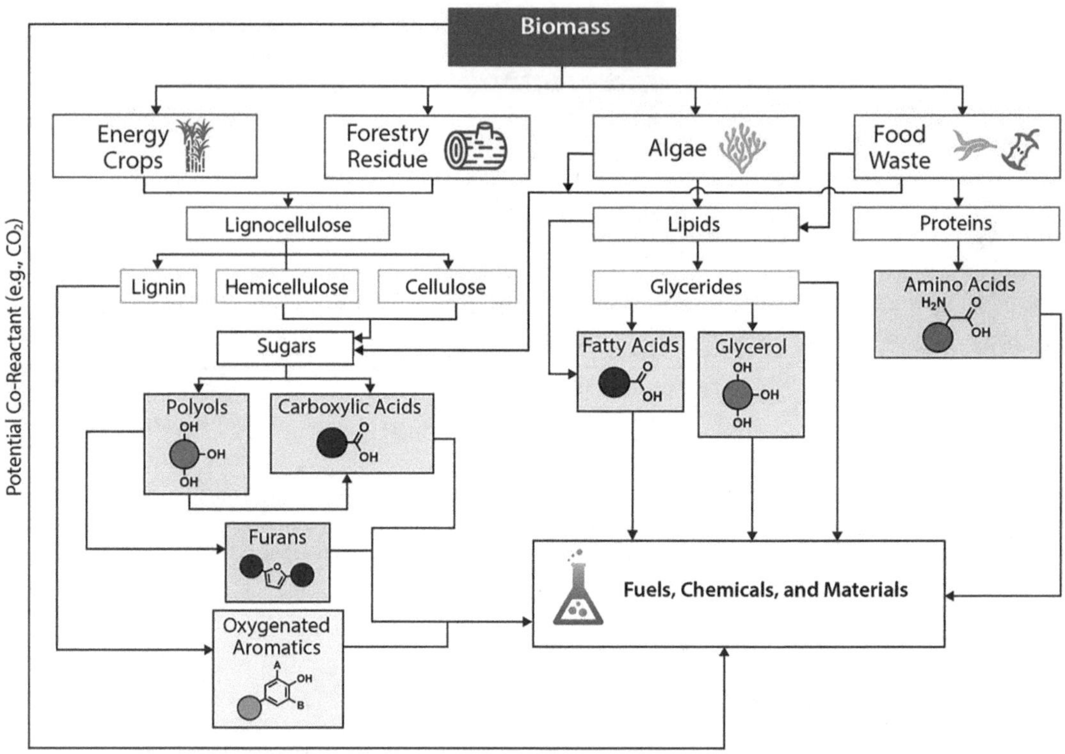

FIGURE 15.15 Overview of biorefining processes involving energy crops, forestry residues, algae, and food waste. (Adapted from ref. [49], with permission. Copyright the American Chemical Society 2021.).

As introduced in Chapter 14, electrochemical processing of biomass can be applied for the generation of biofuels. Among them, anaerobic digestion is being intensively studied [52]. Another way of biomass valorization is the production of biochar and porous carbon products for sensing and energy storage applications. These materials can be used as catalysts, including nanoparticulate materials derived from vegetal biomass [53].

REFERENCES

[1] Kasting, J.F.; Siefert, J.L. 2002. Life and the evolution of Earth's atmosphere. *Science*. 296: 1066–1068.

[2] Gaines, R.H. 1910. Bacterial activity as a corrosive influence in the soil. *The Journal of Industrial and Engineering Chemistry*. 2: 128–130.

[3] Zobell, C.E.; Allen, E.C. 1935. The significance of marine bacteria in the fouling of submerged surfaces. *Journal of Bacteriology*. 29: 239–251.

[4] von Wolzogen Kühr, C.; Van der Vlugt, L. 1934. Graphitization of cast iron as an electrobiochemical pr ocess in anaerobic soils (in Dutch). *Water*. 18: 147–165.

[5] Doménech-Carbó, A.; Doménech-Carbó, M.T. 2023. *Electrochemistry for Cultural Heritage*. Monographs in Electrochemistry series, Scholz, F.; Peter, L., Eds. Springer, Cham. Chapter 10.

[6] Dillmann, P.; Wantkinson, D.; Angelini, E.; Adriaens, A. 2013. *Corrosion and conservation of cultural heritage metallic artefacts*. European Federation of Corrosion, Woodhead, Cambridge.

[7] Perez, N. 2004. *Electrochemistry and Corrosion Science*. Kluwer Academic Publishers, Boston.

[8] Damjanovic, A. 1969. In Bockris, J.O.M.; Conway, B.E., Eds, *Modern Aspects of Electrochemistry*, vol. 5. Plenum Press, New York.

[9] Knisz, J.; Eckert, R.; Gieg, L.M.; Koerdt, A.; Lee, J.S.; Silva, E.R.; Skovhus, T.L.; An Stepec, B.A.; Wade, S.A. 2023. Microbiologically influenced corrosion—more than just microorganisms. FEMS *Microbiology Reviews*. 47: 1–33.

[10] Blackwood, D.J. 2018. An Electrochemist Perspective of Microbiologically Influenced Corrosion. *Corrosion and Materials Degradation*. 1: 59–76.

[11] Little, B.J.; Hinks, J.; Blackwood, D.J. 2020. Microbially influenced corrosion: towards an interdisciplinary perspective on mechanisms. *International Biodeterioration & Biodegradation*. 154: artic. 105062.

[12] Van Loosdrecht, M.C.; Lyklema, N.W.; Zehnder, A.J. 1990. Hydrophobic and electrostatic parameters in bacterial adhesion. *Aquatic Sciences*. 52: 103–114.

[13] Rivas, L.; Fegan, N.; Dykes, G.A. 2005. Physicochemical properties of shiga toxigenic *Escherichia coli*. *Journal of Applied Microbiology*. 99: 716–727.

[14] Pal, M.K.; Lavanya, M. 2022. Microbial Influenced Corrosion: Understanding Bioadhesion and Biofilm Formation. *Journal of Bio- and Tribo-Corrosion*. 8: artic. 76.

[15] Smart, N.R.; Blackwood, D.J.; Werme, L. 2002. Anaerobic corrosion of carbon steel and cast iron in artificial groundwaters: Part 1—Electrochemical aspects. *Corrosion*. 58: 547–559.

[16] Smart, N.R.; Blackwood, D.J.; Werme, L. 2002. Anaerobic corrosion of carbon steel and cast iron in artificial groundwaters: Part 2—Gas generation. *Corrosion*. 58: 627–637.

[17] King, R.A.; Miller, J.D.A.; Smith, J.S. 1973. Corrosion of mild steel by iron sulphides. *British Corrosion Journal*. 8: 137–141.

[18] Costello, J.A. 1974. Cathodic depolarization by sulfate-reducing bacteria. *South Africa Journal of Science*. 70: 202–204.

[19] Rickard, D.; Luther, G.W.III. 2007. Chemistry of Iron Sulfides. *Chemical Reviews*. 107: 514–562.

[20] Almeida, C.M.V.B.; Giannetti, B.F. 2002. Comparative study of electrochemical and thermal oxidation of pyrite. *Journal of Solid State Electrochemistry*. 6: 111–118.

[21] Eghbalnia, M.; Dixon, D.G. 2013. In situ electrochemical characterization of natural pyrite as a galvanic catalyst using single-particle microelectrode technique in ferric sulfate solutions. *Journal of Solid State Electrochemistry*. 17: 235–267.

[22] Li, Y.; Xu, D.; Chen, C.; Li, X.; Jia, R.; Zhang, D.; Sand, W.; Wang, F.; Gu, T. 2018. Anaerobic microbiologically influenced corrosion mechanisms interpreted using bioenergetics and bioelectrochemistry: A review. *Journal of Materials Science & Technology*. 34: 1713–1718.

[23] Li, J.; Du, C.; Liu, Z.; Li, X. 2022. Electrochemical studies of microbiologically influenced corrosion of X80 steel by nitrate-reducing *Bacillus licheniformis* under anaerobic conditions. *Journal of Materials Science & Technology*. 118: 208–217.

[24] Lin, L.; Tsou, C.-H.; Dou, B.; Yan, S.; Zeng, Y.; Gong, M. 2022. Electrochemical corrosion behavior and mechanism of iron-oxidizing bacteria *Thiobacillus ferrooxidans* from acid mine drainage on Q235 carbon steel. *New Journal of Chemistry*. 46: 20279–20291.

[25] Kip, N.; van Veen, J.A. 2015. The dual role of microbes in corrosion. *The ISME Journal*. 9: 542–551.

[26] Little, B.J.; Lee, J.S.; Ray, R.I. 2008. The influence of marine biofilms on corrosion: a concise review. *Electrochimica Acta*. 54: 2–7.

[27] Cai, D.; Wu, J.; Chai, K. 2021. Microbiologically Influenced Corrosion Behavior of Carbon Steel in the Presence of Marine Bacteria *Pseudomonas* sp. and *Vibrio* sp. *ACS Omega*. 6: 3780–3790.

[28] Caniglia, G.; Kranz, C. 2020. Scanning electrochemical microscopy and its potential for studying biofilms and antimicrobial coatings. *Analytical and Bioanalytical Chemistry*. 412: 6133–6148.

[29] Qian, H.C.; Chang, W.W.; Cui, T.Y.; Li, Z.; Guo, D.W.; Kwok, C.T.; Tam, L.M.; Zhang, D.W. 2021. Multi-mode scanning electrochemical microscopic study of microbiologically influenced corrosion mechanism of 304 stainless steel by thermoacidophilic archaea. *Corrosion Science*. 191: artic. 109751.

[30] Liu, L.J.; You, X.Y.; Guo, X.; Liu, S.J.; Jiang, C.Y. 2011. *Metallosphaera cuprina* sp. nov., an acidothermophilic, metal-mobilizing archaeon. *Journal of Systematic. And Evolutionary Microscopy*. 61: 2395–2400.

[31] Xia, D.-H.; Song, S.; Behnamian, Y.; Hu, W.; Frank Cheng, Y.; Luo, J.-L.; Huet, F. 2020. Review—Electrochemical Noise Applied in Corrosion Science: Theoretical and Mathematical Models towards Quantitative Analysis. *Journal of the Electrochemical Society*. 167: artic. 081507.

[32] Chandrasatheesh, C.; Jayapriya, J.; George, R.P.; Kamachi Mudali, U. 2014. Detection and analysis of microbiologically influenced corrosion of 316 L stainless steel with electrochemical noise technique. *Engineering Failure Analysis*. 42: 133.

[33] Homborg, A.M.; Leon Morales, C.F.; Tinga, T.; de Wit, J.H.W.; Mol, J.M.C. 2014. Detection of microbiologically influenced corrosion by electrochemical noise transients. *Electrochimica Acta*. 136: 223–232.
[35] Ortiz-Miranda, A.; Doménech-Carbó, A.; Doménech-Carbó, M.T.; Osete-Cortina, L.; Valle-Algarra, F.; Bolívar-Galiano, F.; Martín-Sánchez, I.; López-Miras, M.M. 2016. Electrochemical characterization of biodeterioration of paint films containing cadmium yellow pigment. *Journal of Solid State Electrochemistry*. 20: 3287–3302.
[36] Ortiz-Miranda, A.; Doménech-Carbó, A.; Doménech-Carbó, M.T.; Osete-Cortina, L.; Bolívar-Galiano, F.; Martín-Sánchez, I. 2017. Analyzing chemical changes in verdigris pictorial specimens upon bacteria and fungi biodeterioration using voltammetry of microparticles. *Heritage Science*. 5: artic. n° 8.
[37] Rao, P.; Mulky, L. 2023. Microbially Influenced Corrosion and its Control Measures: A Critical Review. *Journal of Bio- and Tribo-Corrosion*. 9: artic. 57.
[38] Kip, N.; Jansen, S.; Leite, M.F.A.; de Hollander, M.; Afanasyev, M.; Kuramae, E.E.; Van Veen, J.A. 2017. Methanogens predominate in natural corrosion protective layers on metal sheet piles. *Scientific Reports*. 7: artic. 11899.
[39] Little, B.; Lee, J.; Ray, R. 2007. A review of 'green' strategies to prevent or mitigate microbiologically influenced corrosion. *Biofouling*. 23: 87–97.
[40] Sanni, O.; Iwarere, S.A.; Daramola, M.O. 2022. Evaluation of Corrosion Inhibition of Essential Oil-Based Inhibitors on Aluminum Alloys. *ACS Omega*. 7: 40740–40749.
[41] Doménech-Carbó, A. 2021. *Electrochemistry of Porous Materials*, 2nd edit. Taylor & Francis, Boca Raton, Chapter 16.
[42] Moreira, F.C.; Boaventura, R.A.R.; Brillas, E.; Vilar, V.J.P. 2017. Electrochemical advanced oxidation processes: a review on their application to synthetic and real wastewaters. *Applied Catalysis B*. 202: 217–261.
[43] Martinez-Huitle, C.A.; Ferro, S. 2006. Electrochemical oxidation of organic pollutants for the wastewater treatment: direct and indirect processes. *Chemical Society Reviews*. 35: 1324–1340.
[44] Jiang, D.; Zhao, H.; Zhang, S.; John, R. 2003. Characterization of Photoelectrocatalytic Processes at Nanoporous TiO_2 Film Electrodes: Photocatalytic Oxidation of Glucose. *Journal of Physical Chemistry B*. 107: 12774–12780.
[45] Jiang, Y.; Zhao, H.; Liang, J.; Yue, L.; Li, T.; Luo, Y.; Liu, Q.; Lu, S.; Asiri, A.M.; Gong, Z.; Sun, X. 2021. Anodic oxidation for the degradation of organic pollutants: Anode materials, operating conditions and mechanisms. A mini review. *Electrochemistry Communications*. 123: artic. 106912.
[46] Moradi, M.; Vasseghian, Y.; Khataee, A.; Kobya, M.; Arabzade, H.; Dragoi, E.-N. 2020. Service life and stability of electrodes applied in electrochemical advanced oxidation processes: a comprehensive review. *Journal of Industrial Engineering Chemistry*. 87: 18–39.
[47] Yuan, S.; Li, Z.; Wang, Y. 2013. Effective degradation of methylene blue by a novel electrochemically driven process. *Electrochemistry Communications*. 29: 48–51.
[48] Serrano-Ruiz, J.C.; Luque, R.; Sepúlveda-Escribano, A. 2011. Transformations of Biomass-Derived Platform Molecules: From High Added-Value Chemicals to Fuels via Aqueous-Phase Processing. *Chemical Society Reviews*. 40: 5266–5281.
[49] Lucas, F.W.S.; Grim, R.G.; Tacey, S.A.; Downes, C.A.; Hasse, J.; Roman, A.M.; Farberow, C.A.; Schaidle, J.A.; Holewiinski, A. 2021. Electrochemical Routes for the Valorization of Biomass-Derived Feedstocks: From Chemistry to Application. *ACS Energy Letters*. 6: 1205–1270.
[50] Peduzzi, E.; Boissonnet, G.; Maréchal, F. 2016. Biomass Modelling: Estimating Thermodynamic Properties from the Elemental Composition. *Fuel*. 181: 207–217.
[51] Marken, F.; Atobe, M. Eds. 2019. *Modern Electrosynthetic Methods in Organic Chemistry*. CRC Press, Boca Raton.
[52] Xiao, C.; Fu, Q.; Liao, Q.; Huang, Y.; Xia, A.; Chen, H.; Zhu, X. 2020. Life cycle and economic assessments of biogas production from microalgae biomass with hydrothermal pretreatment via anaerobic digestion. *Renewable Energy*. 151: 70–78.
[53] Kumar, M.; Ambika, S.; Hassani, A.; Nidheesh, P.V. 2023. Waste to catalyst: Role of agricultural waste in water and wastewater treatment. *Science of the Total Environment*. 858: artic. 159762.

ADDITIONAL LITERATURE

Microbiologically Influenced Corrosion

Coetser, S.E.; Cloete, T.E. 2008. Biofouling and biocorrosion in industrial water systems. *Critical Reviews in Microbiology*. 31: 213–232.

Little, B.; Wagner, P.; Mansfeld, F. 1992. An overview of microbiologically influenced corrosion. *Electrochimica Acta*. 37: 2185–2194.

Permeh, S.; Lau, K. 2023. Identification of steel corrosion associated with sulfate-reducing bacteria by electrochemical noise technique. *Materials and Corrosion*. 74: 20–32.

Biofilm Formation

Carniello, V.; Peterson, B.; van der Mei, W.H.C.; Busscher, H.J. 2018. Physico-chemistry from initial bacterial adhesion to surface-programmed biofilm growth. *Advances in Colloid and Interface Science*. 261: 1–14.

Hori, K.; Matsumoto, S. 2010. Bacterial adhesion: from mechanism to control. *Biochemical Engineering Journal*. 48: 424–434.

Procópio, L. 2019. The role of biofilms in the corrosion of steel in marine environments. *World Journal of Microbiology and Biotechnology*. 35:1–8.

Stoodley, P.; Sauer, K.; Davies, D.G.; Costerton, J.W. 2002. Biofilms as complex differentiated communities. *Annual Review of Microbiology*. 56: 187–209.

Biodeterioration of Nonmetallic Materials

Doménech-Carbó, A.; Doménech-Carbó, M.T.; Osete-Cortina, L.; Donnici, M.; Guasch-Ferré, N.; Gasol-Fargas, R.M.; Igelsias-Campos, M.A. 2020. Electrochemical assessment of pigments-binding medium interactions in oil paint deterioration: a case study on indigo and Prussian blue. *Heritage Science*. 8: artic. 71.

Gaylarde, C.C.; Morton, L.H.G.; Loh, K.; Shirakawa, M.A. 2011. Biodeterioration of external architectural paint films–a review. *International Biodeterioration & Biodegradation*. 65: 1189–1198.

Jonkers, H.M.; Thijssen, A.; Muyzer, G.; Copuroglu, O.; Schlangen, E. 2010. Application of bacteria as self-healing agent for the development of sustainable concrete. *Ecological Engineering*. 36: 230–235.

Lamenti, G.; Tiano, P.; Tomaselli, L. 2000. Biodeterioration of ornamental marble statues in the Boboli gardens (Florence, Italy). *Journal of Applied Phycology*. 12: 427–433.

Le Métayer-Levrel, G.; Castanier, S.; Orial, G.; Loubière. J.F.; Perthuisot, J.P. 1999. Applications of bacterial carbonatogenesis to the protection and regeneration of limestones in buildings and historic patrimony. *Sedimentary Geology*. 126: 25–34.

SECM Analysis

Cai, C.; Liu, B.; Mirkin, M.V.; Frank, H.A.; Rusling, J.F. 2002. Scanning electrochemical microscopy of living cells. 3. Rhodobacter sphaeroides. *Analytical Chemistry*. 74: 114–119.

Holt, K.B.; Bard, A.J. 2005. Interaction of silver(I) ions with the respiratory chain of Escherichia coli: an electrochemical and scanning electrochemical microscopy study of the antimicrobial mechanism of micromolar Ag. *Biochemistry*. 44: 13214–13223.

Kaya, T.; Numai, D.; Nagamine, K.; Aoyagi, S.; Shiku, H.; Matsue, T. 2004. Respiration activity of *Escherichia coli* entrapped in a cone-shaped microwell and cylindrical micropore monitored by scanning electrochemical microscopy (SECM). *Analyst*. 129: 529–534.

Li, Z.; Chang, W.; Cui, T.; Xu, D.; Zhang, D.; Lou, Y.; Qian, H.; Song, H.; Mol, A.; Cao, F.; Gu, T.; Li, X. 2021. Adaptive bidirectional extracellular electron transfer during accelerated microbiologically influenced corrosion of stainless steel *Communications Materials*. 2: artic. 67.

Polcari, D.; Dauphin-Ducharme, P.; Mauzeroll, J. 2016. Scanning electrochemical microscopy: a comprehensive review of experimental parameters from 1989 to 2015. *Chemical Reviews*. 116: 13234–13278.

Rudolph, D.; Bates, D.; DiChristina, T.J.; Mizaikoff, B. Kranz, C. 2016. Detection of metal-reducing enzyme complexes by scanning electrochemical microscopy. *Electroanalysis*. 28: 2459–2465.

Electrochemical Noise

Bertocci, U.; Huet, F. 1997. Noise resistance applied to corrosion measurements. III. Influence of the instrumental noise on the measurements. *Journal of the Electrochemical Society*. 144: 2786–2793.

Bertocci, U.; Gabrielli, C.; Huet, F.; Keddam, M. 1997. Noise resistance applied to corrosion measurements. I-Theoretical analysis. *Journal of the Electrochemical Society*. 144: 31–37.

Cottis, R.A.; Al-Awadhi, M.A.A. Al-Mazeedi, H.; Turgoose, S. 2001. Measures for the detection of localized corrosion with electrochemical noise. *Electrochimica Acta*. 46: 3665–3674.

Obot, I.; Onyeachu, I.B.; Zeino, A.; Umoren, S. n.d.. Electrochemical noise (EN) technique: review of recent practical applications to corrosion electrochemistry research. *Journal of Adhesion Science and Technology*. 33: 1453–1496.

Metal Corrosion Inhibition

Kumar, S.; Roy, D.N.; Dey, V. 2021. A comprehensive review on techniques to create the anti-microbial surface of biomaterials to intervene in biofouling. *Colloid and Interface Science Communications*. 43: artic.100464,

Sharma, S.K.; Mudhoo, A.; Khamis, E. 2011. Adsorption studies, modeling, and use of green inhibitors in corrosion inhibition: an overview of recent research. *Journal of Corrosion Science and Engineering*. 11: 319–337.

Videla, H.A.; Herrera, L.K. 2009. Understanding microbial inhibition of corrosion. A comprehensive overview. *International Biodeterioration & Biodegradation*. 63: 896–900.

Zuo, R. 2007. Biofilms: strategies for metal corrosion inhibition employing microorganisms. *Applied Microbiology and Biotechnology*. 76: 1245–1253.

Electrochemical Degradation of Pollutants

Abdullah, F.H.; Abu Bakar, N.H.H.; Abu Bakar, M. 2021. Comparative study of chemically synthesized and low temperature bio-inspired *Musa acuminata* peel extract mediated zinc oxide nanoparticles for enhanced visible-photocatalytic degradation of organic contaminants in wastewater treatment. *Journal of Hazardous Materials*. 406: artic. 124779.

Bishnoi, S.; Kumar, A.; Selvaraj, R. 2018. Facile synthesis of magnetic iron oxide nanoparticles using inedible Cynometra ramiflora fruit extract waste and their photocatalytic degradation of methylene blue dye. *Materials Research Bulletin*. 97: 121–127.

Logan, B.E. 2005. Simultaneous wastewater treatment and biological electricity generation. *Water Science & Technology*. 52: 31–37.

Romero-Cano, L.A.; García-Rosero, H.; Baldenegro-Pérez, L.A.; Marín, F.C.; González-Gutiérrez, L.V. 2020. Coupled Adsorption and Electrochemical Process for Copper Recovery from Wastewater Using Grapefruit Peel. *Journal of Environmental Engineering*. 146: artic. 04020100-1.

Biomass Valorization

Brewer, C.E.; Schmidt-Rohr, K.; Satrio, J.A.; Brown, R.C. 2009. Characterization of biochar from fast pyrolysis and gasification systems. *Environmental Progress & Sustainable Energy*. 28: 386–396.

Bocci, E.; Di Carlo, A.; McPhail, S.J.; Gallucci, K.; Foscolo, P.U.; Moneti, M.; Villarini, M.; Carlini, M. 2014. Biomass to Fuel Cells State of the Art: A Review of the Most Innovative Technology Solutions. *International Journal of Hydrogen Energy*. 39: 21876–21895.

Laufenberg, G.; Kunz, B.; Nystroem, M. 2003. Transformation of vegetable waste into value added products: (A) the upgrading concept; (B) practical implementations. *Bioresource Technology*. 87: 167–198.

Panigrahi, S.; Sharma, H.B.; Tiwari, B.R.; Krishna, N.V.; Ghangrekar, M.M.; Dubey, B.K. 2021. Insight into understanding the performance of electrochemical pretreatment on improving anaerobic biodegradability of yard waste. *Renewable Energy*. 180: 1166–1178.

Paudel, S.R.; Banjara, S.P.; Choi, O.K.; Park, K.Y.; Kim, Y.M.; Lee, J.W. 2017. Pretreatment of agricultural biomass for anaerobic digestion: current state and challenges. *Bioresource Technology*. 245: 1194–1205.

Index

Pages in *italics* refer to figures and pages in **bold** refer to tables

FSC
www.fsc.org
MIX
Papier | Fördert
gute Waldnutzung
FSC® C083411

Kolibri 360 GmbH
Ferdinand-Jühlke-Straße 7
99095 Erfurt, Deutschland
produktsicherheit@kolibri360.de